AF443245

Thermodynamics and Statistical Mechanics

Thermodynamics and Statistical Mechanics

P.V. Panat

Alpha Science International Ltd.

Oxford, U.K.

Thermodynamics and Statistical Mechanics

436 pgs | 1 tbl. | 1 col. photo. | 90 figs.

P.V. Panat
Emeritus Professor
Centre of Modelling and Simulation
University of Pune
Pune, India

ALPHA SCIENCE INTERNATIONAL LTD.
7200 The Quorum, Oxford Business Park North
Garsington Road, Oxford OX4 2JZ, U.K.

www.alphasci.com

Printed from the camera-ready copy provided by the Author.

ISBN 978-1-84265-495-8

Printed in India

Foreword

Professor P. V. Panat of Pune University has written a novel postgraduate textbook on statistical mechanics. The book is distinguished by attention to basic questions, by detailed analyses which do not slur over difficulties (and there are many, both conceptual and mathematical, especially in statistical mechanics), by a number of problems of physical interest discussed and solved separately in the text, and by many important contemporary exemplary applications not found in textbooks. In all this one sees an experienced teacher and the ever inquisitive and acquisitive mind of a researcher in the field at work for the intellectual benefit of students.

As a student of Physics and a teacher, I know the necessity and difficulty of communicating the ideas and methods of this branch of physics. Since it deals with the behavior of a collection of a large number of constituents (particles, molecules, galaxies...) the subject area has direct consequences in perhaps the largest number of areas of human knowledge. An illustration of this is the fact that of the revolutionary contributions of the great Albert Einstein, the one most used by others is his work on Brownian motion, namely the random motion of 'particles' buffeted by the medium (often of the same 'particles') which surrounds them. (This is discussed in chapter 21 of the book).

Most physicists would perhaps consider Einstein's theory of relativity (special, which clarified the notion of relativity of frames of reference or general, which posited an intimate connection between geometry and physics) or his hypothesis of the light quantum to be the contributions which advanced most profoundly our knowledge of the physical world. But Brownian motion directly affects molecules in a gas, colloidal particles which constitute milk, and aerosols in the upper atmosphere. So, the challenge in any exposition of the statistical part of physics is hewing well to the fundamentals while at the same time pointing to its spread. This difficult challenge is well met by this book, uncommonly among those extant, and perhaps uniquely among books easily available to Indian students. This point needs to be emphasized strongly, specially with the realization that disciplinary boundaries are stultifying. In chemistry, chemical engineering, geology, biology and increasingly in the social sciences, statistical physics ideas are seen to be very consequential. For example, a very stimulating recent departure is the analysis of income distributions. Readers of this book will be well equipped with the cause and form of the Maxwell Boltzmann velocity distribution (chapter 9) which has been famously used for describing incomes, and will thus be in a position to both appreciate and create new knowledge far afield from conventional physics.

A comprehensive book on statistical mechanics has to perform many functions that generally do not sit well with each other and so taken together, make the result often uncouth and unwieldy.

The empirical, phenomenological, foundations are rooted in our immediate day to day experience and in the early phase of the industrial revolution (early 19th century). These are enshrined in 'thermodynamics' or the science of motion of heat. In view of its universal bearing, thermodynamics has been dubbed the true 'theory of everything'. It is also the home of a category of physical laws (laws of thermodynamics) qualitatively different from say Newton's laws of motion, and owing their overwhelming strength to the force of large numbers. Emergent patterns of behavior (codified by such laws)s or collective properties are the hallmark of this part of physics, and are both indicators of the nature of physics and counterexamples to the view that Nature can be physically understood only in terms of the behavior of idealized simple components. Panat does excellently here by presenting in the first chapter an account of thermodynamics and its laws. He gives a wealth of example, with a light touch and perceptive awareness. Thermodynamics as a branch of physical science was conceived independently of its statistical mechanical foundations; so its formulation appears often overly abstract and mathematical. In this book Panat takes the student carefully through this, a difficult act, balancing the needs to be correct, comprehensive and independent (of statistical mechanics) while being attractive and accessible to the reader. It is here that his experience and skill show their great value. In the language of thermodynamics (introduced earlier by him), the bugbears of perfect differentials, thermodynamic relations etc. are skillfully and completely dealt with. There is a wealth of physical examples. While every textbook talks about the Carnot cycle as an ideal reversible heat engine, no book discusses the necessary question of limits to the efficiency of the Carnot cycle when power is actually drawn from such a heat engine. This is a contemporary research topic! Here again Panat's book is unique. It goes carefully into this question, comes out with a criterion and applies it to the CANDU reactor. One of the most profound applications of thermodynamics in contemporary physics is to the description and analysis of singular macroscopic behavior near continuous (second order) phase transitions. Two accessible chapters (4 and 5) are devoted to this. A special strength of the book is close attention to the fundamentals in the chapters (6 and 7) on foundations of statistical mechanics I and II, where the latter introduces the quantum mechanical language while the former gives a detailed picture of classical statistical mechanics. The different ensembles are then worked out, and illustrative applications to the perfect classical gas and several problems of physical interest take up in chapters 8, 9 and 10. A very significant application is to the Ising model which is perhaps the most important classical statistical mechanical model system with interacting degrees of freedom (chapter 23).

Quantum statistical mechanics is an area of essential importance to every student of physics both in its fermionic and bosonic aspects, the former because electrons are the basic constituents of all matter and often the essential low energy degrees of freedom determining observed material properties, e.g. in metals, superconductors and semiconductors. Bosons, the other category of many body systems, were earlier believed to have only one simple statistical example -in Nature, namely He. However, the situation changed dramatically in the nineties, with the development of a large number of ultra cold bosonic systems. It is again one of the strengths of this book that the statistical mechanics of the free electron gas model of solids is thoroughly described and the description covers topics unusual for a statistical mechanics book, e.g. magnetism of an electron gas and the quantized Hall effect (Chapters 11-13, 16)! Panat not only develops the statistical mechanics of the Bose gas (Chapters 16 and 18); he goes into the physics of ultra cold bosonic systems giving a detailed description of the experimental setup needed and used, the systems appropriate and the results. This is exceptional in such a textbook, and ties the subject closely with what is possible to do with Nature.

One of the major challenges facing anyone attempting to write high levels text books in India (in science at least) has to do with our diversity. There is first the large university audience of students. For this book, the target audience would be in the M. Sc. classes; and broadly, the content is what is expected in a general (common) paper on statistical mechanics. For such an audience, short on necessary background and on morale, this book can be a godsend. It empowers them. They might have done thermodynamics in the B.Sc. it now provides a clear, strong and principled re-look. Conventional statistical mechanics 'portions' are served up with care, for example by doing the concepts and terminology as well as derivations well. There is a welcome emphasis on solving problems. A clear sense of a lively, growing subject is conveyed.

A fundamental necessity and a major oppression students and teachers face is that a text be not so much subject friendly as examination friendly. Panat's book scores heavily on both counts. The book can be textbook for almost any M. Sc. physics syllabus in an Indian university, because it describes what is 'covered' by such a syllabus properly (the ideas and methods are described as well as derived appropriately). The book may be overcomplete for a particular syllabus but both the teacher and the student can always choose from its contents so as to be well equipped. To me, even more significant is the fact that the student can not only do well at the examinations after reading the necessary parts of the book, but that he can begin to think like a statistical physicist on his own after digesting it. This is the other enabling role of the book.

The second large constituency in India is postgraduate students at a number of institutions which naturally subscribe to the mode of learning which this book illustrates. Such students also will find this book very valuable because of its analytical style, care and detail, and more so because of the rich tapestry of contemporary and traditional applications.

The third major user segment is global and transnational. For such users, the quality and contemporaneity (e.g. in terms of new applications of basics) is an obvious attraction. Even more attractive will be I think the pedagogy, namely the fact that a serious, open eyed and skilled effort is made to obtain carefully and in detail all important results; there is no papering over.

The book is a welcome addition to the recent textbooks on statistical mechanics; welcome both because of its content and style.

T.V. Ramakrishnan, FRS
DAE Homi Bhabha Professor
Physics Department
B.H.U., Varanasi 221005
Also Distinguished Associate
Centre for Condensed Matter Theory, I.I.Sc., Bangalore

Preface

The aim of the present book "Thermodynamics and Statistical Mechanics" is to discuss the physical concepts and methods appropriate for the description of a system of large number of particles.

We have taken a historic approach for our presentation resulting in a careful discussion of thermodynamics. Our experience is that students find thermodynamics as some abstract entity. They also have a notion that thermodynamics means gases, $PV = RT$ and the piston and cylinder only. To erase these notions and bring out the power and generality of thermodynamics, we have discussed many systems (eg. magnetic) right from the beginning. Power of thermodynamics and generality of thermodynamic potentials is illustrated by many examples including those in chapters four and five on phase transitions and critical phenomena. We hope that students will realize the depth of the statement of Einstein given at the start of chapter three. All the macroscopic systems consist of atoms and /or molecules. These entities ultimately obey the laws of Quantum mechanics. In certain conditions they can be treated with the laws of Classical Mechanics. Combination of microscopic concepts for the collection of atoms and molecules and some statistical postulates lead to the general conclusions of macro behavior of the collection irrespective of nature of interaction among them. This discussion is the discussion of foundations of Statistical mechanics. We present this basic but conceptually difficult discussion in as clear and as lucid manner as possible to us. Thus we demystify the thermodynamic function of entropy and consequently other thermodynamic functions. The three ensembles- microcanonical, canonical and grand canonical ensembles are carefully discussed with their limitations. Thus chapters six, seven and eight lay the foundations of Classical Statistical Mechanics, and Quantum Statistical Mechanics.

From chapter Nine onwards, we apply the theory so far developed to various physics problems. Thus we present a detailed analysis of applications of Fermi- Dirac statistics to the standard problems such as (a) specific heat of electron gas (b) Pauli paramagnetism (c) Thermionic emission. Besides these, some new applications which we feel can be grasped by the ambitious students at this level are thoroughly discussed. These are the magnetism of electron gas and integer Quantum Hall effect.

A specific chapter on chemical reactions at a very introductory level is furnished where the process of ionization is treated as a chemical reaction there by deriving Saha's ionization formula. Two chapters are devoted for the Bose system. Bose - I deals with Bose systems of zero chemical potentials. These are the elementary excitations in solids. A specific attention is paid on the problem of Black body radiation. Bose - II deals with Bosons with $\mu \neq 0$. This leads us to Bose- Einstein condensation. Discussion of the experiments due to Cornell et. al. is a specific feature of this chapter.

Historically, Einstein's explanation of Brownian motion paved the way for acceptability of molecular motion ideas as well as a deeper meaning to fluctuation and dissipation. We explore this link in chapter twenty one in a neat way.

Real life is not made up of ideal gases. Interactions are a reality. As a result we discuss classical real gases. We develop Mayer cluster expansion by the method of cumulant, evaluating $B_2(T)$ and $B_3(T)$ in some simple cases.

In chapter 23 we revisit phase transition from statistical mechanical point of view. One dimensional Ising model is discussed. We feel that the ideas of the modern topic of Renormalization group can be introduced for discrete lattice at the introductory level. As a result we essentially elaborate the paper of Maris and Kadanoff.

No book on Statistical Mechanics is complete unless it discusses introductory ideas of non-equilibrium statistical mechanics. This is so, because the response functions here are important to the experimenters. We devote two chapters - one with hand-waving arguments and the other in relaxation time approximation to calculate standard physical quantities like diffusion coefficient, thermal conductivity etc.

We are aware that this book is over complete as far as the general curriculum of M.Sc. programme of most of the Universities. However this book is inclusive, in the sense that a decent 40 lecture course can be tailored to suit the needs. With our experience of teaching for 30 years, following curricula for undergraduate and graduate program are ideal. For an undergraduate course of 40 lectures, consider

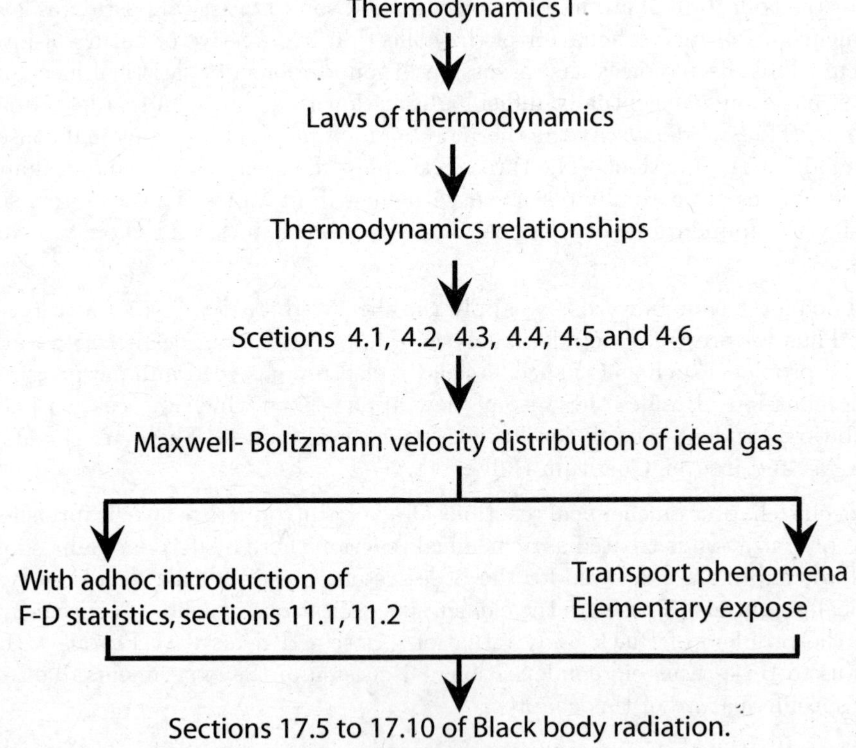

Statistical Mechanics is an integral part of one semester course at a postgraduate level. For such a one semester 40 lecture course, consider,

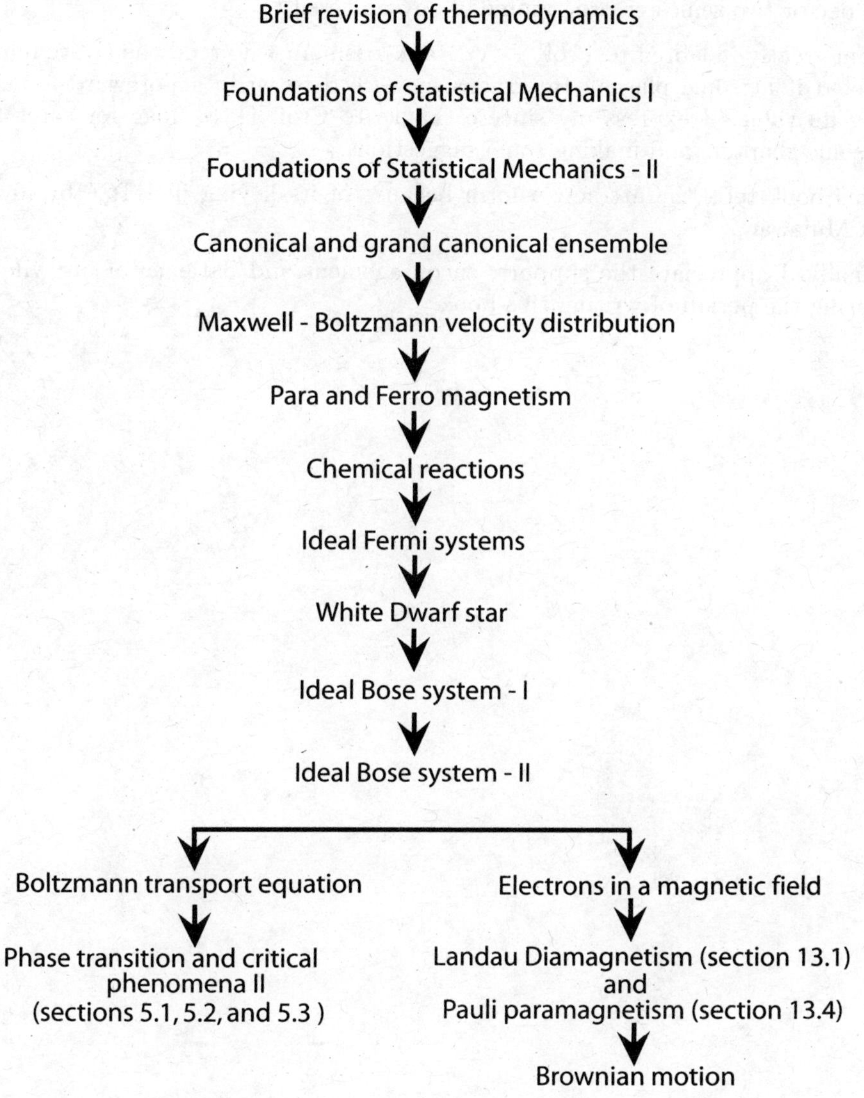

These are only the possible syllabi. The instructor can make many alterations depending on the need of the students. The remaining topics and some other linear combinations of the chapters can form an advanced (specialized) course of statistical mechanics. One special feature of our book is that we have solved large number of problems with variety of applications to various physical phenomena. We feel that the applicability of this basic subject is vast enabling us to take holistic view of physics. Keeping this fact in mind, we have carefully selected the problems in our book. Most of them are solved.

We thus feel that the book is self contained well motivated and has discussed the physical ideas with care. In the process, this book has become verbose and some what longer than might have been otherwise. This book can be useful to undergraduate and graduate one or two semester programme of any university.

I am greatly indebted to Prof. T. V. Ramakrishnan who read the entire manuscript. He corrected it at some places. He graciously agreed to write a Foreward for the book, increasing its value. I express my sincere thanks to Prof. D. S. Joag for carefully going through some chapters and making some suggestions.

This book took the attractive form because of its keying in $\LaTeX$ by my student Mubarak Mujawar

Finally, I appreciate the support, encouragement and patience of my wife Vaishali Panat during the period of writing this book.

P.V. Panat
Pune

Acknowledgement

Author sincerely thanks Prof. Eric Cornell of NIST/JILA, University of Colorado for providing his first original experimental Bose - Einstein condensate pictures appearing in the text. He is also grateful to Prof. Siu A. Chin of Texas A & M University for providing the numerical simulations of the condensates at various temperatures. Prof. Richard Pogge of Astronomy department of Ohio State University allowed us to reprint his H - R diagrams for the chapter on White Dwarf stars. Author sincerely thanks Prof. Pogge.

The author thanks Prof. Arun Jayannavar for bringing the problem of the power output of the Carnot engine to his notice.

I remembered my teachers Prof. Owen Chamberlain, Dr. John Garrison and Dr. Hugh DeWitt of University of California, Berkeley while writing this book. I am grateful to them for creating a deep interest in the subject of Statistical Mechanics by their teaching.

Simulations in statistical Mechanics has played an increasingly important role in recent years. As a matter of fact this field was initiated by Prof. Aneesur Rahman by looking at 32 gas particles on old historic IBM machines like 1620 a long ago. This technique has now grown and has become a prominent tool for studying realistic systems. As a result, I thought it prudent to introduce this technique at a very introductory level. I could not find more competent person to write on this topic than my student Dr. Mihir Arjunwadkar. He contributed an excellent introductory chapter on simulations to this book. I am very much thankful and grateful to Dr. Mihir Arjunwadkar. Needless to say that any errors, if any, are totally owned by the author.

Contents

Foreword   *v*
Preface   *ix*
Acknowledgement   *xiii*

1 Thermodynamics I: Fundamental Notions   **1**

   1.1   Categories of Thermodynamic Systems 2

   1.2   Macroscopic Versus Microscopic Variables 4

   1.3   Heat 5

   1.4   Work 6

       1.4.1   Thermodynamic Equilibrium 6

       1.4.2   Reversible and Irreversible States 6

       1.4.3   Work Done in Various Physical Systems 9

   1.5   Chemical Potential 12

   1.6   Definition of a Mole and the Avogadro Number 12

   1.7   Intensive and Extensive Variables 13

   1.8   Response Functions 13

   1.9   Internal Energy 14

   1.10 Exact and Inexact Differentials 15

   1.11 Useful Mathematical Relations 17

   1.12 Units of Pressure and Volume 19

2 Thermodynamics II: The Laws of Thermodynamics   **23**

   2.1   Zeroth Law of Thermodynamics 23

   2.2   First Law of Thermodynamics 25

   2.3   Ideal Gas 26

   2.4   Carnot's Theorem and the Second Law 30

   2.5   Axiomatic Treatment 35

   2.6   Second Law of Thermodynamics 37

   2.7   Thermodynamic Potentials 38

   2.8   Maxwell's Relations 39

2.9   Third Law of Thermodynamics 40

2.10 First and Second Law with Many Components 40

2.11 Efficiency of an Engine in General and Carnot Engine in Particular 41

2.12 Thermodynamic Temperature Scale 42

2.13 Power Output from Carnot Engine 45

2.14 Equilibrium with a Matter Flow 46

2.15 Entropy, Reversibility and Irreversibility 47

3   Thermodynamic Relationships   51

3.1   Relation between C_P, C_V, β and κ 52

3.2   Internal Energy Relations 53

3.3   Magnetic Response Functions 54

3.4   Minimum Number of Experimental Quantities Necessary to Determine Thermodynamic Functions 56

3.5   Carnot Cycle Ideas to Solve Thermodynamic Problems 57

3.6   Joule-Thomson Process 59

3.7   Adiabatic Demagnetization 63

4   Phase Transitions and Critical Phenomena I   71

4.1 $P - V - T$ Diagrams for Pure Substances 73

4.2   First Order Phase Transitions 76

      4.2.1   Properties of Gibbs Free Energy G 77

4.3   Clausius-Clapeyron Equation 79

4.4   Critical Isotherm 82

4.5   van der Waals Equation 83

4.6   Properties of a Substance Near T_c 85

4.7   Results for Magnetic System 86

4.8   Critical Exponents 86

5   Phase Transitions and Critical Phenomena II   93

5.1   Ginzburg-Landau Theory 93

5.2   Mean Field Theories and Landau's Theory 96

5.3   Fluctuations in Order Parameter 96

5.4   Region Near T_c where Landau Theory is Valid – Ginzburg's Analysis 98

5.5   Prelude to Scaling Hypothesis 100

5.6   Static Scaling Hypothesis 102

6   Foundations of Statistical Mechanics I: Classical Statistical Mechanics   105

6.1   Specification of a Microstate 105

6.2   Ensemble 107

6.3   Microcanonical Ensemble 110

6.4   Ergodic Hypothesis 111

6.5   Entropy 112

7   Foundations of Statistical Mechanics II: Quantum Statistical Mechanics   122

7.1   Introduction to Quantum Ideas 122

7.2   Quantum Statistical Ensemble 127

7.3   Density Matrix 128

7.4   Ideal Quantum Gases – Microcanonical Ensemble 132

7.5   Maxwell-Boltzmann (M-B) Statistics 136

7.6   Entropy of Various Statistics 136

8   Canonical and Grand Canonical Ensembles   140

8.1   Canonical Ensemble 140

8.2   Partition Function and Thermodynamics 143

8.3   Grand Canonical Distribution 145

8.4   Remarks: Microcanonical, Canonical & Grand Canonical Distributions 146

8.5   Fermi and Bose Distributions 149

9   Maxwell-Boltzmann Velocity Distribution of Ideal Gas   156

9.1   Maxwell-Boltzmann Velocity Distribution 156

9.2   Distribution of Component of Velocity 159

9.3   Number of Molecules Hitting a Unit Area of a Container 161

9.4   Pressure Exerted by a Gas 162

9.5   Equipartition Theorem 163

10   Paramagnetism and Ferromagnetism   167

10.1   Paramagnetism 168

10.2 Ferromagnetism: Weiss Theory 170

11 Ideal Fermi System   **176**

11.1 Energy and Pressure of an Ideal Fermi Gas at $T = 0$ 176

11.2 Thermodynamic Quantities of an ideal Fermi gas at finite temperature 180

11.3 Applications of F-D Statistics 184

12 Electrons in Magnetic Field   **190**

12.1 Electron in Magnetic Field 191

12.2 Degeneracy of Levels 191

13 Magnetism of Electron Gas   **195**

13.1 Landau Diamagnetism 195

13.2 Orbital Magnetism at $T = 0K$ 197

13.3 Diamagnetic Susceptibility at All Temperatures in Weak Magnetic Field . . . 201

13.4 Pauli's Paramagnetic Susceptibility 202

13.5 Quantum Hall Effect 205

14 Problems of Physical Interest   **209**

15 Chemical Reactions   **234**

15.1 Law of Mass Action 235

15.2 Heat of Reaction 237

15.3 Discussion of Some Chemical Reactions 237

15.4 Saha's Ionization Formula 238

16 White Dwarf Stars   **242**

16.1 Problem Posed by a White Dwarf Star 245

16.2 Model of a White Dwarf Star 247

16.3 Model Calculations 247

17 Ideal Bose System I   **252**

17.1 Partition Function 253

17.2 Problem of Specific Heat of Solids 254

17.3 Specific Heat Due to Lattice Vibrations: Debye's Theory 255

17.4 Black-Body Radiation 258

17.5 Planck's Law 261

17.6 Mean Total Energy 263

17.7 Pressure Exerted by Radiation 264

17.8 Cosmic Radiation Background 264

17.9 High and Low Frequency Limits of Planck's Law 265

17.10 Einstein's Derivation of Planck's Law 266

18 Ideal Bose System II   **273**

18.1 Thermodynamic Functions of a Bose-Einstein Gas 274

18.2 Illustrative problem 278

18.3 Recent Experiments on BEC 279

19 Transport Phenomena: Elementary Exposé   **289**

19.1 Hand-waving (Elementary) Derivation of Transport Coefficients 289

19.2 Collision Time and Mean Free Path 290

19.3 Calculational Method 292

19.4 Viscosity Calculation 293

19.5 Thermal Conductivity Calculation 294

19.6 Diffusion 294

19.7 Electrical Conductivity 295

19.8 Viscosity and Thermal Conductivity 298

19.9 Thermal Conductivity of Metals 298

20 Boltzmann Transport Equation   **302**

20.1 Derivation of Boltzmann Transport Equation without Collision 302

20.2 Boltzmann Transport Equation with Collision 304

20.3 Relaxation Time (RT) Approximation 308

20.4 Calculation of Relaxation Time 309

20.5 Flux Transported Across a Surface 310

20.6 Calculation of Thermal Conductivity 311

21 Brownian Motion   **316**

21.1 Brownian Motion 317

21.2 Mobility and Diffusion 320

21.3 Correlation Functions and Fluctuation-Dissipation Relations 322

22 Real Gases   **330**

22.1 Mayer Cluster Expansion 331

22.2 Summation of the Diagrams 336

22.3 Calculation of $B_2(T)$ and $B_3(T)$ 338

22.4 Comment About Classical Coulomb Gas 341

22.5 Behavior of $B_2(T)$ with Temperature 343

23 Ising Model, Renormalization Group, Etc.   **348**

23.1 Ising Model in One Dimension 349

23.2 Ising Model in Two Dimensions 351

23.3 Renormalization Group Analysis of 1-D Ising Model 353

23.4 Renormalization Group Analysis of 2-D Ising Model 356

23.5 Epilogue 359

24 Simulation Methods   **363**

24.1 Simulation in Statistical Physics 363

    24.1.1 Reality, Models, Mathematics 363

    24.1.2 Computation and Simulation 364

    24.1.3 Principal Simulation Methods of Statistical Physics 364

24.2 Molecular Dynamics (MD) 364

    24.2.1 Equations of Motion 365

    24.2.2 Integrating Equations of Motion: the Verlet Algorithm 367

    24.2.3 Schematics of a Bare-Basics MD Simulation 371

    24.2.4 Breathing-Mode Oscillations of Highly Symmetric Clusters 372

    24.2.5 Simulating Extended Systems 379

24.3 Monte Carlo (MC) Methods 384

    24.3.1 The Metropolis Algorithm 385

    24.3.2 The Lennard-Jones Fluid Again 387

APPENDICES

A Counting of States   **391**

B Gamma Function and Useful Integrals   394

C Moment Generating Function and Cumulant   396

D Evaluation of Some Fermi-Dirac and Bose-Einstein Integrals   398

E Mean Field Solution of Ising Model   404

BIBLIOGRAPHY   407

INDEX   410

Chapter 1

THERMODYNAMICS I: FUNDAMENTAL NOTIONS

Statistical Mechanics gives a microscopic explanation of certain empirical laws that were crystallized for an assembly of large number of particles confined within a boundary. These laws were formulated in the nineteenth century.The empirical laws are the laws of thermodynamics.

The microscopic explanation of these laws is based upon classical mechanics if the system is classical. Quantum mechanics is used if the system is a quantum system. Usually but not necessarily, large number of particles (say N) are confined within a volume V at a temperature T. The number of particles N could be as large as 10^{23}. Such a situation is called as thermodynamic system. The concept of thermodynamic system is not restricted to the gas above, but is a quite general. Following are the typical examples of various physical thermodynamic systems. They are representative and by no means exhaust all the possibilities.

1. Magnetization of a magnetic solid wherein there are large number atoms (and or molecules) with intrinsic magnetic moment

2. A solid or a liquid system with a surface area

3. A system of two phases such as a liquid and a gas coexisting at a temperature T

4. Electrons in a white dwarf star

5. Electrolytic battery

and so on. It is clear from these examples that the thermodynamic systems are quite general. We must get thoroughly familiarized with the thermodynamic laws and also these systems before we understand them microscopically.

To describe a thermodynamic system, we must discover adequate and minimum number of variables of the system. Moreover these variables must be experimentally measurable.

Identification of such variables is tricky and a most important task, an expression of high creativity. The value of these variables at an instant is called a 'state' of a thermodynamic system. Following are the examples of thermodynamic variables.

1. A single phase gas in volume V has its pressure P, volume V and its temperature T as thermodynamic variables.

2. For a magnetic system, its magnetization $\vec{M}$, external magnetic field $\vec{B}$ and its temperature T are the thermodynamic variables.

3. Two phase system- pressure P,temperature T,chemical potential of two phases and the densities of two phases

4. White Dwarf Star- pressure due to electrons (so called Fermi Pressure), volume V, and inward gravitational contraction and its temperature T are the thermodynamic variables.

In all these examples, and others, a common thermodynamic variable temperature T appears. It is intimately linked with heat . If heat is supplied to a thermodynamic system, then its temperature will generally increase. Its amount will depend upon the system. Thermo means heat and dynamics means motion.Thus thermodynamics means science of a motion of heat. If these thermodynamic variables of a system do not change with time, we say that the system is in thermodynamic equilibrium. In such circumstances, thermodynamic variables are related among each other. This relationship is called an equation of state. For a gas confined in a volume V, its pressure P, volume V and its temperature T are related as

$$f(P, V, T) = 0 \, . \tag{1.1}$$

If a gas is ideal, then,

$$PV = RT \, . \tag{1.2}$$

where R is a gas constant and T is absolute temperature (to be discussed later).The ideal gas is a theoretical construct where in the gas molecules do not interact among themselves. A dilute gas behaves as an almost ideal gas. When the gas is non ideal, simplest equation of state for it, in most cases, is, the van der Waals equation of state. Equation of state for magnetic systems can be written as

$$f(\vec{B}, \vec{M}, T) = 0 \, . \tag{1.3}$$

We will subsequently show this functional relationship to be the Langevin equation in an approximation called the mean field approximation.

1.1   Categories of Thermodynamic Systems

For detailed analysis, we specify thermodynamic systems in to following categories.

1. Isolated System : A thermodynamic system is said to be isolated when it cannot exchange its energy or its number of particles with the surroundings. Its energy and number of particles are strictly conserved.

2. Closed system : Closed system can exchange energy with the surrounding but cannot exchange its matter. Isothermal expansion or isothermal compression can be made to a closed system.Energy is no more conserved. When the surrounding and the system are in equilibrium, the system has average energy $\langle E \rangle$ with fluctuations.

3. Open system : Here the system is confined in a volume V but is in contact with surroundings. It can exchange both, the energy as well as matter.In equilibrium, the system has average energy E and average number of particles N. There are fluctuations in both of these quantities.

We will see that, in statistical mechanics, isolated system is described by the so called microcanonical ensemble, closed system by a canonical ensemble and open system by a grand canonical ensemble .

We have introduced a temperature without really specifying how to measure it. Accurate determination of temperature of a thermodynamic system in a large range of possible temperatures, is a complicated game. Moreover, there is an absolute temperature scale of Kelvin. We have various types of thermometers, such as gas thermometer, platinum resistance thermometer, mercury thermometer, and so on. You have already studied temperature scales such as Celsius (^oC) and Fahrenheit (^oF) and their interrelationships ($\dfrac{F-32}{9} = \dfrac{C}{5}$)

Liquid thermometers are not accurate because two different liquids expand irregularly in temperature ranges. However the dilute gases have an advantage over liquids because their expansion is sizable and almost identical.Their expansion coefficient with temperature is almost same in a wide temperature range during which they do not liquify. The gas thermometers are based on ancient laws of Boyle and that of Charles. Boyle's law states that $P \propto \frac{1}{V}$ at a constant T. Charles law states that $V \propto T$ at constant P. Since interactions between molecules of a dilute gas are weak, it is regarded as a near perfect and therefore we construct a gas thermometer from them. There are two types of thermometer, namely, constant volume gas thermometer and other is constant pressure thermometer. For simplicity, consider a constant pressure gas thermometer. The experiments due to Charles show that, over a range of temperature where the gas in the thermometer does not liquify, volume of the gas is directly proportional to its temperature. Thus, if V_0 is a volume of a thermometric gas at zero degree Celsius, then its volume V at a temperature $T > 0$ is,

$$V(t) = V_0(1 + \alpha T) . \tag{1.4}$$

where α is a coefficient of volume expansion of the dilute gas of thermometer. Plot of equation (1.4) is shown in figure (1.1). value of α for most common gases such as oxygen, nitrogen, Argon etc. is same. Its value is $\alpha = 0.0036608$ per degree Celsius. If we extrapolate the graph of equation (1.4) so that its volume reduces to zero, then corresponding temperature T_0 is $T_0= -273.16^oC$. Since the volume of the gas goes to zero at this temperature, we cannot have a temperature lower than T_0 We take this temperature as an absolute zero. Obviously the temperature corresponding to 0^oC is 273.16 absolute. The unit of absolute temperature is kelvin in honor of Lord Kelvin who is one of the founders of thermodynamics. Thus X^oC is $(X + 273.16)K$. Since the temperature scale developed is valid for almost all dilute gases and for the ideal gas, it is an absolute scale and is independent of

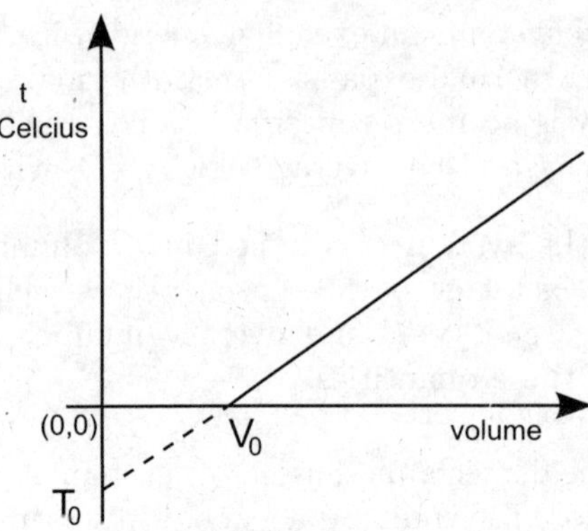

Figure 1.1: Charles's Law

the properties of the gases. The temperature scale thus developed is known as a perfect gas scale or a thermodynamic scale or an absolute scale of temperature. Subsequently, we will present thermodynamic arguments of Lord Kelvin that the absolute temperature scale is independent of the properties of any particular substance. The Kelvin scale of temperature has absolute zero temperature. Experimentalists have devised many ingenious thermometers to determine the temperatures from micro Kelvins to thousands of degrees of Celsius.

From the discussion of temperature, it is clear that this variable is absent in classical mechanics, electrodynamics and also in quantum mechanics. This important variable is statistical in nature. It cannot be defined for one or two particles. It can be defined only when the system has large number of particles ($N \gg 1$).

1.2   Macroscopic Versus Microscopic Variables

In thermodynamics we have a system of large number of particles. Such a system of N particles has a classical state defined by its co ordinates and momenta. Particle 1 has its coordinate $\vec{r}_1$ and momentum $\vec{p}_1$, particle 2 has coordinate $\vec{r}_2$ and momentum $\vec{p}_2$ and so on, and for the Nth particle, coordinate $\vec{r}_N$ and momentum $\vec{p}_N$. This is called *microscopic state or microstate*. Thus this microstate is described by specification $(\vec{r}, \vec{p}) \equiv (\vec{r}_1, \vec{p}_1; \vec{r}_2, \vec{p}_2; ....\vec{r}_N, \vec{p}_N)$. Hamiltonian of this system is a function of $(\vec{r}, \vec{p})$. Thus

$$H = H(\vec{r}_1, \vec{p}_1; \vec{r}_2, \vec{p}_2; ....\vec{r}_N, \vec{p}_N; t) \tag{1.5}$$

Time evolution of the microstate is governed by the Hamilton's equations of motion, which are,

$$\dot{\vec{r}_i} = \nabla_{\vec{p}_i} H(\vec{r}_1, \vec{p}_1; \vec{r}_2, \vec{p}_2; ....\vec{r}_N, \vec{p}_N; t) \tag{1.6}$$

and

$$\dot{\vec{p}_i} = -\nabla_{\vec{r}_i} H(\vec{r}_1, \vec{p}_1; \vec{r}_2, \vec{p}_2; ....\vec{r}_N, \vec{p}_N; t) \tag{1.7}$$

with appropriate values of $\vec{r}_i$ and $\vec{p}_i$ at some specified time say t_o

If the system of N particles is a quantum system, then, its quantum state is its wave function $\psi = \psi_{\vec{p_1}...\vec{p_N}}(\vec{r}_1, \vec{r}_2, ...\vec{r}_N; t)$ which, if initial state at t_o is known, then its time evolution is governed by Schrodinger equation as

$$i\hbar\frac{\partial\psi_{\vec{p_1}...\vec{p_N}}(\vec{r}_1, \vec{r}_2, ...\vec{r}_N; t)}{\partial t}$$

$$= H(\vec{r}_1, \vec{p}_1; \vec{r}_2, \vec{p}_2; ....\vec{r}_N, \vec{p}_N; t)\psi_{\vec{p_1}...\vec{p_N}}(\vec{r}_1, \vec{r}_2, ...\vec{r}_N; t) \qquad (1.8)$$

Clearly, these microscopic equations are impossible to solve and these microscopic variables are impossible to observe.

The macroscopic (sometimes called thermodynamic) variables are easy to observe, e.g. a gas confined in a volume V, having pressure P and temperature T are measurable. Similarly, magnetization $\vec{M}$, external magnetic induction $\vec{B}$ and temperature T for magnetic system are observable. It is very difficult to measure a magnetic moment of a desired particle. [1] For a superconducting state, a detection of an individual electron pair (Cooper pair) and its evolution is very difficult since their number is macroscopic ($N \sim 10^{20}$ close to zero degree kelvin)

Macroscopic, or thermodynamic variables are intuitive in nature, experimentally measurable and do not require any assumption regarding the structure of the matter. Compared to microscopic variables, they are few in number and also are measurable. Temperature is a special variable of a macro state and it is absent in describing micro state. The description of a system in terms of micro state is statistical in nature. If the system is classical, in principle, it is possible to describe its microstate by solving Hamilton's equations with its boundary conditions.This is so because there is no uncertainty principle there. But because it's number of particles ($N \gg 1$) is huge,we are forced to use statistical methods. For the quantum system, we have an inherent probabilistic situation (a la Born). For the quantum system then, we have to double average physical properties e.g. averaging for large numbers and averaging because of intrinsic quantum nature.

The aim of statistical mechanics is to try to understand laws of thermodynamics and calculations of various physical quantities like specific heat, susceptibility, energy and similar other experimentally realizable quantities. This is done from the microscopic variables and is thus system dependent. Understanding of the connection of macroscopic (thermodynamic) variables and state functions such as internal energy E, entropy S (to be discussed later) with the microscopic variables is the aim of statistical mechanics.

1.3   Heat

If the two isolated systems, one at a temperature T_1 and other at temperature T_2 come in contact via a diathermic (or heat conducting) wall, then it is a common experience that, after some time, temperature of the two systems equalizes. We assume that other thermodynamic variable V of both the systems remain unchanged during the process of temperature equalization.We say that during this process "something" flows from the system

---

[1]Local moment and its charge state of a ferromagnetic ion can be determined by Mossbauer effect measurements.

of higher temperature to the the system of lower temperature. This "something" is called heat. If the diathermic wall is not rigid, then, along with the transfer of heat, an external work will be done by the movement of the wall. Volumes will change.

1.4   Work

In mechanics, we know the definition of work as $\int \vec{F} \cdot d\vec{r}$ where $\vec{F}$ is an external force. If the thermodynamic system as a <u>whole</u> exerts a force on the surrounding, and this results in displacements, then the work is said to be done either by the system or on the system. This work is called external work. Simplest example is an expansion of a gas in cylinder with piston at constant pressure. In this process, piston moves, and the work is done. There is another type of work called internal work. Example of internal work is, molecular rearrangement in a gas because of mutual interaction among its molecules. In thermodynamics, we deal only with the external work and henceforth we will simply refor it as work.

1.4.1   Thermodynamic Equilibrium

A thermodynamic system is said to be in equilibrium provided

1. it is in mechanical equilibrium. i.e. there are no external forces or torques acting on the system, and

2. it is in thermal equilibrium which means, there are no temperature variations between different parts of the system and between the system and its surroundings, and

3. it is in chemical equilibrium, means there are no chemical reactions within the system. It means that the concentration of various constituents of the system does not change with time.

1.4.2   Reversible and Irreversible States

Irreversible processes are of common occurrence. Consider a gas confined in a cylinder in volume V_1, at temperature T_1 and having a pressure P_1. Suppose it comes in contact with a surrounding at temperature T_0 less than T_1. Let the pressure of the surrounding be $P_0 < P_1$. Clearly the piston of the cylinder will expand till the pressure and temperature of the gas in the cylinder equalizes with that of surrounding. Even though it is energetically possible, on its own, the gas in the cylinder will not get compressed to go to its original thermodynamic state viz. (P_1, V_1, T_1). Thus we have irreversibility. In an irreversible process, the change in thermodynamic variables occur so fast that the thermodynamic variables like P, V, T cannot be defined during the process. This is so because, there may be turbulence near the boundaries, gradients in pressure and temperature and all these may vary from time to time and position to position. That is why we cannot define a thermodynamic state uniquely during the irreversible process. Any process that takes a finite amount of time for its completion is irreversible. Friction between wall of the cylinder and piston contributes to irreversibility and so also the viscosity of the gas. Finally, when the system comes to a thermodynamic equilibrium, thermodynamic variables can be defined. Consider a $P - V$ diagram of a gaseous system. Here one plots P against V at a constant temperature,

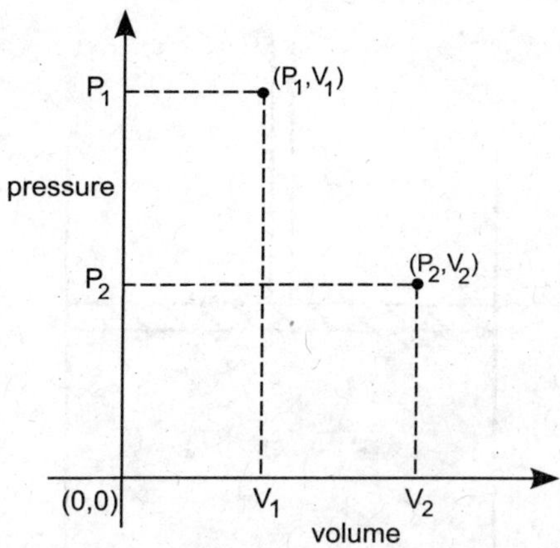

Figure 1.2: Irreversible States

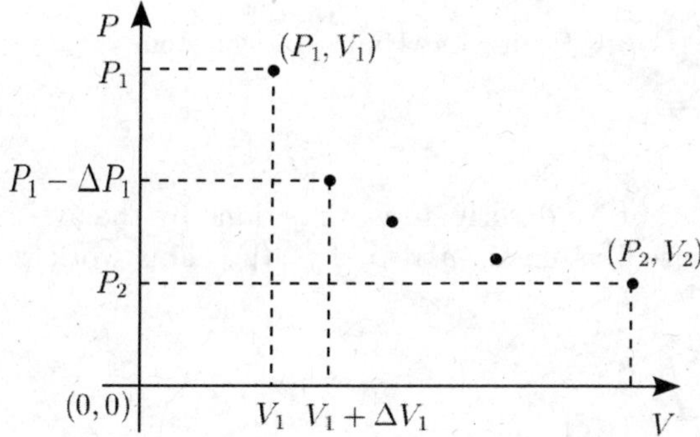

Figure 1.3: Quasi static process

which is determined by the equation of state. For the irreversible process, only the initial state (P_1, V_1) and the final state (P_2, V_2) are shown in figure (1.2).We cannot know the intermediate states because there is no equilibrium. On the other hand, suppose, we let expand volume V_1 to $(V_1 + \Delta V_1)$ i.e. by an infinitesimal amount. Correspondingly, the pressure will decrease to $(P - \Delta P_1)$.We then wait long enough so that turbulence etc dies down and equilibrium state is reached. Then increase volume infinitesimally and decrease P infinitesimally. Wait long enough till an equilibrium is reached. This process can be continued until the state (P_2, V_2) is reached $(V_2 > V_1)$. Since every intermediate state is an equilibrium state, we can represent it on $P - V$ diagram as shown in the figure (1.3). Such a process is called quasi static process. We cannot go from one equilibrium state to another one, which may be very close, by the static process. When we connect all the intermediate points, we get a curve. Assuming every point on the curve drawn is an equilibrium point, we call the process as reversible process. This idealized process is never realized and quasi static process comes close to it. Cylinder with piston and gas in it is called "Hydrostatic

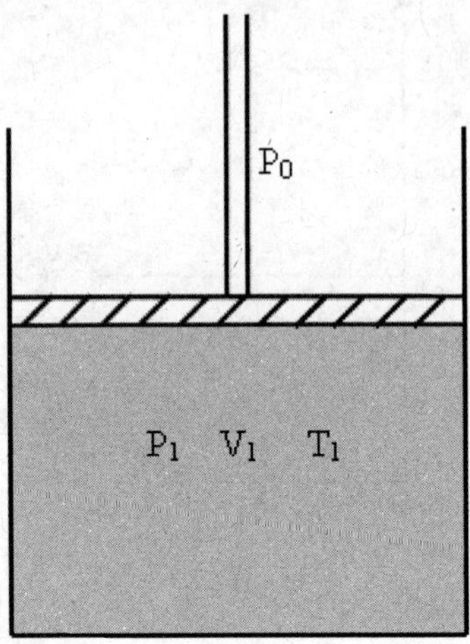

Figure 1.4: Hydrostatic system

system". In the process that we described, work is done by the system against surrounding. This work has its numerical value as, $\delta W = \int P \, dV$. This work is taken to be positive if

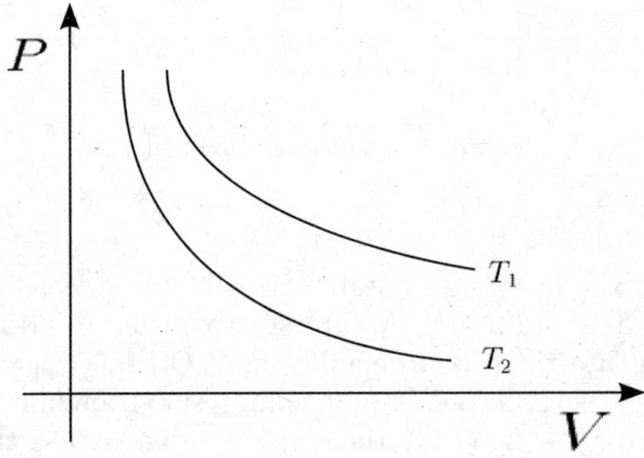

Figure 1.5: Isotherms

the gas is compressed $(V_2 > V_1)$ i.e. work is done against the system. It is,

$$\delta W = - \int_{V_2}^{V_1} P \, dV \tag{1.9}$$

On the other hand, when the gas in the cylinder expands, work is done against surroundings and is taken as negative

$$\delta W = \int_{V_2}^{V_1} P\, dV \tag{1.10}$$

We wrote δW rather than dW because going from initial state (P_1, V_1) to the final state (P_2, V_2) depends upon the path chosen. It must be said that there are many curves which connect two points. Each of this curve is called as a path. Area under the curve or path is amount of work done. The curve P versus V at a constant temperature T is called "isotherm". For every isotherm, there is a fixed temperature as shown in figure (1.5). Two isotherms do not intersect.

1.4.3 Work Done in Various Physical Systems

In all the illustrations of this section we will assume quasi static process.

1. Expansion of a film: Film is a two dimensional surface. Let σ be the surface tension of a film. If an area of the film expands by an amount dA, then work done is,

$$\delta W = \sigma\, dA \tag{1.11}$$

   Here a surface tension is measured in N/m and area in m^2 and work in Joules.

2. Stretching of a wire: Here we have a one dimensional system where a wire is loaded in surrounding atmosphere. Young's modulus Y of the wire is defined to be

$$Y = \frac{stress}{strain} = \frac{FL}{A\,\Delta L} \tag{1.12}$$

   Here, the force F is acting along a wire of length L with ΔL as an increment in its length. A is its cross sectional area. It is assumed that in the process of stretching, cross section of the wire practically remains same. (i.e. its Poisson ratio is negligible.) We make a usual assumption that the stretching is quasi static. Moreover it is isothermal because the wire is in contact of surroundings. Young's modulus depends upon temperature but not on F. Thus in quasi static isothermal stretching, equation (1.12) can be written as

$$Y = \frac{L}{A}\left(\frac{\partial F}{\partial L}\right)_T \tag{1.13}$$

   Force is a tension itself. It is related to the extension of the wire by Hooke's law as,

$$\text{Tension} = -k\,\Delta L \tag{1.14}$$

   where k is spring or Hooke constant. Thermodynamic infinitesimal work done is

$$\delta W = FdL \tag{1.15}$$

Illustrative Problems

Problem 1: In a quasi static isothermal process ,a wire of length L and cross sectional area A is stretched. Its initial tension was F_1 and the final tension is F_2. Neglect the Poisson ratio. Find the work done in the stretching.

Solution: Infinitesimal work done is $\delta W = F dL$. By equation (1.13),

$$dL = \frac{L}{A\,Y} dF \tag{1.16}$$

Thus,

$$W = \int_{F_1}^{F_2} F\, dL = \frac{L}{2A\,Y}(F_2^2 - F_1^2) \tag{1.17}$$

Problem 2: Find the work done when a spherical soap bubble is formed quasi statically and in an isothermal condition. σ is the surface tension and r is a radius of the bubble.

Solution: The bubble has two sides and thus has an area $8\pi r^2$. The work done is $W = 8\pi\sigma r^2$.

Problem 3: Find a work done in changing polarization of a dielectric:

Solution: Consider a slab of a isotropic dielectric material placed between two parallel plates of the parallel plate capacitor[2]. The work done per unit volume due to electric field $\vec{E}$ in moving a charge in dielectric with electric displacement $\vec{D}$ as,

$$\delta W = \frac{1}{2} \int \vec{E} \cdot \delta\vec{D} \tag{1.18}$$

In dielectric, both the positive and negative charges move appropriately and are equal. Then in a quasi static condition,

$$W = \int \int \vec{E} \cdot \delta\vec{D} \, d^3x.$$

But $\vec{D} = \epsilon_0 \vec{E} + \vec{P}$, where $\vec{P}$ is the total polarization per unit volume (MKSI system). Thus,

$$W = \frac{\epsilon_0 E^2 V}{2} + \int \vec{E} \cdot d\vec{P} \, d^3x \tag{1.19}$$

In this equation a first term is a field term which exists even when the dielectric is absent.Therefore the net work done on dielectric is

---

[2]Introduction to electrodynamics by Capri and Panat p-154 and equation 4.120 (Narosa)

$$W = \int \vec{E} \cdot d\vec{P} \, d^3x \tag{1.20}$$

$\therefore$ Infinitesimal work done per unit volume $= \delta W = \vec{E}.d\vec{P}$.

Temperature dependence of a polarization density, or that of dielectric constant is found for a dielectric material as

$$P = \left(a + \frac{b}{T} \right) E \tag{1.21}$$

where a and b are constants. This equation is clearly an equation of state for dielectric.

Problem 4: Find the magnetic work done.

Solution: Consider a paramagnetic sample in external magnetic field H. Then the sample gets magnetized. Following the argument similar to that of dielectric[3], we get the quasi static work done as the sample gets infinitesimal magnetization dM to be

$$\delta W = \vec{H} \cdot d\vec{M} \tag{1.22}$$

For a paramagnetic sample, the equation of state is

$$\vec{M} = \frac{C}{T}\vec{H} \tag{1.23}$$

This is a Curie's law

Problem 5: Find work done, when a van der Waals gas expands from volume V_1 to V_2 in a quasi static and isothermal fashion

solution: van der Waals gas obeys equation of state as,

$$\left(P + \frac{a}{V^2} \right)(V - b) = R\,T \tag{1.24}$$

Clearly,

$$W = -\int_{V_1}^{V_2} P\,dV$$

But

$$P = \frac{RT}{V - b} - \frac{a}{V^2}$$

Work done then is,

$$W = R\,T \ln\left(\frac{V_1 - b}{V_2 - b} \right) + a\left(\frac{1}{V_1} - \frac{1}{V_2} \right) \tag{1.25}$$

---

[3]See Introduction to Electrodynamics by Capri and Panat p-263(Narosa)

1.5   Chemical Potential

Chemical potential μ is an amount of energy necessary to add a single particle to a thermodynamic system. The system is initially in equilibrium. The process of addition of the particle is such that it remains in equilibrium even after the addition. One can get a misconception that this energy should be zero. But this is not so, since the system should remain in equilibrium after an addition of the particle. This is possible if the added particle has an energy comparable to the average energy of a particle in the system. Thus if the particle number of the system changes by dN, then the work done is,

$$\delta W = \mu\, dN \qquad (1.26)$$

If a system under consideration has many species, then,

$$\delta W = \mu_1\, dN_1 + \mu_2\, dN_2 + \mu_3\, dN_3 + ... \qquad (1.27)$$

where dN_i is a change in particle number of ith specie with a chemical potential μ_i. Important point regarding these various work to be noted is that they can be converted in to each other without any restriction.

1.6   Definition of a Mole and the Avogadro Number

One atomic mass unit u is defined as $\frac{1}{12}^{th}$ mass of Carbon isotope $^{12}C_6$. Atomic masses are very accurately measured with mass spectrometer. Mass of this Carbon isotope is taken as a standard for calibration purposes. Avogadro number N_A is defined as the number of $^{12}C_6$ atoms in one gram of $^{12}C_6$. Thus,

$$N_A = \frac{1\ gm}{1u} \cong 6.022 \times 10^{23}\ \text{particles.} \qquad (1.28)$$

The quantity N_A particle is called 1 mole of particles. If mass of atom is m, then the mass of one mole is

$$M = mN_A \qquad (1.29)$$

A mass of one mole of a substance is called as molar mass and is denoted by M. If we have ν moles of a substance of molar mass M then, its mass is simply νM. The equation of an ideal gas is $PV = RT$. This equation is valid for 1 mole of a gas confined in a volume V. R is a gas constant and has a value

$$R \cong 8.314\ J/moleK \qquad (1.30)$$

For an ideal gas with ν moles, the equation of state is,

$$PV = \nu RT \qquad (1.31)$$

1.7   Intensive and Extensive Variables

Consider two systems of volumes V_1 and V_2 of same temperature T and pressure P. Let these two systems come in thermal as well as mechanical contact with each other. Then obviously they will come in thermodynamic equilibrium with each other. The new system will have a volume $V_1 + V_2$ but its pressure and temperature will not change. We thus have two kinds of variables. If a size of a thermodynamic system is scaled by a factor say α some thermodynamic variables do not scale at all. Such variables are called *intensive variables*. Those thermodynamic variables which are scaled by a factor α are called as *extensive variables*.

From the thermodynamic systems mentioned earlier, pressure P, temperature T, tension F, surface tension σ electric field $\vec{E}$ and the magnetic field $\vec{H}$ are intensive quantities.

On the other hand, volume V, length L, area A, total polarization $\vec{P}$, and total magnetization $\vec{M}$ are extensive quantities.

1.8   Response Functions

It is interesting to note that there is a lot of similarity between various thermodynamic variables. By applying a pressure to a confined gas, its volume and temperature changes. A paramagnetic system responds to the application of magnetic field $\vec{H}$ by developing its magnetization. By applying a heat to the thermodynamic system, its temperature goes up. Thus, there is some kind of an external controllable quantity , when applied to the thermodynamic system, the system responds appropriately. The measure of this response is known as a response function. Specific heat is one such response function. It is an amount of heat necessary to raise the temperature of the body by one degree. When heat is supplied at a constant volume to the body then it is called as a specific heat at constant volume , denoted as C_V. Process of supplying the heat at constant pressure can also be realized and corresponding response function is the specific heat at constant pressure C_P.

$$C_V = \left(\frac{\delta Q}{dT}\right)_V \tag{1.32}$$

and

$$C_P = \left(\frac{\delta Q}{dT}\right)_P \tag{1.33}$$

Other response function is compressibility , again evaluated in isothermal situation and is,

$$\kappa = -\frac{1}{V}\left(\frac{\partial V}{\partial P}\right)_T \tag{1.34}$$

Volume expansion coefficient at constant pressure is denoted by β Thus,

$$\beta = \frac{1}{V}\left(\frac{\partial V}{\partial T}\right)_P \tag{1.35}$$

For the magnetic system, response function is susceptibility . Isothermal susceptibility χ for any magnetic system is defined as,

$$\chi = \left(\frac{\partial M}{\partial H}\right)_T \tag{1.36}$$

where M is magnetization. In a similar way, we can define electrical susceptibility.

All these response functions are important because they are experimentally measurable. During the phase transition, their behavior is drastic. As far as thermodynamics is concerned,the response functions with their temperature behaviors is provided. Thermodynamics tells their inter relationship but does not tell you how they depend upon system parameters. Their calculations for various systems is a topic of statistical mechanics.

1.9   Internal Energy

A macroscopic system consists of large number of atoms, molecules, charged particles etc. interacting in a complex manner. For example, charged particles will interact with each other via Lorentz force. Neutral atoms interact among themselves by van der Waals interaction and so on. In spite of such a complex situation for a macroscopic system, we can still define its energy called internal energy of the system. If the macroscopic system has particles which do not interact, then, its energy is sum of energies of each particle. On the other hand if there are interactions, then we have to add potential energy and kinetic energy of each particle of the system. For a quantum macroscopic system, we have to add energy eigen values of the particles. Thus there is a well defined energy of a thermodynamic system. Thermodynamics was developed before microscopic view. Over a period of long time, it was realized that energy is conserved and also, heat is a form of energy. It was Joule who experimentally conclusively demonstrated that the mechanical work and heat are equivalent. His famous experiment was as follows. A highly insulated pot (see figure (1.6)) was filled with water of mass m. Arrangement inside was such that a rod with fins inside could freely churn the water inside. The rod was attached by falling weights. As the weight fell from a height H, work done is $m\,g\,H$. Due to this work, the water inside the pot was churned. The temperature of the water inside the pot rose. The heat generated in this process, equals, $m \times T$ where T is a temperature rise of the water in the pot and specific heat of water is 1. Joule conclusively established that the energy lost because of falling weights (i.e. the work done by the gravity) bears a constant ratio to the amount of heat generated in the pot. Since this ratio is constant, it means heat and mechanical work are different forms of energy and their units are different because of historic reasons.The ratio is called " mechanical equivalent of heat". Since the pot is completely insulated, all work done because of falling weights goes in increasing of the heat content of the water, i.e. increasing energy of the water.Thus, on the macroscopic scale, energy function can be defined. This means macroscopic systems have precise and measurable energies. The macroscopic energies obey energy conservation principle. Only differences in energy (rather than its absolute value)

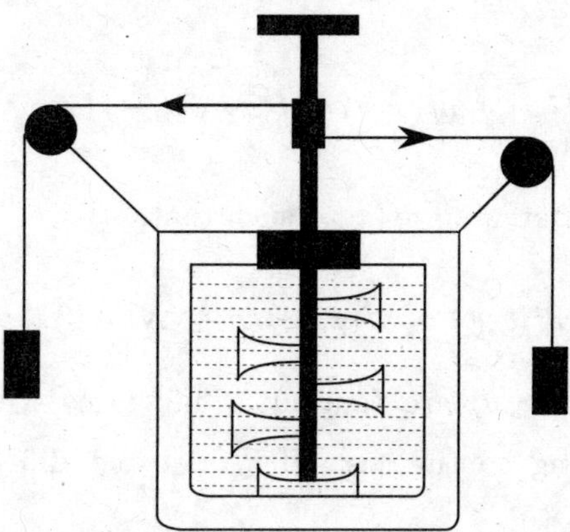

Figure 1.6: Joule's Apparatus

have a physical significance.We then choose a particular state of the system and assign its energy to be zero. We call the energy of macroscopic system as an internal energy and will denote it by E. Internal energy is an extensive quantity. It is a function of a temperature T, volume V of the system and of mole number of chemical constituent. Some other books on thermodynamics denote internal energy by U. Historically, the unit of heat is calories (abbreviated as " cal "). It is amount of heat necessary to raise temperature of one gram of water by one degree Celsius. However, specific heat of water varies slowly with temperature. This necessitated redefining calorie. It is an amount of heat necessary to raise a temperature of one gram of water from $14.5\ ^0C$ to $15.5\ ^0C$. In this unit the mechanical equivalent of heat is 4.1860 J/cal. Even now, chemists and physicists use calorie as a unit of heat.

In conclusion, we see that there exist a thermodynamic function called internal energy E. If an adiabatically sealed thermodynamic system passes through states 1 and 2, doing a work W_{12} adiabatically, then, clearly this work done is equal the difference in internal energy between the states 2 and 1. Thus, $W_{12} = E_2 - E_1$.

1.10   Exact and Inexact Differentials

Exact and inexact differentials play an important role in thermodynamics. This is mathematical but very important topic. Consider amount of work $\delta W = -P\,dV$. Going from a state (P_1, V_1, T_1) to (P_2, V_2, T_2), the work done $\int \delta W$ depends upon the path chosen and not on starting and end points alone. Thus, if δf is an inexact differential, then,

$$\int_{(x_1,y_1)}^{(x_2,y_2)} \delta f(x,y) \neq f(x_2,y_2) - f(x_1,y_1) \tag{1.37}$$

but its value will depend upon the path taken to reach (x_2, y_2) from (x_1, y_1). On the other hand, if $df(x,y)$ is an exact differential , then,

$$\int_{(x_1,y_1)}^{(x_2,y_2)} df(x,y) = f(x_2,y_2) - f(x_1,y_1) \tag{1.38}$$

Consider an exact differential $df(x,y)$ such that,

$$df(x,y) = M(x,y)\,dx \,+\, N(x,y)\,dy \tag{1.39}$$

Here $M(x,y)$ and $N(x,y)$ are continuous functions with their partial derivatives $\dfrac{\partial M}{\partial x}, \dfrac{\partial M}{\partial y}, \dfrac{\partial N}{\partial x}, \dfrac{\partial N}{\partial y}$, being continuous. If df is exact differential, then, $M(x,y)$ and $N(x,y)$ satisfy,

$$\left(\frac{\partial M}{\partial y}\right)_x = \left(\frac{\partial N}{\partial x}\right)_y \tag{1.40}$$

Clearly,

$$\left(\frac{\partial f}{\partial x}\right)_y = M(x,y)$$

$$\left(\frac{\partial f}{\partial y}\right)_x = N(x,y)$$

With these equations, we get,

$$\left(\frac{\partial^2 f}{\partial y \partial x}\right) = \left(\frac{\partial^2 f}{\partial x \partial y}\right) \tag{1.41}$$

If the equation (1.38) can be written as $df = \nabla f \cdot \vec{dl}$, then df is an exact differential. Here $\vec{dl} = \hat{i}dx + \hat{j}dy$. Let $\vec{F} = \nabla f$. Then clearly, $\nabla \times F = 0$. This implies $\left(\dfrac{\partial^2 f}{\partial y \partial z}\right) = \left(\dfrac{\partial^2 f}{\partial z \partial y}\right)$ and two similar equations of second derivative of f w.r.t. x and y and x and z. In case $\vec{F} \cdot \vec{dl}$ is not exact differential, we can make an exact differential by multiplying it with another function $g(x,y)$ This function g is called as integrating factor. Thus $[g(x,y)\vec{F} \cdot \vec{dl}]$ is a perfect differential. Then,

$$\frac{\partial[gM]}{\partial y} = \frac{\partial[gN]}{\partial x} \tag{1.42}$$

To solve the partial differential equation (1.41), we seek a form of g as $g(x,y) = g_1(x)\,g_2(y)$. After some simplification, we get,

$$M(x,y)\frac{d\ln g_2(y)}{dy} + \frac{\partial M}{\partial y} = N(x,y)\frac{d\ln g_1(x)}{dx} + \frac{\partial N}{\partial x} \qquad (1.43)$$

Equation (1.43) can be solved if explicit forms of M and N are known. Clearly $g(x,y)$ is an integrating factor.

An exact differential is also called as a perfect differential.

1.11   Useful Mathematical Relations

If we have a functional relationship $f(x,y,z) = 0$ then x can be thought as a function of y and z. Then,

$$dx = \left(\frac{\partial x}{\partial y}\right)_z dy + \left(\frac{\partial x}{\partial z}\right)_y dz. \qquad (1.44)$$

Similarly y can be thought as a function of x and z. Thus,

$$dy = \left(\frac{\partial y}{\partial x}\right)_z dx + \left(\frac{\partial y}{\partial z}\right)_x dz. \qquad (1.45)$$

Then,

$$dx = \left(\frac{\partial x}{\partial y}\right)_z \left(\frac{\partial y}{\partial x}\right)_z dx + \left(\frac{\partial x}{\partial y}\right)_z \left(\frac{\partial y}{\partial z}\right)_x dz + \left(\frac{\partial x}{\partial z}\right)_y dz. \qquad (1.46)$$

Because of a functional relationship between x, y, and z exists, only two of them are independent. We choose x and z to be independent variables. We also choose $dz = 0$ and $dx \neq 0$. Then,

$$\left(\frac{\partial x}{\partial y}\right)_z \left(\frac{\partial y}{\partial x}\right)_z = 1. \qquad (1.47)$$

or,

$$\left(\frac{\partial x}{\partial y}\right)_z = 1./(\partial y/\partial x)_z \qquad (1.48)$$

If we choose $dx=0$, and use equation (1.48), then

$$\left(\frac{\partial x}{\partial y}\right)_z \left(\frac{\partial y}{\partial z}\right)_x \left(\frac{\partial z}{\partial x}\right)_y = -1. \qquad (1.49)$$

If x,y,z are P, V, and T variables, then,

$$\left(\frac{\partial P}{\partial V}\right)_T \left(\frac{\partial V}{\partial T}\right)_P \left(\frac{\partial T}{\partial P}\right)_V = -1. \qquad (1.50)$$

But from a definition of response functions, volume expansion coefficient β and isothermal compressibility (see equations 1.35 and 1.36) κ, we get,

$$\left(\frac{\partial P}{\partial T}\right)_V = \frac{\beta}{\kappa} \qquad (1.51)$$

Clearly, at constant volume, if we increase the temperature from T_1 to T_2, then a change in pressure is,

$$P_2 - P_1 = \int_{T_1}^{T_2} \frac{\beta}{\kappa}\, dT \qquad (1.52)$$

It may so happen that T_1 and T_2 differ by a small amount like , say, $T_1 = 20^oC$ and $T_2 = 30^oC$. Also there is no abnormality such as a phase transition of any kind in the temperature range, then we get,

$$P_2 - P_1 \cong \frac{\beta}{\kappa}(T_2 - T_1) \qquad (1.53)$$

Illustrative Problems

Problem 6: Given, $dF = 2xy^3 dx + 3x^2 y^2 dy$, show that it is a perfect differential. Show also that $\int dF$ evaluated along a path $(0,0) \to (1,1)$ is same if evaluated along $(0,0) \to (0,1) \to (1,1)$.

Solution: $M = 2xy^3$ and $N = 3x^2 y^2$. Then, $\left(\frac{\partial M}{\partial y}\right)_x = \left(\frac{\partial N}{\partial x}\right)_y$. Thus df is a perfect differential. We integrate along S_1.

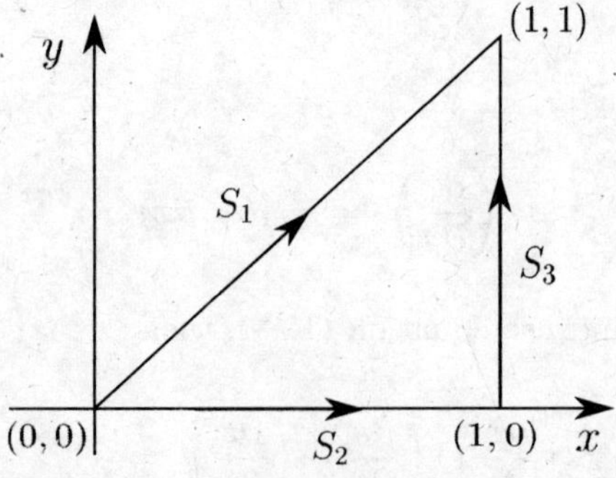

Figure 1.7: Figure for problem 6

$$\int_{S_1} (2xy^3 dx + 3x^2 y^2 dy) = \int_{S_1} (2x^4 + 3x^4)dx \text{ which integrates to 1 from 0 to 1.}$$

On the other hand, along S_2, from $(0,0)$ to $(1,0)$, $y = 0$ implying $dy = 0$. For integration along $(1,0)$ to $(1,1)$, $x = 1$ and $dx = 0$. It is then easy to see that

$$\int_{S_2} dF = \int_1^1 3y^2 dy = 1$$

Problem 7: Find integrating factor for $\delta F = 2x^2 y^3 dx + 3x^3 y^2 dy$

Solution: By taking appropriate derivatives $M_y = 6x^2 y^2$ and $N_x = 9x^2 y^2$. We see that dF is not exact. Let, $g(x,y) = g_1(x)g_2(y)$. Then, after some simplifications,

$$2x^2 y^3 \, g_1(x)\frac{dg_2(y)}{dy} + 6x^2 y^2 g_1(x)g_2(y) = 9x^2 y^2 g_1(x)g_2(y) + 3x^3 y^2 g_2(y)\frac{dg_1(x)}{dx}.$$

Dividing by $g_1 g_2$ and simplifying, we get,

$$\frac{2y}{g_2(y)}\frac{dg_2(y)}{dy} = c = 3 + \frac{3x}{g_1(x)}\frac{dg_1(x)}{dx}.$$

We have equated each side to a constant c because x and y are independent variables. These differential equations can be solved immediately to give,

$$g(x,y) = g_1(x)g_2(y) = x^{c/3-1}y^{c/2}$$

Then $g(x,y)\delta F$ in the problem becomes exact differential.

Convention of Notation Regarding d and δ

We will now make a convention throughout this book that symbols like dF, dE, dH etc will represent perfect differentials. Symbols like δF, δW etc will not be perfect differentials but merely small quantities of F, W etc.

1.12 Units of Pressure and Volume

Strictly speaking pressure is a tensor. We normally take it as a diagonalized form. Pressure is force per unit area. Here a perpendicular component is taken per unit area. MKSI unit of pressure is pascal. Pascal is $Newton/meter^2$. There are other units of pressure. One of the most popular one is atmosphere. It is a pressure exerted by a column of 76 centimeter of Mercury. Units of volume are either cm^3 or m^3.

Short Questions

1. What is meant by the equation of state? Write equation of state for an ideal gas. Write van der Waals equation of state explaining the meaning of each term.

2. Write equation of state for a paramagnetic salt.

3. Why a thermodynamic variable temperature T has absolute zero?

4. For the following systems state the microscopic and macroscopic variables.

    (a) Magnetic system.

    (b) A gas confined in a cylinder.

    (c) Ferro and Para electric systems.

    (d) Surface of any material.

    (e) Super conducting system.

5. Define what is thermodynamic equilibrium?

6. Explain what is meant by a quasi static process?

7. Is the work done by or on the thermodynamic system a perfect differential? Explain.

8. Find the work done in isothermal process if the volume of a one mole of an ideal gas expands from V_1 to V_2.

$$\text{Ans:} R \ln \frac{V_2}{V_1}$$

    What would be its value if the gas has ν moles?

9. What is meant by a chemical potential? If a system has two species of N_1 and N_2 molecules, and one particle of each specie is added to the system, what is the work done?

10. State what is meant by Avogadro's number.

11. Give the examples of extensive and intensive quantities of various thermodynamic systems.

12. Explain what are response functions? State some of them with concrete examples.

13. What is a micro variable and a macro variable for an electro chemical cell (battery)?

14. What is mechanical equivalent of heat? Describe Joule's experiment stating its implications.

15. What is the difference between exact and inexact differential? Give their examples from various thermodynamic systems.

16. Draw the isotherms for paramagnetic systems explaining their shapes.

17. Electrochemical cell with electrodes provide a voltage for a circuit. When the circuit is complete, electro chemical reaction in the battery starts and the charge is transferred from one electrode to the other. Guess the correct thermodynamic variable for the cell.

    Answer: 1)Transferred charge Z Coulomb. 2)Voltage across the terminals (Emf) in Volts. 3) Absolute temperature T in Kelvins.

18. Why the change in internal energy is a measure of the work done by thermodynamic system in an adiabatic process?

19. Why is dE exact?

20. Work done during the adiabatic process in going from a state 1 to state 2 is

$$W_{12} = E_2 - E_1$$

claim: In the above adiabatic process, whether it is quasi static or irreversible, W_{12} is given by the above expression. Justify this claim.

21. We make following statements. Justify them.

a)Internal energy of the magnetic system is a function of absolute temperature and its magnetization.

b)Internal energy of a stretched wire is a function of an absolute temperature and a strain in the wire.

c)Internal energy of a thin film is a function of an absolute temperature and its area.

22. If dq is a charge that flows through an external circuit connected to the battery of emf E; argue that the thermodynamic work done is

$\delta W = Edq = EI\,dt$ on the surrounding during the discharge process.

Problems

Problem 1. Show that $\delta F = yx\,dx + x^2\,dy$ is not exact. Show that $g(x, y) = 1/x$ makes $g(x, y)\,\delta F$ exact.

Problem 2. Points A, B, C have co ordinates (1,1); (2,1); and (2,2) respectively. Integrate $\delta F = (x^2 - y)\,dx + x\,dy$ along the line AC and also along the line ABC. Find its integrating factor. (See figure (1.8))

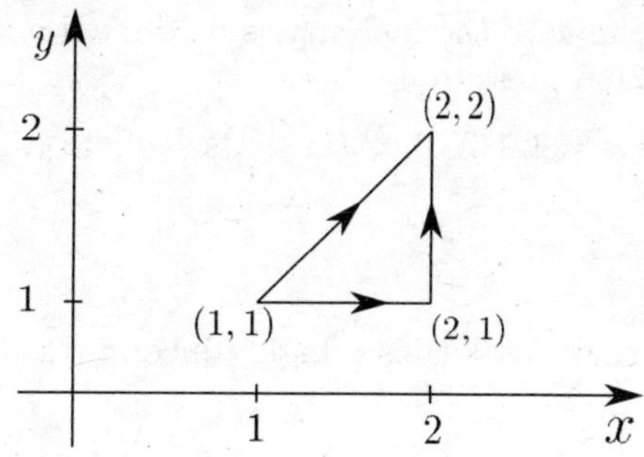

Figure 1.8: Figure for problem 2

Ans: Integral along $AC = 7/3$. Along ABC $10/3$ and $g(x, y) = 1/x^2$.

Problem 3. Suppose a relationship exists between x, y and z and we are also given a function $f(x, y, z)$. Show that,

$$\left(\frac{\partial x}{\partial y}\right)_f \left(\frac{\partial y}{\partial z}\right)_f \left(\frac{\partial z}{\partial x}\right)_f = 1$$

and,

$$\left(\frac{\partial x}{\partial f}\right)_z = \left(\frac{\partial x}{\partial f}\right)_y + \left(\frac{\partial x}{\partial y}\right)_f \left(\frac{\partial y}{\partial f}\right)_z$$

Problem 4. During a quasi static expansion of a gas in an adiabatic container, following relationships exists.

$$P V^\gamma = \text{constant}; \ \gamma = C_p/C_v$$

Find the work done in expansion of a gas from (P_1, V_1) to (P_2, V_2) during an adiabatic process.

$$\text{Ans: } - \frac{P_1 V_1 - P_2 V_2}{\gamma - 1}$$

Problem 5. Let σ_W be a surface tension of a clear water surface. Let σ be a surface tension of water covered by the oil film. Then experimentally it is found that, in a restricted range of area values A, the equation of state is given by

$(\sigma - \sigma_W) A = aT$. where a is a constant. Find the work done when the area of the oil film increases from A_1 to A_2 in an isothermal manner.

$$\text{Ans: } W = \sigma_W (A_2 - A_1) + aT \ln(A_2/A_1)$$

Problem 6. A mass of mercury is held in a container at one atmospheric pressure $(10^5 Pa)$ at a temperature $20\,^\circ C$. Volume of the mercury is held fixed and its temperature is raised from $10^\circ C$ to $20\,^\circ C$. Find final pressure.

Data: $\beta = 1.8 \times 10^{-4} K^{-1}$; $\kappa = 4.0 \times 10^{-11} Pa^{-1}$ Use equation (1.51).

$$\text{Ans: } 451 \text{atm}$$

This problem shows that mercury is a highly incompressible substance.

Chapter 2

THERMODYNAMICS II: THE LAWS OF THERMODYNAMICS

After the preliminary discussion of various concepts useful for understanding of thermal Physics in chapter 1, we will try to understand inter conversion of various forms of work with heat and vice versa. This understanding is based upon four empirical laws of thermodynamics which are postulates and can be understood completely using microscopic and statistical ideas. The four laws are named as

1. Zeroth law: This law deals with the equilibrium of thermodynamic systems.

2. First law: This law deals with the conservation of energy.

3. Second law: The second law relates change of a state function called entropy with a transfer of heat in to the system at some absolute temperature.

4. Third law (Nernst's Theorem): The third law relates a value of entropy at absolute zero which is either zero (in most cases) or some constant value depending upon the system.

2.1 Zeroth Law of Thermodynamics

Thermodynamic state is defined by macroscopic variables. These variables can be defined only in an equilibrium state. In equilibrium state, these variables do not change with time.

Thermometer or any temperature sensitive sensor can be used to find temperature of the system. It is a well known experience that if the two thermodynamic systems are brought in contact with each other, then there is a transfer of heat if the two have different temperature. This transfer continues until the temperatures of the two system become equal. Then we say that the two systems are in equilibrium. Consider now three systems A, B and C. Suppose that A and B are in equilibrium, and, so also B and C are in equilibrium. Then in that case A and C are also in equilibrium with a common thermodynamic property of all the three systems. This common property is their temperature. This empirical fact is known as "Zeroth law of thermodynamics".

We define generalized coordinates x_i with generalized forces X_i such that the work done due to change in i^{th} coordinate is $dW_i = X_i dx_i$. In general, total work done is then

$$dW = \sum_{i=1}^{s} X_i dx_i \qquad (2.1)$$

where s are generalized coordinates. x_i are sometimes called as deformation coordinates. The deformation coordinates would be volume V, area A, magnetization M, strain etc. Corresponding generalized forces are X_i as pressure P, surface tension σ, magnetic field H, stress in the wire etc.

We define a simple thermodynamic system as the one which is described by two variables (so called parameters of the state). We will for example consider a typical system of a gas confined in a cylinder. The parameters of the state are pressure P and its volume V.

Postulate 1: Equilibrium state of a simple thermodynamic system is defined by its two thermodynamic parameters.

Thus for a gas enclosed in a cylinder, its volume can not specify equilibrium. We have to also know corresponding pressure.

Postulate 2: If two simple systems enclosed in a adiabatic vessels are brought in contact with each other through a diathermic wall, then in equilibrium, the four parameters of the two systems (e.g. $P_1, V_1; P_2, V_2$) cannot take arbitrary values but are related with each other.

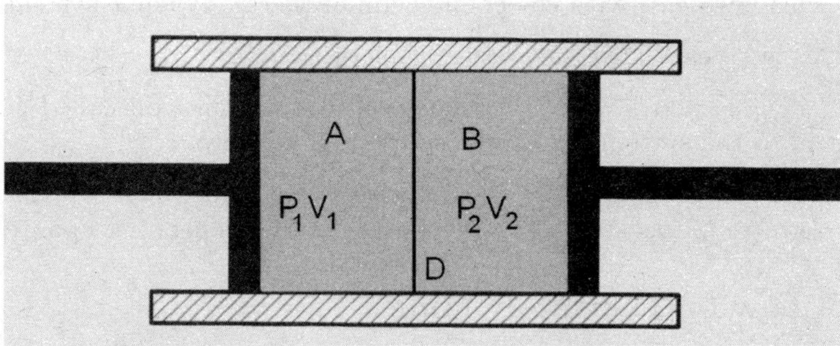

Figure 2.1: Two isolated systems separated by a diathermic wall.

To illustrate the postulate 2, we have two otherwise isolated systems (see figure 1) that are brought in contact with each other by a diathermic wall D. Then the second postulate states that

$$F(P_1, V_1; P_2, V_2) = 0 \qquad (2.2)$$

This function F can be found experimentally.

Postulate 3: If two simple systems are each in thermal equilibrium with a third simple system, then they are in thermal equilibrium with each other.

Suppose systems 1 and 2 have parameters P_1, V_1 and P_2, V_2 respectively and each is in equilibrium with the third system of parameters P_3 and V_3; then postulate 2 imply that

$$F_2(P_1, V_1; P_3, V_3) = 0 \text{ and } F_1(P_2, V_2; P_3, V_3) = 0 \qquad (2.3)$$

Postulate 3 then implies

$$F_3(P_1 V_1; P_2, V_2) = 0 \qquad (2.4)$$

By solving equation (3), we get a general functional relation as ,

$$P_3 = f_2(P_1, V_1; V_3) = f_1(P_2, V_2; V_3). \qquad (2.5)$$

Equation (5) is valid for any V_3 and is equivalent to equation (4). As V_3 does not appear in equation (4) it will not appear in equation (5) and we thus get

$$f_1(P_1, V_1) = f_2(P_2, V_2).$$

We write the 'temperature' θ_1 of the system 1 as $\theta_1 = f_1(P_1, V_1)$ and 'temperature' θ_2 of the system 2 as $\theta_2 = f_2(P_2, V_2)$. Then for thermal equilibrium, $\theta_1 = \theta_2$. By zeroth law, if $\theta_1 = \theta_3$ and $\theta_2 = \theta_3$ then $\theta_1 = \theta_2$. This quantity θ is called empirical temperature.

Simplest example is ideal gas where $PV = RT$ holds. The functional relationship $f(P, V) = 0$ is $PV - RT = 0$.

2.2   First Law of Thermodynamics

Consider an adiabatic process in which initial state i changes to final state f. Then the change in internal energy is $(E_f - E_i)$, which is equal to the work required to change the state of the system.

On the other hand, if we change same initial state i to the same final state f but in an isothermal manner, then the change in internal energy $(E_f - E_i)$ is not a work done to achieve this change of state. To resolve this dilemma, we are forced to accept a transfer of energy (so called heat) from the surrounding of the system during the process, provided the principle of conservation of energy holds. Joule in his experiments have shown the equivalence of heat and energy. We denote the transfer of heat energy Q during isothermal change of state from the state i to the state f. The conservation of energy then gives,

$$E_f - E_i = W + Q \qquad (2.6)$$

where W is isothermal work done. Equation (6) is first law of thermodynamics. In a differential form the first law of thermodynamics can be written as,

$$dE = \delta Q + \delta W \qquad (2.7)$$

Here dE is a perfect differential. It is a change in the internal energy during infinitesimal change of thermodynamic system. δQ is small transfer of energy during the same infinitesimal change of state where the work done is δW. The quantities δW and δQ are not perfect differentials. They are sometimes called inexact differentials because, they depend upon the path taken by the system to go from its initial state to its final state infinitesimally close to it. For various systems equation (7) is written as

$$
\begin{array}{llll}
(i) & \delta Q = dE + PdV & & \text{for hydrostatic system} \\
(ii) & \delta Q = dE + PdV - HdM & & \text{for paramagnetic system} \\
(iii) & \delta Q = dE + PdV - \mu_1 dN_1 - \mu_2 dN_2... & \text{Multicomponent hydrostatic system} \\
(iv) & \delta Q = dE - \sigma dA & & \text{surface system}
\end{array}
\tag{2.8}
$$

It is important to note that, equation (7) is nothing but an energy conservation principle. Change in internal energy is an exact differential. It therefore depends upon the initial and final thermodynamic state and does not depend upon how the final state is reached from initial state. It is therefore irrelevant whether the changes δW and δQ occur through a reversible or irreversible process. Thus,

$$
dE = \delta W)_{rev} + \delta Q)_{rev} = \delta W)_{irr} + \delta Q)_{irr}.
\tag{2.9}
$$

where the subscript 'rev' represents reversible ($\cong$ quasistatic) process and 'irr' represent an irreversible process. In general $\delta W)_{irr}$ is larger than $\delta W)_{rev}$. Thus,

$$
\delta W)_{irr} > \delta W)_{rev} = -PdV
\tag{2.10}
$$

Then, from equation (9) we get

$$
\delta Q)_{rev} > \delta Q)_{irr}
$$

It thus appears that part of the work is always converted in to heat in irreversible process. This heat is radiated out.

Since an internal energy of the system depend only upon its thermodynamics state we realize that for a cyclical process,

$$
\oint dE = 0
$$

The integral is path (contour) independent.

2.3   Ideal Gas

Ideal gas, as defined earlier is the one where its atomic and/or molecular constituents do not interact with each other. As a result its constituents have a kinetic energy only. We will show subsequently that the internal energy of an ideal gas is nothing but is a product of the average kinetic energy of the constituent and total number of the constituents (atoms and/or molecules). We will show that the average kinetic energy of the constituent (hereafter referred as the molecule in general) is related with absolute temperature T. It is further shown that for an each degree of freedom of molecule an average energy is $\dfrac{k_B T}{2}$ where k_B is known as the Boltzmann constant with a value

$$
\begin{aligned}
k_B &= 1.381 \times 10^{-23} J/K \\
&= 1.382 \times 10^{-16} ergs/degree
\end{aligned}
\tag{2.11}
$$

When we have a 1 gram-mole of the gas then it contains the number of molecules equal to the Avogadro number N_A. The value of N_A is estimated to be $N_A = 6.022 \times 10^{23}$. The value of the gas constant R is

$$
\begin{aligned}
R &= N_A k_B = 6.022 \times 1.381 \times 10^{-23} \times 10^{23} \\
&= 8.314 \ J/molK
\end{aligned}
$$

For one gram-mole of a gas, the ideal gas equation is

$$PV = N_A k_B T = RT \tag{2.12}$$

The internal energy of an ideal gas is purely a function of temperature. It does not depend upon its volume V. On the other hand in a real gas (where gas - gas molecules are interacting amongst each other) internal energy is a function of both T and V. Thus

$$
\begin{aligned}
E(T,V) &= E(T) \qquad \text{for ideal gas} \\
E(T,V) &= E(T,V) \qquad \text{for real gas}
\end{aligned}
$$

Thus for an ideal gas

$$\left(\frac{\partial E}{\partial V}\right)_T = 0 \ . \tag{2.13}$$

and

$$\left(\frac{\partial E}{\partial P}\right)_T = 0 \ . \tag{2.14}$$

By first law of thermodynamics,

$$\delta Q = dE + PdV \ . \tag{2.15}$$

Clearly,

$$\delta Q = C_V dT + PdV \tag{2.16}$$

for an ideal gas. The equilibrium states of an ideal gas are represented by an ideal gas equation

$$PV = \nu RT$$

where ν are the number of moles of the gas.

Simple Ideal Gas Relations

1. $C_P - C_V = \nu R$
This simple relation can be proved as follows.
Differentiating the gas law we get,

$$PdV + VdP = \nu RdT \ .$$

Using equation (15) we get

$$\delta Q = (C_V + \nu R)\, dT - VdP \tag{2.17}$$

The specific heat C_P at constant pressure is defined as

$$C_P = \left(\frac{\delta Q}{\delta T}\right)_P .$$

Using equation (16) we obtain

$$C_P - C_V = \nu R \tag{2.18}$$

Like equation (15) we obtained another useful relation

$$\delta Q = C_P dT - V dP. \tag{2.19}$$

2. $PV^\gamma = constant$

The ratio C_P/C_V is an important parameter for any gas and is denoted by γ. It is useful in discussing in adiabatic processes . In adiabatic processes, $\delta Q = 0$.

$$C_P dT = V dP . \tag{2.20}$$

From equation (16)

$$C_V dT = -P dV . \tag{2.21}$$

Thus,

$$\frac{dP}{P} = -\frac{C_P}{C_V}\frac{dV}{V} = -\gamma\frac{dV}{V} \tag{2.22}$$

Strictly speaking γ depend upon the temperature for real gases. For an ideal gas, γ is constant with a value 5/3 for mono-atomic dilute gases. Its value is 7/5 for the di-atomic dilute gases such as H_2, O_2, N_2 etc. For the ideal gas then equation (22) can immediately be integrated for the quasi static, adiabatic processes to give

$$PV^\gamma = constant \tag{2.23}$$

which is equivalent to

$$\frac{T}{P^{(\gamma-1)/\gamma}} = constant \tag{2.24}$$

Illustrative Problems

Problem 1: Consider a real gas where its internal energy E can be considered as a function of T and P or of T and V. Derive:

$$1. \quad \delta Q = \left[\left(\frac{\partial E}{\partial T}\right)_P + P\left(\frac{\partial V}{\partial T}\right)_P\right] dT + \left[\left(\frac{\partial E}{\partial P}\right)_T + P\left(\frac{\partial V}{\partial P}\right)_T\right] dP$$

$$2. \quad \delta Q = \left(\frac{\partial E}{\partial P}\right)_V dP + \left[\left(\frac{\partial E}{\partial V}\right)_P + P\right] dV$$

$$3. \quad \left(\frac{\partial E}{\partial T}\right)_P = C_P - \beta PV$$

$$4. \quad \left(\frac{\partial E}{\partial P}\right)_V = \frac{C_V \kappa}{\beta}$$

$$5. \quad \left(\frac{\partial E}{\partial P}\right)_T = PV\kappa - (C_P - C_V)\kappa/\beta$$

$$6. \quad \left(\frac{\partial E}{\partial V}\right)_P = \frac{C_P}{V\beta} - P \tag{2.25}$$

Solution: Given $E = E(T, P)$ and $V = V(T, P)$,

$$dE = \left(\frac{\partial E}{\partial T}\right)_P dT + \left(\frac{\partial E}{\partial P}\right)_T dP$$

$$dV = \left(\frac{\partial V}{\partial T}\right)_P dT + \left(\frac{\partial V}{\partial P}\right)_T dP.$$

1. We write $\delta Q = dE + PdV$, substitute and get the result.

2. Regard $E = E(P, V)$

$$dE = \left(\frac{\partial E}{\partial P}\right)_V dP + \left(\frac{\partial E}{\partial V}\right)_P dV$$

Put in first law and get the result.

3. Using the response functions defined earlier

$$\left(\frac{\partial V}{\partial T}\right)_P = \beta V, \quad \left(\frac{\partial V}{\partial P}\right)_T = -\kappa V \quad \text{and} \quad \left(\frac{\partial P}{\partial V}\right)_T = -\frac{B}{V}$$

From (1),

$$\left(\frac{\delta Q}{dT}\right)_P = C_P = \left(\frac{\partial E}{\partial T}\right)_P + \beta PV$$

4. From (2)

$$\left(\frac{\delta Q}{dT}\right)_V = C_V = \left(\frac{\partial E}{\partial P}\right)_V \left(\frac{\partial P}{\partial T}\right)_V.$$

But

$$\left(\frac{\partial P}{\partial T}\right)_V = \frac{\beta}{\kappa}$$

Therefore

$$\left(\frac{\partial E}{\partial P}\right)_V = \frac{C_V}{\beta}\kappa.$$

From this it is clear that by measurement of various response functions we can determine the behavior of an internal energy with pressure.

5.   Problem (1) can be written as

$$\delta Q = \left[\left(\frac{\partial E}{\partial T}\right)_P + \beta PV\right] dT + \left[\left(\frac{\partial E}{\partial P}\right)_T - \kappa PV\right] dP$$

Thus,

$$\left(\frac{\delta Q}{dT}\right)_V = C_V = \left(\frac{\partial E}{\partial T}\right)_P + \beta PV + \left[\left(\frac{\partial E}{\partial P}\right)_T - \kappa PV\right]\left(\frac{\partial P}{\partial T}\right)_V$$

But using (3)

$$C_V = C_P + \left[\left(\frac{\partial E}{\partial P}\right)_T - \kappa PV\right]\frac{\beta}{\kappa}$$

or

$$\left(\frac{\partial E}{\partial P}\right)_T = \kappa PV - \frac{(C_P - C_V)\kappa}{\beta}$$

6.   From (2)

$$\left(\frac{\delta Q}{dT}\right)_P = \left[\left(\frac{\partial E}{\partial V}\right)_P + P\right]\left(\frac{\partial V}{\partial T}\right)_P = C_P$$

Thus

$$\frac{C_P}{\beta V} - P = \left(\frac{\partial E}{\partial V}\right)_P$$

Problem 2. In Joule's experiment, one mole of water is put in the pot of his experiment where the pedals dip completely in the water. The pot is insulated. Weights of $0.427\ kg$ rotate pedals and fall through a one meter height. The temperature of the water increases from 15.5^oC to 16.5^oC. Find how much heat is created.

Answer: Work done $= mgh = 0.427 \times 9.8 \times 1\ J = 4.1846\ J$. This is a unit of heat called 'calorie'. This unit of heat was used earlier where it was defined as an amount of heat necessary to raise a temperature of one gram of water by one degree Celsius. This is a C.G.S. unit of heat and is still commonly used by chemists and physicists.

2.4   Carnot's Theorem and the Second Law

Reversible and Irreversible Processes

So far, we have come across a state function namely the internal energy. A state function depends on equilibrium values of the state coordinates of the thermodynamic system such as P, V, T, H, M, σ, A etc. We need to know one more state function which will turn out to be the entropy S. We have already seen in Chapter (1) and figure (2) that we can have equilibrium states on $P - V$ diagram very close to each other. But while going from one equilibrium state to another closely spaced equilibrium state, there are non-equilibrium states. It is not possible to go from one equilibrium state to the other closely spaced equilibrium state in a continuous manner. This is because a slight change in pressure (or a generalized force in general) causes violent changes like turbulence near the walls of the piston and cylinder which propagates in the gas. Collisions among the molecules of the gas restores a new equilibrium state. This takes some finite time called relaxation time τ. If our time of measurement t is such that $t >> \tau$ we see only the equilibrium states. On the other hand, if $t < \tau$, we see a non-equilibrium situation where the molecules readjust themselves to reach an equilibrium. Figure (2) of chapter 1 clearly shows a quasi-static process. There is an idealized process called a reversible process which is obtained by smoothly joining the points on quasi-static curve. In reality, this reversible process is not realized. Quasi static processes can come as close to the reversible process as possible. We are thus joining two closely space equilibrium points for our own convenience. *This implies that there is no curve* $df(P, V, T) = 0$ *which can connect two closely spaced equilibrium points by going through all intermediate equilibrium points in a continuous manner.*

All processes performed by the macro system in nature are not reversible. They are irreversible. e.g. a compressed spring when set free in air oscillates. The oscillations of spring die down. The potential energy in a compressed spring is released in the surrounding air which gets heated. In a reverse situation even if we cools the surrounding air, the spring never starts oscillating and finally gets compressed on its own. Even though in the reverse process like the spring getting compressed, there is no violation of the principle of conservation of energy the reverse process does not occur. Moreover, there is a friction, with frictional heat losses.

Carnot's Theorem

In a reversible cyclic process,
$\oint \dfrac{\delta Q}{T} = 0$ *where δQ is infinitesimal amount of heat taken in or given out by the system at a temperature T when the system undergoes a cyclic process.*

We first state the necessity of a cyclic process. A state of a thermodynamic system is defined by specifying generalized force and corresponding deforming coordinate (e.g. a pair (P, V), (H, M) etc.). The third coordinate T is decided by the equation of the state. If by some process like isothermal or adiabatic or isochoric, initial state i of the system (P_i, V_i) changes to final state (P_f, V_f) and then by any of the above processes the system comes back to its initial state i, we have the cyclic process. Cyclic processes are important because the engine works with cyclic process. We can convert a rotational motion in to a linear motion. Rotational motion to the wheel can be given by a cyclic process. This motion can be converted to a linear motion by an off axis shaft attached to the wheel and to the piston of the engine.

Simplest but most important cyclic process was discovered by Carnot. Working substance of the Carnot engine is an ideal gas. The engine operates between two reservoirs, one at a high temperature T_H and the other at low temperature T_L. Four processes are performed in the following sequence. All processes described are idealized and therefore reversible. Consider the Carnot cycle picturised in figure (2). Here we have two heat

reservoirs, one at a temperature T_H and the other at temperature T_L such that $T_H > T_L$.
(1) The engine starts with a initial state of (P_1, V_1) at A. This means ideal gas enclosed in

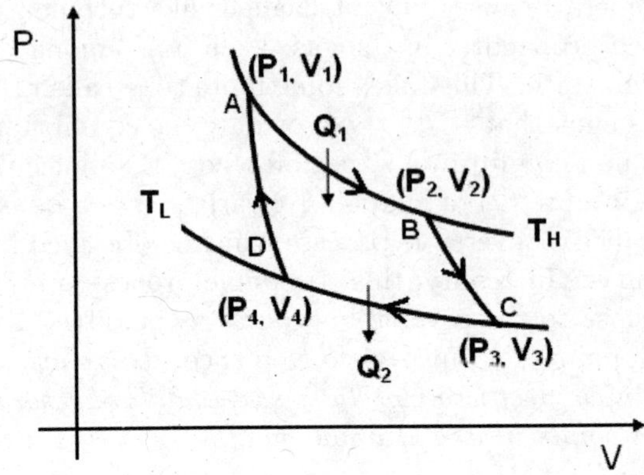

Figure 2.2: Carnot cycle with ideal gas.

a cylinder with piston has a volume V_1 and is at pressure P_1. The gas is in thermodynamic equilibrium state whose temperature is T_H, given by $P_1 V_1 = \nu R T_H$. This cylinder is kept in contact with reservoir at a temperature T_H.

(2) The gas in the cylinder expands isothermally till its volume becomes V_2 with corresponding pressure P_2.

(3) The cylinder is detached from the reservoir, isolated and the gas is allowed to expand adiabatically from its volume V_2 to V_3 with pressure P_3. During the isothermal process $A \rightarrow B$, the temperature of the gas is T_H. During adiabatic process $B \rightarrow C$ the temperature of the gas in cylinder falls to T_L.

(4) The gas in a state (P_3, V_3) at C is isothermally compressed at a temperature T_L to a state (P_4, V_4). Clearly, the cylinder is in thermal contact with a reservoir at a temperature T_L.

(5) Finally the gas is compressed adiabatically to to its original or the initial state making a cyclic process complete.

Since, for the system going from A to B is isothermal and the gas is ideal,

$$\frac{P_1}{P_2} = \frac{V_2}{V_1} . \tag{2.26}$$

During this isothermal expansion process, a heat Q_1 is taken in from the reservoir at temperature T_H. As the working substance is an ideal gas, there is no change in its internal energy. Using the first law, we get the amount of heat Q_1 absorbed during the isothermal process A to B as,

$$Q_1 = \int_{V_1}^{V_2} P \, dV = \nu R T_H \int_{V_1}^{V_2} \frac{dV}{V} = \nu R T_H \ln \frac{V_2}{V_1} . \tag{2.27}$$

Clearly, as $V_2 > V_1$, $Q_1 > 0$.

During the B to C, no heat is transferred to the gas. Then using equation (22) and the equation of state,

$$\frac{V_3}{V_2} = \left(\frac{T_H}{T_L}\right)^{\frac{1}{\gamma-1}} \tag{2.28}$$

During the isothermal process C to D heat is given out to the reservoir at a temperature T_L. Its amount is

$$Q_2 = \nu R T_L \ln\left(\frac{V_4}{V_3}\right) < 0 \ . \tag{2.29}$$

This is obvious because $V_4 < V_3$.

In an adiabatic compression from V_4 to V_1 we have

$$\frac{V_4}{V_1} = \left(\frac{T_H}{T_L}\right)^{\frac{1}{\gamma-1}} \tag{2.30}$$

From equation (27), (28) and (29) we see that

$$\frac{Q_1}{T_H} + \frac{Q_2}{T_L} = 0 \tag{2.31}$$

This equation is very important and is a very general result valid for any reversible cyclic process. We can decompose the Carnot cycle to be made up of the combination of infinitesimal paths. Then equation equivalent to (30) can be written as

$$\oint \frac{\delta Q_{rev}}{T} = 0 \tag{2.32}$$

Here δQ_{rev} is a heat and can be positive or negative during the infinitesimal reversible path at a temperature T. The equation (31) is a most general result. To prove its validity we must show this result to be true for any closed contour. Consider an arbitrary cycle divided into large number N of a sequence of Carnot cycle as shown in figure (3). The lines in smaller cycles are such that in the ith cycle it represents compression whereas in the $(i+1^{th})$ cycle the same line represents expansion by exactly the same amount. Thus the two effects cancel. When N becomes large, the cycles become thinner and finally taken together approximates the given arbitrary cycle. For each of the Carnot process equation (31) is valid. Since all these Carnot processes add to give the arbitrary cycle, we have $\oint \frac{\delta Q_{rev}}{T} = 0$ in general.

Since the equation (30) is quiet general we immediately see that $\frac{\delta Q_{rev}}{T}$ is an exact differential. We know that δQ is not an exact differential and $1/T$ is integrating factor to make $\delta Q/T$ an exact differential. We write

$$dS = \frac{\delta Q_{rev}}{T} \ . \tag{2.33}$$

S is called entropy. We now have besides an internal energy E, there is an entropy S which is also a state function. This means dE and dS are exact differentials. We draw

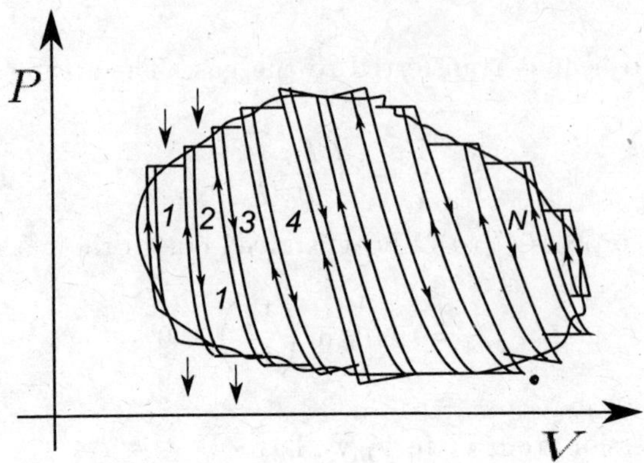

Figure 2.3: A general cycle subdivided into many Carnot cycles.

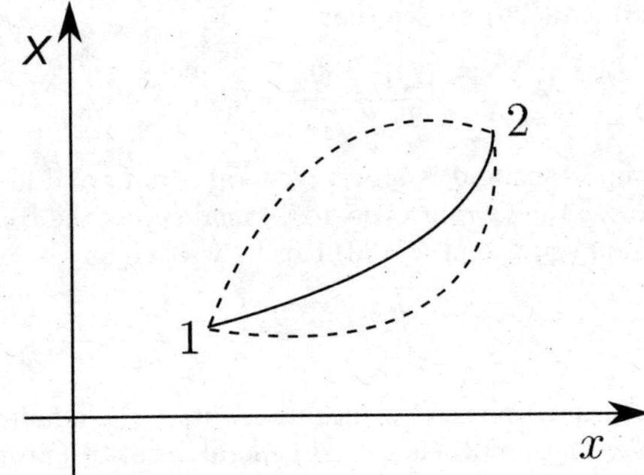

Figure 2.4: Two state points 1 and 2 where the entropy difference is evaluated.

a graph of generalized force X versus a generalized coordinate x. (Say, P versus V or H versus M etc.) Consider two state points 1 and 2 as shown in figure (4). The entropy difference is

$$S(2) - S(1) = \int_1^2 \frac{\delta Q_{rev}}{T} \qquad (2.34)$$

This is independent of path taken from state 1 to state 2 by the system. With equation (32) we see that if we have an isolated system then the entropy S of the system is constant. The state function entropy was discovered by Clausius in 1850. We thus see that

$$\delta Q)_{irr} < \delta Q)_{rev} = T dS$$

For an isolated system, $\delta Q)_{rev} = 0$ which implies $dS = 0$ and S is maximum. At equilibrium, all irreversible processes for isolated systems are such that the entropy is maximum.

2.5   Axiomatic Treatment

We have seen that with infinitesimal quasi static process the heat absorbed can be divided into infinitesimal internal energy and infinitesimal work done. Mathematical structure of the theory of thermodynamics is given by the theory of differential forms.[1] We study some of their properties.

Consider a linear differential form

$$\begin{aligned} \delta f &= A_x(x,y,z)dx + A_y(x,y,z)dy + A_z(x,y,z)dz \\ &= \vec{A}.\vec{dl} \end{aligned} \tag{2.35}$$

The quantity δf may not be an exact differential. But it is possible to make δf exact by multiplying it with a factor, $M(\vec{r})\delta f = d\phi(\vec{r})$. Then the solution of a differential equation $\delta f = 0$ implies $\phi(\vec{r})$=constant. Then, it can easily be shown that the necessary condition for integrating factor for $\delta f = 0$ to exist is

$$\vec{A}.(\nabla \times \vec{A}) = 0 \tag{2.36}$$

[Use $MA_x = \partial\phi/\partial x$, $\dfrac{\partial(MA_y)}{\partial x} = \dfrac{\partial(MA_x)}{\partial y}$ and eliminate M from three conditions.]

The condition can be shown to be sufficient also. This condition can be generalized to a linear differential form $\delta f = \sum_{i}^{n} X_i dx_i$ of n variables. For $n = 2$ the integrating factor always exists. We will state a theorem on differential forms without proof.

Theorem: Given continuous functions A_x, A_y, A_z,... of variables $x, y, z, ...$, a differential form is given as

$$\delta f = A_x dx + A_y dy + A_z dz + ...$$

then, every neighborhood of an arbitrary point P, another general point P_0 exists which can not be reach by a curve satisfying $\delta f = 0$. Under such circumstances, the differential form $\delta f = 0$ has an integrating factor.

Clearly, this theorem has a similarity with thermodynamics. We can imagine $\delta f = 0$ means $\delta Q = 0$ for an adiabatic process. Also, we can not reach one equilibrium state from another by a continuous manner. We therefore have a quasi static process. (This is equivalent to an inability to reach $P \to P_0$ via $\delta f = 0$.)

Consider two systems of two gases confined in volumes V_1 and V_2 at pressures P_1 and P_2 at a temperature θ. Clearly, the gases are in equilibrium (since θ is same). Then for an infinitesimal quasi static process

$$\delta Q = dE_1 + P_1 dV_1 + dE_2 + P_2 dV_2 \tag{2.37}$$

---

[1] The treatment given here is an axiomatic treatment due to Caratheodary. This section may be skipped by the beginner.

Here θ is an empirical temperature with a status as an independent variable. It is related with an absolute temperature. Taking $E = E(V, \theta)$;

$$
\delta Q = \left(\frac{\partial E_1}{\partial \theta}\right)_{V_1} d\theta + \left(\frac{\partial E_1}{\partial V_1}\right)_{\theta} dV_1 + \left(\frac{\partial E_2}{\partial \theta}\right)_{V_2} d\theta + \left(\frac{\partial E_2}{\partial V_2}\right)_{\theta} dV_2
$$
$$
+ P_1 dV_1 + P_2 dV_2 \tag{2.38}
$$

Clearly, by the theorem, it has an integrating factor which can be written as

$$
\delta Q = M d\sigma \tag{2.39}
$$

where the integrating factor M^{-1} and a variable σ are functions of V_1, V_2 and θ. We can also write for two systems as

$$
\delta Q_1 = \left(\frac{\partial E_1}{\partial \theta}\right)_{V_1} d\theta + \left\{ P_1 + \left(\frac{\partial E_1}{\partial V_1}\right)_{\theta} \right\} dV_1
$$
$$
\equiv M_1 d\sigma_1. \tag{2.40}
$$

Similarly,

$$
\delta Q_2 = \left(\frac{\partial E_2}{\partial \theta}\right)_{V_2} d\theta + \left\{ P_2 + \left(\frac{\partial E_2}{\partial V_2}\right)_{\theta} \right\} dV_2
$$
$$
\equiv M_2 d\sigma_2. \tag{2.41}
$$

From above equations, M_1 is a function of V_1 and θ and M_2 is a function of V_2 and θ. Thus,

$$
M(\theta, \sigma_1, \sigma_2) d\sigma(\theta, \sigma_1, \sigma_2) = M_1(\theta, \sigma_1) d\sigma_1 + M_2(\theta, \sigma_2) d\sigma_2 \tag{2.42}
$$

Clearly,

$$
\left.\begin{aligned}
\left(\frac{\partial \sigma}{\partial \theta}\right)_{\sigma_1, \sigma_2} &= 0 \\
\left(\frac{\partial \sigma}{\partial \sigma_1}\right)_{\sigma_2, \theta} &= \frac{M_1}{M} \\
\left(\frac{\partial \sigma}{\partial \sigma_2}\right)_{\sigma_1, \theta} &= \frac{M_2}{M}
\end{aligned}\right\} \tag{2.43}
$$

Since the empirical temperature is held fixed, from equation (42) we get

$$
\frac{\partial}{\partial \theta}\left(\frac{M_1}{M}\right) = \frac{\partial}{\partial \theta}\left(\frac{M_2}{M}\right) = 0
$$

or,

$$
\frac{1}{M}\frac{\partial M}{\partial \theta} = \frac{1}{M_1}\frac{\partial M_1}{\partial \theta}
$$

we then get

$$
\frac{1}{M_1}\frac{\partial M_1}{\partial \theta} = \frac{1}{M_2}\frac{\partial M_2}{\partial \theta} = \frac{1}{M}\frac{\partial M}{\partial \theta} \tag{2.44}
$$

But M_1 is a function of θ and σ_1 and M_2 is a function of θ and σ_2. This is possible if all three quantities in equation (43) are functions of θ only. We call this function $u(\theta)$. We can integrate equation(43) as,

$$\left. \begin{aligned} M_1(\theta,\sigma_1) &= \Sigma_1(\sigma_1)\exp[\textstyle\int u(\theta)d\theta] \\ M_2(\theta,\sigma_2) &= \Sigma_2(\sigma_2)\exp[\textstyle\int u(\theta)d\theta] \\ M(\theta,\sigma_1,\sigma_2) &= \Sigma(\sigma_1,\sigma_2)\exp[\textstyle\int u(\theta)d\theta] \end{aligned} \right\} \qquad (2.45)$$

Here, Σ ^{s} are constants of integration and are functions of σ ^{s}. We now replace $\exp[\int u(\theta)d\theta]$ in terms of absolute temperature T as

$$T = C\exp\left[\int u(\theta)d\theta\right]$$

where C is another constant.

With these considerations

$$\left. \begin{aligned} \delta Q &= T\Sigma(\sigma_1,\sigma_2)d\sigma/C \\ \delta Q_1 &= T\Sigma_1(\sigma_1,)d\sigma_1/C \\ \text{and} \quad \delta Q_2 &= T\Sigma_2(\sigma_2)d\sigma_2/C \end{aligned} \right\} \qquad (2.46)$$

We define a function entropy to be such that

$$S = \frac{1}{C}\int \Sigma(\sigma)d\sigma \qquad (2.47)$$

This leads us to

$$\delta Q = TdS, \quad \delta Q_1 = TdS_1, \quad \delta Q_2 = TdS_2. \qquad (2.48)$$

From this discussion, it is clear that

$$S = S_1 + S_2.$$

We have arrived at a state function of entropy S because it is not possible to go from equilibrium state to another equilibrium state by continuous manner.

This conclusion was arrived at because of properties of differential forms and the general theorem which assures an integrating factor of absolute temperature T and a state function S. We call S as entropy. It is function of internal energy E, volume V, area A, magnetization M etc.

2.6 Second Law of Thermodynamics

We have seen a general engine(Carnot engine) wherein a heat is taken from a hot reservoir at a temperature T_H and some heat is rejected to a cold reservoir at a temperature T_L. The difference in two heats is converted in to a work. This is a general process and no exception in any type of a cycle found so far. Second law of thermodynamics is statement of this observation. Lard Kelvin stated this observation as a law known as 'Second Law of Thermodynamics'. It states *"It is impossible to construct an engine, which operating in a cycle, will produce no effect other than extraction of a heat from a reservoir and performance of an equivalent work."*

Clausius stated the second law as *"It is impossible to design a refrigerator (or an engine) which will extract a heat from a reservoir at temperature $T_L < T_H$ and gives it to a reservoir at a temperature T_H without an extra input of energy from outside"*

Equivalence of these two statements of the second law of thermodynamics is discussed in Zemansky's book.

We have shown that $\dfrac{\delta Q_{rev}}{T}$ is an exact differential. Or

$$\delta Q)_{rev} = TdS \tag{2.49}$$

where S is an entropy function. Equation(48) is a statement of second law of thermodynamic and is equivalent to a statement of Kelvin and Clausius. Combining this with first law of thermodynamics for hydrostatic systems we write,

$$\delta Q)_{rev} = TdS = dE + PdV - \mu dN$$

A similar statement can be made for other systems with appropriate variables like σ, A, H, M etc. This is a combination of first and second law of thermodynamics. Here, first law conserves energy and second law gives a relation of heat, entropy and absolute temperature.

Once we have two state functions(internal energy E and entropy S), we construct other state functions by combination of E, S, T, P, V etc. These state functions have definitive properties. They are called thermodynamic potentials.

They are (a) Helmholtz free energy $F = E - TS$ (b) Enthalpy $H = E + PV = TS + \mu N$ (c) Gibbs free energy $G = E - TS + PV$. It is also called as Gibbs potential. All are for one component system. All these potentials are related with each other as a Legendre transforms of each other.[2]

2.7 Thermodynamic Potentials

(a) Free energy is defined as $F = E - TS$ for a fixed number of particle in a single specie. Here,

$$\begin{aligned} dF &= dE - TdS - SdT \\ &= -PdV - SdT \end{aligned} \tag{2.50}$$

This shows that the work done in isothermal process is equal to the change in free energy.

(b) Enthalpy H is defined as $H = E + PV$ for a fixed number of particle in a single specie. Here,

$$\begin{aligned} dH &= dE + PdV + VdP \\ &= TdS + VdP \end{aligned} \tag{2.51}$$

---

[2]See 'Classical Mechanics' by P.V.Panat (p 205, Narosa Publishers).

Thus, for the processes where the pressure is constant (isobaric processes) change in enthalpy is equal to the heat intake or a heat release in the system.

(c) Gibbs free energy G (sometimes also called as Gibbs potential) is defined for a fixed number of particle in a single specie as,

$$G = E - TdS + PV$$

or,

$$dG = -SdT + VdP \qquad (2.52)$$

For a process, where there is no change of temperature and pressure, G remains constant. In first order phase transitions (e.g. boiling of water), G remains constant.

2.8 Maxwell's Relations

Because the thermodynamic functions are state functions, there derivatives are exact differentials. As a result we obtain the relationship of derivatives of entropy with the derivatives of experimentally measurable quantities like pressure, volume, temperature, magnetic field etc. From equation (49)

$$\left(\frac{\partial F}{\partial V}\right)_T = -P$$

and

$$\left(\frac{\partial F}{\partial T}\right)_V = -S$$

Because dF is exact,

$$\left(\frac{\partial^2 F}{\partial V \partial T}\right) = \left(\frac{\partial S}{\partial V}\right)_T = \left(\frac{\partial P}{\partial T}\right)_V \qquad (2.53)$$

Similarly,

$$\left(\frac{\partial H}{\partial S}\right)_P = T; \quad \left(\frac{\partial H}{\partial P}\right)_S = V$$

or,

$$\left(\frac{\partial^2 H}{\partial S \partial P}\right) = \left(\frac{\partial^2 H}{\partial P \partial S}\right) = \left(\frac{\partial T}{\partial P}\right)_S = \left(\frac{\partial V}{\partial S}\right)_P \qquad (2.54)$$

and,

$$\left(\frac{\partial G}{\partial T}\right)_P = -S; \quad \left(\frac{\partial G}{\partial P}\right)_T = V$$

or

$$\left(\frac{\partial S}{\partial P}\right)_T = \left(-\frac{\partial V}{\partial T}\right)_P \qquad (2.55)$$

Similarly, using $dE = TdS - PdV$, we get

$$\left(\frac{\partial T}{\partial V}\right)_S = -\left(\frac{\partial P}{\partial S}\right)_V \qquad (2.56)$$

The equations (52) to (55) are known as Maxwell's relations holding for any hydrostatic system. These equations are so general that they apply for any system. For magnetic systems similar relations can be obtained by replacing $P \to H$, $V \to -M$, $T \to T$ and $S \to S$

We have defined a state function entropy S. Still its physics is not clear. Salute to Clausius but it was Boltzmann who clarified the relationship of entropy S, a macroscopic concept with microscopic world which will be studied subsequently in statistical mechanics.

2.9   Third Law of Thermodynamics

This is also called as Nernst theorem . It states that *The entropy of any system at absolute zero is either zero or has a constant value S_0. The constant S_0 is independent of all parameters of the system.* Since we deal with entropy differences the question of its value at absolute zero does not arise. However, in many situations the Nernst theorem is useful in the study of various phases of materials at low temperatures.

Consequences of the third law are that most of the response functions go to zero as $T \to 0$. Consider a specific heat C_P at constant pressure. Then if the temperature changes by an amount dT the heat absorbed will be $\delta Q = C_P dT$. Entropy of the system at temperature T is

$$S = \int_0^T \frac{C_P dT}{T} \tag{2.57}$$

This equation suggest that $C_P(T = 0)$ and $C_V(T = 0)$ must be zero. This is so because (a) $C_P > C_V$ at all temperatures; (b) The integral for entropy will diverge if $C_P(T = 0)$ is not zero. This result is in complete agreement with experiments. Most important examples are those of Debye's specific heat and specific heat of electrons in the metals. In both cases the specific heat tends to zero as absolute temperature $T \to 0$.

2.10   First and Second Law with Many Components

If we have only one component thermodynamic system then the first and the second law together give

$$T dS = dE + P dV. \tag{2.58}$$

Here we have two natural variables. In general there may be $(\alpha + 1)$ natural variables then, there are $\alpha(\alpha + 1)/2$ pairs of variables. Since each thermodynamic potential has a perfect differentials (dE, dF, dH, dG) there are $\alpha(\alpha + 1)/2$ mixed derivatives. This gives $\alpha(\alpha + 1)/2$

Maxwell relations. The thermodynamic potential differentials are

$$\left.\begin{aligned}
dE &= TdS - PdV + \sum_{i=1}^{\alpha-1} \mu_i dN_i \\
dF &= -SdT - PdV + \sum_{i=1}^{\alpha-1} \mu_i dN_i \\
dH &= TdS + VdP + \sum_{i=1}^{\alpha-1} \mu_i dN_i \\
dG &= -SdT + VdP + \sum_{i=1}^{\alpha-1} \mu_i dN_i
\end{aligned}\right\} \tag{2.59}$$

Here μ_i is a chemical potential of ith specie with N_i particles.

Maxwell relations are similar to the one obtained in section 2.9. Besides the Maxwell's relations there, additional relations are derived because of the μdN terms. They are

$$\left.\begin{aligned}
\frac{\partial T}{\partial N_i}\bigg)_{S,V} &= \frac{\partial \mu_i}{\partial S}\bigg)_{V,N_i} \\
\frac{\partial P}{\partial N_i}\bigg)_{S,V} &= -\frac{\partial \mu_i}{\partial V}\bigg)_{S,N_i}
\end{aligned}\right\} \text{ from } dE \text{ equation} \tag{2.60}$$

$$\left.\begin{aligned}
\frac{\partial S}{\partial N_i}\bigg)_{T,V} &= -\frac{\partial \mu_i}{\partial T}\bigg)_{V,N_i} \\
\frac{\partial P}{\partial N_i}\bigg)_{T,V} &= -\frac{\partial \mu_i}{\partial V}\bigg)_{T,N_i}
\end{aligned}\right\} \text{ from } dF \text{ equation} \tag{2.61}$$

$$\left.\begin{aligned}
\frac{\partial T}{\partial N_i}\bigg)_{S,P} &= \frac{\partial \mu_i}{\partial S}\bigg)_{P,N_i} \\
\frac{\partial V}{\partial N_i}\bigg)_{S,P} &= \frac{\partial \mu_i}{\partial P}\bigg)_{S,N_i}
\end{aligned}\right\} \text{ from } dH \text{ equation} \tag{2.62}$$

$$\left.\begin{aligned}
-\frac{\partial S}{\partial N_i}\bigg)_{T,P} &= \frac{\partial \mu_i}{\partial T}\bigg)_{P,N_i} \\
\frac{\partial V}{\partial N_i}\bigg)_{T,P} &= \frac{\partial \mu_i}{\partial P}\bigg)_{T,N_i}
\end{aligned}\right\} \text{ from } dG \text{ equation} \tag{2.63}$$

These Maxwell's relations are in addition to those given in section (9)

2.11 Efficiency of an Engine in General and Carnot Engine in Particular

The thermal efficiency η of an engine is defined as

$$\eta = \frac{\text{Work output}}{\text{Heat input}}$$

In case of a heat engine, heat input Q_1 is given by a hot reservoir at a temperature T_1, heat Q_2 is rejected to the cold reservoir of temperature T_2. The work done by it is $W = Q_1 - Q_2$
Thus

$$\eta = \frac{W}{Q_1} = 1 - \frac{Q_2}{Q_1} \tag{2.64}$$

From equation (30) for Carnot engine it is clear that Q_2/Q_1 is T_2/T_1
Thus,

$$\eta = 1 - \frac{T_2}{T_1} \tag{2.65}$$

is an efficiency of Carnot engine.

There is an important property of Carnot engine. It is, that the Carnot engine working between two heat reservoirs at temperatures T_1 and T_2 has higher efficiency than any other engine working between the same two reservoirs.

The other important property is that all Carnot engines operating between same two reservoirs have same efficiency.

Moreover, efficiency of the Carnot engine is independent of the nature of working substance. In the chapter on Bose statistics, we will operate Carnot engine with a black body radiation and arrive the same efficiency formula of equation (64)

2.12   Thermodynamic Temperature Scale

In section (2) of chapter (1) we have mentioned that the absolute scale of temperature is identical with a perfect gas scale. The argument was given by Lord Kelvin, based upon efficiency of the Carnot engine, which is independent of its working substance and depends only upon the temperatures of its reservoirs.

Clearly the efficiency of the Carnot engine is $\eta = 1 - \dfrac{Q_2}{Q_1}$ and is a function of T_1 and T_2
Thus

$$\frac{Q_1}{Q_2} = f(T_1, T_2) \tag{2.66}$$

Consider three Carnot reversible engines as shown in figure (5).

1. The engine R_A operates between T_1 and T_2
2. The engine R_B operates between T_1 and $T_3 > T_2$
3. The engine R_C operates between T_3 and T_2.

The engine R_B rejects heat Q_3 to the reservoir at T_3. The engine R_C sucks heat Q_3 from the same reservoir and rejects heat Q_2 to the reservoir at T_2.

It is easy to see that because all the engines are Carnot engines, heat absorbed by R_B is Q_1 and heat rejected by R_C is Q_2. (Think why?) With arguments similar to those used for arriving at equation (65) we get

$$\frac{Q_1}{Q_3} = f(T_1, T_3) \text{ and } \frac{Q_3}{Q_2} = f(T_3, T_2). \tag{2.67}$$

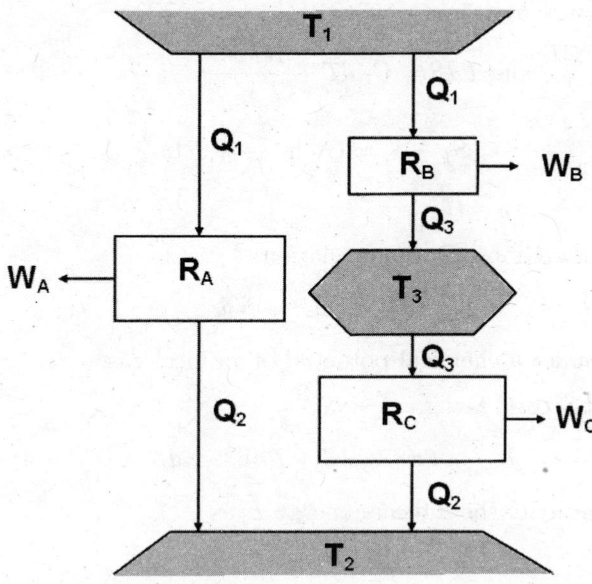

Figure 2.5: Three coupled Carnot engines.

Combining equation (65) and (66) we get

$$f(T_1, T_2) = \frac{f(T_1, T_3)}{f(T_2, T_3)} \qquad (2.68)$$

With the intermediate reservoir the temperature T_3 is arbitrary and does not appear on the left. This means T_3 must drop out. Therefore

$$f(T_1, T_2) = \frac{Q_1}{Q_2} = \frac{f(T_1, T_3)}{f(T_2, T_3)} = \frac{\psi(T_1)}{\psi(T_2)} \qquad (2.69)$$

where ψ is yet another unknown function. The ratio $\dfrac{\psi(T_1)}{\psi(T_2)}$ is defined to be a ratio of two thermodynamic temperatures $\dfrac{T_1}{T_2}$. This ratio is absolute because the Carnot engine has its efficiency independent of its working substance. The temperature scale is fixed by assigning 273.16 degrees to freezing point of water. This scale is absolute and its unit is kelvin. Thus the lowest limit of a temperature on absolute scale is $0K$. There is no upper bound to the absolute scale of the temperature. The discovery of absolute scale is due to Lord Kelvin. At absolute zero, the efficiency of Carnot engine is 100% and therefore it is not possible to have a negative absolute temperature.

Illustrative problems

Problem 3: Calculate the change in entropy where 1 mole of an ideal gas changes its volume and temperature from initial value V_i, T_i to V_f, T_f.

Solution: Use $P = \dfrac{RT}{V}$ and $TdS = C_V dT + \dfrac{RT dV}{V}$

$$S_f - S_i = C_V \ln \frac{T_f}{T_i} + R \ln \frac{V_f}{V_i} \tag{2.70}$$

Problem 4: Prove the Gibbs- Durham relation

$$SdT - VdP + Nd\mu = 0$$

and obtain expression for a chemical potential of an ideal gas.

Solution: First and Second law gives

$$TdS = dE + PdV - \mu dN \tag{2.71}$$

From this relation we write the internal energy E as

$$E = E(S, V, N)$$

The variables S, V and N are extensive variables and so also is the internal energy E. Let's scale S, V and N by $1 + \epsilon$. E is then scaled accordingly. Thus

$$E((1+\epsilon)S, (1+\epsilon)V, (1+\epsilon)N) = E(S, V, N) + \epsilon \times E(S, V, N) \tag{2.72}$$

However, from equation (70), $\left. \dfrac{\partial E}{\partial S} \right)_{V,N} = T$ etc. Equation (71) then becomes,

$$E(S, V, N) = TS - PV + \mu N \tag{2.73}$$

Differentiating equation (72) and using equation (70) we get the Gibbs Durham relation as

$$SdT - VdP + Nd\mu = 0 \tag{2.74}$$

Using the equation (73) we write

$$d\mu = -\frac{S}{N}dT + \frac{V}{N}dP.$$

Then using equation (69) we get

$$d\mu = \frac{k_B T}{P}dP - \left\{ \frac{S_i}{N} + \frac{C_V}{N} \ln\left(\frac{T}{T_i}\right) + \frac{R}{N} \ln\left(\frac{V}{V_i}\right) \right\} dT$$

It is to be noted that the chemical potential is a state function. Therefore, its integration is path independent. We choose a path in $P - T$ plane such that we keep $P = P_i$ and change the temperature from T_0 to T. Here, $dP = 0$. Then we go from (P_i, T) to (P, T) where $dT = 0$. The final result is

$$\mu(P, T) = \mu(P_i, T_i) + \frac{(T - T_i)}{N}(C_P - S_i) + k_B T \ln\left(\frac{P}{P_i}\right) - \frac{C_P}{N} T \ln\left(\frac{T}{T_i}\right).$$

2.13 Power Output from Carnot Engine

We know that the Carnot engine has highest efficiency among heat engines. The other engines are less efficient because of irreversibility, friction and heat leakages.

Interesting to observe is that the power output of the Carnot engine is "zero". The two isothermal and two adiabatic processes of Carnot engine are quasi static and hence require an infinite time to complete the cycle. The power output thus is zero. Only a cycle with finite time will give a nonzero power.

While discussing the Carnot cycle, we did not bother about a temperature of the working substance and took it to be same as the temperature of the hot reservoir when the heat was absorbed. We also delivered heat to the sink at colder temperature of the working substance. These processes require infinite time with quasi static process.

To make the cycle in finite time, heat has to flow from hot reservoir to the working substance. This requires temperature of the working substance to be lower than that of the hot reservoir during the isothermal expansion of the cycle. Similarly, the temperature of the sink is lower than that of the working substance during the isothermal compression of the Carnot cycle. This heat flow process is irreversible. An important result of maximum power output, limited by the rate of heat transfer to and from the working substance of the Carnot engine is obtained by Curzon and Ahlborn[3]
We produce their derivation here.

Stage 1: T_1- temperature of a hot reservoir
T_{1w} is temperature of the working substance during the isothermal expansion.
F_1- heat flux such that

$$F_1 = \alpha(T_1 - T_{1w}). \tag{2.75}$$

Here α is constant depending upon thermal conductivity and the geometry of the wall of cylinder. Let t_1 be the time for the isothermal expansion to last. Input energy Q_1 is

$$Q_1 = F_1 t_1 = \alpha t_1 (T_1 - T_{1w}). \tag{2.76}$$

For the sake of simplicity, we assume that the two adiabats, binding two isothermals in the Carnot cycle to be reversible.

Like equation (74) isothermal compression leads to the rejection of heat to the sink to be,

$$Q_2 = \beta t_2 (T_{2w} - T_2). \tag{2.77}$$

Here, t_2 is a time taken for the isothermal compression of the working gas at temperature T_{2w}. T_2 is temperature of the sink. β is a constant similar to α.

Because the adiabats are reversible, we have

$$\frac{Q_1}{T_{1w}} = \frac{Q_2}{T_{2w}}. \tag{2.78}$$

---

[3] "Efficiency of a Carnot Engine at Maximum Power output" by F.L. Curzon and B.Ahlborn, Am. Journ. Phys. **43**, 22(1975). This is very readable paper. See also, a very general derivation in "Thermodynamic Efficiency at Maximum Power" by C.Van den Broeck, Phys. Rev. Letts. **95**, 190602 (2005)

Time spent in the two isothermals is $(t_1 + t_2)$. We assume that the total time expended in the Carnot cycle to be proportional to $(t_1 + t_2)$ as $\gamma(t_1 + t_2)$. Thus, power of our Carnot engine is

$$P = \frac{(Q_1 - Q_2)}{\gamma(t_1 + t_2)}. \tag{2.79}$$

Let us write $T_1 - T_{1w} = x$, $T_2 - T_{2w} = y$. Then, using equation(74) to (78), we get

$$P = \frac{\alpha\beta xy(T_1 - T_2 - x - y)}{\gamma[\beta T_1 y + \alpha T_2 x + xy(\alpha - \beta)]}. \tag{2.80}$$

The power is maximized requiring

$$\frac{\partial P}{\partial x} = 0 = \frac{\partial P}{\partial y}. \tag{2.81}$$

$\dfrac{\partial P}{\partial y} = 0$ gives, $y = (\alpha T_2/\beta T_1)^{1/2} x$ after some very elementary algebra. Putting this value

of y in $\dfrac{\partial P}{\partial x} = 0$ we get a quadratic equation for $\xi = x/T_1$ as

$$[1 - \alpha/\beta]\,\xi^2 - 2\left[(\alpha T_2/\beta T_1)^{1/2} + 1\right]\xi + (1 - T_2/T_1) = 0 \tag{2.82}$$

Since $\xi < 1$, physically relevant ξ is

$$\xi = \frac{x}{T_1} = \frac{1 - (T_2/T_1)^{1/2}}{1 + (\alpha/\beta)^{1/2}} \tag{2.83}$$

With the values of x and y that optimize power P, we have the efficiency of Carnot engine as

$$\begin{aligned}
\eta &= \frac{Q_1 - Q_2}{Q_1} \\
&= 1 - T_{2w}/T_{1w} \\
&= 1 - (T_2 + y)/(T_1 - x) \\
&= 1 - (T_2/T_1)^{1/2} \tag{2.84}
\end{aligned}$$

Equation (83) is a principal result of Curzon and Ahlborn. The results are compared with observed performance of real heat engines e.g. consider CANDU(Canada) PHW Nuclear Reactor. Here $T_1 = 300^\circ C$, $T_2 = 25^\circ C$. This gives $\eta(Carnot) = 48\%$, η from equation (83) is 28% and observed efficiency is 30%. The result of Carnot efficiency at optimal power has triggered a new field of investigation of finite time thermodynamics (Or endoreversible thermodynamics) and is valid quiet generally.

2.14 Equilibrium with a Matter Flow

There are many thermodynamic systems, which when come in contact with each other, there is a flow of energy, adjustments of pressures and also a flow of matter, so that the equilibrium is reached. Interesting example of mass flow is p-n junction in semiconductor diode. Here, when p type silicon is fused with n type silicon, at the junction flow of electrons from n type to p type takes place to reach an equilibrium. This is besides temperature equalization.

Consider two systems with parameters (E_1, S_1, N_1, V_1) and (E_2, S_2, N_2, V_2) in contact with each other. Assume that the composite system is isolated. Then

$$
\begin{aligned}
E_1 + E_2 &= E \quad Constant \\
N_1 + N_2 &= N \quad Constant \\
V_1 + V_2 &= V \quad Constant
\end{aligned}
$$

This means, $dE_1 = -dE_2$, $dV_1 = -dV_2$, $dN_1 = -dN_2$. At equilibrium, $dS = 0$ because S is maximum. Also, by additivity of entropy, $dS = dS_1 + dS_2$.
We use the thermodynamic law,

$$
dE = TdS - PdV + \mu dN
$$

Thus

$$
dS = 0 = \left.\frac{\partial S_1}{\partial E_1}\right)_{V_1, N_1} dE_1 + \left.\frac{\partial S_1}{\partial V_1}\right)_{E_1, N_1} dV_1 + \left.\frac{\partial S_1}{\partial N_1}\right)_{E_1, V_1} dN_1
$$

$$
+ \left.\frac{\partial S_2}{\partial E_2}\right)_{V_2, N_2} dE_2 + \left.\frac{\partial S_2}{\partial V_2}\right)_{E_2, N_2} dV_2 + \left.\frac{\partial S_2}{\partial N_2}\right)_{E_2, V_2} dN_2
$$

or,

$$
dS = 0 = \left(\frac{1}{T_1} - \frac{1}{T_2}\right) dE_1 + \left(\frac{P_1}{T_1} - \frac{P_2}{T_2}\right) dV_1
$$

$$
+ \left(\frac{\mu_2}{T_2} - \frac{\mu_1}{T_1}\right) dN_1
$$

At equilibrium, we get $T_1 = T_2 = T$, $P_1 = P_2 = P$ and $\mu_1 = \mu_2$

If there is an equal temperature of both systems and are interacting via a wall that can allow transfer of mass, then

$$
\Delta S = \left(\frac{\mu_2 - \mu_1}{T}\right) \Delta N_1 > 0
$$

Thus, if $\mu_1 > \mu_2$, then $\Delta N_1 < 0$ It means, in matter flow, mass flows from higher chemical potential to the lower chemical potential. The flow occurs until μ_1 equal μ_2. Then the flow stops and the equilibrium is reached.

In the example of p-n junction that we gave earlier, an electron transfer takes place until the chemical potential of the entire sample is same.

2.15   Entropy, Reversibility and Irreversibility

Let a heat δQ be exchanged in a reversible manner to a thermodynamic system. This exchange changes entropy of the system by dS at temperature T. Since the heat δQ is exchanged in a reversible manner, we denote it with subscript R and we know that

$$
\delta Q_R = TdS .
$$

We know that a thermodynamic system exchanges heat with surroundings. We call system plus the surrounding as a universe. The surrounding usually is also called as reservoir. Compared to the system, reservoir is very big. As a result, temperature of the reservoir does not change (to an excellent approximation) even if the heat δQ is exchanged between the system and the reservoir. Thus the reservoir undergoes non dissipative changes, entirely determined by exchanged heat δQ. Thus it does not matter whether δQ is exchanged reversibly or irreversibly by the reservoir. Therefore, by whatever process, the reservoir absorbs heat δQ from any system at a temperature T, the entropy change of the reservoir is $\delta Q/T$.

If δQ_R is a heat absorbed by the system from the reservoir of temperature T, in a reversible manner, then,

$$dS \text{ of the system } = \delta Q_R/T$$

and

$$dS \text{ of the reservoir } = -\delta Q_R/T$$

Entropy change of the universe = Entropy change of the system plus entropy change of the reservoir = 0.

Thus, the entropy change of the universe is zero in the reversible process.

In case of irreversible process, entropy of the universe increases. Consider, two thermodynamic equilibrium states connected by the irreversible process. Since these two states i (initial) and f (final) are equilibrium states, $(S_f - S_i)$ can be found by replacing the irreversible process by a reversible one. Once the value of $\Delta S = S_f - S_i$ is calculated, it will always be found that (ΔS) for universe is positive or zero. That means,

$$\Delta S(Universe) \geq 0.$$

Here equality refers to reversible process and inequality to irreversible process. To illustrate this important fact, we solve some problems.

Problem 5. Consider a long wire of unstretched length L_0, stretched to a length $L = L_0 + \Delta L$ $(\Delta L << L_0)$, by application of a tension τ. The wire is enclosed in a cylinder so that it doesn't exchange heat with surrounding. The equation of state of the wire is $\tau = CT\left(\dfrac{L}{L_0} - \dfrac{L_0^2}{L^2}\right)$ where C is constant. Find (a) Change in entropy during stretching to L. (b) The wire is cut at a point $L_0 + \Delta L/2$ Find entropy of snapping.

Solution: Clearly, the work done $= \displaystyle\int \tau dL$ is a heat input.

$$\therefore \frac{\delta Q}{T} = dS = C\left\{\frac{L}{L_0} - \frac{L_0^2}{L^2}\right\} dL .$$

(a) Change in entropy when the string is stretched from L_0 to $L_0 + \Delta L$ is

$$\Delta S = C \int_{L_0}^{L_0+\Delta L} \left(\frac{L}{L_0} - \frac{L_0^2}{L^2} \right) dL$$

$$= \frac{3}{2} C \frac{\Delta L^2}{L_0} .$$

(b) To find entropy change of snapping, we find entropy change ΔS_1 in stretching the string from L_0 to $L_0 + \dfrac{\Delta L}{2}$ and ΔS_2 in stretching the string from $L_0 + \dfrac{\Delta L}{2}$ to $L_0 + \Delta L$. Clearly, entropy change of cutting the wire is

$$\Delta S_1 + \Delta S_2 - \Delta S.$$

$$\Delta S_1 = C \int_{L_0}^{L_0+\Delta L/2} \left(\frac{L}{L_0} - \frac{L_0^2}{L^2} \right) dL = \frac{5}{8} C \frac{\Delta L^2}{L_0}.$$

Similarly,

$$\Delta S_2 = C \int_{L_0+\frac{\Delta L}{2}}^{L_0+\Delta L} \left(\frac{L}{L_0} - \frac{L_0^2}{L^2} \right) dL = \frac{9}{8} C \frac{\Delta L^2}{L_0}.$$

$\therefore$ Change of entropy in the process of cutting the wire is

$$C \frac{\Delta L^2}{L_0} \left(\frac{9}{8} + \frac{5}{8} - \frac{3}{2} \right) = \frac{1}{4} C \frac{\Delta L^2}{L_0} .$$

Change of entropy of the surrounding is zero.

$\therefore$ Change of entropy of the universe $= \dfrac{1}{4} C \dfrac{\Delta L^2}{L_0} > 0$

Problem 6. Calculate entropy change when an ideal gas rushes in to a vacuum (Joule's free expansion)

Solution: There is entropy change of the system due to irreversible process of free expansion. We replace free expansion process by a reversible process that takes a gas from its initial volume V_i at a temperature T to its final state of volume V_f at a temperature T. Clearly, this is reversible isothermal expansion of an ideal gas of ν moles. The entropy change ΔS of the system is

$$\Delta S = \int_{V_i}^{V_f} \frac{\delta Q_R}{T} = \nu R \ln \left(\frac{V_f}{V_i} \right) .$$

Entropy change of surrounding is zero because the free expansion takes place for an isolated system. The entropy of universe is $\nu R \ln \left(\frac{V_f}{V_i} \right) > 0$ since $V_f > V_i$. Thus entropy of the universe increased even when no heat entered or left the system.

So far, it is found that, whenever an irreversible process occurs, entropy of the universe increases. This is known as entropy principle.

Here, in all processes, we find that there are internal constraints (i.e. restriction) on the parameters of the thermodynamic system. For example, in the problem of free expansion of the gas, gas is isolated in a volume V_i (Here V is parameter). After the free expansion, it stays in an isolated volume $V_f > V_i$. Thus in the process of free expansion, we say that, an initial constraint of volume on the gas is removed and a new constraint on volume is imposed. In all such processes, a thermodynamic equilibrium state is reached for which the entropy of the system is maximum. In general, then, for any thermodynamic system we postulate,

1. There exist an entropy function of extensive variables defined for all equilibrium states of the system. The equilibrium state takes the values of the parameters such that, entropy attains maximum value upon the removal of internal constraints. Thus, $dS = 0$ for the process. Moreover, $d^2S < 0$ signifying the maximum of entropy.

2. Entropy is a continuous, differentiable and monotonically increasing function of internal energy. We thus write

   (a) $S = S(E, V, N_1, N_2...)$ for gases

   (b) $S = S(E, M, N_1, N_2...)$ for magnetic system

   (c) $S = S(E, \Delta, ...)$ for superconducting system, where Δ is superconducting gap and

so on.

Monoatomic increasing property means

$$\left(\frac{\partial S}{\partial E}\right)_{V, N_1, N_2...} > 0$$

We know that $\left(\dfrac{\partial S}{\partial E}\right) = \dfrac{1}{T}$

Thus this implies that the absolute temperature is non negative

3. Entropy of the composite system is additive. Thus, if we have two systems, system 1 and system 2, then the entropy of the composite system S is

$$S = S_1 + S_2$$

4. There exists a state called ground state of the system. This state is attained at absolute zero and has a property that entropy of the state is zero. Sometimes, entropy of this state is constant and can be taken as zero. This is Nernst theorem. Better understanding of this statement will come when we develop quantum statistical mechanics.

Chapter 3

THERMODYNAMIC RELATIONSHIPS

A theory is the more impressive the greater the simplicity of its premises is, the more different kinds of things it relates, and the more extended is its area of applicability. Therefore the deep impression which classical thermodynamics made upon me. It is the only physical theory of universal content concerning which I am convinced that, within the framework of the applicability of its basic concepts, it will never be overthrown. ¬ Albert Einstein

We have, so far emphasized the fact that the thermodynamics has a large domain of applicability. With thermodynamic functions entropy S, internal energy E, Helmholtz free energy F, Kelvin's enthalpy H and Gibbs free energy G, we can study magnetic systems, electrical and electrochemical systems, gas-liquid-solid systems, superconductivity etc. The intensive variables and extensive variables may vary from system to system.

We have seen in the last chapter that if an external stimulus is applied to the system, then the response of the system is governed by its response function.Since we are using them extensively in this chapter, we revise them. Following are the examples.

(a). If gaseous system is given a heat at either a constant pressure or at a constant volume, its temperature rises. Corresponding response function is specific heat C_P or C_V.

(b). If an external magnetic field H is applied to the magnetic system at temperature T, the system responds by acquiring a magnetic moment M. Response function then is isothermal magnetic susceptibility χ_T and is a function of T and H. It is

$$\chi_T(T, H) = \left(\frac{\partial M}{\partial H}\right)_T \tag{3.1}$$

(c). If a temperature of a gaseous system is raised, then under isobaric condition the gas expands and the corresponding response function is coefficient of volume expansion. It is given by

$$\beta = \frac{1}{V}\left(\frac{\partial V}{\partial T}\right)_P \tag{3.2}$$

Similarly, if the gas is compressed isothermally, the response function is isothermal compressibility. It is given by

$$\kappa_T = -\frac{1}{V}\left(\frac{\partial V}{\partial P}\right)_T.$$
(3.3)

If the gas compressed adiabatically, then we define isentropic compressibility κ_S, where

$$\kappa_S = -\frac{1}{V}\left(\frac{\partial V}{\partial P}\right)_S$$
(3.4)

We can give many such examples of response function for different systems. They will be mentioned from time to time. As far as thermodynamics is concerned, there is an inter-relationship between these response functions. These relationships are <u>exact</u>. These response functions can not be calculated in terms of microscopic quantities of the system by thermodynamics alone. To calculate them, we have to use statistical mechanics. None the less after their calculations, inter-relationships among them, as given by thermodynamics, must be satisfied. This is the strength of thermodynamics. Inability of thermodynamics to calculate them for the system in terms of its microscopic parameters is its weakness.

3.1 Relation between C_P, C_V, β and κ

To get various thermodynamics relations, we will have to invoke Maxwell's relations again and again. This is because, Maxwell's relations connect derivatives of entropy with derivatives of pressure, volume and temperature.

Consider the entropy of a pure substance to be a functions of V and T. Equation of state gives connection of P, V and T. Then

$$TdS = T\left[\left(\frac{\partial S}{\partial V}\right)_T dV + \left(\frac{\partial S}{\partial T}\right)_V dT\right]$$
(3.5)

For a general process, $T\left(\frac{\partial S}{\partial T}\right)_V = C_V$.

By Maxwell's relation,

$$\left(\frac{\partial S}{\partial V}\right)_T = \left(\frac{\partial P}{\partial T}\right)_V$$
(3.6)

Thus,

$$TdS = C_V dT + T\left(\frac{\partial P}{\partial T}\right)_V dV.$$
(3.7)

In a similar manner, considering $S = S(P,T)$ we get,

$$TdS = C_P dT - T\left(\frac{\partial V}{\partial T}\right)_P dP.$$
(3.8)

Equating, we get

$$dT = \frac{T\left(\frac{\partial P}{\partial T}\right)_V dV}{C_P - C_V} + \frac{T\left(\frac{\partial V}{\partial T}\right)_P dP}{C_P - C_V}.$$

From equation of states, $T = T(P, V)$. Thus,

$$dT = \left(\frac{\partial T}{\partial P}\right)_V dP + \left(\frac{\partial T}{\partial V}\right)_P dV.$$

We get after simplification,

$$C_P - C_V = T\left(\frac{\partial V}{\partial T}\right)_P \left(\frac{\partial P}{\partial T}\right)_V \tag{3.9}$$

But

$$\left(\frac{\partial P}{\partial T}\right)_V = -\left(\frac{\partial V}{\partial T}\right)_P \left(\frac{\partial P}{\partial V}\right)_T.$$

Thus,

$$C_P - C_V = -T\left(\frac{\partial V}{\partial T}\right)_P^2 \left(\frac{\partial P}{\partial V}\right)_T \tag{3.10}$$

For all known substances, $(\partial P/\partial V)_T$ is always negative, leading to $C_P > C_V$. Putting $C_P - C_V$ in terms of β and κ we get,

$$C_P - C_V = VT\beta^2/\kappa_T \tag{3.11}$$

For an ideal gas, evaluating β and κ; we get a well known result

$$C_P - C_V = R. \tag{3.12}$$

3.2 Internal Energy Relations

If a system with single component undergoes an infinitesimal reversible process, then

$$dE = TdS - PdV. \tag{3.13}$$

Since we are free to take E, S and P as a function of T and V, we can write the above equation, by keeping T constant as,

$$\left(\frac{\partial E}{\partial V}\right)_T = T\left(\frac{\partial S}{\partial V}\right)_T - P. \tag{3.14}$$

But by Maxwell's relation, $(\partial S/\partial V)_T = (\partial P/\partial T)_V$.

Thus,

$$\left(\frac{\partial E}{\partial V}\right)_T = T\left(\frac{\partial P}{\partial T}\right)_V - P. \tag{3.15}$$

In an exactly similar manner, taking a derivative of E with respect to P we get,

$$\left(\frac{\partial E}{\partial P}\right)_T = -T\left(\frac{\partial V}{\partial T}\right)_P - P\left(\frac{\partial V}{\partial P}\right)_T. \tag{3.16}$$

Illustrative Problem

Problem 1. Suppose that a gas of photons is enclosed in a volume at a temperature of T. If u is the energy density of the photon gas, then, it is given that, the pressure exerted by the photon gas is $P = u/3$. Show that $u \propto T^4$

solution: Because u is the energy density at a given equilibrium temperature, $(\partial E/\partial V)_T = u$. Black body volume is fixed. Thus

$$(\partial P/\partial T)_V = \frac{1}{3}\frac{du}{dT}$$

From the internal energy equation (15),

$$u = \frac{T}{3}\frac{du}{dT} - \frac{u}{3}$$

or,

$$4\frac{dT}{T} = du/u$$

or,

$$u = \text{ constant } T^4$$

Clearly, the specific heat of the Black body radiation gas is proportional to T^3. The coefficient of proportionality can not be derived from thermodynamics. We will derive it in terms of fundamental constants k_B, c, and $\hbar$ when we deal with Bose-statistics.

3.3 Magnetic Response Functions

While applying thermodynamics to the magnetic system (assumed to be solid form), we ignore changes in volume of the system due to changes in pressure. It means we ignore PdV term in the thermodynamic law. We will restrict a discussion to the magnetic phenomena where experiments are performed at low pressure. Thus, no significant error will result by ignoring PdV term.

Role played by the pressure for a fluid system is similar to the role played by an external magnetic field H for a magnetic system. Similarly the role played by volume V is similar to the role played by the negative of magnetization $(-M)$ for the magnetic system. (As P increases, volume of fluid decreases. On the other hand as H increases, the magnetization of the para or ferromagnetic system increases). Corresponding thermodynamic potentials are $F = E - TS$, $H_{en} = E - MH$ and $G = E - TS - MH$. Here we wrote for enthalpy H_{en} since H is used for the field.

Fundamental thermodynamic relation is,

$$T\,dS = dE - H\,dM \tag{3.17}$$

Thus, the differentials of these potentials are

$$
\begin{aligned}
dE &= T\,dS + H\,dM \\
dF &= H\,dM - S\,dT \\
dG &= -S\,dT - M\,dH \\
dH_{en} &= T\,dS - M\,dH
\end{aligned}
\tag{3.18}
$$

Maxwell's relation corresponding to equation (2.51??) to (2.54??) are

$$
\begin{aligned}
\left(\frac{\partial H}{\partial T}\right)_M &= -\left(\frac{\partial S}{\partial M}\right)_T \\
\left(\frac{\partial M}{\partial S}\right)_H &= -\left(\frac{\partial T}{\partial H}\right)_S \\
\left(\frac{\partial M}{\partial T}\right)_H &= \left(\frac{\partial S}{\partial H}\right)_T
\end{aligned}
\tag{3.19}
$$

and

$$
\left(\frac{\partial H}{\partial S}\right)_M = \left(\frac{\partial T}{\partial M}\right)_S
$$

Corresponding to C_V and C_P we have C_M and C_H. We define them as

$$
C_M = \left(\frac{\delta Q}{\delta T}\right)_M = T\left(\frac{\partial S}{\partial T}\right)_M = -T\left(\frac{\partial^2 F}{\partial T^2}\right)_M
\tag{3.20}
$$

and

$$
C_H = \left(\frac{\delta Q}{\delta T}\right)_H = T\left(\frac{\partial S}{\partial T}\right)_H = T\left(\frac{\partial H_{en}}{dT}\right)_H = -T\left(\frac{\partial^2 G}{\partial T^2}\right)_H
\tag{3.21}
$$

For the fluid systems, we have isothermal and adiabatic compressibility κ_T and κ_S defined as

$$
\kappa_T = -\frac{1}{V}\left(\frac{\partial V}{\partial P}\right)_T \quad \text{and} \quad \kappa_S = -\frac{1}{V}\left(\frac{\partial V}{\partial P}\right)_S
\tag{3.22}
$$

Corresponding functions for magnetic system are susceptibilities. Thus, the isothermal susceptibility χ_T is

$$
\chi_T = \left(\frac{\partial M}{\partial H}\right)_T = -\left(\frac{\partial^2 G}{\partial H^2}\right)_T
\tag{3.23}
$$

Similarly, the adiabatic susceptibility is

$$
\chi_S = \left(\frac{\partial M}{\partial H}\right)_S = -\left(\frac{\partial^2 H_{en}}{\partial H^2}\right)_S
\tag{3.24}
$$

Like the coefficients of volume expansion, we have

$$
\beta_H = \left(\frac{\partial M}{\partial T}\right)_H
$$

Thus, if we can calculate the thermodynamic potentials using statistical mechanics, we can theoretically obtain all the response functions.

Illustrative Problem

Problem 2. Show that, for a magnetic system,

$$C_H - C_M = T\beta_H^2/\chi_T \tag{3.25}$$

analogous to the fluid system.

Solution: Following exactly the steps to obtain equation(7) and (8),

$$TdS = C_M dT - T\left(\frac{\partial H}{\partial T}\right)_M dM = C_H dT + T\left(\frac{\partial M}{\partial T}\right)_H dH. \tag{3.26}$$

Use $T = T(M, H)$ as the equation of state.

Also, with

$$\left(\frac{\partial H}{\partial T}\right)_M = \left(\frac{\partial M}{\partial T}\right)_H \left(\frac{\partial H}{\partial M}\right)_T \tag{3.27}$$

and various response functions, we get

$$C_H - C_M = T\beta_H^2/\chi_T. \tag{3.28}$$

Clearly, just like $C_P \geq T\beta^2/\kappa_T$, we have

$$C_H \geq T\left(\frac{\partial M}{\partial T}\right)_M^2 /\chi_T. \tag{3.29}$$

At a critical point, this inequality is used to obtain Rushbrooke inequality.

3.4   Minimum Number of Experimental Quantities Necessary to Determine Thermodynamic Functions

From the number of thermodynamic functions and response functions, it may be thought that enormous experimental efforts would be necessary to determine them. We will consider a particular case of fluids with two parameters. Generalization of other systems is immediate. For this system, we count their measurable thermal properties. (1) Equation of state (2) Isothermal compressibility κ_T (3) Isothermal volume expansion coefficient (4) Adiabatic equation, i.e. relation between any two variables between P, V and T in adiabatic process (5) Specific heats C_P and C_V which are the functions of parameters (6) Joule coefficient $(\partial T/\partial V)_E$ which involves free expansion of gas (7) Joule- Thompson coefficient $(\partial T/\partial P)_{enthalpy}$ which occurs in the expansion of a gas through a throttle.

Experimentally to measure all these quantities in a wide temperature range is an enormous task. However we will show that the complete specification of a thermodynamic state can be gotten by the measurement of (a) Equation of state and (b) one of the specific heats (C_P or C_V) in a relevant temperature range. This assertion can be seen as follows.

Consider the evaluation of internal energy E. From the relation

$$dE = TdS - PdV$$

we can find dE for an infinitesimal change in S and V. To find dS, consider $S \equiv S(V,T)$. Then

$$dS = \frac{\partial S}{\partial V}\bigg)_T dV + \frac{\partial S}{\partial T}\bigg)_V dT \tag{3.30}$$

or,

$$
\begin{aligned}
TdS &= T\left(\frac{\partial S}{\partial T}\right)_V dT + T\left(\frac{\partial S}{\partial V}\right)_T dV. \\
&= C_V dT + T\left(\frac{\partial P}{\partial T}\right)_V dT.
\end{aligned}
$$

(Use Maxwell relation for $(\partial S/\partial V)_T$)
Then,

$$dE = C_V dT + \left[T\left(\frac{\partial P}{\partial T}\right)_V - P\right] dV.$$

Integrating between state (T_0, V_0) and (T, V); we get

$$E(T,V) - E(T_0, V_0) = \int_{(T_0,V_0)}^{(T,V_0)} C_V dT + \int_{(T,V_0)}^{(T,V)} \left[T\left(\frac{\partial P}{\partial T}\right)_V - P\right] dV. \tag{3.31}$$

With the experimental knowledge or otherwise, $C_V = C_V(T)$ between T_0 and T at constant volume V_0 is known and therefore the first integral can be evaluated. Then, the second integral is evaluated at a temperature T from volume V_0 to V provided we know the equation of state $f(P,V,T) = 0$.

We can also show that

$$S(T,V) - S(T_0, V_0) = \int_{(T_0,V_0)}^{(T,V_0)} \frac{C_V}{T} dT + \int_{(T,V_0)}^{(T,V)} \left(\frac{\partial P}{\partial T}\right)_V dV \tag{3.32}$$

We thus see that the minimum data necessary to find various thermodynamic functions is the knowledge of equation of state and C_V (or C_P) in a temperature range of interest. This is the reason why measurements of specific heats as a function of T is so important. Once we know E and S, other thermodynamic functions $F = E - TS$; $H = E + PV$ and $G = F + PV$ can be found.

3.5   Carnot Cycle Ideas to Solve Thermodynamic Problems

We use Carnot cycle method, useful sometimes to solve elementary problem. Here the gas of the problem acts as a working substance of the Carnot cycle. The cycle is run between two isotherms, one at absolute temperature T and other at $T - \delta T$, bounded by two adiabats as shown in figure (1). A heat in Q is an intake at temperature T. Amount of work done during the cycle is an area δA of the cycle. Hence the heat rejected to the reservoir at temperature $T - \delta T$ is $Q - \delta A$. Then using equation (2.65), we get

$$\frac{Q - \delta A}{Q} = \frac{T - \delta T}{T} \quad \text{or,} \quad \frac{\delta A}{Q} = \frac{\delta T}{T}. \tag{3.33}$$

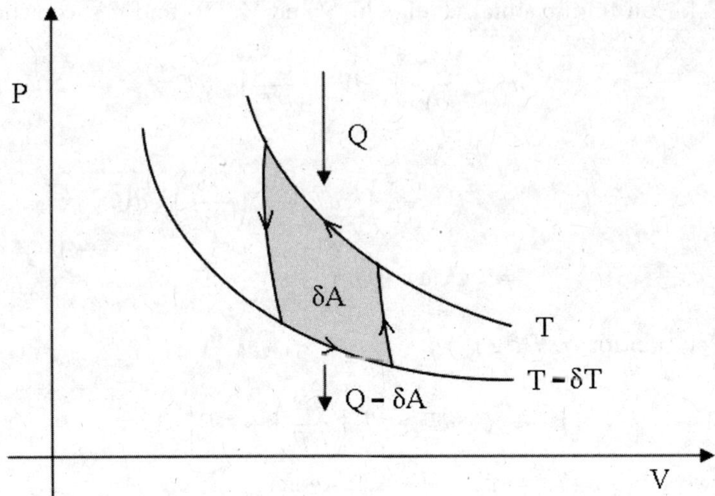

Figure 3.1: Carnot cycle

This is the equation when applied to the system, gives a useful result. To illustrate the method, we rederive Stefen's law. Following is data. For black-body, pressure exerted by the radiation is $P = u(T)/3$ where $u(T)$ is energy density that depends upon T only. Consider now a black-body expanded isothermally from volume V_1 to V_2 at an absolute temperature T. Then, expand it still further adiabatically to V_3 at temperature $T - \delta T$. Compress it isothermally from V_3 to V_4 and finally compress it adiabatically to its original volume V_1 as shown in figure(2) with corresponding Carnot cycle. Internal energy is $E(T, V) = u(T)V$.

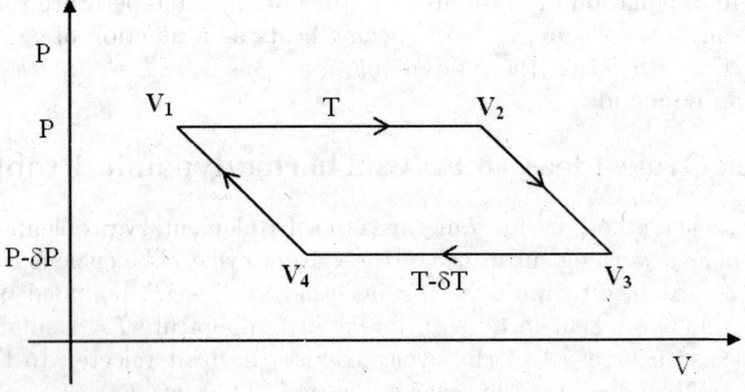

Figure 3.2: Carnot cycle for black body radiation

Work done by Carnot engine is

$$\delta A = \delta P(V_2 - V_1)$$

Heat input Q in the engine is,

$$\begin{aligned}
Q &= \int_{V_1}^{V_2} dE(T,V) + \int_{V_1}^{V_2} PdV \\
&= E(V_2,T) - E(V_1,T) + \frac{u}{3}(V_2 - V_1) \\
&= \frac{4}{3}u(T)(V_2 - V_1).
\end{aligned}$$

Then, equation (33) gives

$$\frac{\delta P(V_2 - V_1)}{\frac{4}{3}u(T)(V_2 - V_1)} = \frac{\delta u(T)}{4u(T)} = \frac{\delta T}{T}$$

Or, clearly, $u(T) = \text{constant} \times T^4$
Which is Stefan's law. The derivative given here is due to Boltzmann and hence it is also called as Stefen-Boltzmann law.

3.6   Joule-Thomson Process

This is a practical process, still used for liquifying the gases. Problem of liquification of gases was an important problem in the last century. It not only was of scientific importance but also of practical value, because the gases were required for welding and therefore required to be transported in a convenient and efficient manner.

In a Joule-Thomson process, a long insulated pipe is divided into two parts by a porous plug. Porous plug provides an obstruction to a flow of a gas. Constant pressure P_0 is maintained on the left side of the plug at a temperature T_0. On the right hand side of the porous plug the gas comes at a constant pressure P_1 and temperature T_1. Gas is assumed to flow from left to right when a steady state is reached. This means $P_0 > P_1$. Our aim is to find condition so that $T_1 < T_0$.

Consider a mass M of a gas confined by two imaginary planes X and Y as shown in figure (3). Mass M is initially confined within a volume V_0. Porous plug has a negligible volume compared with V_0. After some time, the gas of mass M occupies a larger volume V_1 on the right as shown figure (3). During a process of the flow, no heat flows in or out of system because of insulation. It should be noted that , then,

$$dE + PdV = 0.$$

Or,

$$E_0 + P_0 V_0 = E_1 + P_1 V_1 \tag{3.34}$$

where E_0 and E_1 are internal energies of the mass M, before and after the passage of the gas. Clearly, we see that an enthalpy $H(T,P)$ of the gas during Joule-Thomson process (sometimes also called throttling process) is constant.

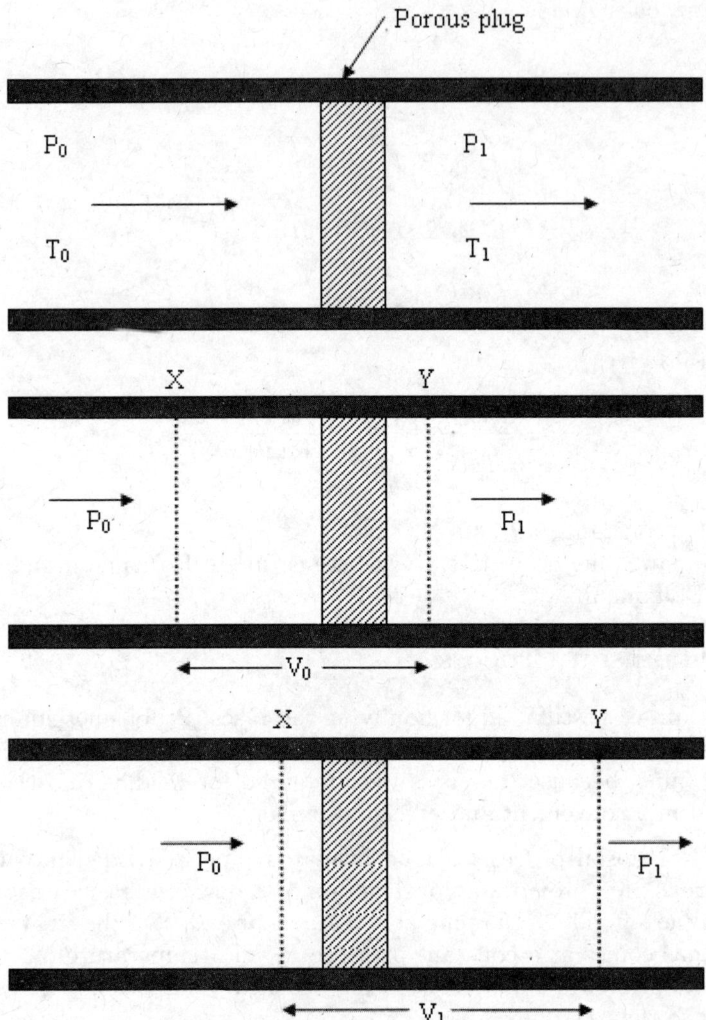

Figure 3.3: Mass M of a gas after passing through a porous plug in a Joule Thomson process.

This process is such that the initial and final states of flowing gas are equilibrium states on left and on right of the porous plug. The flow process of gas through the porous plug region is a highly convective non-equilibrium process. Enthalpy function is defined for equilibrium state. In general, enthalpy is a function of absolute temperature and pressure. i.e. $H(T,P)$. Then,

$$H(T_0, P_0) = H(T_1, P_1) \qquad (3.35)$$

It should be noted that if an ideal gas undergoes a throttling process, then, H is a function of absolute temperature T only. Under such condition, there is no change of temperature of the gas. Equation (34) determines T_1 in terms of P_0, T_0 and P_1. The method is to plot temperature versus pressure for various constant values of enthalpy - the so called constant enthalpy curves in $P-T$ plane. We draw typical $T-P$ curves for different values of enthalpies in figure (4). Locus of maxima on the curve is called inversion curve. We want a region,

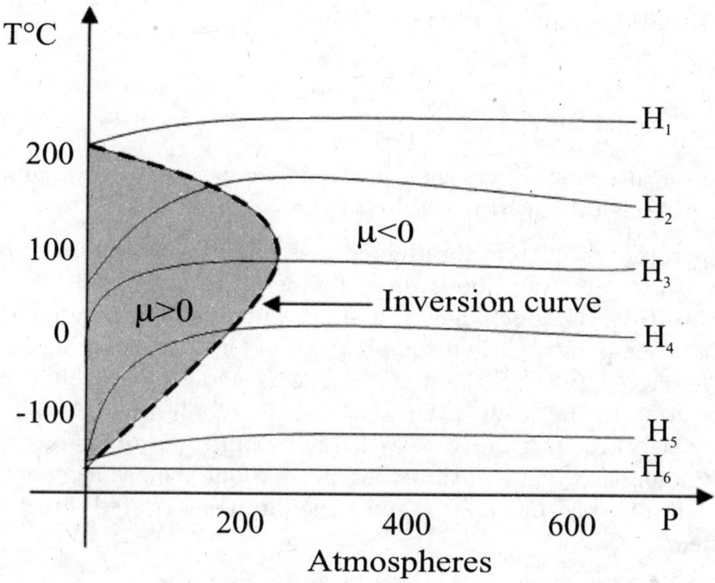

Figure 3.4: Curves of constant enthalpy in $P - T$ plane for a gas.

where, upon decrease of pressure, a decrease of temperature is needed. Correspondingly, we define Joule-Thomson coefficient μ such that

$$\mu = \left(\frac{\partial T}{\partial P}\right)_H \tag{3.36}$$

If $\mu < 0$, gas will be heated during the throttling process and vice versa for $\mu > 0$. Inversion curve corresponds to the points of $\mu = 0$.

This means during a throttling process, one descends to the lower temperature side on constant H curve in $\mu > 0$ region. To find μ, we note,

$$dH = TdS + VdP = 0 \tag{3.37}$$

Also, taking $S = S(T, P)$, we get

$$T\left(\frac{\partial S}{\partial T}\right)_P dT = -\left[V + \left(\frac{\partial S}{\partial P}\right)_T\right]dP = C_P dT$$

Or,

$$\mu = \left(\frac{\partial T}{\partial P}\right)_H = -\frac{V + T\left(\frac{\partial S}{\partial P}\right)_T}{C_P}$$

but by Maxwell's relations,

$$\left(\frac{\partial S}{\partial P}\right)_T = -\left(\frac{\partial V}{\partial T}\right)_P = -V\beta$$

where β is coefficient of volume expansion. Thus,

$$\mu = \frac{V}{C_P}\left(\beta T - 1\right)$$

All points lying on inversion curves give inversion temperatures. Maximum of inversion temperature for gases is tabulated in Table (1).

This temperature is different for different gases. It is seen that most of the maximum inversion temperatures are way above room temperature and hence do not require pre-cooling before throttling process. Thus, if hydrogen and helium undergo a Joule-Thomson expansion at room temperature, final temperature of these gases is higher than the room temperature. We have to pre-cool them ($T_H = 204^o K$ and $T_{He} = 43^o K$) and then undergo a throttling process to further cool it. On the other hand, all other gases given in the table can be cooled by the throttling process carried out from room temperature. One may not have a liquification in one cycle of a throttling process but the gas has a lower temperature after that cycle. In that case, one has to send a gas through repeated throttling cycles before the gas is liquified.

Table 1: Inversion Temperature for Common Gases.

	Gas	Inversion Temperature degree K
1	He	43
2	Xe	1486
3	Kr	1079
4	Ar	794
5	H_2	204
6	N_2	607
7	CO	644
8	CO_2	1275

Free expansion of a gas is a very similar process. Here a thermally insulated vessel with rigid walls is divided in to two compartments by a partition. A gas is confined in one compartment and other compartment is evacuated. If the partition is removed, the gas will undergo a free expansion. Since no mechanical work is done and the system is isolated, internal energy E of the gas remains unchanged during a free expansion. Obvious quantity to measure in the free expansion is so called Joule's coefficient $\left(\frac{\partial T}{\partial V}\right)_E$. For an ideal gas, there is no change in the temperature during free expansion.

3.7 Adiabatic Demagnetization

This is an effect where there is a decrease in temperature produced by adiabatic decrease in applied magnetic field. This is a very convenient method of cooling below about $1^o K$.

To make the physics of this method clear, we break the method in to steps and compare it with cooling of a gas in a cylinder. It should be noted that PdV is mechanical work whereas $-MdH$ is a magnetic work. We also assume that there are no magnetostriction effects and therefore PdV term is ignored from magnetic equation.

Step 1.
Gas cylinder system: Consider a gas in a cylinder at temperature T_i, in contact with a heat reservoir of temperature T_i. Compress the gas to volume V_i. Gas still will have some temperature T_i since this is an isothermal process.

Magnetic system: Take some paramagnetic salt such as cerium magnesium nitrate (CMM) in a solenoid (or magnetic coil) that is dipped in liquid Helium at about $1^o K$. Liquid Helium acts as a heat bath. Now switch on magnetic field from 0 to H_i so that the magnetic work is done in an isothermal manner (This is equivalent to isothermal compression of the gas).

Step 2.
Gas cylinder system: The gas is then thermally isolated and allowed to expand. This is adiabatic expansion. Here, part of the internal energy goes in doing work. Thus there is a reduction in internal energy of the gas, leading to a decrease in its temperature.

Magnetic system: Magnetic sample is thermally insulated by pumping out the Helium. Then external magnetic field H_i is reduced to H_f which is almost zero. Clearly, this process is adiabatic, and, because the field is reduced to zero, is also demagnetizing. In this process we get a reduction of temperature of the sample from T_i to T_f. Many a times T_f is about $1/100 \, ^{th} T_i$

For magnetic system we draw a $T - S$ diagram as shown in figure (5). It is easy to see that the entropy decreases monotonically with a decreases of temperature. Also, the entropy decreases with an increase of the magnetic field.

Clearly $a \rightarrow b$ is isothermal step and $b \rightarrow c$ is an adiabatic step where a field is reduced from H_1 to H_0 at fixed S. This is adiabatic demagnetization.

Mathematical Analysis: Entropy $S = S(H, T)$. For adiabatic process,

$$dS = 0 = \left(\frac{\partial S}{\partial H}\right)_T dH + \left(\frac{\partial S}{\partial T}\right)_H dT$$

Then,

$$\left(\frac{\partial T}{\partial H}\right)_S = -\frac{(\partial S/\partial H)_T}{(\partial S/\partial T)_H}$$

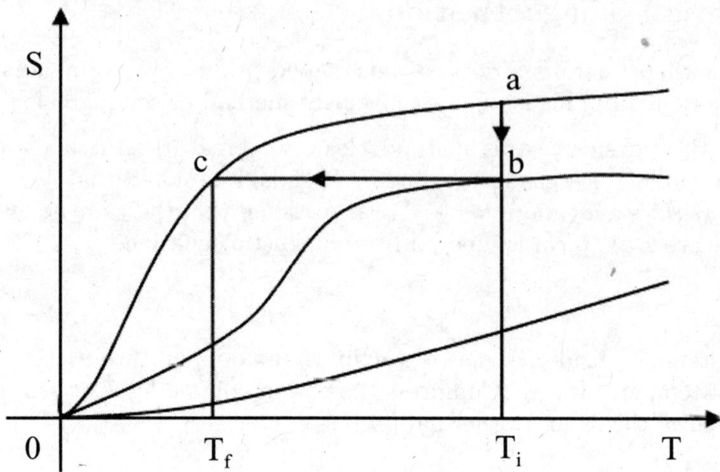

Figure 3.5: Behavior of an entropy as a function of a temperature for different values of H for a paramagnetic salt used in adiabatic demagnetization.

But magnetic specific heat at constant H is $C_H(T, H) = T \left(\frac{\partial S}{\partial T} \right)_H$.
Also, from Maxwell's relations,

$$\left(\frac{\partial S}{\partial H} \right)_T = \left(\frac{\partial M}{\partial T} \right)_H$$

Let $\chi(T, H)$ be magnetic susceptibility/volume of the paramagnetic salt.. Then

$$M = V\chi H.$$

We get,

$$\left(\frac{\partial T}{\partial H} \right)_S = -\frac{VTH}{C_H} \left(\frac{\partial \chi}{\partial T} \right)_H.$$

Clearly $\chi(T, H)$ and $C_H(T, H)$ can be measured for the paramagnetic salt for different T and H. Then simple integration gives T_f in terms of T_i, H_i and H_f.

Illustrative Problems

Problem 3. The quantity of interest in Joule's free expansion is $\eta = (\partial T / \partial V)_E$. Show that

$$\eta = \left(\frac{\partial T}{\partial V} \right)_E = \frac{1}{C_V} \left[P - \frac{\beta T}{\kappa} \right]$$

Solution: Let $E = E(T, V)$. In Joule's expansion E is unchanged.

$$\therefore \quad dE = 0 \;=\; \left(\frac{\partial E}{\partial T}\right)_V dT + \left(\frac{\partial E}{\partial V}\right)_T dV$$

$$=\; C_V dT + \left(\frac{\partial E}{\partial V}\right)_T dV$$

Using Maxwell's relations $\left(\dfrac{\partial S}{\partial V}\right)_T = \left(\dfrac{\partial P}{\partial T}\right)_V$; definitions of β and κ, we get

$$\left(\frac{\partial T}{\partial V}\right)_E = \frac{1}{C_V}\left[P - T\left(\frac{\partial P}{\partial T}\right)_V\right] = \frac{1}{C_V}\left[P - T\beta/\kappa\right]$$

q.e.d.

Problem 4. $P - V$ diagram for a petrol engine is shown in figure (6). The complete cycle $ABCD$ is as follows.

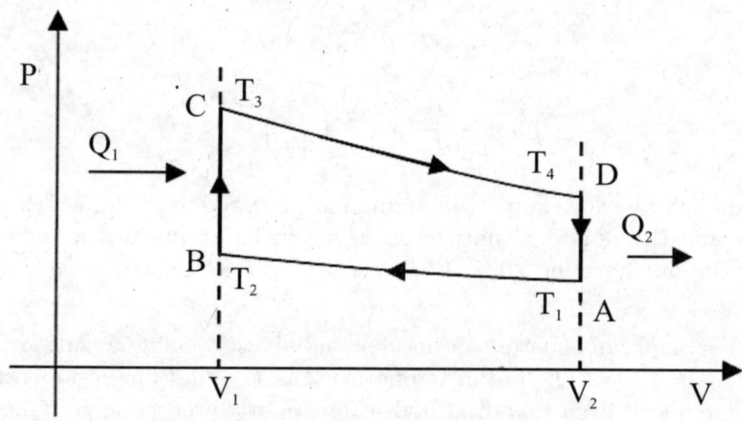

Figure 3.6: Carnot cycle for petrol engine.

$A \to B$ is adiabatic compression from volume V_2 to V_1. The compressed vapor of air fuel mixture is exploded at B when pressure increases from P_B to P_C at constant volume. It is shown as $B \to C$. From C to D we have adiabatic expansion. Here engine does a useful work. From $D \to A$, a final cooling at constant volume takes place. Find efficiency of this cycle in terms of V_1, V_2 and γ. Make a drastic assumption of an ideal gas as a working substance.

Solution:

$$\eta = \frac{Q_1 - Q_2}{Q_1}.$$

Let T_1 be the temperature at A, T_2 at B, T_3 at C and T_4 at D.

For the adiabatic process, $A \rightarrow B$,

$$P_A V_2 = RT_1, P_B V_1 = RT_2$$

$$P_A V_2^\gamma = P_B V_1^\gamma$$

This gives

$$\frac{T_1}{T_2} = \frac{V_1^{\gamma-1}}{V_2^{\gamma-1}}$$

Also, $Q_1 = C_V(T_3 - T_2)$, $Q_2 = C_V(T_4 - T_1)$.

Then

$$\eta = 1 - \frac{T_4 - T_1}{T_3 - T_2}.$$

But by similar arguments,

$$\frac{T_3}{T_4} = \left(\frac{V_2}{V_1}\right)^{\gamma-1}$$

or

$$\frac{T_1}{T_2} = \frac{T_4}{T_3}$$

Then

$$\eta = 1 - \frac{T_4}{T_3} = 1 - \left(\frac{V_1}{V_2}\right)^{\gamma-1}$$

The ideal efficiency for the petrol engine may be around 40 to 50 %. However with frictional losses and the non-ideal nature of an air fuel mixture, actual efficiency for the petrol engine is obtained around 20 to 30 %

Problem 5. For a paramagnetic salt used in adiabatic demagnetization, $C_H(H,T) = V(b + aH^2)/T^2$. Also, $\chi = a/T$. Initial temperature is T_i. Find final temperature T_f if the magnetic field is reduced from its initial high value H_i to lower value H_f adiabatically.

Solution: In adiabatic demagnetization, $dS = 0$.

$$dS = 0 = C_H dT + \left(\frac{\partial S}{\partial H}\right)_T dH$$

But by Maxwell's relations, from equation (19)

$$\left(\frac{\partial S}{\partial H}\right)_T = \left(\frac{\partial M}{\partial T}\right)_H$$

Also $\qquad M = V\chi H = VaH/T$

$$\therefore \quad \left(\frac{\partial M}{\partial T}\right)_H = -\frac{VaH}{T^2}$$

Then,

$$dS = 0 \;\;=\;\; C_H dT + \left(\frac{\partial M}{\partial T}\right)_H dH$$

$$=\;\; \frac{V(b + aH^2)}{T^2}dT - \frac{VaH}{T^2}dH$$

Integrate from T_i to T_f and H_i to H_f and simplify,

$$\frac{T_f}{T_i} = \left(\frac{b + aH_f^2}{b + aH_i^2}\right)^{1/2}$$

q.e.d.

Short Questions

1. Elaborate the Einstein's statement about Thermodynamics, given at the beginning of this chapter.

2. List as many response functions that you know giving their significance in short.

3. In actual gases, the internal energy is a function of both, the temperature of the gas and its volume. On the other hand, internal energy of ideal gas is a function of absolute temperature only. Explain these results.

4. Obtain Maxwell's relations for magnetic system.

5. Show,

$$S(T, V) - S(T_0, V_0) = \int_{T_0, V_0}^{T, V_0} \frac{C_V dT}{T} + \int_{T, V_0}^{T, V} \left(\frac{\partial P}{\partial T}\right)_V dV.$$

6. Show that the enthalpy for an ideal gas is purely a function of T.

7. What is inversion temperature? Explain its significance. Can one reduce a temperature of an ideal gas by throttling process?

8. It is observed that if hydrogen and helium are passed through a porous plug of Joule-Thomson apparatus at room temperature ($\approx 300K$), then their temperature rises rather than is reduced. Explain this result.

9. Explain qualitatively the process of adiabatic demagnetization for cooling.

Problems

Problem 1. Show that for a van der Waals gas C_V is independent of V. Show also that the enthalpy $H = E + PV$ is a function of absolute temperature T for an ideal gas.

(Hint:

$$P = \frac{RT}{V - b} - \frac{a}{V^2}$$

$$\frac{\partial C_V}{\partial V}\Big)_T = \frac{\partial}{\partial V}\Big)_T \left(T\frac{\partial S}{\partial T}\right)_V = T\frac{\partial^2 S}{\partial V \partial T} = \frac{\partial}{\partial T}\Big)_V \left(\frac{\partial S}{\partial V}\right)_T$$

$$= T\frac{\partial}{\partial T}\left(\frac{\partial P}{\partial T}\right)_V = T\frac{\partial^2 P}{\partial T^2}\Big)_V = 0 \quad)$$

Problem 2. Dependence of pressure on V and T for real gases is given by Virial expansion. Lowest Virial coefficient is B_2 and is a function of T alone i.e. $B_2(T)$. With n as a density,

$$P = nk_BT(1 + nB_2(T))$$

Show that

$$(1) \qquad\qquad K_T = \frac{1}{P + n^2 k_B T\, B_2(T)}$$

$$(2) \qquad\qquad \left(\frac{\partial E}{\partial V}\right)_T = \frac{(nk_BT)^2}{k_B}\frac{dB_2}{dT} > 0$$

(Hint: Show

$$P\left(\frac{\partial V}{\partial P}\right)_T + V = -\frac{N^2}{V^2}k_B T\, B_2 \left(\frac{\partial V}{\partial P}\right)_T$$

$$\therefore \left(\frac{\partial V}{\partial P}\right)_T = -\frac{V}{p + n^2 k_B T\, B_2(T)}. \quad \text{Hence get } K_T.$$

Moreover,

$$\left(\frac{\partial E}{\partial V}\right)_T = T\left(\frac{\partial P}{\partial T}\right)_V - P = P + n^2 k_B T^2 \frac{\partial B_2}{\partial T} - P.$$

But, $\dfrac{\partial B_2}{\partial T}$ is > 0 and hence $\left(\dfrac{\partial E}{\partial V}\right)_T > 0$)

Problem 3. An inventor says that he has constructed an engine with a heat intake of 10^5 Joule at a temperature 1000 K. Heat rejected by his engine is 2×10^4 Joule at 500 K. Will you invest your hard earned money in his factory to manufacture his engine? Why? Explain in details.

Ans.: No!

Problem 4. Justify that the Carnot cycle in figure 7 represents the cycle for paramagnetic salt. Identify as to which lines are adiabatics and which ones are isothermals?

Problem 5. (a) Assume that, for a certain system, volume expansion coefficient β is temperature independent and volume is pressure independent. Show that the amount of heat Q flows out of the system upon compression if $\beta > 0$ and vice versa. Show that

$$Q = -TV\beta(P_f - P_i)$$

Find Q when mercury is compressed from 1 atmosphere ($\approx 10^5$ Pa) to 10^8 Pa. $\beta_{Hg} = 1.8 \times 10^{-4}K$, $T = 300K$ and volume of Hg is $V = 10^{-5}m^3$.

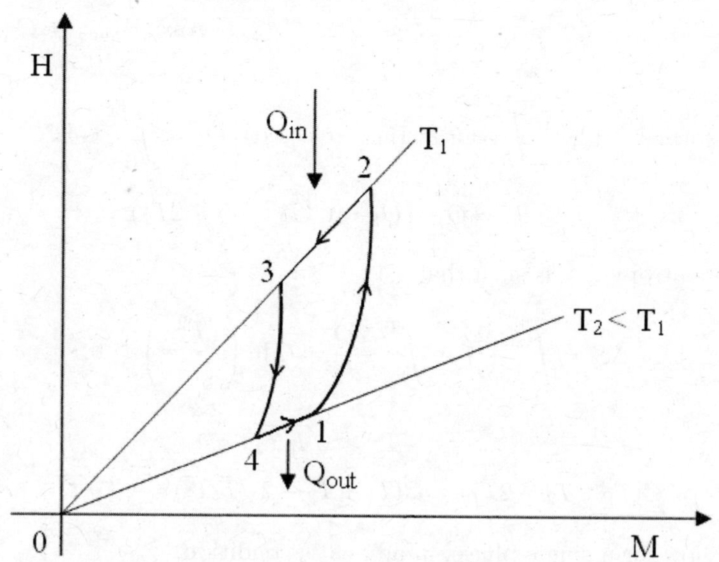

Figure 3.7: Carnot cycle for a paramagnetic salt.

$$\text{Ans.: } Q = -54\text{J}$$

(b) Show that the isothermal work done during compression is

$$W = \frac{1}{2} V k_T \left(P_f^2 - P_i^2 \right).$$

Consider the case of mercury in (a). Take $\kappa_{Hg} = 4 \times 10^{-11} Pa^{-1}$. Estimate W.

$$\text{Ans: } W = 2J$$

(c) Find total change in internal energy.

$$(\Delta E = Q + W = -54 + 2 = -52J.)$$

Problem 6. You are given two identical bodies A_1 and A_2 at initial temperatures T_1 and T_2 respectively. C_P is their heat capacity and is almost independent of temperature. The bodies are large enough to act as reservoirs. A heat engine is run between the two bodies with $A_1(T_1 > T_2)$ as a hot reservoir and A_2 as cold one. As a result of operation of heat engine, a final common temperature T_f is attained at the end. Throughout the process, pressure on A_1 and A_2 is kept constant.

Find,
(i) Total amount of work done by the engine.
(ii) Using $\Delta S \geq 0$, show that $T_f = \sqrt{T_1 T_2}$.
(iii) Maximum work done by the engine.

$$\text{Ans: } W_{max} = C\left(\sqrt{T_1} - \sqrt{T_2}\right)^2.$$

(Hint: Heat absorbed $= Q_1 = \displaystyle\int_{T_f}^{T_1} C\delta T$, Heat rejected$= Q_2 = \displaystyle\int_{T_2}^{T_f} C\delta T$

$$\therefore \quad W = Q_1 - Q_2 = C(T_1 + T_2 - 2T_f)$$

Total change of entropy ΔS is such that

$$\Delta S = \int_{T_1}^{T_f} \frac{\delta Q}{T} + \int_{T_2}^{T_f} \frac{\delta Q}{T} = C \ln\left(\frac{T_f^2}{T_1 T_2}\right) \geq 0.$$

$$\therefore \quad T_f \geq \sqrt{T_1 T_2}$$

Maximum work $= C(T_1 + T_2 - 2T_f) = C(T_1 + T_2 - 2\sqrt{T_1 T_2}) = C(\sqrt{T_1} - \sqrt{T_2})^2.)$

Problem 7. Show for a single phase, in adiabatic condition,

$$\left(\frac{\partial P}{\partial T}\right)_S = \frac{C_P}{TV\beta}.$$

Here V is specific volume=volume/mass. Calculate $(\partial P/\partial T)_S$ for Hg at $300K$, $V_{Hg} = 7.4 \times 10^{-5} m^3/kg$, $C_P = 140 J/kg.K$ and $\beta = 1.8 \times 10^{-4}/K$.

$$\text{Ans: For Hg, Slope} \simeq 3 \times 10^7 Pa/K$$

Chapter 4

PHASE TRANSITIONS AND CRITICAL PHENOMENA I

Study of phase transition s began with classic experiments of Amagat and Andrews on phase properties of carbon dioxide in the late 19th century. Its theoretical explanation was offered by van der Waals who proposed an equation of states for real gases. This was for solid–liquid–gas transformation. A similar theory for paramagnetic to ferromagnetic transition was given by Weiss.

A phase is an equilibrium state of a substance for certain range of thermodynamic variables. At certain values of these parameters, suddenly, another phase (another thermodynamical equilibrium state) is created if a phase transition occurs. Some examples of Phase transitions are

1. normal metal to superconducting metal,
2. para electric to ferroelectric transition of $BaTiO_3$,
3. various phases of liquid crystals,
4. normal to super fluid state of liquid helium,
5. various structural phase transitions e.g. cubic phase of brass and its b.c.c. structure.

All these phase transitions occur at certain fixed thermodynamic parameters. Most important parameter in phase transition phenomenon is a fixed temperature T_c called critical temperature. A range of temperatures around T_c is important. The ranges are characterized by a dimensionless parameter $\dfrac{T - T_c}{T_c} = t$ for $T > T_c$ and $\dfrac{T_c - T}{T_c} = -t$ for $T < T_c$.

It is observed that for $t > 0$, only one phase exists. For $t < 0$, phases can coexists in equilibrium. Example is that of a gas and a liquid coexisting at $T < T_c$. Only gaseous phase exists for $T > T_c$. This happens at a certain pressure called critical pressure P_c. A critical volume V_c is automatically decided by the equation of state. For water, critical temperature T_c has value of about $648K$ and corresponding critical pressure P_c is about 219 atmosphere.

In case of ferromagnetic systems such as iron, there is a critical temperature T_c below which its magnetization M is nonzero even if the external magnetic field H is switched off.

The value of M increases as T decreases below T_c and has a maximum value at $0K$. On the other hand, $M = 0$ when $H = 0$ for $T > T_c$.

We give another example - example of co-existence of a normal fluid (viscosity $\neq$ 0) and of a superfluid (viscosity = 0) in case of Helium II below $T_c = 1.27K$. As the temperature decreases from T_c, the density of superfluid component increases. This is modeled by Tisza in his two fluid model of Helium.

It was observed by van der Waals (vdW) that if the equation is written in terms of $\widetilde{P} = \dfrac{P}{P_c}$, $\widetilde{V} = \dfrac{V}{V_c}$, and $\widetilde{T} = \dfrac{T}{T_c}$ for various substances(like Methane, Carbon dioxide, water etc.), the equation becomes universal i.e. all points fall on one surface. This surface has projections in various planes such as $P - T, T - V$ and $P - V$. The curve in a $P - V$ plane is a parabola if vdW equation is used. Experimentally one obtains the universal curve but it is not a parabola. This curve is called co-existence curve. It was discovered by Guggenheim that the co-existence curve can best be fitted by a cubic equation(1945). Another landmark in the theory was by Onsagar's exact solution of two dimensional Ising model in 1944. A better understanding of phase transitions was due to Landau's phenomenological theory of second order phase transitions(1937). He introduced a very important concept called 'Order Parameter'. This Order parameter is different for different systems. It has a property that the order parameter tends to zero as $T \to T_c$ from below. It has a small but a nonzero value for $t = \dfrac{T_c - T}{T_c} << 1$. Identification of this order parameter is obvious in many usual situations but is a skillful job in 'not so obvious situation'.
Some examples of order parameters are
(i) $\rho_L - \rho_G$ where ρ_L is liquid density and ρ_G is gas density in co existence region.
(ii) $\vec{M}(T)$, the magnetization of ferromagnetic materials.
(iii) $\Delta(T)$, the super conducting gap [1].

Next important advance was made by Kadanoff who introduced the idea of scaling and showed that, on macroscopic scale, all these phase transitions were very similar and many of their properties could be calculated [2]. This theory of scaling was extended to behavior of non-equilibrium steady state situations (Diffusion, thermal conductivity etc.) by Ferrell and also by Hohenberg and Halperin in 1969.

Most important advance, which encompasses all aspects of phase transitions was made by K. G. Wilson in 1970 by using the renormalization group ideas of field theory due to Bogoliubov. The theory of Wilson also relied upon Kadanoff's ideas (loc.cit.). Wilson got Nobel prize in 1982. We will discuss the essentials of all these ideas which we feel are important.

---

[1] Normal metals have all electrons stacked in energy levels according to Pauli principle up to maximum energy level (called Fermi level E_f). When this metal becomes superconductor below $T < T_c$ there is a temperature dependent energy gap Δ in the energy spectrum. For details see any standard book on solid state physics like that of Kittel.
[2] Kadanoff et.al. Rev. Mod. Phys. <u>39</u>, 395(1967). This long paper is well written and enjoyable to read.

4.1 $P - V - T$ Diagrams for Pure Substances

Equation of state is

$$f(P, V, T) = 0$$

and relates pressure, volume and temperature in an equilibrium thermodynamic state and is a surface. This kind of relationship is true for any substance. The form of the equation may change from system to system e.g. for magnetic systems an equation of states is a relation between M, H and T.

We will discuss the gas-liquid-solid system for the time being. Projection of the surface is easier to visualize on $P - V$ plane or $P - T$ plane. Fig.1(a) shows the $P - V - T$ surface of water and its projection in $P - T$ plane.

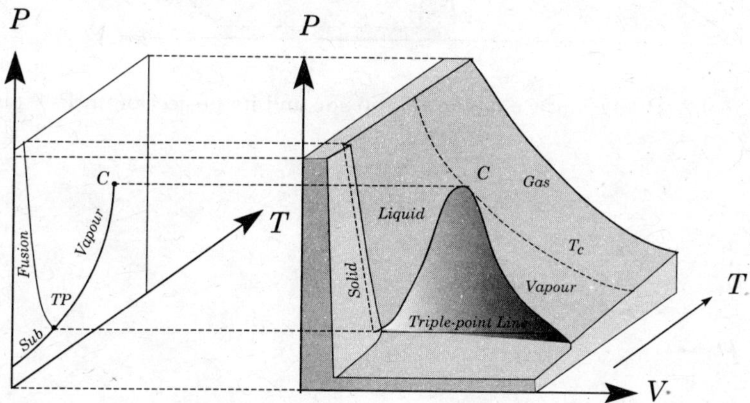

Figure 4.1: P-V-T surface of water and its projection in P-T plane

Fig.(2) shows the $P - V - T$ surface for CO_2 . Note the difference between the fusion curves of water and that of CO_2 in the two figures.

The projection of $P - V - T$ surface in the $P - T$ plane for water is shown in Fig.(3)

All the substances may not have same type of $P - V - T$ surface, because equation of state differs from substance to substance. However, the diagrams that we discuss are typical of many substances. A careful look at the phase diagram of water indicates that there are two important points (a) triple point T_{Tp}, (b) critical point T_c. There are three important lines (a)sublimation curve (b)vaporization curve and (c) fusion curve. All these curves meet at the triple point. At the triple point, all the three phases, namely, liquid, vapor and solid coexist. The value for the triple point temperature for water $T_{T_p} = 273.16K$ and corresponding pressure $P_{T_p} = 0.6MPa$ or about six atmospheric pressure (1 atmospheric pressure, atm. is, $1.013 \times 10^5 Pa$). Corresponding liquid density is $\rho_c \approx 1000 \ kg/m^3$. Clearly the triple point temperature is approximately $0^o C$.

It is well known that at NTP, water boils at $100^o C$. As the pressure is reduced, the boiling point of water decreases. All boiling point temperatures and corresponding pressures lie on a vaporization curve.

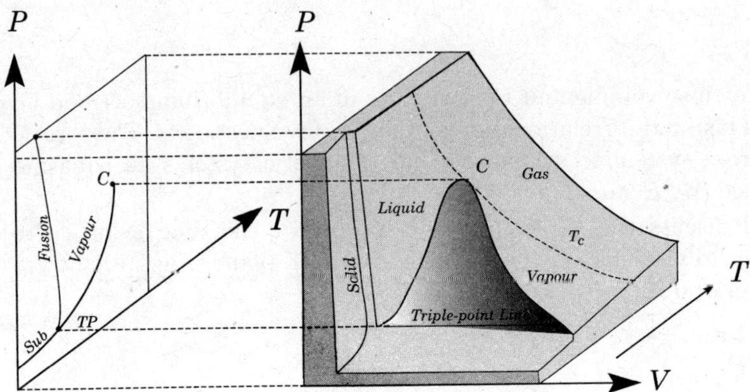

Figure 4.2: P-V-T surface of carbon dioxide and its projection in P-T plane

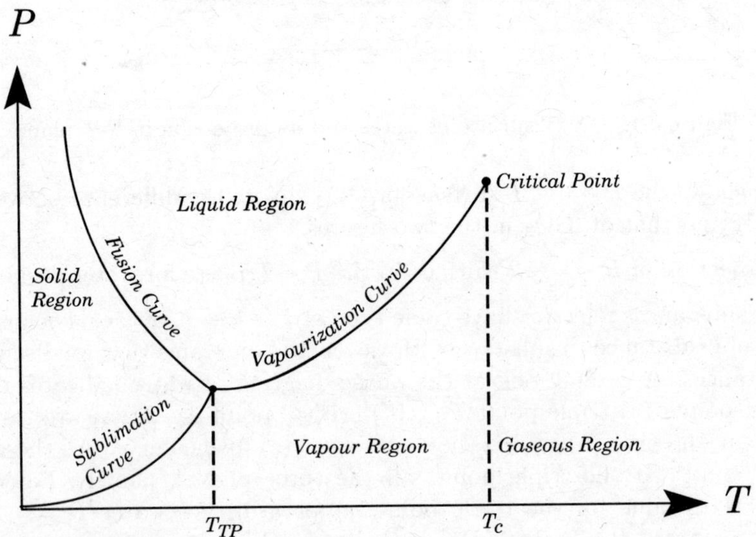

Figure 4.3: Phase diagram for water in P-T plane

Solid and liquid (here ice and water) coexists on a fusion curve which is unbounded and starts from triple point. For water and some other substances (like iron, bismuth, antimony etc.) the slope of fusion curve is negative. It is well known that as the pressure increases melting point decreases for the substances whose fusion curve has negative slope. Good castings can be made from only those substances whose fusion curve has a negative slope. The melting points and corresponding pressures lie on a fusion curve. Generally, the fusion curve has a positive slope for most of the substances. These substances contract upon freezing. Most of the metals, paraffin wax, CO_2 have a fusion curve with a positive slope.

We discuss the curve for H_2O in the $P - T$ plane as shown in figure 3. Region near sublimation curve is interesting. If we have an ice (solid region) at a temperature T (obviously less that the triple point temperature) its pressure value must lie above the sublimation curve. If we increase the pressure, keeping its temperature fixed, we cross the fusion curve at some value of a pressure. Beyond this value we will melt the ice. On the other hand if we decrease the pressure and cross the sublimation curve we have a water vapor and no water.

Vaporization curve starts from the triple point temperature T_{T_p} and ends at the critical point temperature T_c.

We define gas in a following way. A substance with no free surface (sometimes called miniscus) and with a volume determined by that of its container is called a gas provided its temperature is lager than T_c. Otherwise it is called as a vapor. Thus a vapor is a gas in equilibrium with its liquid and has always a temperature less than T_c.

At a temperature $T < T_c$, a vapor can be condensed in its liquid form by isothermally increasing the pressure. On the other hand a gas at $T > T_c$ can not be condensed in its liquid by any pressure, how so ever high!.

> The critical temperature for Helium is $T_c = 5.2K$. It is very low and therefore all the attempts to liquify it before K.Onnes failed. K.Onnes succeeded in liquifying Helium in 1908. By careful study of isotherms of Helium down to liquid hydrogen temperature(14 K) he estimated T_c to be about $5.25K$. By pre cooling Helium gas in Hydrogen vapors generated from liquid Hydrogen and then allowing it to throttle (Joule-Thomson Process) Helium got liquified.

It must be noted that the slope of the vaporization curve and sublimation curve is always positive. On the other hand the fusion curve may have a positive or a negative slope.It is obvious that the triple point or the critical point in $P - T$ plane are actually lines in $P - V - T$ surface.

This discussion of $P - V - T$ space will not be complete without understanding of isotherms in $P - V$ plane. Typical experimental $P - V$ plot for different isotherms is shown in figure(4).

From this figure it is seen that, all isotherms with temperature larger than T_c are hyperbolas and are approximately given by the equation $P V = constant$. Isotherm at critical temperature T_c is called the critical isotherm. All isotherms with $T < T_c$ have a flat portion. Edges of the flat portions can be joined by dotted line. A region below the dotted

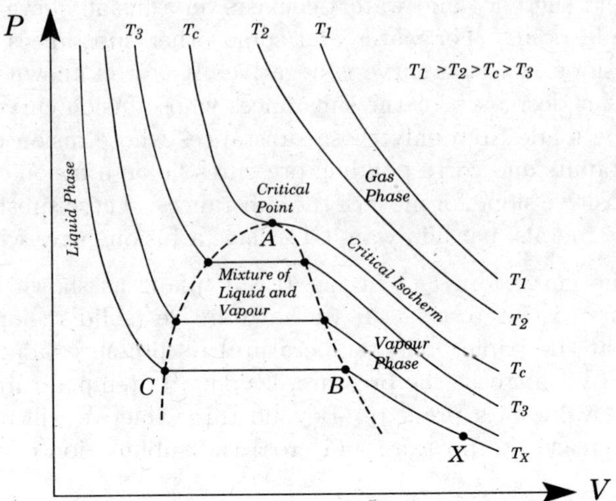

Figure 4.4: Phase diagram in P-V plane for liquid-gas transition.(Experimental)

line has a mixture of vapor and liquid. A region on the right of AB is a vapor phase region. The region on left of AC is a liquid region.

Suppose, we have a water vapor at pressure P_X and volume V_X at $T_X < T_c$. It is represented by a point X on the isotherm as shown in figure 3. We now compress this vapor isothermally so that its volume decreases and pressure rises till point B is reached.

If the compression is continued, condensation occurs forming the droplets of water. There is no corresponding increase of pressure . Region $B - C$ is isobaric and isothermal region. Along the $B - C$ line, liquid and vapor co-exist. At point C, all the vapor becomes liquid and is hard to compress indicating sharp rise in pressure for relatively low change in volume. Point B represents the state of saturated vapor. As the temperature T of the isothermals approaches T_c $(T < T_c)$; the region $B - C$ goes on shrinking and is zero at T_c.

4.2   First Order Phase Transitions

In solid-liquid-gas type of phase transition, it is observed that temperature and pressure remain constant. What changes is the volume and the entropy between two coexisting phases. Consider now the changes in the four thermodynamic potentials for a single specie

$$dE = TdS - PdV$$

$$dH = TdS + VdP$$

$$dF = -PdV - SdT$$

and

$$dG = VdP - SdT \tag{4.1}$$

Clearly for $dP = 0 = dT$, only $dG = 0$ and all other potentials change as the phase change. Thus during the process of this phase transition, Gibbs free energy remains invariant. Since thermodynamic potentials are extensive quantities, we can define $g = \frac{G}{N}$ as Gibbs free energy per particle of the N particle system. We can also define $g = \frac{G}{\nu}$ where ν is number of moles of a substance in the system as a molar Gibbs Potential. Gibbs free energy per particle is same as the chemical potential. This can be seen by writing equation 1, when the numbers change. Then

$$dG = -SdT + VdP + \sum_k \mu_k dN_k \tag{4.2}$$

which, for single component becomes,

$$dG = -SdT + VdP + \mu dN \tag{4.3}$$

If at constant P and T, N changes by some proportion, i.e. $dN = Nd\lambda$, then $dG = Gd\lambda$. Clearly then,

$$G = \mu N \text{ and } \mu = \frac{G}{N} \tag{4.4}$$

4.2.1 Properties of Gibbs Free Energy G

Gibbs free energy is also known as Gibbs potential because of its connection with the chemical potential μ. Free energy F and Gibbs potential G has certain geometrical properties.

We define a convexity and a concavity of a function $f(x)$ of a real variable x. Referring to the figure (5) of a convex function we see that a chord joining two points $y_1 = f(x_1)$ and $y_2 = f(x_2)$ is above the curve.

We define a function to be convex when every chord of it lies above the curve. When the derivative exists $f(x)$ has a tangent. This tangent always lie below the curve $f(x)$. This also implies that $f''(x) \geq 0$ for all values of x in the domain of f.
If $f(x)$ is a convex function, $-f(x)$ is a concave function.

Theorem: The Gibbs potential is a concave function of both T and V.

Proof: We have to show that $\left(\frac{\partial^2 G}{\partial T^2}\right)_P$ and $\left(\frac{\partial^2 G}{\partial P^2}\right)_T$ are negative definite. Clearly,

$$S = -\left(\frac{\partial G}{\partial T}\right)_P$$

$$\left(\frac{\partial^2 G}{\partial T^2}\right)_P = -\left(\frac{\partial S}{\partial T}\right)_P = -\frac{C_P}{T} \leq 0$$

and,

$$\left(\frac{\partial^2 G}{\partial P^2}\right)_T = \left(\frac{\partial V}{\partial P}\right)_T = -VK_T \leq 0$$

q.e.d.

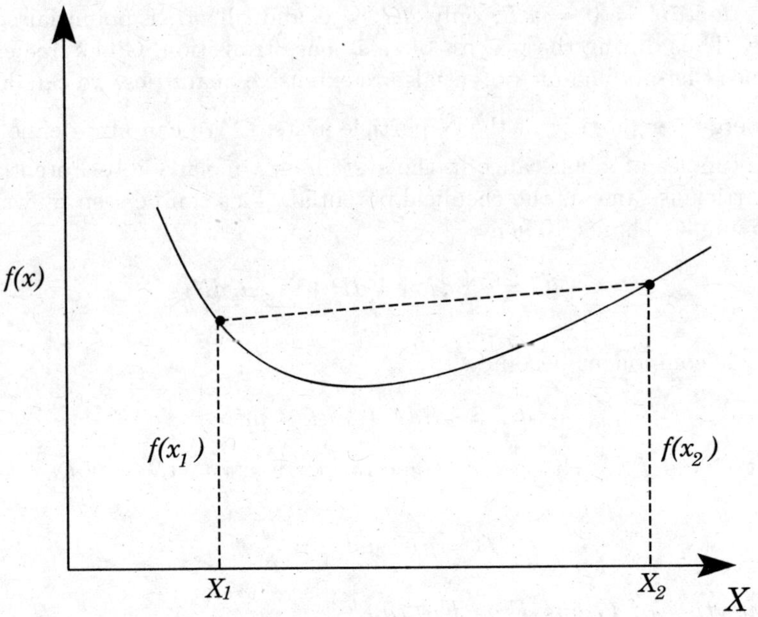

Figure 4.5: Convex function

We plot g as a function of absolute temperature as shown in figure 6. At the transition temperature, g is continuous but $\left(\dfrac{\partial g}{\partial T}\right)$ is discontinuous. This derivative is $-S$ i.e. negative of entropy per mole. We thus see that entropies of two phases differ. This implies that, when the two phases co-exists at some temperature T, there is a different heat content in them. The difference in heat content is $T|S_I - S_{II}| = T|\Delta S|$. From equation 1; moments reflection will reveal that this is nothing but a change in enthalpy ΔH. This ΔH is called latent heat L. L is a weak function of T. L is given out when a transition from lower entropy phase to a higher entropy phase takes phase and is absorbed vice versa. Correspondingly, there is a change in a molar volume between different phases.

We <u>define</u> a first order phase transition as the one where a latent heat L is involved during the transition process. Incidentally, in a first order phase transitions the first derivative of G with respect to T is discontinuous.

Ehrenfest classified various orders of the phase transitions according to the derivatives

$$G^{(n)} \equiv \left(\frac{\partial^n G}{\partial T^n}\right)_{P,N}.$$

A phase transition, according to Ehrenfest, is of nth order if $G^{(n)}$ is discontinuous but all the lower derivatives of G are continuous. Clearly, according to this scheme, a transition,

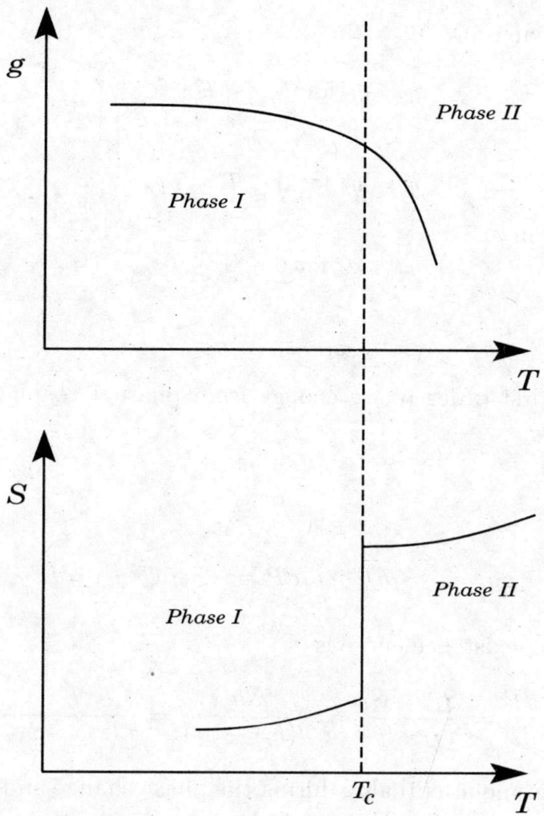

Figure 4.6: g and S as a function of T

where there is a continuous change in entropy S, is of second order, (about which we will talk subsequently). The solid- liquid-gas transition is of first order because, G is continuous but $G^{(1)} \sim S$ is discontinuous. Now a days, though, this type of scheme of classification in terms of derivatives of G is no more used because all types of phase transition can not be fitted in to this scheme. For example, we can not classify the logarithmic singularity in the specific heat of liquid Helium II. This is the reason why we 'defined' first order phase transition in terms of the latent heat.

4.3   Clausius-Clapeyron Equation

Clausius-Clapeyron equation (Here after referred as CCE) is an equation for vaporization curve, fusion curve and a sublimation curve. These curves meet at the triple point T_{T_p}.
As seen in the last section, g is same for two co-existing phases. Thus, if g_s, g_l and g_v denote mole Gibbs energies for solid, liquid and the vapor phase respectively, then,

along the vaporization curve,

$$g_l = g_v \text{ for } T_{T_p} \le T \le T_c,$$

along the fusion curve

$$g_l = g_s \text{ for } 0 \le T \le T_{T_P},$$

along the sublimation curve

$$g_v = g_s \text{ for } 0 \le T \le T_{T_p}.$$

At T_{T_P},

$$g_l = g_v = g_s \tag{4.5}$$

Thus for any general first order phase change from phase I to phase II, there will be a coexistence curve where

$$g_I = g_{II} \tag{4.6}$$

or

$$dg_I = -S_I dT + v_I dP = -S_{II} dT + v_{II} dP$$

Then the slope of the coexistence curve is

$$\frac{dP}{dT} = \frac{S_{II} - S_I}{v_{II} - v_I} = \frac{\Delta h}{T(v_{II} - v_I)} = \frac{L}{T(v_{II} - v_I)} \tag{4.7}$$

where Δh is a change in molar enthalpy during the phase change and v is the molar volume. Equation(7) is known as Clausius-Clapeyron Equation and is one of the most important equations of classical thermodynamics. It has extensive applications in Cloud Physics. From CCE; it is clear that the latent heat at the triple point T_p is not defined.

Illustrative Problems

Problem 1: Derive CCE using Carnot Engine.

Solution: Consider two isotherms differing in temperature by δT as shown in figure(7). Let us run Carnot Engine with reservoir at T and at $T - \delta T$. Let AB be isothermal expansion, BC be adiabatic expansion and similarly CD as isothermal compression and DA be adiabatic compression. While expanding from A to B, evaporation takes place and heat energy L per unit mass is absorbed. During infinitesimal adiabats AD and BC, pressure change is

$$\delta P = \frac{\delta P}{\delta T} \delta T$$

The work done by the engine= area $ABCD = \delta A$.

$$\delta A = \frac{\delta P}{\delta T} \delta T(v_B - v_A)$$

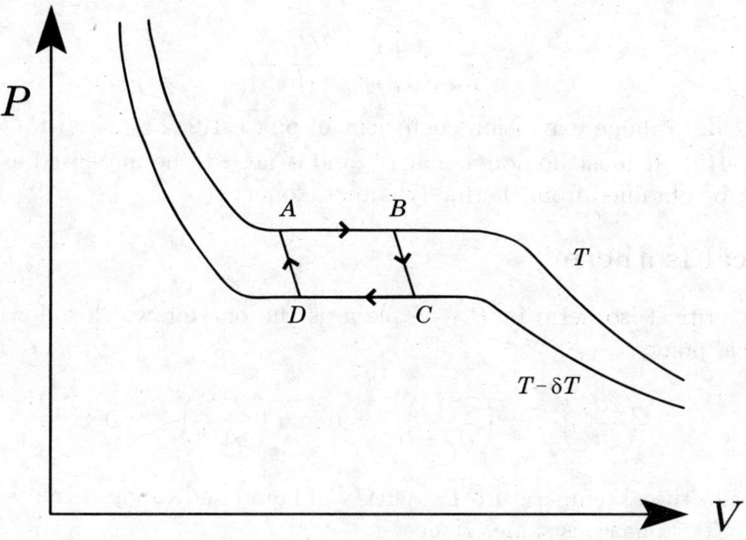

Figure 4.7: Carnot cycle for deriving CCE.

Thus using equation for efficiency for Carnot engine , we get

$$\frac{\delta P}{\delta T}\frac{\delta T(v_B - v_A)}{L} = \frac{\delta T}{T}$$

or

$$\frac{dP}{dT} = \frac{L}{T(v_g - v_l)}$$

q.e.d.

Problem 2: Derive an equation similar to CCE for second order phase transition.

Solution:
In a second order phase transition , entropy and volume are continuous. Thus latent heat is zero. This means, while evaluating dP/dT , we encounter 0/0 situation. This is remedied by applying L'Hospitals's rule. Because the transition under consideration is second order, S is continuous but C_P is discontinuous. Thus $(\partial S/\partial T)_P$ is not continuous. Then,

$$\frac{dP}{dT} = \frac{\left(\dfrac{\partial s_2}{\partial T}\right)_P - \left(\dfrac{\partial s_1}{\partial T}\right)_P}{\left(\dfrac{\partial v_2}{\partial T}\right)_P - \left(\dfrac{\partial v_1}{\partial T}\right)_P}$$

$$= \frac{1}{T} \frac{(C_P)_2 - (C_P)_1}{v_2(\alpha_P)_2 - v_1(\alpha_P)_1} \tag{4.8}$$

where $(\alpha_P)_{1(2)}$ are volume expansion coefficient of phase 1(2) and so are $(C_P)_{1(2)}$ specific heats of phase 1(2). It must be noted that C_p and α need to be measured as a function of T and can not be obtained from thermodynamics alone.

4.4   Critical Isotherm

Theorem: A critical isotherm in $P - V$ plane is the one for which following conditions hold. At critical point,

$$\left(\frac{\partial P}{\partial V}\right)_{T_c} = \left(\frac{\partial^2 P}{\partial V^2}\right)_{T_c} = 0 \text{ and } \left(\frac{\partial^3 P}{\partial V^3}\right)_{T_c} < 0 \tag{4.9}$$

Proof: Near its critical temperature T_c, density of liquid and vapor are close to each other. The pressure of two phases is same. Hence,

$$\begin{aligned} P(T_c, V) &= P(T_c, V + \delta V) \\ &= P(T_c, V) + \left(\frac{\partial P}{\partial V}\right)_{T_c} \delta V + \frac{1}{2!} \left(\frac{\partial^2 P}{\partial V^2}\right)_{T_c} (\delta V)^2 + \ldots \end{aligned}$$

or

$$\left(\frac{\partial P}{\partial V}\right)_{T_c} = 0 \tag{4.10}$$

Now, $G = F + PV$ is minimum along a transition. Thus, any change in G, i.e. $\delta G > 0$ at constant P and T. This change can come about because of changing V isothermally. Thus,

$$\delta F + P \delta V > 0$$

i.e.

$$\left(\frac{\partial F}{\partial V}\right)_T \delta V + \frac{1}{2!} \left(\frac{\partial^2 F}{\partial V^2}\right)_T (\delta V)^2 + \frac{1}{3!} \left(\frac{\partial^3 F}{\partial V^3}\right)_T (\delta V)^3 + \ldots + P \delta V > 0$$

but

$$\left(\frac{\partial F}{\partial V}\right)_T = -P$$

Thus

$$-\frac{1}{2!} \left(\frac{\partial P}{\partial V}\right)_T (\delta V)^2 + \left(-\frac{1}{3!}\right) \left(\frac{\partial^2 P}{\partial V^2}\right)_T (\delta V)^3 + \left(-\frac{1}{4!}\right) \left(\frac{\partial^3 P}{\partial V^3}\right)_T (\delta V)^4 \ldots > 0$$

Using equation (10), we get

$$-\frac{1}{6}\left(\frac{\partial^2 P}{\partial V^2}\right)_T - \frac{1}{24}\left(\frac{\partial^3 P}{\partial V^3}\right)\delta V + \ldots > 0$$

In order this inequality to hold for any δV, we have

$$\left(\frac{\partial^2 P}{\partial V^2}\right)_{T_c} = 0, \quad \left(\frac{\partial^3 P}{\partial V^3}\right)_{T_c} < 0 \tag{4.11}$$

Equations (10) and (11) determine critical isotherm completely.

4.5 van der Waals Equation

The equation of state was proposed by Johannes Diderik van der Waals (vdW) in his Ph.D. thesis. The equation was meant for real gases. Its shear simplicity makes it useful even today. It must be noted that no ideal gas ($PV = RT$) will lead to vapor-liquid transition(phase transition). vdW was the first equation of state to be useful to understand phase transition. This equation for one mole is

$$\left(P + \frac{a}{v^2}\right)(v - b) = RT \tag{4.12}$$

Clearly, for small P and large v, it reduces to the ideal gas equation. Also, v is always greater than b, otherwise the pressure will be negative. 'a' and 'b'are empirical constants. After this equation, many equations of states were empirically proposed for different temperature ranges. We will not discuss them. We will derive vdW equation subsequently using statistical mechanics.

The vdW equation is cubic in volume. Its diagram from fig.(8) can be compared with experimental fig.(4). We see that there is a drastic deviation in the coexistence region. Because of a cubic nature, curve has maxima and minima. Consider a typical curve c_1 at temperature T_1 well below T_c. Up to a point E, the curve has a physical behavior, i.e., as pressure decreases, volume increases. Obviously, this physical behavior is given by $\left(\frac{\partial P}{\partial V}\right)_T < 0$. At the points A,B,C,D,E; $\left(\frac{\partial P}{\partial V}\right)_T = 0$. In the region AbE, as volume increases, pressure also increases, which is unphysical. The states in this unphysical region are collapsible by a slight perturbation. This is the reason why experimental curves do not show this part of curve. The curve c_1 in experiments is observed as $a - b - c$ and not as $a - A - b - E - c$.

The portion $c - E$ of c_1 represents supercooled vapor. Portion $a - A$ of c_1 represents superheated liquid. The dotted curve$A - B - C - D - E$ is obtained by joining the points with $\left(\frac{\partial P}{\partial V}\right)_T = 0$. This is why the critical point C has $\left(\frac{\partial P}{\partial V}\right)_T = 0$. It should be noted that the experimental curve c_1 starts becoming horizontal from point a. This point 'a' is away from A. This horizontal line ends at 'c' which is far away from E. The points a and c are located in such a manner that area $a - A - b - a$ is equal to $b - c - E - b$. The points 'a', 'c' are the stable equilibrium points.

Value of pressure, volume and temperature at the critical point (P_c, v_c, T_c) can be

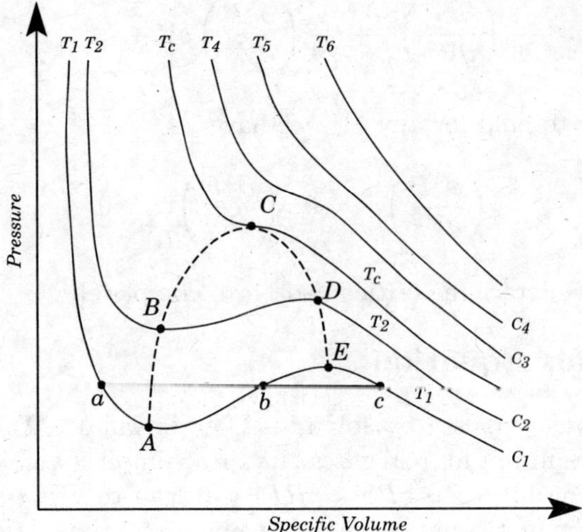

Figure 4.8: P-V diagram according to van der Waals equation

obtained in terms of a and b using van der Waals equation, Here we use (i) vdW equation, (ii) condition $\left(\dfrac{\partial P}{\partial V}\right)_{T_c} = 0$ and the condition (iii) $\left(\dfrac{\partial^2 P}{\partial V^2}\right)_{T_c} = 0$.

After some simple but tedious algebra, we obtain

$$T_c = \frac{8a}{27bR}, \quad v_c = 3b, \text{ and } P_c = \frac{a}{27b^2} \tag{4.13}$$

It is interesting to note that vdW equation can be scaled in terms of P_c, v_c, and T_c. The new equation thus obtained is same for <u>all</u> substances.

Let $\tilde{P} = P/P_c$; $\tilde{v} = v/v_c$ and $\tilde{T} = T/T_c$.

Then the vdW equation becomes,

$$\left(\tilde{P}\, P_c + \frac{a}{(\tilde{v}\, v_c)^2}\right) (\tilde{v}\, v_c - b) = R\tilde{T}T_c$$

or

$$\left(\tilde{P} + \frac{3}{\tilde{v}^2}\right) (3\tilde{v} - 1) = 8\tilde{T} \tag{4.14}$$

Thus, if P, V and T of any substance are measured in terms of their critical values, then , the equation of state satisfied by any substance is same. This is called 'law of corresponding states'. The relation between P_c, v_c and T_c can be obtained using equation (13) for vdW gas. It is,

$$Z_c = \frac{P_c v_c}{RT_c} = \frac{a}{27b^2}\frac{3b}{8a}27b = \frac{3}{8} = 0.375$$

For the ideal gas $Z = \dfrac{Pv}{RT} = 1$

For real fluids, Z_c is less than $3/8$.

4.6   Properties of a Substance Near T_c

We obtain these properties using vdW equation. We use equation(14) near T_c. Let,

$$\tilde{P} = 1 + P; \ \tilde{v} = 1 + v \ \text{and} \ \tilde{T} = 1 + t. \tag{4.15}$$

Then equation (14) becomes,

$$\left[1 + P + \frac{3}{(1 + v)^2} \right] [3v + 2] = 8(1 + t) \tag{4.16}$$

Clearly, close to the critical point,

$$P = \frac{P - P_c}{P_c}, \ \ v = \frac{v - v_c}{v_c}, \ \text{and} \ t = \frac{T - T_c}{T_c}$$

are all far less than a one and are close to zero. Thus equation(16) can be simplified by binomial expansion as

$$P = -6tv + 4t - \frac{3}{2}v^3 + 9tv^2 + ... \tag{4.17}$$

Equation(17) is very important because it immediately gives behavior of various physical quantities of interest near the critical point.

Case1: Suppose we want to find behavior P with v on a critical curve. On this curve,

$$t = 0 \ \ \text{and} \ \ P \approx -\frac{3}{2}v^3 \tag{4.18}$$

Case2: Behavior of isothermal compressibility is obtained from

$$K_T = -\frac{1}{V} \left(\frac{\partial V}{\partial P} \right)_T$$

Thus

$$K_T^{-1} \ \propto \ \left(\frac{\partial P}{\partial V} \right)_T \ \ \text{and} \ \ K_T \simeq -t^{-1} \sim \frac{T_c}{T_c - T} \tag{4.19}$$

Case3: Behavior of order parameter $\rho_g - \rho_l$ can be found by looking the behavior at 'a' and 'c' of figure(7). At a fixed t, the pressure at a and at c is same. Moreover, since points a and c are symmetric about V_c, $v_a = -v_c$ with reference to the critical volume. Then $\rho_g - \rho_l \propto (v_a + v_c)$ which is the width ac.

Then, clearly, from equation(17) we get

$$\rho_g - \rho_l \ \propto \ \left(\frac{T_c - T}{T_c} \right)^{1/2} \ \ \ \ T < T_c$$

$$= \ \ \ \ \ 0 \ \ \ \ \ \ \ \ \ \ T \geq T_c \tag{4.20}$$

Case4: Behavior of the latent heat near T_c is given by

$$L = T(S_2 - S_1) \approx T_c \left(\frac{\partial S}{\partial v}\right)_t (v_2 - v_1) \tag{4.21}$$

$$\propto \sqrt{(T_c - T)} \qquad (T < T_c)$$

At the critical point, latent heat is zero, which indicates that only, at the critical point, the phase transition is of second order. At all other points on the co-existence curve, i.e. along vaporization curve (sans the critical point), $L \neq 0$ and the transition is of first order.

4.7   Results for Magnetic System

Like van der Waals equation for liquid-vapor transition there is Weiss molecular theory for ferromagnetic to paramagnetic transition. In Weiss theory also we get a similar behaviour near critical point. Here, M, the magnetization behaves as

$$M \propto \sqrt{T_c - T} \qquad \text{for } T < T_c$$
$$= 0 \qquad \text{for } T \geq T_c \tag{4.22}$$

compare this with equation(20). Clearly, M is order parameter in ferro-para transition. In critical isotherm, Weiss theory gives

$$H \propto M^3 \tag{4.23}$$

which is very similar behavior as given in equation(18). Like isothermal compressibility K_T in liquid-gas transition, we have isothermal susceptibility χ_T in magnetic system. Thus,

$$\chi_T = \left(\frac{\partial M}{\partial H}\right)_T \tag{4.24}$$

Weiss theory gives behaviour of χ_T as,

$$\chi_T \propto |T - T_c|^{-1} \tag{4.25}$$

Molar specific heat at constant H has a jump at T_c for spin $\frac{1}{2}$ system. It is,

$$\Delta C_H = \frac{3}{2} R \tag{4.26}$$

4.8   Critical Exponents

It is observed that phase transitions of different systems behave similarly near their critical points. Usually a physical quantity depends upon some fractional power of t. This is a non analytical behavior. The fractional power, or exponent is called critical exponent. Thus
(i) Order parameter

$$O \sim (-t)^{\beta} \qquad \text{for } T < T_c$$
$$= \quad 0 \qquad \text{for } T \geq T_c \tag{4.27}$$

From earlier discussion $\beta = \frac{1}{2}$ in van der Waals case and Weiss model.

(ii) Specific heat at constant volume (or at constant H for magnetic system). Thus

$$C_v \sim (-t)^{-\alpha'} \qquad \text{for } T < T_c$$
$$\sim (t)^{-\alpha} \qquad \text{for } T > T_c \tag{4.28}$$

Here we study behaviour of a response function as $T \rightarrow T_c$ from above and below the transition temperature. Exponents α and α' turn out to be same at temperatures sufficiently close to T_c.

(iii) Isothermal compressibility K_T and zero field magnetic susceptibility χ_T are similar. Near the critical point, they behave as

$$K_T \sim (-t)^{-\gamma'} \qquad \text{for } T < T_c$$
$$\sim (t)^{-\gamma} \qquad \text{for } T > T_c \tag{4.29}$$

Similarly

$$\chi_T \sim (-t)^{-\gamma'} \qquad \text{for } T < T_c$$
$$\sim (t)^{-\gamma} \qquad \text{for } T > T_c \tag{4.30}$$

Exponents $\gamma = \gamma'$ at sufficiently close to T_c.
vdW and Weiss theory give $\gamma = 1$

(iv) On the critical isotherm,

$$H \sim |M|^{\delta} \, \text{Sgn} \, M$$

for the magnetic system. For fluid system,

$$P - P_c \sim |\rho_L - \rho_G|^{\delta} \, \text{Sgn} \, (\rho_L - \rho_G)$$

Value of δ in case of Weiss and vdW theory is 3.

Near the critical point, there are significant fluctuations in density. They are small for the temperature range far away from T_c. We illustrate this concept with reference to fluid system. Generalization for other systems is immediate. Let n be the average fluid density and $n(\vec{r})$ be a local density at $\vec{r}$. The quantity

$$\delta n(\vec{r}) = n(\vec{r}) - n$$

is known as deviation. Correlation function $G(\vec{r}, \vec{r'})$ is defined as

$$
\begin{aligned}
G(\vec{r}, \vec{r'}) &= \ <\delta n(\vec{r})\delta n(\vec{r'})> \\[2mm]
&= \ <n(\vec{r})n(\vec{r'})> -n^2
\end{aligned}
\tag{4.31}
$$

Here $<\ >$ is a thermal average. These correlations are such that when $|\vec{r}-\vec{r'}|$ is large, $<n(\vec{r})n(\vec{r'})> = n^2$ and $G \to 0$.

Detailed discussion of $G(\vec{r}, \vec{r'}) = G(|\vec{r}-\vec{r'}|) \equiv G(r)$ is beyond our scope. However, this important function is related with many experimental quantities For example, the space integral of $G(r)$ is related with isothermal compressibility. Intensity of the light scattered by liquid-vapor mixture near T_c is proportional to the fourier transform of $G(r)$.

Behavior of $G(r)$ near T_c is important. Asymptotic form for large r behaves as,

$$
G(r) \propto \frac{e^{-r/\xi}}{r^{1+\eta}}
\tag{4.32}
$$

The quantity ξ is called correlation length and it diverges at T_c. Its behavior near T_c is given by

$$
\begin{aligned}
\xi \ &\sim \ (-t)^{-\nu'} \qquad \text{for } T < T_c \\[2mm]
&\sim \ (t)^{-\nu} \qquad \text{for } T > T_c
\end{aligned}
\tag{4.33}
$$

and

$$
G(r) \sim \frac{1}{r^{1+\eta}} \quad \text{at } T_c
\tag{4.34}
$$

The experimental values of exponents are given below. Their values are approximate and are

$$
\begin{aligned}
\alpha \ &\approx \ \alpha' \approx 0.1 \\
\beta \ &\approx \ \tfrac{1}{3} \\
\delta \ &\approx \ 4.3 \\
\nu' \ &\approx \ 0.57 \\
\gamma' \ &\approx \ 1.2
\end{aligned}
\tag{4.35}
$$

The values obtained by using van der Waals equation differ considerably from above experimental values.

Values obtained using van der Waals theory or Weiss theory are called mean field values. These theories are called mean field theories. Mean field results capture essential physics and are of qualitative guidance.

Further more, we mention here that these critical exponents depend upon the dimensionality of the phase transiting system. All above critical exponents are quoted in equation (35) are for three dimensions. Hence forth, we will assume our system as three dimensional.

Short Questions

1. Identify the order of transition in Ehrenfest's sense with reason. C_P is given in the units of cal/gram/^oC

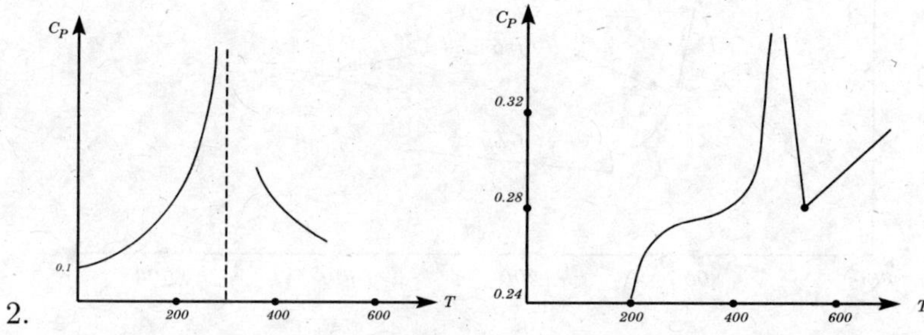

2.

Figure 4.9: (a) Specific heat of β brass   (b) Specific heat of crystalline quartz

Ans: λ transition

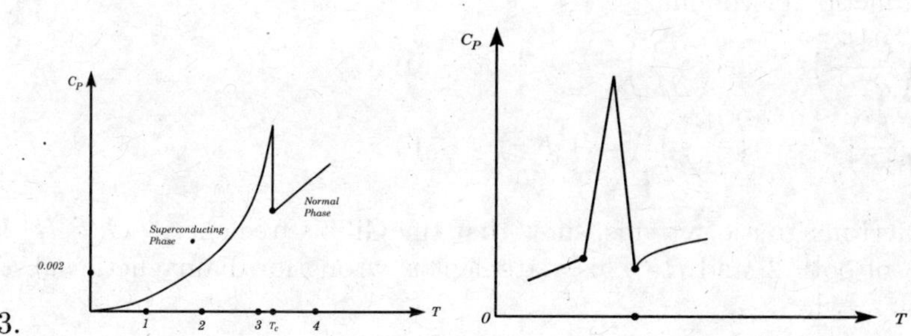

3.

Figure 4.10: (a) Specific heat of tin   (b) Specific heat of a certain gas

Ans: a- Second order
b- first order

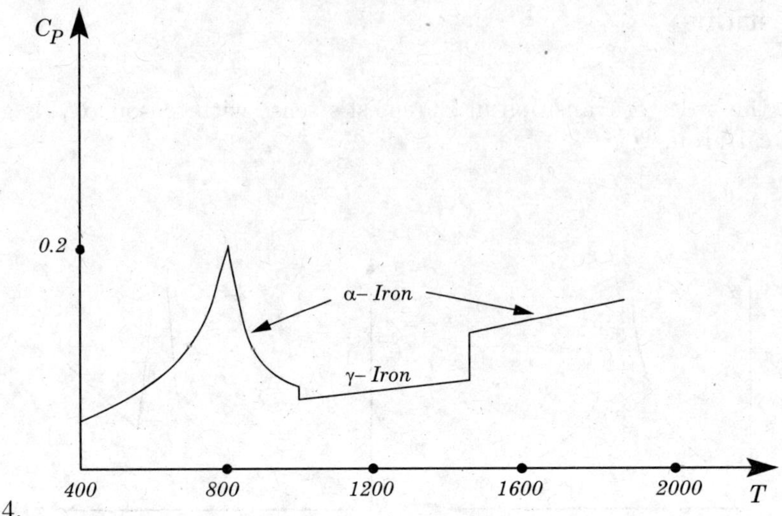

4.

Figure 4.11: Specific heat of iron

Ans: Third order phase transition

5. Which type of phase transition has entropy function continuous with respect to temperature? Give examples.

6. Show that Helmholtz free energy $F(T, V)$ is a concave function of temperature and convex function of volume.

(Hint: $\left(\dfrac{\partial^2 F}{\partial T^2}\right)_V = -\left(\dfrac{\partial S}{\partial T}\right)_V = -\dfrac{1}{T}C_V \leq 0$

and $\left(\dfrac{\partial^2 F}{\partial V^2}\right)_T = -\left(\dfrac{\partial P}{\partial V}\right)_T = (VK_T)^{-1} \geq 0$)

7. For the ferromagnetic systems, show that the Gibb's free energy $G(T, H)$ is a concave function of both T and H. This statement is wrong for diamagnetic substances. Can you explain why?

8. Substances such as ice, bismuth and iron have a negative slope of its fusion curve. Show that they contract upon melting.

9. It is observed that, for ice at a pressure larger than pressure at triple point, the melting temperature is less than the triple point temperature. Justify this observation.

Problems

Problem 1
Derive CCE from Maxwell relation

$$\left(\frac{\partial P}{\partial T}\right)_V = \left(\frac{\partial S}{\partial V}\right)_T$$

(Hint: Along $P - V$ diagram, in a coexistence region $\left(\frac{\partial P}{\partial T}\right)_V$ is independent of V. We take it $\frac{dP}{dT}$. $\Delta S = \frac{L}{T}$, $\Delta V = v_2 - v_1$. Thus $\frac{dP}{dT} = \frac{L}{T(v_2 - v_1)}$)

Problem 2

Show that the Latent heat of vaporization of van der Waals gas is given by

$$L = -a\left(\frac{1}{v_g} - \frac{1}{v_l}\right) + P\left(v_g - v_l\right)$$

(Hint: Use a) $\left(\frac{\partial E}{\partial V}\right)_T = T\left(\frac{\partial P}{\partial T}\right)_V - P$

b) $P = \frac{RT}{v - b} - \frac{a}{v^2} \Rightarrow \left(\frac{\partial P}{\partial T}\right)_V = \frac{R}{v - b}$

c) Since process takes place at constant P and T,

$$L = \int_{v_l}^{v_g} T dS = \int dE + \int P dv = -a\left(\frac{1}{v_g} - \frac{1}{v_l}\right) + P(v_g - v_l))$$

Problem 3

Latent heat is a slow function of T along vaporization curve and is given as

$$L = L_0 - \alpha T$$

where L_0 is a value of the latent heat at some specified temperature and α is a small amount of energy per degree kelvin. Show that the equation of vaporization curve is

$$\ln \frac{P(T)}{P_0} = B \ln T + \frac{C}{T}$$

Find B and C.

(Hint: In reality, $v_l << v_g \rightarrow \frac{dP}{dT} \approx \frac{L}{Tv_g} \simeq \frac{ML}{RT^2}P$

Here M is molecular weight and we have used ideal gas law $Pv_g = \frac{RT}{M}$.
Integrate, P_0 as a reference pressure.)

$$\text{Ans: } B = -\frac{M\alpha}{R}$$
$$C = -\frac{ML_0}{RT}$$

Problem 4

Find the equation of a curve $ABCDE$ in the fig.(7), which connects the last metastable point of supersaturation or supercooling. Use van der Waals equation.

$$\text{Ans: } P = \frac{a(v - 2b)}{v^3}$$

(Hint: At all the points $ABCDE$, $\left(\dfrac{\partial P}{\partial V}\right)_T = 0$. Use vdW equation to get $T = \dfrac{2a(v - b)^2}{Rv^3}$ at these points. Use vdW to get $Pv^3 = a(v - 2b)$ for these points.)

Problem 5

Find the rate of change of melting temperature per unit pressure from the following data. Initial pressure$=1.01 \times 10^5$ Pa, $T = 273K$ and Latent heat of fusion of ice$=3.34 \times 10^5 J/kg$. Consider a small change in specific volume on melting to be $-9.05 \times 10^{-5} \ m^3/kg$

$$\text{Ans: } \approx 7.5 \times 10^{-8} \ K/Pa$$

Problem 6

For two phase system, from figure(3) it is clear that, in a co-existence region, P is independent of V and is a function of T alone. Prove,

$$\left(\frac{\partial P}{\partial T}\right)^2 = \frac{C_v}{TV\kappa_s}$$

(Hint: $\dfrac{dP}{dT} = \dfrac{T(S_2 - S_1)}{T(v_2 - v_1)} = \dfrac{T\left(\dfrac{\partial S}{\partial T}\right)_v dT}{T\left(\dfrac{\partial v}{\partial P}\right)_S dP}$. Use Definitions of C_v and κ_s and get desired result.)

Chapter 5

PHASE TRANSITIONS AND CRITICAL PHENOMENA II

In the chapter I of the critical phenomena, we studied solid-liquid-vapor phase transition in great details and made some comments about magnetic system. We also mentioned mean field theories(vdW and Weiss theories). In this chapter, we study Landau-Ginzburg theory(GLT). GLT is a generalized mean field theory (MFT). We will then consider fluctuations in order parameter and then the scaling hypothesis. Most of the discussion in this chapter pertains to a second order phase transition.

5.1   Ginzburg-Landau Theory

We know that the phase transiting system has order parameter O. It is asymmetric around a transition temperature T_c, and has the property,

$$
\begin{aligned}
O \;\neq\; 0 \qquad &\text{for } T < T_c \\[2mm]
=\; 0 \qquad &\text{for } T > T_c
\end{aligned}
\tag{5.1}
$$

Order parameter is like a coordinate. Landau assumed that the free energy F, besides being a function of volume and temperature is also a function of the order parameter. Average value of the order parameter is small near but below T_c; and goes on increasing as $T \to 0$. Landau then postulates that the free energy F can be expanded in terms of powers of order parameter and its gradients near T_c.

The major defect of this expansion is that F diverges at and very close to T_c. One can obtain the range of temperatures where the Landau theory is applicable. For example, it turns out that for $|T - T_c|$ more than about 10^{-10}, the expansion is convergent for old superconductors. Bardeen Cooper and Schrieffer theory of old superconductors is a mean field theory. It has an vast applicability of mentioned above. For new superconductors - the so called high T_c superconductors , we do not have such vast region of convergence. Originally, Landau expanded F in powers of the order parameter but did not include its gradients.

Ginzburg applied this theory to old superconductors wherein he included the gradient term in the free energy expansion, known as Ginzburg -Landau (GL) expansion of free energy. We will discuss a ferromagnetic system with magnetization M as the order parameter. For a general theory, replace M by appropriate order parameter O (e.g. superconducting gap Δ, electric polarization $\vec{P}$ etc.) Thus the GL expansion for the ferromagnetic system is,

$$F = \int \mathcal{F}(\vec{r}, T) d^3 r \tag{5.2}$$

where $\mathcal{F}(\vec{r}, T)$ is a free energy density. Then,

$$\mathcal{F}(\vec{r}, T) = \mathcal{F}_0 - \vec{B}.\vec{M} + a(T)|M_z(\vec{r})|^2 + b(t)|M_z(\vec{r})|^4 + c(T)|\nabla M_z(\vec{r})|^2 + \ldots\ldots \tag{5.3}$$

Here $\mathcal{F}_0$ is free energy of a normal (uncondensed) phase. $\vec{B} = B_z \hat{k}$ is an external magnetic field. It is used for aligning the spins of ferromagnetic material. Z-direction is taken as a preferred direction along which the magnetization is aligned. Following are the features of this expansion,

(a) only the even powers of the order parameter M occur. This is so because F should remain same even if we flip the sign of $\vec{B}$.

(b) $|\nabla M|^2$ makes free energy larger when the order parameter varies in space - e.g. near the boundaries. It also represents the transverse magnetic fluctuations.

    Thermodynamically, the most probable value of M is the one that minimizes F. Note that F is a functional of M and stationarity of F demands[1]

$$\frac{\delta F}{\delta M} = 0 = -B_z + 2a(T)M_z(\vec{r}) + 4b(T)M_z{}^3(\vec{r}) - 2c\nabla^2 M_z(\vec{r}) \tag{5.4}$$

Thus

$$\left\{ 2c\nabla^2 - (2a(T) + 4bM_z{}^2) \right\} M_z(\vec{r}) = -B_z. \tag{5.5}$$

Landau then asserts that the solution of equation (5) is an average value, or a most probable value. Moreover, as F is a convex function of M, $(\equiv \chi_T > 0)$, we expect $b > 0$. Also one can never get a uniform solution of M from equation (5) if $c < 0$. Therefore, $c > 0$. We now discuss various particular cases of equation (5).

Case 1: B is uniform, implying M is uniform, $\nabla^2 M = 0$
Then

$$\left[2a(T) + 4bM^2 \right] M = B$$

$$\text{If } B = 0, \qquad M = 0 \qquad \text{corresponding to } T > T_c$$

$$\tag{5.6}$$

$$\text{and} \qquad M = \pm\sqrt{\frac{-a(T)}{2b}} \qquad \text{corresponding to } T < T_c$$

---

[1] For the definition of functional and its derivative, see chapter (8) of Classical Mechanics by P.V.Panat(Narosa, 2005)

Landau chooses $a > 0$ for $T > T_c$ and $a < 0$ for $T < T_c$ with temperature dependence as,

$$a(T) = a'(T - T_c) \tag{5.7}$$

where a' is independent of T.

It is clear that, at $T = T_c$, $\mathcal{F} \sim O(M^4)$ and we take $b(T) = b(T_c) > 0$ and $c(T_c) > 0$. Because of M^4 term in $\mathcal{F}$ at T_c, there are large scale fluctuations in M i.e. in the order parameter.

Here, from equation (6) and (7); we get

$$M \propto (T_c - T)^{1/2} \equiv (T_c - T)^\beta; \ \ \beta = 1/2. \tag{5.8}$$

This result is same as what we get from Weiss theory.

Case 2: On the critical isotherm, $a(T_c) = 0$ and

$$B \propto M^3 \ \text{ or } \delta = 3 \tag{5.9}$$

Case 3:

$$\chi_T = \left(\frac{\partial M}{\partial B}\right)_T$$

$$\Rightarrow \ (2a + 12M^2)\chi_T = 1 \ \text{ or } \ \chi_T \propto \frac{1}{T_c - T}. \tag{5.10}$$

But $\chi_T \propto t^{-\gamma}$ implying $\gamma = 1$

Because of the nature of $a > 0$ for $T > T_c$, we see that $\chi_T \propto t^{-\gamma'}$ and $\gamma' = \gamma = 1$ in Landau theory.

Illustrative Problem 1:

Find the value of a jump in the specific heat at T_c according to 'LG' theory.

Solution:

$$-\frac{C_B}{T} = \left(\frac{\partial^2 F}{\partial T^2}\right)_{B=0}$$

Then, $F = \int \mathcal{F}_0(T) d^3r$ for $T > T_c$.

For $T < T_c$, $F = \int \left(\mathcal{F}_0 + aM^2 + bM^4\right) d^3r$.

Use $M^2 = -\dfrac{a}{2b}$. Then

$$F = \int \left(\mathcal{F}_0 - \tfrac{a^2}{2b}\right) d^3r \ \text{ for } T < T_c$$

Now $a^2 = a'^2(T - T_c)^2$, b is independent of T,

$$\therefore \frac{\partial^2 F}{\partial T^2} \propto -2a'^2$$

Thus,

$$\Delta C_b = \frac{a'^2}{b} \tag{5.11}$$

5.2   Mean Field Theories and Landau's Theory

We have seen that the values of the critical exponents as found by vdW theory or Weiss theory are same as those given by the Landau theory. This is not surprising. The derivation of vdW equation or that of Curie-Weiss law will be presented subsequently. Thermodynamics will not be able to provide it. In these derivations we observe the reasons for the name of 'mean field' theories.

Consider the Weiss case. Ferromagnetic materials have spins at their lattice site and are fluctuating. In the model, nearest neighbor spins strongly interact among each other. In an external magnetic field, these spins align along the field. Aligned spins produce local magnetization. This local magnetization is averaged to give rise to magnetic field which adds to the external magnetic field. Thus, we have cause and effect intermingled. There is a competition between thermal agitation of the spins and their alignment. At a certain temperature and below, thermal agitations lose out. This leaves the spins aligned. Weiss theory gives the equation of state $H = H(T, M)$ and is a power series in M. This should be compared with equation (4) without ∇^2 term. As far as vdW equation goes, n, the density is similar to M. The virial expansion for pressure is actually a power series expansion in n for a pressure P. One can derive vdW equation by retaining first virial coefficient $B(T)$. In its derivation, because of intermolecular interactions (which are attractive at large distances), the pressure is reduced from its ideal gas value. This reduction is argued to be proportional to n^2. Here n is a mean value of a density of the gas. Hence, pressure changes mean density which in turn corrects the pressure. Thus, because the mean values are used in vdW or Weiss theory, and the equation of order parameter, we call these theories as mean field theories.

The form of Landau theory has a similar structure as an equation of state. Therefore the Landau theory is also a mean field theory. This kind of ideas are common in physics. Hartree's theory of correcting Coulomb potential on an electron in multi-electron atom, due to its other electrons can identically be called as a mean field theory[2]

5.3   Fluctuations in Order Parameter

Landau [3] theory takes average value of the order parameter in evaluating free energy functional. Near T close to T_c, there are large fluctuations which can be estimated using a following theorem from statistical mechanics.

Theorem: If the energy function (usually Hamiltonian) of the system has a term of the type $-\int h(\vec{r})O(\vec{r})d^3r$, then, if $h(\vec{r}) \rightarrow h(\vec{r}) + \delta h(\vec{r})$ the average value of $O(\vec{r})$ changes to $\left\langle \hat{O}(\vec{r}) \right\rangle \rightarrow \langle O(\vec{r}) \rangle + \delta \langle O(\vec{r}) \rangle$ where,

$$\delta \left\langle O(\vec{r}) \right\rangle = \frac{1}{k_B T} \int d^3 r' \left\langle (O(\vec{r}) - \langle O \rangle) \right\rangle \left\langle \left(O(\vec{r'}) - \langle O \rangle \right) \right\rangle \delta h. \tag{5.12}$$

Here, $h(\vec{r})$ is a conjugate parameter of $O(\vec{r})$. We omit the proof [4].

---

[2]'Quantum theory of atomic structure' Vol I, by John Slater(McGraw-Hill 1960)
[3]This discussion can be skipped by the beginner
[4]See 'Principles of condensed matter physics' by Chaikin and Lubensky (Cambridge University Press,

Clearly, M is order parameter, $\hat{O}$ and B is corresponding conjugate variable. This theorem tells us how a fluctuation in order parameter M can be found.
When $\langle M \rangle \rightarrow \langle M \rangle + \delta \langle M \rangle$ in equation(4), we get

$$\delta B \cong 2a(T)\,\delta\langle M \rangle + 12b\,\langle M \rangle^2\,\delta\langle M \rangle - 2c\nabla^2\delta\langle M \rangle \qquad (5.13)$$

Putting equation(13) in to (12) and calling the mean square deviation of M as

$$\delta\langle M \rangle = \frac{1}{k_B T}\int g(\vec{r}, \vec{r'})\,d^3r'\,\delta B,$$

we get,

$$\int\left\{\left[2c\nabla^2 - 2a - 12b\langle M^2 \rangle\right] g(\vec{r} - \vec{r'}) - k_B T \delta(\vec{r} - \vec{r'})\right\}\delta B = 0 \qquad (5.14)$$

Since δB is arbitrary, we write equation for the correlation function using Landau theory as

$$\left(2c\nabla^2 - 2a - 12b\langle M \rangle^2\right) g(\vec{r} - \vec{r'}) = k_B T \delta(\vec{r} - \vec{r'}) \qquad (5.15)$$

Case 1 Above T_c,

$$\langle M \rangle = 0$$

and,

$$(2c\nabla^2 - 2a)g = k_B T \delta(\vec{r} - \vec{r'}) \qquad (5.16)$$

We write $a = a'(T - T_c)$ and get the solution as,

$$g(r) = g(|\vec{r} - \vec{r'}|) = \frac{k_B T}{8\pi c}\,\frac{e^{-r/\xi}}{r} \qquad T > T_c \qquad (5.17)$$

with $\xi = \sqrt{\dfrac{c}{a'}}(T - T_c)^{-1/2}$.

Case2 Below T_c, $\langle M \rangle \neq 0$ but

$$\langle M \rangle^2 = \frac{a'(T_c - T)}{2b}$$

and equation (14) becomes,

$$[2c\nabla^2 - 8a'(T_c - T)]g(r) = k_B T \delta(r)$$

It has a solution

$$g(r) = \frac{e^{-r/\xi}}{r}\frac{k_B T}{8\pi c} \tag{5.18}$$

where $\xi = \sqrt{\dfrac{c}{4a'}}(T_c - T)^{-1/2}$.

We see then, as $\xi \sim t^{-\nu}$, $\nu = +\frac{1}{2}$.

It is seen that, the correlation length ξ diverges close to T_c. This leads to a divergence in the thermodynamic functions. From the theorem we stated, it is clear that

$$\chi = \frac{\delta\langle M\rangle}{\delta h} = \frac{1}{k_B T}\int g(r)d^3r$$

$$\propto \int \frac{e^{-r/\xi}}{r}4\pi r^2 dr$$

$$\propto \xi^2$$

Thus

$$\chi_{T\to T_c} \propto \xi^2 \propto |T - T_c|^{-1}$$

which is same as what we get in equation (10). Because of the fluctuations becoming larger and larger as T approaches T_c; writing the average value of the order parameter in F is wrong in a close temperature range near T_c. That is the reason why Landau's theory does not work in the temperature range and we get wrong critical exponents.

5.4　Region Near T_c where Landau Theory is Valid – Ginzburg's Analysis

Landau's theory of expanding the free energy in terms of powers of average order parameter would work when the fluctuations [5]in the order parameter are small. We quantify the 'smallness' of fluctuations in a following way. We consider magnetic system. Identical discussion follows when we replace M by $\langle O\rangle$ in general.

$$\delta\langle M(\vec{r})\rangle = \frac{1}{k_B T}\int g(\vec{r},\vec{r'})d^3r'\delta B$$

---

[5]This section may be omitted by the beginner

We average $g(\vec{r})$ over a coherence volume ξ^3. Thus

$$
\begin{aligned}
\langle \delta \langle M(\vec{r}) \rangle \rangle_{coh} &= \frac{1}{k_B T} \frac{1}{\xi^3} \int_{v_\xi} g(\vec{r}) d^3 r \\[2mm]
&= \frac{4\pi}{k_B T} \frac{1}{\xi^3} \int_0^\xi r^2 g(r) dr \\[2mm]
&= \frac{1}{2c\xi} \int_0^1 x e^{-x} dx \\[2mm]
&= \frac{A}{2c\xi}
\end{aligned}
\tag{5.19}
$$

where $A = \frac{(e-2)}{e}$. Here we used

$$
g(r) = \frac{k_B T}{8\pi c} \frac{e^{-r/\xi}}{r}
$$

We also know that $\langle M \rangle^2 = \frac{a(T)}{2b}$. The fluctuations are said to be 'small' when,

$$
\delta \langle M(\vec{r}) \rangle < \langle M \rangle^2
\tag{5.20}
$$

or,

$$
\frac{A}{2c\xi} < \frac{a(T)}{2b}
\tag{5.21}
$$

Equation (21) is Ginzburg criterion. We can put it in more physical form in a following way.

We define a bare correlation length as,

$$
\xi_0 = \sqrt{\frac{c}{4a(T=0)}} = \sqrt{\frac{c}{4a'T_c}}
\tag{5.22}
$$

From equation (17), we also have

$$
\xi = \sqrt{\frac{c}{a(T)}}
\tag{5.23}
$$

From equation (11) we have

$$
\Delta C_B = \frac{a'^2}{b} \quad \text{and} \quad a(T) = a'|T - T_c|.
\tag{5.24}
$$

Combining equation (21) with (22) to (24), we get Ginzburg criterion as,

$$
\left| \frac{T - T_c}{T_c} \right|^{1/2} > \frac{A}{4T_c^2 \Delta C_B \xi_0^{3}}
\tag{5.25}
$$

Thus, so long as T is away from T_c, satisfying Ginzburg criterion, mean field theory is valid. We call T_G, the Ginzburg temperature when fluctuations become important. T_G is then given by

$$\left|\frac{T_G - T_c}{T_c}\right| = \left(\frac{A}{4T_c{}^2 \Delta C_B \xi_0{}^3}\right)^2 \tag{5.26}$$

We can evaluate Ginzburg temperature by a more careful evaluation of A and write $t_G = \left|\frac{T_G - T_c}{T_c}\right|$ to give

$$t_G = \frac{k_B{}^2}{32\pi^2 (\Delta C_B)^2 \xi_0{}^6} \tag{5.27}$$

in c.g.s. units.

In all our discussion of phase transition, we have restricted our calculations in three dimensions. A general discussion of phase transition in D dimensions can be done. Then equation (25) modifies to

$$\left(\frac{\xi}{\xi_0}\right)^{D-4} = \left|\frac{T - T_c}{T_c}\right|^{\frac{4-D}{2}} > \frac{8A_D}{\Delta C_B \xi_0{}^D} \tag{5.28}$$

From this equation, it is clear that, for $D > 4$, the Ginzburg criterion is always satisfied. For $D < 4$, the Ginzburg criterion is never satisfied since $\xi^{D-4} \to 0$ as $T \to T_c$. It is clear that, the mean field theory provides a consistent description of the second order phase transition (of the order $\langle O \rangle^4$) when $D > 4$. The dimension D_c below which the mean field theory breaks down is called upper critical dimension. Thus, $D_c = 4$ when the free energy is dominated up to $\langle O \rangle^4$ in Ginzburg-Landau expansion.

5.5   Prelude to Scaling Hypothesis

We have defined various critical exponents We have evaluated them in the framework of mean field theory in Landau sense. These critical exponents correspond to the behavior of various physical quantities (like magnetization, $\rho_L - \rho_G$, superconducting gap, $|t| = \frac{|T - T_c|}{T_c}$, H, P etc.) neat T_c. Since thermodynamics describes the physical quantities as derivatives of thermodynamic potentials, we see that these exponents are related among each other. We get inequalities among exponents. To see this point, consider, from equation(29) of a chapter 'Thermodynamic relationships'. The specific heat of the magnetic system at constant magnetic field is

$$C_H \geq T \left(\frac{\partial M}{\partial T}\right)^2 / \chi_T \tag{5.29}$$

By definition of critical point exponents, for $T < T_c$, $C_H \sim (-t)^{-\alpha'}$; $\chi_T \sim (-t)^{-\gamma'}$ and $\left(\frac{\partial M}{\partial T}\right)_H \sim (-t)^{\beta-1}$.

Then putting this behaviour in equation(30) we get

$$(-t)^{-\alpha'} \geq T(-t)^{2\beta-2+\gamma'}$$

or

$$\alpha' + 2\beta + \gamma' \geq 2 \tag{5.30}$$

This inequality is known as Rushbrooke inequality. There are many such inequalities derived from thermodynamic relations. It is found experimentally that in most cases the inequalities are actually equalities. This fact is made plausible by the so called scaling hypothesis due to Widom and its subsequent elaboration based upon heuristic arguments by Kadanoff. Fruitful use of the scaling hypothesis requires some properties of homogeneous function given below.

A) A function $f(x)$ is said to be a homogeneous function of degree k if

$$f(\lambda x) = \lambda^k f(x) \tag{5.31}$$

e.g. $f(x) = \text{constant} \times x^k$

B) A function $f(x_1, x_2, ...x_n)$ of n variables is a homogeneous function of degree k if

$$f(\lambda x_1, \lambda x_2, ...\lambda x_n) = \lambda^k f(x_1, x_2, ...x_n)$$

e.g. $f(x_1, x_2) = Ax_1{}^k + Bx_1{}^{k-m}x_2{}^m + cx_2{}^k$

C) A generalized homogeneous function is defined as

$$f(\lambda^a x_1, \lambda^b x_2) = \lambda f(x_1, x_2) \tag{5.32}$$

or equivalently,

$$f(\lambda x_1, \lambda^b x_2) = \lambda^p f(x_1, x_2)$$
$$f(\lambda^a x_1, \lambda x_2) = \lambda^p f(x_1, x_2) \tag{5.33}$$

It is easy to see that these forms are equivalent. For, let $\Lambda^{1/p} = \lambda$. Then,

$$f(\lambda x_1, \lambda^b x_2) = f(\Lambda^{1/p} x_1, \Lambda^{b/p} x_2) = \Lambda f(x_1, x_2).$$

D) If a homogeneous function is of the form

$$f(\lambda x_1, \lambda x_2, ...\lambda x_n) = \lambda^p f(x_1, x_2, ...x_n)$$

then,

$$f(x_1, x_2, ...x_n) = x_n{}^{-p} f\left(\frac{x_1}{x_n}, \frac{x_2}{x_n}, ...\frac{x_{n-1}}{x_n}; 1\right)$$
$$\equiv x_n{}^{-p} F\left(\frac{x_1}{x_n}, \frac{x_2}{x_n}, ...\frac{x_{n-1}}{x_n}\right)$$

Clearly, for a function of two variables,

$$f(\lambda x, \lambda y) = \lambda^p f(x, y)$$

and

$$f(x, y) = y^p F(x/y) = y^p f\left(\frac{x}{y}, 1\right)$$

5.6   Static Scaling Hypothesis

The static scaling hypothesis states that

Any of the thermodynamic potentials, namely the Helmholtz free energy F, internal energy E, Gibbs free energy G and enthalpy H are generalized homogeneous functions of their respective arguments near the critical point temperature T_c provided the system undergoes a second order phase transition.

Thus according to this hypothesis, typically for the magnetic system, with $t = (T - T_c)/T_c$;

$$\lambda\, G(t, H_{mag}) = G(\lambda^{a_t} t, \lambda^{a_H} H_{mag}) \tag{5.34}$$

$$\lambda\, F(t, M) = F(\lambda^{a_t} t, \lambda^{a_M} M) \tag{5.35}$$

$$\lambda\, E(S, M) = E(\lambda^{a_S} S, \lambda^{a_M} M) \tag{5.36}$$

$$\lambda\, H(S, H_{mag}) = H(\lambda^{a_S} S, \lambda^{a_H} H_{mag}) \tag{5.37}$$

where H is usual enthalpy function and H_{mag} is external magnetic field inducing the magnetization M.

This is a hypothesis and can be made plausible by heuristic arguments of Kadanoff. (loc.cit.)

We are now in a position to obtain the critical exponents in terms of two basic exponents. If we analyze $G(t, H_{mag})$; then the two basic exponents are a_t and a_H. Since we are not using enthalpy function, we will write $H_{mag} = H$ for brevity. The scaling hypothesis and thermodynamics can not provide the values of a_t and a_H. They can be found by the help of models or by renormalization group. With,

$$G(\lambda^{a_t} t, \lambda^{a_H} H) = \lambda G(t, H);$$

we take a derivative w.r.t. H. From our earlier discussion it is the magnetization M. Thus,

$$\lambda^{a_H} \frac{\partial G(\lambda^{a_t} t, \lambda^{a_H} H)}{\partial(\lambda^{a_H} H)} = \lambda\frac{\partial G}{\partial H} = \lambda M(t, H)$$

$$\therefore \quad \lambda^{a_H - 1} M(\lambda^{a_t} t, \lambda^{a_H} H) = M(t, H) \tag{5.38}$$

In the limit $H \to 0$, $M \sim t^\beta$. Similarly at $t = 0$, $H \sim M^\delta$. We first choose $\lambda = \left(-\frac{1}{t}\right)^{1/a_t}$ and $H = 0$

Then, equation (38) becomes,

$$M(t,0) \; = \; (-t)^{(1-a_H)/a_t} \, M(-1,0)$$

$$\text{or } \beta \; = \; (1-a_H)/a_t.$$

Similarly, at $t=0$ and $\lambda = H^{-1/a_H}$, we find

$$\delta = \frac{a_H}{1-a_H}$$

Solving,

$$a_t = (\beta\delta + \beta)^{-1} \text{ and } a_H = \delta(1+\delta)^{-1} \tag{5.39}$$

Isothermal susceptibility $\chi_T = -\dfrac{\partial^2 G}{\partial H^2}\bigg)_t$

$$\lambda^{2a_H} \chi_T \left(\lambda^{a_t} t, \lambda^{a_H} H\right) = \lambda \, \chi_T(t,H) \tag{5.40}$$

Put $H=0$ and put $\lambda = (-t)^{-1/a_t}$ for $T < T_c$. Then,

$$\chi_T(t,0) \; = \; \lambda^{2a_H - 1} \chi(-1,0)$$

$$= \; (-t)^{-\dfrac{2a_H - 1}{a_t}} \chi(-1,0)$$

$$\sim \; (-t)^{-\gamma'}$$

Thus,

$$\gamma' = \frac{2a_H - 1}{a_t} \tag{5.41}$$

With a_H and a_t expressed in terms of β and δ, we get Widom equality as

$$\gamma' = \beta(\delta - 1) \tag{5.42}$$

For $T > T_c$, choose $\lambda = (t)^{-1/a_t}$ in equation(40) to get γ. Then,

$$\gamma = \frac{2a_H - 1}{a_t} \tag{5.43}$$

The scaling hypothesis gives $\gamma = \gamma'$.

Specific heat of a magnetic system at constant magnetic field H is given by

$$C_H = -T\left(\frac{\partial^2 G}{\partial T^2}\right)_H \tag{5.44}$$

$$\lambda^{a_t} \frac{\partial G\left(\lambda^{a_t} t, \lambda^{a_H} H\right)}{\partial\left(\lambda^{a_t} t\right)} = \lambda \frac{\partial G(t,H)}{\partial t}$$

With one more derivative w.r.t. t,

$$\lambda^{2a_t - 1} C_H \left(\lambda^{a_t} t, \lambda^{a_H} H \right) = C_H(t, H)$$

Thus, with $\lambda = 1/(-t)^{1/a_t}$ for $T < T_c$,

$$C_H(t, 0) = (-t)^{2 - 1/a_t} C_H(-1, 0)$$

Critical exponent α' corresponds to behavior of the specific heat neat $T < T_c$.

Thus,

$$\alpha' \;\; = \;\; 2 - \frac{1}{a_t}$$

$$= \;\; 2 - \beta\delta - \beta$$

which gives Griffith's equality. It is ,

$$\alpha' + \beta(\delta + 1) = 2. \tag{5.45}$$

Combining Griffith and Widom equality, we see that the Rushbrook inequality (equation(30)) becomes equality as

$$\alpha' + 2\beta + \gamma' = 2.$$

We thus see that from the scaling hypothesis, two scaling exponents and the properties of the thermodynamic potentials, we obtain various inequalities/equalities between the critical exponents. This probably as far as one can understand the second order phase transition using thermodynamics alone.

Chapter 6

Foundations of Statistical Mechanics I: Classical Statistical Mechanics

We have seen in chapter 1 that a thermodynamic state is defined by few variables like P, V, T, M, B etc. Such a state is called macrostate. We know that the macroscopic system has large number of particle, of the order of 10^{22} at N.T.P. Classical Mechanics predicts future values of $\{q_i\}$ and $\{p_j\}$, if they are initially known, via Hamilton's equations. If the classical system has f degrees of freedom, a microscopic state is defined by $2f$ values of $\{q_k\}$ and $\{p_k\}$ at every time t. A microstate of a quantum system is specified by N particle wave function with f labels of quantum number. Evolution of this wavefunction is governed by Schrödinger equation. Aim of this chapter is to understand an inter relation between the microstate and macrostate of a thermodynamic system. We will first demystify entropy function and then will study various types of ensembles. In the process, partition and grand partition function is defined. There inter relation with free energy F and pressure P is established respectively.

Finally using these ensembles, the distribution function is obtained. This programme is divided in three parts. First part deals with classical Statistical Mechanics

6.1   Specification of a Microstate

a. Classical System

Consider a system of N atoms with $3N$ coordinates $(\vec{r_1}(t), \vec{r_2}(t), ...\vec{r_N}(t))$. We will take generalized coordinates $q_1, ...q_f$ where f is number of degrees of freedom. Corresponding canonical momenta are $(p_1.........p_f)$. Time evolution is governed by Hamilton's equation of motion

$$\frac{\partial H}{\partial p_i} = \dot{q_i} \quad \text{and} \quad \frac{\partial H}{\partial q_i} = -\dot{p_i} \tag{6.1}$$

$H = H(q_1, q_2...q_f; p_1....p_f; t)$ is a hamiltonian of the system

$$H = \sum_{k=1}^{f} \left(\frac{p_k^2}{2m} + \frac{1}{2} \sum_j V(\vec{r}_{jk}) \right) \qquad (6.2)$$

Microscopic state, called microstate, of the system is described by statement of the values of f coordinates and corresponding f momenta at every instant t. For brevity, we write this state as

$$(q_1(t), q_2(t), ...q_f(t); \; p_1(t), p_2(t), ...p_f(t)) \equiv (q(t), p(t)) \qquad (6.3)$$

Clearly, $(q(t), p(t))$ is known if $(q(0), p(0))$ are provided and $H(q, p; t)$ is known.

To represent the 'path' i.e. a microstate $(q(t), p(t))$ at every instant t, we consider a $2f$ dimensional space, where for each coordinate q_k, we have an axis and 'perpendicular' to it is an axis for p_k. This is a generalization of our standard three dimensional space. A microstate (q, p) is given by one point in this $2f$ dimensional space. The point $(q(t), p(t))$ is called phase point and the $2f$ dimensional space in which the phase point moves, is called 'phase space'. We denote the phase space by Γ space. We will consider conservative systems. As a result, during an entire motion of a phase point, its total energy will be conserved. Thus $H = E$. But H is the relation between $\{q_k\}$ and $\{p_k\}$. As result, a motion of the phase point will be restricted on a surface of $(2f - 1)$ dimensions and is characterized by $H = E$. (In three dimensions, a surface has dimensions $3 - 1 = 2$; where as, in two dimensions, "surface" has dimension $2 - 1 = 1$ and is a line.)

A velocity in Γ space is <u>defined</u> as $\vec{v} = (\dot{q}_1, \dot{q}_2....; \dot{p}_1....\dot{p}_f)$. In the Γ space, a gradient has components $\left(\dfrac{\partial}{\partial q_1}, \dfrac{\partial}{\partial q_2}....\dfrac{\partial}{\partial q_f}; \; \dfrac{\partial}{\partial p_1} \dfrac{\partial}{\partial p_2}....\dfrac{\partial}{\partial p_f} \right)$. Clearly,

$$|\vec{v}| = |\nabla H| \qquad (6.4)$$

Since the direction of the gradient is normal to the surface, $\vec{v}$ and ∇v are at right angles to each other in space. Thus

$$\vec{v} \cdot \nabla H = 0 \qquad (6.5)$$

We see that the Γ space is a graph of p_1 versus q_1; p_2 versus q_2;......p_f versus q_f. According to quantum mechanics, $\delta q \, \delta p \sim \hbar$. As result, smallest volume of a Γ space that can be probed is $(\hbar)^f$

b. Quantum System

The microscopic system of collection of atoms and molecule is described by quantum mechanics. Here, a state $|\psi\rangle$ is a vector ('strictly a ray') in a Hilbert space. Corresponding to every measurable physical quantity, there is a Hermitian operator which operates on the Hilbert space corresponding to the system. If $|q\rangle$ is a position eigenvector, then, Schrödinger wave function is $\psi(q) = \langle q|\psi\rangle$. For microscopic system under consideration, the position eigenvector $|q>$ is $|q_1, q_2, ...q_f>$. There will be f quantum numbers for f degrees of freedom. A microscopic quantum state is completely specified by furnishing these quantum numbers.

To illustrate this point, we give following two examples

i) Consider a system consisting of N non interacting spins each with $S = 1/2\ \hbar$. The quantum state of this system is very simple. Since each spin has $S_z = 1/2\ \hbar$ or $-1/2\ \hbar$ i.e. its m value, quantum state of the system will simply be given by specifying values of $m_1, m_2, ...m_N$. Thus $|\uparrow, \downarrow, \uparrow, \uparrow, \downarrow, \downarrow, \uparrow, ...\rangle$ is a typical quantum state of this system.

ii) Consider N quantum particles. Each of these is confined in a 1-dimensional box of size L. A typical particle will have an energy

$$E_n = \frac{\hbar^2}{2m} \frac{\pi^2 n^2}{L^2} \tag{6.6}$$

Corresponding wave function is $\langle x|n \rangle = \psi_n(x) = \sin \dfrac{n\pi x}{L}$.

For N particles, we will have to specify corresponding values of n. Typically, say n_k be the value for quantum number of kth particle. The N particle state is $|n_1, n_2, n_3, ...n_k... >$ and corresponding wavefunction is

$$\psi_{n_1, n_2, n_3 ....n_N}(q_1, q_2, q_3, ...q_k) = \prod_{k=1}^{N} sin \left(\frac{n_k \pi q_k}{L} \right) \tag{6.7}$$

Here, we have treated the particles as distinguishable.

6.2   Ensemble

First we consider classical situation. We know that the phase point of the system moves on a surface of constant energy $H = E$. But we really do not know initial condition of 10^{23} particles at $t = 0$. As a result, we consider many, many copies of the N particle system with a constraint of $H = E$ as shown in figure(1).

This ignorance of initial condition forces us to resort to statistical considerations. As a result, we study the concept of ensemble. On a constant energy surface $H = E$, there will be a phase point corresponding to each system. These phase points will be distributed on the constant energy surface. We do not know the initial condition $(q(0), p(0))$ of each of these system. Since we assume large number of systems as members of the ensemble, then points on $H = E$ surface in Γ space are very dense. We define their density ρ in the following manner.

Let

$$\rho(q_1, q_2....q_f;\ p_1, p_2....p_f;\ t)\ dq_1....dq_f\ dp_1...dp_f$$

be number of systems contained within a volume element q_1 and $(q_1 + dq_1)$; q_2 and $(q_2 + dq_2)$....p_f and $p_f + dp_f$ of Γ space at time t. In a short notation we write $\rho(q, p; t)\ dq\ dp$ as

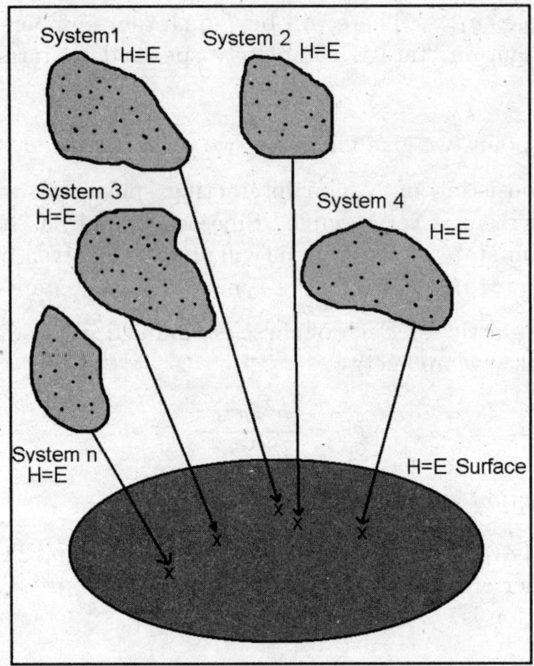

Figure 6.1: $H = E$ surface with phase points corresponding to different members of ensemble. $\times$ represents phase point of corresponding ensemble..

number of systems in a phase volume $(dq\ dp) = dq_1...dq_n, dp_1...dp_n$. Here ρ is called as a density function in Γ space.

Clearly, total number of system in ensemble is M.

Therefore,

$$M = \int \rho(q, p;\ t)\ dq\ dp \tag{6.8}$$

Since the total number of systems in the ensemble do not change with time t,

$$\frac{d\rho(q, p;\ t)}{dt} = 0$$

This means

$$\frac{d\rho}{dt} = \frac{\partial\rho}{\partial t} + \sum_{k=1}^{f}\left(\frac{\partial\rho}{\partial q_k}\dot{q}_k + \frac{\partial\rho}{\partial p_k}\dot{p}_k\right)$$

$$= \frac{\partial\rho}{\partial t} + \sum_{k=1}^{f}\left(\frac{\partial\rho}{\partial q_k}\frac{\partial H}{\partial p_k} - \frac{\partial\rho}{\partial p_k}\frac{\partial H}{\partial q_k}\right)$$

$$= \frac{\partial\rho}{\partial t} + [\rho, H] = 0 \tag{6.9}$$

where $[\rho, H]$ is a Poisson Bracket [1]. We know that if the Poisson Bracket of a physical quantity with Hamiltonian is zero, then that quantity is a constant of motion. For an equilibrium, the physical quantities should not depend upon time. This means the density function in Γ space should not explicitly depend upon time t. Thus for equilibrium, $\partial\rho/\partial t = 0$. This means the Poisson Bracket of ρ with H is zero, implying that the ρ is a constant of motion. This means ρ is a function of constant of motion - namely the energy E.

Moreover, we prove that the flow of density function is like that of incompressible fluid. Velocity defined here is such that

$$\nabla.\vec{v} = 0.$$

For,

$$\nabla.\vec{v} = \sum_{k}\frac{\partial\dot{q}_k}{\partial q_k} + \frac{\partial\dot{p}_k}{\partial p_k}$$

$$= \sum_{k}\left(\frac{\partial^2 H}{\partial q_k\partial p_k} - \frac{\partial^2 H}{\partial p_k\partial q_k}\right) = 0 \tag{6.10}$$

That means the flow is incompressible. Using equation (9) we see that

$$\frac{d\rho}{dt} = 0 = \frac{\partial\rho}{\partial t} + \sum_{k}\frac{\partial\rho}{\partial q_k}\dot{q}_k + \frac{\partial\rho}{\partial p_k}\dot{p}_k$$

$$= \frac{\partial\rho}{\partial t} + \vec{v}.\nabla\rho$$

But since $\nabla.\vec{v} = 0$, we rewrite this equation as,

$$\frac{d\rho}{dt} = 0 = \frac{\partial\rho}{\partial t} + \vec{v}.\nabla\rho + \rho\nabla.\vec{v}$$

$$= \frac{\partial\rho}{\partial t} + \nabla.(\rho\vec{v}) \tag{6.11}$$

This is an equation of continuity for the flow of phase point density.

---

[1] For details of properties and theory of Poisson Bracket see Ch.21 P-236 Classical Mechanics by P. V. Panat (Narosa Publisher)

Let Φ be the volume of the Γ space whose surface is a constant energy surface $H = E$. Then, integral version of equation (11) is

$$\frac{d}{dt} \int \rho \, dp \, dq = 0.$$

Or,

$$\int_\phi \frac{\partial \rho}{\partial t} dp \, dq = - \int \nabla.(\rho \vec{v}) \, dq \, dp = - \oint \rho \vec{v}.d\vec{s}$$

where $d\vec{s}$ is an area element of the constant energy surface. This means, the number of systems contained in the volume Φ of Γ space decrease with time. Exactly same number of systems leave the surface 'S' of Φ. The current of the 'flow' of the system is $\vec{j} = \rho \vec{v}$.

We now introduce an operator that corresponds to co-moving derivative

$$\frac{D}{Dt} = \frac{\partial}{\partial t} + \vec{v}.\nabla \tag{6.12}$$

We also note that $\nabla.\vec{v} = 0$. Then, constancy of the density function can be written as,

$$\frac{D\rho}{Dt} = 0 \tag{6.13}$$

or, its equivalent form

$$\frac{d\rho}{dt} = \frac{\partial \rho}{\partial t} + [\rho, H] \tag{6.14}$$

is known as Liouville's theorem.

6.3   Microcanonical Ensemble

With large number of identical systems as members of the ensemble, and the density function $\rho(q, p; t)$ defined earlier, we discuss micro-canonical ensemble. First we note that we should not talk about $(2f - 1)$ dimensional zero thickness surface, but we must consider a shell of width δE at $H = E$. This is consistent with quantum mechanics. To measure the energy exactly, one will require an infinite amount of time. The measurements take a finite time. This gives energy a width δE. It should be noted that in an equilibrium situation, ρ is constant. It means every phase point in the shell of thickness δE around E, is equally likely. We postulate that every state of the system lying the energy shell between E and $E + \delta E$ is *a priori*, equally probable when the closed system is in equilibrium.

The ensemble, where in, in equilibrium,

$$\rho(q, p) = \begin{cases} \text{Constant} & \text{for} \quad E < H < E + \delta E \\ \\ 0 & \text{outside the shell} \end{cases} \tag{6.15}$$

is called microcanonical ensemble.

This can be understood as follows. In the volume element, $(dp \, dq)$ of the Γ space, $\rho \, dp \, dq$ are the number of states, and each is equally probable. This is due to our postulate of equal *a priori* probability. Therefore, the probability of the classical system to have co-ordinates lying between

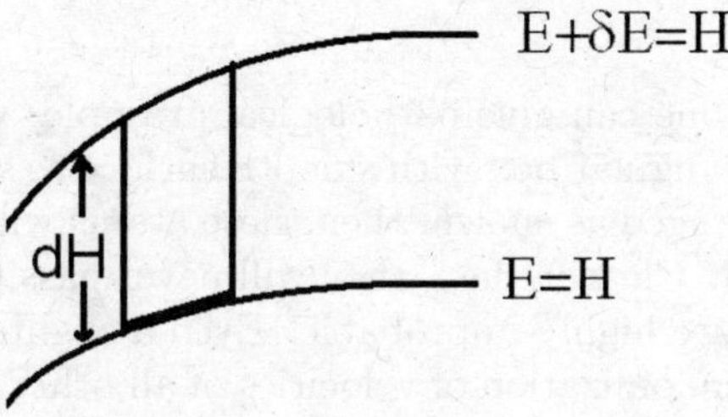

Figure 6.2: Probability density for energy at temperature T.

$$[(q_1 \text{ and } q_1 + dq_1), (p_1 \text{ and } p_1 + dp_1)]; \dots [(q_f \text{ and } q_f + dq_f), (p_f \text{ and } p_f + dp_f)] \text{ is}$$

$$\omega(q,p) = \frac{\rho(q,p)dq\ dp}{\int \rho(q,p)dq\ dp} \qquad (6.16)$$

If $f(q,p)$ is a physical quantity, the microcanonical average is,

$$\langle f \rangle_m = \int \omega(q,p)\ f(q,p)\ dq\ dp = \frac{\int f(q,p)\ \rho(q,p)\ dq\ dp}{\int \rho(q,p)\ dq\ dp} \qquad (6.17)$$

6.4   Ergodic Hypothesis

We consider a single thermodynamic system. It has single phase point in Γ space. This phase point moves on a constant energy surface $H = E$ of the system. Boltzmann postulated ergodic hypothesis. Ergodic hypothesis states that, the phase point moves on a constant energy surface $H = E$ and it visits all the points on $H = E$ surface. Boltzmann does not specify the time taken by the phase point to visit all the points.

Ergodic Hypothesis has many difficulties from pure mathematics point of view. To overcome these difficulties, P. and T. Ehrenfest postulated quasi-ergodic hypothesis. It states that, in a course of time the phase point of the system will come arbitrarily close with every other point on $H = E$ surface. This was proved by Birkoff under certain conditions. Then the time average of a physical quantity $f(q(t), p(t))$ is,

$$\langle f \rangle_t = \lim_{\tau \to \infty} \frac{1}{\tau} \int_0^{\tau} f(q(t), p(t))dt \qquad (6.18)$$

If the ergodic hypothesis is true, we prove,

$$\langle f \rangle_t = \langle f \rangle_m \qquad (6.19)$$

According to ergodic hypothesis, every point on $H = E$ surface is reached by the phase point of the system. Then the time average $< f >_t$ is independent of the starting point $(q(t=0), p(t=0))$ on $H = E$ surface. Thus, every point on $H = E$ surface of microcanonical ensemble can act as a position of the phase point of the system at $t = 0$. Then let's take time average of the microcanonical average as $\langle < f >_m \rangle_t$. By Liouville's theorem, in equilibrium, $\langle f \rangle_m$ is independent of time t. Obviously, in this case, $\langle f \rangle_m = \langle \langle f \rangle_m \rangle_t = \langle f \rangle_t$ This implies that we can take ensemble average of $f(q, p)$, which has same answer as with its time average.

Strictly speaking, one can give pathological examples where ergodic hypothesis does not hold. Consider a rectangular box with smooth surface as shown in figure (2). Let N non interacting atoms be arranged as shown, then these atoms will simply be reflected back and forth from sides A and B. Clearly, then, they will never pass through all other points in the box. But such situations are highly improbable. Even a slightest disturbance to the velocity of any atom will cause randomization of velocities of all other atoms through collisions. The

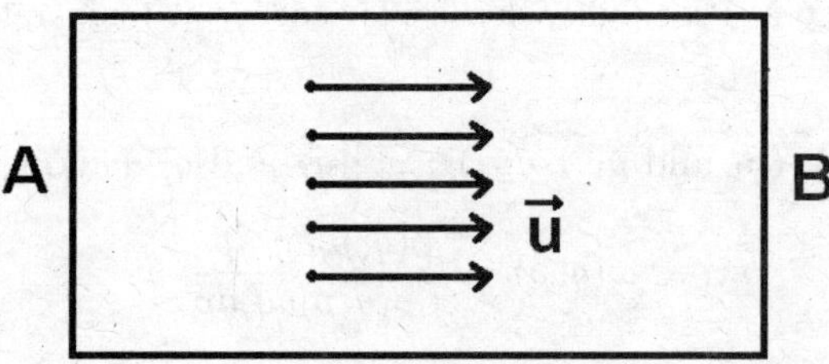

Figure 6.3: Unusual nonergodic situation.

question is which average quantity of f is experimentally observable? It is the time average of f which is observable. The time τ is long enough, much larger than the relaxation time of the system over which the average is taken.

6.5   Entropy

We first understand clearly some of the ideas before we discuss intimate relation between microstate of physical system and its entropy. We define $\Phi = \Phi(E, N, V)$ as number of states contained in a volume enclosed by a constant energy surface $H = E$ for the system.

Density of states $\omega(E)$ is defined as a derivative of $\Phi(E, V, N)$ with respect to energy E. Thus,

$$\omega(E) = \left[\frac{\partial \Phi(E, V, N)}{\partial E} \right]_{V,N} \tag{6.20}$$

In microcanonical ensemble, we consider an energy shell of width δE at $H = E$ surface. Total number of states in the shell of microcanonical ensemble are denoted by $\Omega(E, N, V)$ and it is,

$$\Omega(E, N, V) = \omega(E)\delta E \tag{6.21}$$

Next Quantity of interest is volume of a shell of width δR in n dimensions where $n \gg 1$. By dimensional arguments, volume of a sphere of radius R in n dimension is,

$$V = C_n R^n \tag{6.22}$$

C_n can be calculated but is not important for present discussion. Area of the surface of the volume V is

$$\text{Area} = nC_n R^{n-1}$$

Volume of shell of thickness δR is

$$\delta V = C_n \left[R^n - (R - \delta R)^n \right]$$
$$= V \left[1 - \left(1 - \tfrac{\delta R}{R}\right)^n \right]$$

But we know that

$$\lim_{n \to \infty} \left[1 + \frac{x}{n} \right]^n = e^x$$

Thus, when $n \gg 1$,

$$\delta V \simeq V \left(1 - e^{-n\delta R/R} \right) \tag{6.23}$$

It therefore follows that, for a phase space of large number of particle, volume of entire sphere enclosed by the surface $H = E$ is almost same as a volume of the shell between two surfaces of $H = E$ and $H = E + \delta E$.

To get the feel for the numbers, we estimate Ω and Φ for a free particle system. We will present the detailed calculation subsequently. Here,

$$\sum_{k=1}^{N} \frac{p_k^2}{2m} = E$$

The radius of the sphere $H = E$ is proportional to

$$\sqrt{\sum_{k=1}^{N} p_k^2} = \sqrt{2mE}$$

Thus

$$\Phi(E, V, N) = \left(\frac{1}{h^f} \right) \int_{H=E} dp\, dq \propto E^{f/2} V^f / h^f$$

and

$$\Omega(E, V, N) \propto \frac{1}{h^f} E^{f/2} V^f \tag{6.24}$$

where f are number of degrees of freedom. In case of the free particle system, $f = 3N$. This makes

$$\Omega(E, V, N) \approx \Phi(E, V, N) \propto \frac{1}{h^{3N}} E^{3N/2} V^{3N},$$

which is a huge number when $N \sim 10^{22}/c.c.$.

It is clear that for one thermodynamic state represented by (E, V, N), there are huge number of microstates.

Boltzmann discovered that entropy S that appears in thermodynamics is proportional to $\ln \Omega$. Planck explicitly wrote down

$$\boxed{S = k_B \ln \Omega} \tag{6.25}$$

where k_B is universal constant known as Boltzmann constant. This equation essentially connects thermodynamics with statistical ideas. Value of k_B is $k_B = 1.38 \times 10^{-16} ergs/deg$ ($= 1.38 \times 10^{-23} Joule/degree$). Note that in the above formula for entropy, there is no initial entropy S_0. Thus value of S provided by above formula is absolute value. It is also significant that Nernst theorem or third law of thermodynamics is already present in above formula. For, at $T = 0$, in most of the systems, its ground state is attained. It is only one state and $\ln 1 = 0$. For some systems, at $T = 0$, $\Omega = constant$. Therefore, $S = constant = S_0$ at $T = 0$ indicating the inclusion of the third law.

We have to prove that entropy so stated has properties of thermodynamic entropy.

To get the further understanding of entropy formula and emphasize correctness of Planck's assertion, we consider two closed systems A_1 and A_2 as shown in figure (3). A gas of $N_1(N_2)$ atoms, at a temperature $T_1(T_2)$ is confined in a volume $V_1(V_2)$ constituting a system $A_1(A_2)$. We first bring A_1 and A_2 in constant with each other forming a composite system A_0. A_1 and A_2 can exchange energy with each other, however $N_i, V_i (i = 1, 2)$ do not change. A_0 system is closed. Number of accessible states of $A_1(A_2)$

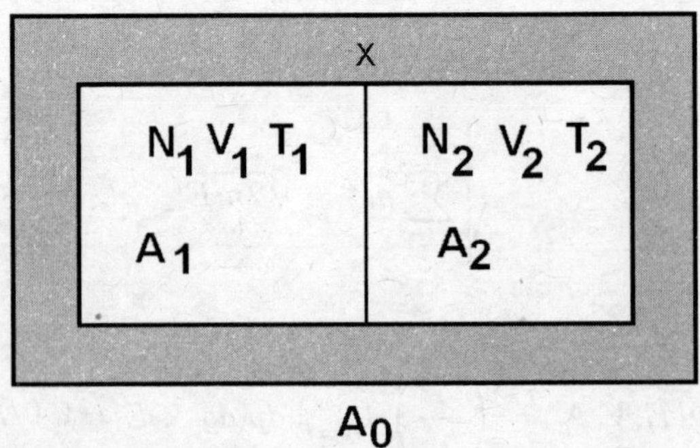

Figure 6.4: Two systems A_1 and A_2 exchanging energy through X.

are $\Omega_1(E_1, V_1, T_1)$ ($\Omega_2(E_2, V_2, T_2)$). For A_0, its energy E_0 is such that, $E_0 = E_1 + E_2$ as constant even though E_1 and E_2 may readjust. Question is, what are number of accessible state $\Omega_0(E_1, E_2)$ of A_0 in its equilibrium? They are

$$\Omega_0(E_1, E_2) = \Omega_1(E_1)\Omega_2(E_2) = \Omega_1(E_1)\Omega_2(E_0 - E_1) \tag{6.26}$$

Hence, how E_1 is adjusted? We assert that E_1 takes a value that maximizes Ω_0. This is required because the equilibrium state is that state, which maximizes entropy. We also take logarithm of Ω_0 since it is a huge number.

Let $V_0 = V_1 + V_2$, $N_0 = N_1 + N_2$.
Therefore $\ln \Omega_0(E_0, E_1) = \ln \Omega_1(E_1) + \ln \Omega_2(E_0 - E_1)$
and

$$\left. \frac{\partial \ln \Omega_0}{\partial E_1} \right)_{E_0, V_0, N_0} = \left. \frac{\partial \ln \Omega_1}{\partial E_1} \right)_{E_0, V_0, N_0} - \left. \frac{\partial \ln \Omega_2(E_0 - E_1)}{\partial E_1} \right)_{E_0, V_0, N_0} = 0$$

Compare this with the equation

$$T dS = dE + P dV - \mu dN \tag{6.27}$$

Clearly, with the discovery $S = k_B \ln \Omega$, $\dfrac{\partial \ln \Omega}{\partial E} = \dfrac{1}{k_B} \dfrac{\partial S}{\partial E}$.
From thermodynamic equation,

$$\left. \frac{\partial S}{\partial E} \right)_{V, N} = \frac{1}{T}. \tag{6.28}$$

This means, E_1 gets adjusted in such a manner that the entropy of the total system is maximum.
Consequently,

$$\frac{1}{k_B T_1} = \frac{1}{k_B T_2}$$

or

$$T_1 = T_2 \tag{6.29}$$

which is a well known thermodynamic condition for equilibrium.

We now relax a constraint of $V_1(V_2)$ being fixed. That means the boundary X can slide when A_1 and A_2 equilibrate. Thus A_1 and A_2 not only exchange energy but also adjust volumes V_1 and V_2 during the process of equilibrium, keeping V_0 constant. Then $\Omega_0(E_0, V_1, N_0)$ is such that the entropy of A_0 is maximum. This means

$$\left. \frac{\partial \ln \Omega_0}{\partial V_1} \right)_{\overline{E}_1, N_1} = \left. \frac{\partial \ln \Omega_1}{\partial V_1} \right)_{\overline{E}_1, N_1} + \left. \frac{\partial \ln \Omega_2}{\partial V_2} \right)_{\bar{E}, N_1} = 0.$$

Or,

$$\left. \frac{\partial \ln \Omega_1}{\partial V_1} \right)_{\overline{E}_1, N_1} - \left. \frac{\partial \ln \Omega_2}{\partial V_1} \right)_{\overline{E}_1, N_1} = 0 \tag{6.30}$$

From the thermodynamic equation

$$\frac{\partial \ln \Omega}{\partial V} = \frac{1}{k_B} \frac{\partial S}{\partial V} = \frac{P}{k_B T}.$$

Therefore, (30) imply

$$\frac{P_1}{k_B T_1} = \frac{P_2}{k_B T_2}.$$

But $T_1 = T_2$ as a result of an energy exchange. This means V_1 and V_2 get readjusted in such a manner that $P_1 = P_2$. This is same relation we derived in thermodynamics.

Thus,

$$\left. \frac{\partial \ln \Omega}{\partial E} \right)_{V,N} = \frac{1}{k_B T} \equiv \beta \tag{6.31}$$

and

$$\left. \frac{\partial \ln \Omega}{\partial V} \right)_{E,N} = \frac{P}{k_B T} \equiv \beta P \tag{6.32}$$

It is easy to see that, if the divider X in fig. (3) acts as a permeable membrane, i.e. allows the exchange of particles as well, then N_1 and N_2 along with V_1 and V_2 and T_1 and T_2 get readjusted. Similar analysis shows that, in equilibrium, $\mu_1 = \mu_2$.
$\ln \Omega$ is connected with μ as

$$\left. \frac{\partial \ln \Omega}{\partial N} \right)_{E,V} = -\beta \mu = -\frac{\mu}{k_B T} \tag{6.33}$$

From this analysis, it supports the validity of Boltzmann assertion that

$$S = k_B \ln \Omega \tag{6.34}$$

Illustrative Problems

Problem 1: Show that, in microcanonical ensemble, we can write

$$S = -k_B \sum_r P_r \ln P_r$$

where P_r is equal *a priori* probability.

Solution: In microcanonical ensemble, we consider a system energy between E and $E + \delta E$. All states in this range are equally probable. If there are total of Ω states in the energy range , each state has equal *a priori* probability $= \frac{1}{\Omega}$. Therefore, a state in this range has $P_r = \frac{1}{\Omega}$.

$$\therefore P_r \ln P_r = \frac{1}{\Omega} \ln \left(\frac{1}{\Omega} \right)$$

$$\therefore -k_B \sum_r P_r \ln P_r = -k_B \sum_r \frac{1}{\Omega} \ln\left(\frac{1}{\Omega}\right) = -k_B \cdot \Omega \frac{1}{\Omega} \ln \frac{1}{\Omega} = k_B \ln \Omega = S$$

q.e.d

Problem 2: Consider an isolated system of $N(>> 1)$ spins of $\frac{1}{2}\hbar$. Each has magnetic moment μ. Energy E of the system is $E = -(N_+ - N_-)\mu H$. Here $N_+(N_-)$ is number of spins aligned parallel (antiparallel) to H.

(i) Consider $\delta E << E$ but $\delta E >> \mu H$. Find $\Omega(E)$, the total number of accessible states lying between E and $E + \delta E$.

(ii) Find $S = k_B \ln \Omega$. Use Sterling's formula for simplification.

Solution:

(i) Here $N = N_+ + N_-$ and let $n = N_+ - N_-$.

Total energy is $E = -n\mu H$. Clearly, then,

$$N_{\pm} = \frac{N}{2} \pm \frac{n}{2} = \frac{N}{2} \mp \frac{E}{2\mu H}$$

We can arrange N spins among themselves in $N!$ ways. Many ways do not give independent distinguishable arrangements into groups $\{N_+\}$ or $\{N_-\}$. Corresponding number of interchanges are $N_+!$ and $N_-!$. Then the microstates density $\omega(E)$ is,

$$\omega(E) = \frac{N!}{N_+!N_-!} = \frac{N!}{\left(\frac{N}{2} - \frac{E}{2\mu H}\right)!\left(\frac{N}{2} + \frac{E}{2\mu H}\right)!}$$

But energy range is δE and level difference is $2\mu H$. Thus total number of states in the shell is

$$\Omega(E) = \frac{N!}{\left(\frac{N}{2} - \frac{E}{2\mu H}\right)!\left(\frac{N}{2} + \frac{E}{2\mu H}\right)!} \cdot \frac{\delta E}{2\mu H}$$

(ii)

$$\ln \omega(E) = \ln N! - \ln(N_+)! - \ln(N_-)!$$

Use Sterling's formula

$$\ln N! = N \ln N - N$$

After some simplification

$$\ln \omega(E) \cong - \left\{ \frac{N}{2}\left(1 - \frac{n}{N}\right) \ln \frac{1}{2}\left(1 - \frac{n}{N}\right) + \frac{N}{2}\left(1 + \frac{n}{N}\right) \ln \frac{1}{2}\left(1 + \frac{n}{N}\right) \right\}$$

But, for $0 < x < 1$,

$$\ln(1 \pm x) \simeq \pm x - \frac{x^2}{2} \pm \frac{x^3}{3} - \ldots\ldots$$

Putting $n = -E/\mu II$,

$$\ln \omega(E) \simeq N \ln 2 - \frac{E^2}{2N\mu^2 H^2}$$

Entropy of the system $= k_B \ln \Omega = S$
Or,

$$S = N k_B \ln 2 - \frac{k_B E^2}{2N\mu^2 H^2} + k_B \ln \delta E - k_B \ln(2\mu H)$$

The first term is entropy of random orientation of N spins in the absence of H.
Thus,

$$\omega(E) \quad \sim \quad e^{-\dfrac{(E/\mu H)^2}{2N}}$$
$$\sim \quad e^{-\dfrac{(N_+ - N_-)^2}{2N}}$$

which is a Gaussian distribution, with a spread of $\sqrt{N}$ as shown in figure(5).

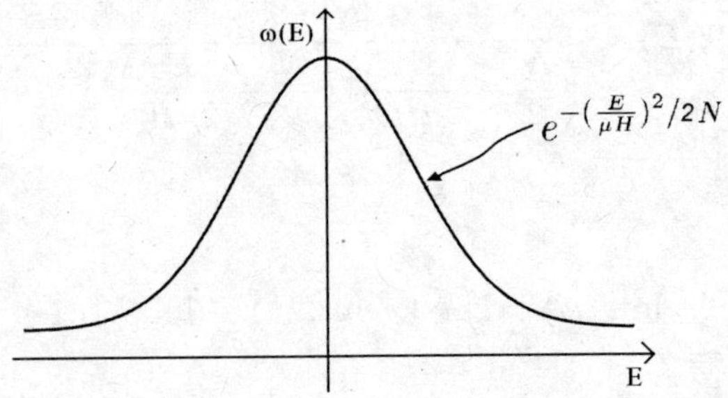

Figure 6.5: Gaussian distribution.

Short Questions

1. Draw a phase space of free particle moving along x-axis.

2. Distinguish between the microstate and a macrostate. Give two example of each.

3. How a microstate is specified in case of classical system? Give example.

4. How a microstate is specified in case of quantum system? Give example.

5. State ergodic hypothesis.

6. Define time average of $f(p, q)$.

7. Define ensemble average of $f(p, q)$.

8. Explain how quasi-ergodic hypothesis makes 'time average=ensemble average' plausible?

9. State and prove Liouville's theorem. Explain its physical significance.

10. Give an example, even though pathological, where ergodic hypothesis is not true.

11. Show that, when the dimension of space $N >> 1$, volume of sphere of radius R in this space is almost same as that of a shell enclosed by the spheres of radius R and radius $R - \delta R$ where $\delta R << R$.

12. State what is microcanonical ensemble. Define entropy S in a microcanonical ensemble and show how this definition is consistent with laws of thermodynamics.

13. Draw a phase space diagram of a rigid rotator.

Problems

Problem 1. For an ideal non relativistic gas of N atoms, show
(a) $P\,V^{5/3}$ for reversible adiabatic processes. (b) $\gamma = C_p/C_V = 5/3$

(Hint: From the text, $\Omega =$ constant $E^{3N/2}V^N$

$\therefore S = S_0 + Nk_B \ln V + \dfrac{3N}{2} k_B \ln E.$

Use $\left(\dfrac{\partial S}{\partial E}\right)_{N,V} = \dfrac{1}{T} = \dfrac{3Nk_B}{2E},\ \left(\dfrac{\partial S}{\partial V}\right)_{N,E} = \dfrac{P}{T} = \dfrac{Nk_B}{V}$

Using above, $\frac{2}{3}P\,V = E$; for adiabatic, $V E^{3/2} =$ constant, get $PV^{5/3} =$ constant.

$C_V = \dfrac{\partial E}{\partial T}\bigg)_{V,N} = \dfrac{3}{2}Nk_B,\ \left(C_p = \dfrac{\partial H}{\partial T} = \dfrac{\partial}{\partial T}(E + pV) = \dfrac{5}{2}C_V.\right)$

$\therefore \gamma = C_p/C_V = 5/3)$

Problem 2. Show that in an adiabatic process for ideal extreme relativistic gas with single
particle energy $\varepsilon = \dfrac{hc}{2L}(n_x{}^2 + n_y{}^2 + n_z{}^2)^{1/2} = \dfrac{hc}{2L}n,$

a) $PV^{4/3} =$ constant and b) $\gamma = 4/3$

(Hint: $\Omega \propto E^{3N/2} V^N$; $\dfrac{3Nk_B}{E} = \dfrac{1}{T}$; $PV = \dfrac{E}{3}$; $C_V = 3Nk_B, C_p = 4/3\, C_V$)

Problem 3. Consider a gas of N molecules confined in a volume V. Let v be a small fraction of V. Then the probability of the molecule lying in v is $v/V = p$. Probability of its being in $(V - v)$ is $\left(1 - \frac{v}{V}\right) = (1 - p) = q$

(i) Show that the probability of n molecules in volume v is

$$\omega(n) = \frac{N!}{n!(N-n)!} \left(\frac{v}{V}\right)^n \left(1 - \frac{v}{V}\right)^{N-n}$$

(ii) show

$$\sum \omega(n) = 1$$

and

$$\overline{n} = \sum n\omega(n) = Np = \frac{Nv}{V} = \rho V$$

where $\rho = N/V$ is density of the gas

(iii) For large $N \gg 1$, using Stirling approximation $\ln N! \cong N \ln N - N$, Show

$$\omega(n) \simeq \frac{e^{-\dfrac{(n - \overline{n})^2}{2Npq}}}{\sqrt{2\pi Npq}}$$

(Hint: Number of possibilities of picking n different molecules out of N is $\begin{pmatrix} N \\ n \end{pmatrix} = \dfrac{N!}{n!(N-n)!}$. Probability of molecules in v is p. Thus Probability of n atoms in v and not in $V - v$ is,

$$\omega(n) = \frac{N!}{n!(N-n)!} p^n q^{N-n}$$

(ii) $(p + q)^N = \sum_n \begin{pmatrix} N \\ n \end{pmatrix} p^n q^{N-n}$ (Revise binomial theorem and its properties).
$p + q = 1$ and hence the result.

$$\begin{aligned}
\overline{n} &= \sum n\omega(n) = \sum \frac{\partial}{\partial p}\left[\begin{pmatrix} N \\ n \end{pmatrix} p^n q^{N-n}\right] \\
&= \sum n\omega(n) = p\, \frac{\partial}{\partial p} (p + q)^N = Np
\end{aligned}$$

$$\therefore \overline{n} = N\frac{v}{V} = \rho v.$$

· (iii) $\omega(n)$ has a sharp maximum. Show that the maximum value occurs at $n_m = Np = \overline{n} = \rho v$. We expand $\ln \omega(n)$ because it is easier to handle logarithms of large numbers. Expand around maximum,

$$\ln \omega(n) = \ln \omega(\overline{n}) + \frac{1}{2}(n - \overline{n})^2 \frac{d^2\omega(n)}{dn^2}\bigg|_{n=\overline{n}}$$

Show $\dfrac{d^2\omega(n)}{dn^2}\bigg|_{n=\overline{n}} = \dfrac{-1}{Npq}$ Then

$$\omega(n) = \omega(\overline{n})e^{-\frac{(n-\overline{n})^2}{2Npq}}$$

and $w(\overline{n})$ can be obtained by normalizing

$$\int_{-\infty}^{\infty} \omega(n)dn = 1 \text{ or } \omega(n) = \frac{1}{\sqrt{2\pi Npq}}.$$

q.e.d.

Following facts are learned from this example.

(a) Distribution of number of particles is very sharply peaked around $\overline{n} = \rho v$. It means the distribution of the molecules is uniform.

(b) Estimations are as follows. Typically for $1mm^3$ in $1cc$, average number of molecules is $\overline{n} \cong 10^{23} \times 10^{-3} = 10^{20}$. Suppose we want a deviation of $n = \overline{n} \pm \delta$ where $\delta \cong 10^{10}$, then we find values of $\omega(n)$ for $\delta = 2\sqrt{2} \times 10^{10}$, $3\sqrt{2} \times 10^{10}$ and $4\sqrt{2} \times 10^{10}$ decreasing vary fast from approximately 10^{-4} to 10^{-9} indicating sharpness of $\omega(n)$ around $n = \overline{n} = vN/V$)

Problem 4. For a pendulum, generalised coordinate is its angular displacement θ and the generalized momentum is $p_\theta = ml^2\dot{\theta}$. Find trajectory in phase space and show that the area A of enclosed trajectory is $E\tau$ where E is total energy of the pendulum and τ is its period.

(Hint: $\theta \sim A\sin\omega t$, $p_\theta = ml^2\omega A\cos\omega t$. Show

$$\left(\frac{\theta}{A}\right)^2 + \left(\frac{p_\theta}{ml^2 A\omega}\right)^2 = 1; \tau = 2\pi\sqrt{\frac{l}{g}}$$

Area $= \pi ml^2 A^2\omega$. Find E, substitute.)

Chapter 7

Foundations of Statistical Mechanics II: Quantum Statistical Mechanics

7.1 Introduction to Quantum Ideas

We state here some important aspects of quantum mechanics, necessary for our development of the subject. There are many excellent text books on quantum mechanics. To mention one without the other would be unjust. But from our opinion, Quantum Mechanics by Landau and Lifshitz (Non relativistic theory) is best. You may skip this entire section if you are already knowledgeable of quantum mechanics.

In quantum mechanics, every observable quantity is represented by a Hermitian operator. It is the eigen value of the operator that is observable. Hermitian operator has always real eigen values. Since an observed physical quantity is real, corresponding operator is Hermitian.

In a coordinate representation, position vector is represented by $\hat{r} = \vec{r}$. We will put a cap on the top to emphasize operator nature. Momentum in the same representation is $\hat{p} = -i\hbar\nabla$. Corresponding to energy, the operator is Hamiltonian H. It is

$$
\begin{aligned}
\hat{H} &= \hat{T} + \hat{V} \\
&= \frac{\hat{p}^2}{2m} + V(\hat{r}) \\
&= -\frac{\hbar^2}{2m}\nabla^2 + V(\vec{r})
\end{aligned}
\tag{7.1}
$$

It is also represented by

$$
\hat{H} = -\frac{\hbar^2}{2m}\nabla^2 + V(\vec{r}) = i\hbar\frac{\partial}{\partial t}.
$$

Angular momentum is $\vec{L} = \vec{r} \times \vec{p}$ and its operator form in coordinate representation is $\hat{\vec{L}} = (\vec{r} \times (-i\hbar\nabla))$. There are certain physical quantities which are purely of quantum nature. They have no classical analogue. One such example is spin angular momentum. Corresponding operators need to be discovered as were by Pauli in the form of matrices named after him.

The operators operate on vector space (more formally as a 'ray space') the Hilbert space. A physical system is described by a complex vector $|\psi(t)\rangle$ in the abstract Hilbert space. Its coordinate representation is a well known Schrödinger wave function as $\psi(q,t) = \langle q | \psi(t) \rangle$. Quantum description of a physical system is probabilistic. $|\psi(q,t)|^2$ is a probability density. The probability that a quantum particle is between q and $q + dq$ is $|\psi(q,t)|^2 dq$

The process of measurement in quantum mechanics is such that its apparatus is classical. Process of a measurement of physical quantity is mathematically represented by the operation of the corresponding operator on the quantum state. What we observe is an average value of the operator. Consider an operator $\hat{O}$. Its average value will be obtained in the process of measurement. It is

$$\langle \hat{O} \rangle = \frac{\int \psi^*(q,t)\, \hat{O}\, \psi(q,t)\, dq}{\int \psi^*(q,t)\, \psi(q,t)\, dq} \tag{7.2}$$

If the quantum state $\psi(q,t)$ is normalized, then

$$\int \psi^*(q,t)\psi(q,t)dq = \int |\psi(q,t)|^2\, dq = 1 \tag{7.3}$$

If not, it can always be normalized.

Quantum mechanics is a linear theory. Principle of superposition works. Thus, if $\psi_1(\vec{r},t)$ and $\psi_2(\vec{r},t)$ represent two different states, then $\psi(\vec{r},t) = a\psi_1 + b\psi_2$ is also a state. In general, for an operator $\hat{O}$, there will be a set of eigenvalues $\{o_n\}$ and corresponding eigenfunctions $\{\phi_n(\vec{r})\}$ in a Hilbert space. The eigenfunctions $\{\phi_n\}$ form a basis set in the Hilbert space. Thus, any vector $\psi(\vec{r},t)$ can be expanded as a superposition of basis.

We write

$$\psi(\vec{r},t) = \sum_n a_n(t)\phi_n(\vec{r}) \tag{7.4}$$

and

$$\hat{O}\,\phi_n(\vec{r}) = o_n\,\phi_n(\vec{r}) \tag{7.5}$$

A process of measurement of a physical quantity $\hat{O}$ on a state $\psi(\vec{r},t)$ is such that one gets o_n with probability $|a_n|^2$. we say that, after the measurement, a state ψ collapses in to an eigen state $\phi_n(\vec{r})$. If one makes a repeated measurement of $\hat{O}$ on $\phi_n(\vec{r})$, one gets o_n with certainty. (This is a consequence of the fact that the basis vectors are linearly independent of each other.)

Time evolution of a state $|\psi(t)\rangle$ is governed by a Schödinger equation

$$i\hbar\frac{\partial |\psi(t)\rangle}{\partial t} = \hat{H}\,|\psi(t)\rangle \tag{7.6}$$

Here, the state evolves with time and can uniquely be found if $|\psi(t_0)\rangle$ for the system is known. In this picture, the operators are independent of time.

There is another picture known as Heisenberg picture where the operators are time dependent and the states are time independent. Thus the operator $\hat{O}(t)$ evolves in time according to Heisenberg equation as,

$$i\hbar\frac{d\hat{O}}{dt} = i\hbar\frac{\partial\hat{O}}{\partial t} + [\hat{O}, \hat{H}] \tag{7.7}$$

where

$$[\hat{O}, \hat{H}] = \hat{O}\hat{H} - \hat{H}\hat{O}$$

is known as commutation bracket.

Process of measurement in quantum mechanics is such that not all physical quantities can be measured simultaneously. Well known example is x and p_x which cannot be measured simultaneously. Another example is components of angular momenta . They cannot be measured simultaneously. Manifestation of this fact in quantum mechanics, is that corresponding operators do not commute. Thus

$$[x, p_x] = xp_x - p_xx \neq 0$$

It is

$$[x, p_x] = i\hbar \tag{7.8}$$

Similarly

$$[L_x, L_y] = i\hbar L_z$$

There are many such pairs of operators in quantum mechanics. Uncertainty principle is a consequence of that.

There is a great deal of importance of commuting observables. Those observables whose corresponding operators commute, can be measured simultaneously. It means, they have a common set of eigenfunctions. Moreover, those operators which are time independent and commute with $\hat{H}$ are constants of motion. Knowledge of the constant of motion simplifies solution of the problem considerably.

The set $\{\phi_n(\vec{r})\}$ is said to be complete in a Hilbert space, if any vector ψ belonging to the same space can be uniquely expanded as

$$\psi(\vec{r}) = \sum_n a_n\, \phi_n(\vec{r}) \tag{7.9}$$

If ψ is time dependent, i.e., $\psi(t)$, then a_n become time dependent.

The eigenfunctions $\{\phi_n\}$ form orthonormal (o.n.) set . It means

$$\int \phi_n^*(\vec{r})\, \phi_m(\vec{r})d^3r = \delta_{mn} \tag{7.10}$$

Here δ_{mn} is Kronecker delta s.t.

$$\delta_{mn} = \begin{cases} 1 & if \quad m = n \\ \\ 0 & if \quad m \neq n \end{cases} \tag{7.11}$$

Clearly, then, $a_m = \int \phi_m^* \, \psi(\vec{r}) d^3 r \equiv \langle m | \psi \rangle$

A set $\{\phi_n(\vec{r})\}$ is complete means, it satisfies following property

$$\sum_n \phi_n^*(\vec{r}) \, \phi_n(\vec{r'}) = \delta(\vec{r} - \vec{r'})$$

or,

$$\sum_n |n\rangle \langle n| = 1 \tag{7.12}$$

where $\delta(\vec{r})$ is a three dimensional Dirac[1] delta function.

Any operator $\hat{O}$ has a matrix representation

$$\hat{O}_{mn} = \int \phi_m^*(\vec{r}) \, \hat{O} \, \phi_n(\vec{r}) d^3 r \tag{7.13}$$

This matrix will be denoted by $[O]$ and may or may not be diagonal. It is easy to see that if $\{\phi_n\}$ are eigenfunctions of $\hat{O}$, then $\hat{O}\phi_n = o_n\phi_n$ and $\hat{O}_{mn}$ is diagonal. If a matrix $[O]$ is not diagonal, then it can be made diagonal by a similarity transformation. Then the diagonal elements are the eigenvalues of $\hat{O}$.

We note that the Trace (sum of diagonal elements) of a matrix $\hat{O}$ is invariant under similarity transformation and is sum of all eigenvalues of $\hat{O}$.

Transpose of a matrix O is defined by $\tilde{O}$ and in the process of transposing, the rows and columns are exchanged.

Thus

$$O_{mn} = \tilde{O}_{nm} \tag{7.14}$$

A Hermitian matrix O_H is the one which has real eigenvalues and is such that

$$O_H = \tilde{O}_H^* \equiv O_H^\dagger \tag{7.15}$$

where $*$ means complex conjugation. $O_H^\dagger$ is called adjoint of O_H. Thus, Hermitian operator is one whose adjoint is same as the operator himself.

A projection operator on a state $|n\rangle$ is $P_n = |n\rangle \langle n|$ which projects any state $|\psi\rangle$ on $|n\rangle$.

---

[1] For a reasonable discussion of Dirac delta function, see p-19, section 2.3 Introduction to Electrodynamics by Capri and Panat (Narosa Publishers).

Thus, $P_n \left| \psi \right\rangle = \left| n \right\rangle \left\langle n \right| \psi \rangle = a_n \left| n \right\rangle$. It is easy to see that the projection operator has its eigenvalue zero or one. More over

$$P_n{}^2 = \left| n \right\rangle \left\langle n | n \right\rangle \left\langle n \right| = \left| n \right\rangle \left\langle n \right| = P_n. \tag{7.16}$$

Energy eigenfunctions are eigenfunctions of H and have a special place in quantum mechanics.

$$\hat{H}\psi_n(\vec{r}) = E_n \psi_n(\vec{r}) \tag{7.17}$$

Then, time dependence of $\psi_n(r', t)$ is

$$\psi_n(\vec{r}, t) = e^{-iHt/\hbar} \ \psi_n(\vec{r}) = e^{-iE_n t/\hbar} \ \psi_n(\vec{r}) \tag{7.18}$$

This is a stationary state and will remain in energy E_n since no transitions are induced.

It is to be noted that if the operator $\hat{O}$ is Hermitian, then $e^{iO} = \sum\limits_{n=0}^{\infty} \dfrac{(i)^n}{n!} \hat{O}^n$ is unitary. Unitary operator U has a property that

$$U^{-1} = \tilde{U}^* = U^\dagger \text{ and } UU^\dagger = 1$$

Clearly, $e^{iHt/\hbar}$ is a unitary operator. Thus a unitary transformation is,

$$\psi(t) = U(t)\psi(0) \tag{7.19}$$

Unitary transformation preserves the norm. Thus

$$\left\langle \psi(t) \left| \psi(t) \right\rangle = \left\langle \psi(0) \right| U^\dagger U \left| \psi(0) \right\rangle = \left\langle \psi(0) \left| \psi(0) \right\rangle \right.$$

and

$$UU^\dagger = 1 \tag{7.20}$$

Finally, the last part that we will need is Fermi's golden rule. If the Hamiltonian is such that

$$H = H_0 + H'(t)$$

where $H'(t)$ is "small" time dependent part, then it will induce transitions from a state $\left| m \right\rangle$ to $\left| n \right\rangle$ of H_0 at the rate of

$$w_{mn} = \frac{2\pi}{\hbar} \left| \left\langle m \left| H' \right| n \right\rangle \right|^2 \delta(E_m - E_n) \tag{7.21}$$

7.2   Quantum Statistical Ensemble

While discussing the classical system, we recognized the necessity of ensemble of many similar systems, with the phase point of each, on constant energy surface in its phase space.

In a system of large number of quantum particles, the information about the state is given by Schödinger wave function

$$\psi(q,t) = \psi(q_1, q_2.....q_f, t)$$

This $\psi(q,t)$ is equivalent to a phase point in Γ space for classical statistical system. For quantum ensemble, we consider many many identical systems with the same state function $\psi(q,t)$ whose time evolution given by

$$i\hbar\frac{\partial \psi(q,t)}{\partial t} = \hat{H}\psi(q,t) \tag{7.22}$$

This $\psi(q,t)$ can be expanded in a complete O.N. set

$$\psi(q,t) = \sum a_n(t)\phi_n(q)$$

It must be noted that all members of ensemble are independent. Therefore, we assume that measurement of one of the members of the ensemble is not going to affect other members. Now we make a drastic statement to describe the mixture as

$$
\begin{array}{lll}
N_1 & \text{members are in state} & \phi_1 \\
N_2 & \text{members are in state} & \phi_2 \\
\cdot & \quad " & \cdot \\
\cdot & \quad " & \cdot \\
\cdot & \quad " & \cdot \\
N_k & \text{members are in state} & \phi_k
\end{array}
$$

If there are N systems as members of the ensemble, then clearly,

$$\sum_{k=1} N_k = N \tag{7.23}$$

This way of looking at the ensemble imply that we are simultaneously making a same measurement on all the members of the ensemble. After the measurement is over we simply count, how many of the members collapse in various eigenstates.

Then $W_k = \dfrac{N_k}{N}$ with $\sum W_k = 1$ is probability of finding a state ϕ_k in the ensemble.

The average of a physical quantity $\hat{O}$ in a state $|k>$ is defined as

$$\hat{O}_{kk} = \left\langle k \left| \hat{O} \right| k \right\rangle = \int \phi_k^*(\vec{r})\ \hat{O}\ \phi_k(\vec{r})d^3r$$

The average of $\hat{O}_{kk}$ on ensemble is then given by

$$\langle \hat{O} \rangle = \sum \frac{N_j}{N} \langle j | \hat{O} | j \rangle = \sum W_j \hat{O}_{jj} \tag{7.24}$$

Effectively, we then regard the wave function ψ of the system as

$$\psi = \sum b_n \phi_n(q)$$

and the average value of an operator is,

$$\langle \hat{O} \rangle = \frac{\sum\limits_{n} |b_n|^2 \langle n | \hat{O} | n \rangle}{\sum\limits_{n} |b_n|^2}$$

A microcanonical ensemble is constructed as,

$$|b_n|^2 = \begin{cases} 1 & \text{if} \quad E < E_n < E + \delta E \\ \\ 0 & \text{otherwise} \end{cases} \tag{7.25}$$

The phases of complex numbers $\{b_n\}$ are random. This randomness means, the time averages of the cross products is zero. Let $b_n = |b_n| e^{i\delta_n}$. Then, denoting time average by a bar on top,

$$\overline{b_n{}^* b_m} = |b_n| |b_m| \overline{e^{i(\delta_m - \delta_n)}} = \delta_{mn} |b_n|^2 \tag{7.26}$$

Thus, the assumption of random phases come naturally for an ensemble system. This assumption means destruction of quantum coherence. Thus, the postulate of random phases implies that state of a system in equilibrium can be considered as an incoherent superposition of the eigenstates, or is a mixture of various eigenstates.

If there are Ω_k states accessible to the system, then in a usual langauge, each one of them have an equal *a priori* probability so long as corresponding E_k is within E and $E + \delta E$. Let total number of quantum states be $\Omega(E)$ in the energy range δE around energy E.

Then, the same microcanonical ensemble can be stated as

$$W_k = \begin{cases} \dfrac{\Omega_k}{\Omega} & \text{for} \quad E \leq E_k \leq E + \delta E \\ \\ 0 & \text{otherwise} \end{cases} \tag{7.27}$$

7.3 Density Matrix

We have considered N_j members in the state ϕ_j. We then see that

$$\langle \hat{O} \rangle = \sum_j \frac{N_j}{N} \langle j | \hat{O} | j \rangle = \sum_j W_j \hat{O}_{jj} \tag{7.28}$$

The quantity $< \hat{O} >$ has two averages - one due to quantum mechanics as given by $< j|\hat{O}|j >$ and an ensemble average of $< j|\hat{O}|j >$ as $\sum_j W_j \langle j|\hat{O}|j \rangle$. Clearly W_j is probability that the state $|j >$ is realized (i.e. $\hat{O}_{jj}$ is measured). W_j is called diagonal element of a density matrix $\rho_{jj} = W_{jj} = \sum_{n=1}^{j} |b_n|^2/N$.

We thus see that the quantum ensemble is an incoherent mixture of various states $\{\phi_n\}$ with corresponding realization probability $\{|b_n|^2\}$.

This can be formalized by introducing a density operator $\hat{\rho}$ as

$$\hat{\rho} = \sum_j W_j |j\rangle\langle j| \tag{7.29}$$

Clearly,

$$\begin{aligned}
\langle j|\hat{\rho}|j'\rangle &= \sum_j W_j \langle j|j\rangle \langle j|j'\rangle \\
&= W_j
\end{aligned}$$

This is a diagonal representation of the density operator.

Consider a following quantity

$$\mathrm{Tr}\left(\hat{\rho}\,\hat{O}\right) = \sum_k \left\langle k \left|\hat{\rho}\hat{O}\right| k\right\rangle \tag{7.30}$$

But

$$\sum_l |l\rangle\langle l| = 1$$

Thus,

$$\begin{aligned}
\mathrm{Tr}\left(\hat{\rho}\,\hat{O}\right) &= \sum_k \sum_l \langle k|\hat{\rho}|l\rangle \left\langle l\left|\hat{O}\right| k\right\rangle \\
&= \sum_k \sum_l W_k \delta_{kl} O_{lk} = \sum_k W_k O_{kk} = \langle O \rangle
\end{aligned}$$

If the density operator is not normalized, then,

$$\left\langle \hat{O} \right\rangle = \frac{\mathrm{Tr}\left(\hat{\rho}\hat{O}\right)}{\mathrm{Tr}\left(\hat{\rho}\right)} \tag{7.31}$$

The form of density matrix we gave, viz $\rho = \sum_j W_j |j><j|$ is a diagonal form. Its diagonal elements are $\{W_j\}$. W_j is a probability that a state $|j >$ is realized upon measurement on ensemble. Clearly,

$$W_j \geq 0 \quad \forall j$$
$$\sum_j W_j = 1$$

In general, we can have a representation of ρ such that it is non diagonal.

If all but one of the W_j are zero, then we say that the system is in pure state; otherwise, it is in a mixed state. This means, a state, which can interfere with itself is a 'pure state', others are simply mixture or a 'mixed state'.

Thus for a pure state,

$$\rho = |j_{pure}\rangle \langle j_{pure}| \tag{7.32}$$

For a pure state,

$$\begin{aligned}
\rho^2 &= |j_{pure}\rangle \langle j_{pure}|j_{pure}\rangle \langle j_{pure}| \\
&= |j_{pure}\rangle \langle j_{pure}| \\
&= \rho
\end{aligned}$$

We can easily see that the necessary and sufficient condition for ρ of a pure state is $\rho^2 = \rho$.

Density matrix can be written in terms of wave functions. Thus,

$$\begin{aligned}
\rho(x',x) &= \sum_j W_j \langle x'|j\rangle \langle j|x\rangle \\
&= \sum_j W_j \phi_j(x')\phi_j^*(x) \tag{7.33}
\end{aligned}$$

In this representation, we can write averages as

$$\begin{aligned}
\langle \hat{O} \rangle &= \mathrm{Tr}\, \rho\, \hat{O} = \int d^3x \left\langle \vec{x} \left| \rho\, \hat{O} \right| \vec{x} \right\rangle \\
&= \int\int d^3x d^3x' \langle \vec{x}|\rho|\vec{x}'\rangle \left\langle \vec{x}' \left| \hat{O} \right| \vec{x} \right\rangle \\
&= \int \rho(x,x')\hat{O}(x',x) d^3x' d^3x \tag{7.34}
\end{aligned}$$

We usually encounter a situation where we study a small system X, interacting with a large system Y called reservoir. The two together form an isolated system. Let the coordinates of X be denoted by x and that of Y by y. The total system $X \oplus Y$ is isolated. Then the state ψ of the total system is a function of x and y. Thus $\psi(x,y) = <x| <y|\psi>$. To get the density matrix of the system X we will have to integrate out y coordinates. Thus,

$$\rho(x,x') = \int \psi(x,y)\psi^*(x',y)dy$$

and we will again arrive

$$\langle \hat{O} \rangle = \int \rho(xx')\hat{O}(x',x)\, dx'dx$$

Illustrative Problem

Problem 1. Consider a spin 1/2 system with a pure state

$$|\alpha\rangle = \begin{pmatrix} 1/\sqrt{2} \\ 1/\sqrt{2} \end{pmatrix} = 1/\sqrt{2} \begin{pmatrix} 1 \\ 0 \end{pmatrix} + 1/\sqrt{2} \begin{pmatrix} 0 \\ 1 \end{pmatrix}$$

a) Find corresponding density matrix.

b) Find density matrix for 'up' spin state.

c) Find density matrix for a mixture of 50% up spin and 50% down spin.

Solution:

a) $\rho = |\alpha\rangle \langle\alpha| = \begin{pmatrix} 1/\sqrt{2} \\ 1/\sqrt{2} \end{pmatrix} \begin{pmatrix} 1/\sqrt{2} & 1/\sqrt{2} \end{pmatrix} = \begin{pmatrix} 1/2 & 1/2 \\ 1/2 & 1/2 \end{pmatrix}$

b) For up spin state, $|\alpha\rangle = \begin{pmatrix} 1 \\ 0 \end{pmatrix}$ and $\rho = \begin{pmatrix} 1 \\ 0 \end{pmatrix} \begin{pmatrix} 1 & 0 \end{pmatrix} = \begin{pmatrix} 1 & 0 \\ 0 & 0 \end{pmatrix}$

c) For the mixture,

$$\rho = \frac{1}{2}\rho_{up} + \frac{1}{2}\rho_{down} = \begin{pmatrix} 1/2 & 0 \\ 0 & 1/2 \end{pmatrix}$$

q.e.d.

Density operator satisfies the equation of motion as

$$i\hbar \frac{\partial \hat{\rho}}{\partial t} = - \left[\hat{\rho}, \hat{H} \right] \tag{7.35}$$

where $[\hat{\rho}, \hat{H}]$ is a commutation bracket. This should be compared with Liouville's theorem in classical statistical mechanics. There ρ is a classical phase space density and $\left(-\frac{1}{i\hbar} \right) \left[\hat{\rho}, \hat{H} \right]$ is equivalent to Poisson bracket. This equation of motion for the operator $\hat{\rho}$ has a negative sign as compared to a general operator evolution equation of Heisenberg.

In terms of the density operator, we write $\Omega(E)$ total number of states between E and $E + \delta E$ of microcanonical ensemble as

$$\Omega(E) = \text{Tr}\hat{\rho} \tag{7.36}$$

where the density operator is defined as,

$$\hat{\rho} = \sum_{E \leq E_k \leq E+\delta E} W_k |k\rangle \langle k|$$

Then the entropy is

$$S = k_B \ln \Omega(E) \tag{7.37}$$

7.4   Ideal Quantum Gases – Microcanonical Ensemble

A collection of N electrons form N identical particles. But N electrons and N positrons do not form $2N$ identical particles. Similarly N protons will form a collection of N identical particles. Without going in to a very precise meaning of what identical particles are, we will understand them notionally.

Nature has made two types of identical particles. Those that obey Pauli principle, have a half integer spin. They have antisymmetric wave function ψ upon exchange of two particles of this group (i.e. it changes sign of ψ). Major consequence of this is that a single particle state $|\vec{p}\rangle$ is occupied by either none of the particle or only one particle. Such particles are called Fermions. Theoretically, one can consider spinless Fermions as those with above property. In such case, spin interactions are unimportant. The other types of identical particles in nature are those with integer intrinsic spin. The wave function of the system of such particles is symmetric upon the exchange of two such particles (sign does not change). Consequence of this is that a state $|\vec{p}\rangle$ can accommodate any number of these type of particles. Such particles are known as Bosons.

We consider N identical noninteracting quantum particles confined in a volume V.

Accordingly we have to take into account Fermionic or Bosonic nature of particles while evaluating $\Omega(E)$ in microcanonical ensemble. This is clearly a counting exercise.

The Hamiltonian of N particle system is

$$\hat{H} = \sum_{i=1}^{N} \frac{p_i^2}{2m} = -\frac{\hbar^2}{2m} \sum_{i=1}^{N} \nabla_i^2$$

Since the particles are noninteracting, the energy eigenvalue will be sum of single particle energy levels $p^2/2m$. Since the particles are confined in volume V, $\vec{p}$ is quantized.

$$\vec{p} = \hbar \left\{ \hat{i} \frac{n_x}{L_x} + \hat{j} \frac{n_y}{L_y} + \hat{k} \frac{n_z}{L_z} \right\}$$

where L_x, L_y and L_z are lengths of the sides of V. Clearly, as V increases, $\vec{p}$ takes continuum values. We now specify state of entire ideal system by the occupancy of a single particle state $|\vec{p}\rangle$. Thus, there may be n_1 number of particles occupying a state $|\vec{p_1}\rangle$, n_2 number of particles occupying state $|\vec{p_2}\rangle$,..... n_p number of particles occupying state $|\vec{p_n}\rangle$ and so on. Such a state is obviously denoted in Dirac notation as,

$$\left| \psi_{n_1, n_2, \ldots n_p \ldots} \right\rangle = \left| n_1, n_2, \ldots n_p \ldots \right\rangle$$

Technically, such a state is called occupation number state or Fock state. This state is equivalent to a Schrödinger wavefunction. We ask our reader to refer to a standard book on quantum mechanics for this kind of equivalence.

The occupancy n_p is such that

$$
n_p = \begin{cases} 0, 1 & (Fermions) \\ 0, 1, 2, \ldots & (Bosons) \end{cases}
$$

Let the single particle energy be denoted by ϵ_p as

$$\epsilon_p = p^2/2m = \hbar^2 k^2/2m$$

Then the total energy will be

$$E = \sum_p \epsilon_p n_p$$

and total number of particles as,

$$N = \sum_p n_p$$

Let g_1, g_2, $g_3 \ldots .g_p \ldots$ be the degeneracy of the levels with momenta $\vec{p_1}$, $\vec{p_2} \ldots .. \ \vec{p} \ldots$etc. Here degeneracy means number of levels which may or may not be occupied but are technically available for occupancy. We call these degeneracies as a cell. Let the g_i cell have average energy ϵ_i and let its occupation number be denoted by n_i. We have the values of $\vec{p_1}$, $\vec{p_2} \ldots$ etc chosen to be such that they give total energy E in the shell of δE for our isolated system represented by the microcanonical ensemble. Let $W\{n_i\}$ be number of states of the system corresponding to the set of occupation number $\{n_i\}$. We emphasize again, the occupancy n_i of g_i is not fixed.

There are many possible combinations n_i. Only requirements are,

$$\sum \epsilon_i n_i = E$$

and

$$\sum n_i = N$$

Clearly E can remain fixed for various values of n_i and ϵ_i and so is N. We have to consider all such possibilities and each possibility is a compatible state. All such compatible states, when added together, give $\Omega(E)$. Thus

$$\Omega(E) = \sum_{\{n_i\}} W\{n_i\}$$

Let w_i be the number of ways of putting n_i particles in g_i levels of ith cell. Let w_j be a similar number for the jth cell. Then total number of ways of putting $(n_i + n_j)$ particles in i^{th} and j^{th} cell is $w_i w_j$. For entire system,

$$W\{n_i\} = w_i w_j ....$$
$$= \prod_j w_j$$

Thus, the problem now is reduced to counting of w_j for Bose and Fermi system.

We first consider Bose system.

Bose System: Here we fix our attention to i^{th} cell with degeneracy g_i. Any number of particles can occupy i^{th} cell. Let us partition the cell in to g_i subshells as shown in figure (1).

Let cross denote number of particles in each subshell. Total number of crosses gives n_i. If we permute(exchange) the particles within a subshell or the subshells themselves, then number of distinct possibilities are

$$w_i = \frac{(n_i + g_i - 1)!}{n_i!(g_i - 1)!}$$

Then

$$W\{n_i\} = \prod_i \frac{(n_i + g_i - 1)!}{n_i!(g_i - 1)!} \qquad (Bose) \tag{7.38}$$

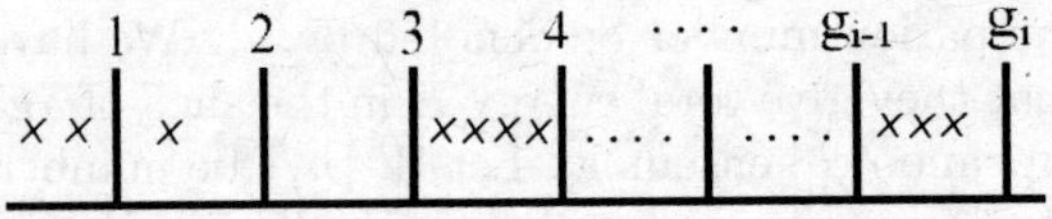

Figure 7.1: One possible Bosonic partition of g_i cell.

Fermi System: Here g_i cell is occupied by 0 or 1. Then w_i is number of ways in which n_i identical particles can be chosen from g_i subshells. Thus,

$$W_i = \prod_i \frac{g_i!}{n_i!(g_i - n_i)!} \qquad (Fermi) \tag{7.39}$$

Getting $\Omega(E) = \sum_{\{n_i\}} W\{n_i\}$ is a formidable task. We want to find entropy S such that

$$S = k_B \ln \Omega(E) = k_B \ln \left(\sum_{\{n_i\}} W\{n_i\} \right). \tag{7.40}$$

We will approximate it with a set $\{\overline{n_i}\}$. We will thus write

$$\sum_{\{n_i\}} W\{n_i\} = W\{\overline{n_i}\} \tag{7.41}$$

where a set $\{\overline{n_i}\}$ maximizes W_i subject to the condition that E and N remain constant. This is done best by the Lagrange's method of undetermined multipliers[2].

We will perform this job in case of Bose system and Fermi system is left as an exercise. We will use

$$\ln N! = N \ln N - N \quad for \ N >> 1$$

We write (with $n_i >> 1$),

$$\ln W_i = \ln \ (n_i + g_i - 1)! - \ln n_i! - \ln(g_i - 1)!$$

$$\approx (n_i + g_i)\ln(n_i + g_i) - (n_i + g_i) - n_i \ln n_i + n_i - (g_i - 1)\ln(g_i - 1) + g_i - 1.$$

We maximize $\ln W_i$ subject to the condition

$$N = \sum n_i$$

and

$$E = \sum n_i \epsilon_i \ .$$

Let α and β be two Lagrange multipliers for N and E respectively. Then the functional whose maximum is to be found by Lagrange's method is,

$$\begin{aligned}
F\,[n_i, g_i] \ = \ & (n_i + g_i)\ln(n_i + g_i) \\
- \ & (n_i + g_i) - n_i \ln n_i + n_i \\
- \ & (g_i - 1)\ln(g_i - 1) + g_i - 1 \\
- \ & \alpha n_i - \beta n_i \epsilon_i \ .
\end{aligned}$$

Thus

$$\frac{\delta F}{\delta n_i} = 0 = 1 + \ln(\overline{n_i} + g_i) - 1 - \ln \overline{n_i} - \alpha - \beta \epsilon_i \ . \tag{7.42}$$

or

$$\ln \left(1 + \frac{g_i}{n_i} \right) = \alpha + \beta \epsilon_i$$

or

$$\overline{n_i} = \frac{g_i}{e^{\beta \epsilon_i + \alpha} - 1} \quad \text{(Bose)} \tag{7.43}$$

For Fermi case,

$$\overline{n_i} = \frac{g_i}{e^{\beta \epsilon_i + \alpha} + 1} \quad \text{(Fermi)} \tag{7.44}$$

---

[2]See Classical Mechanics by P.V.Panat p-100 Ch-9 (Narosa Publishers)

The Lagrange's undetermined multipliers are determined by two equations

$$N = \sum_p \frac{g_p}{e^{\beta \epsilon_p + \alpha} \pm 1} \tag{7.45}$$

and

$$E = \sum_p \frac{g_p \epsilon_p}{e^{\beta \epsilon_p + \alpha} \pm 1} \tag{7.46}$$

Since E and N are given, the two undetermined multipliers α and β can be found.

Subsequently, it will be shown that, $e^{\alpha} = e^{-\beta \mu}$ where μ is chemical potential and $\beta = 1/k_B T$. Thus, Fermi-Dirac and Bose-Einstein distributions are

$$
\begin{aligned}
n_{\vec{p}} &= \frac{g_p}{e^{(\epsilon_p - \mu)/k_B T} + 1} \quad \text{(F-D)} \\
&= \frac{g_p}{e^{(\epsilon_p - \mu)/k_B T} - 1} \quad \text{(B-E)}
\end{aligned}
\tag{7.47}
$$

7.5 Maxwell-Boltzmann (M-B) Statistics

This statistics can be obtained as an asymptotic case of Fermi or Bose statistics. It is sometimes called Boltzmann statistics also. Here the gas is so rare that the particles could be distinguished and therefore labeled. This means de-Broglie wavelength of each particle scarcely overlap over each other. At large T, the chemical potential of an ideal gas is large and negative. Under this situation, in the denominator of F-D or B-E distribution one is negligible. Thus,

$$\overline{n_p} \approx g_p e^{-\beta(\epsilon_p - \mu)} \tag{7.48}$$

which is Boltzmann distribution. The states available for Boltzmann statistics are many more than Bose-Einstein case. This is so because, two identical particles arranged as AB and BA are counted as two states, whereas this is counted as a single state in B-E or F-D case. Formally, the quantity to be maximized for getting Boltzmann statistics subject to total energy E and total number of particles N to be constant is,

$$W\{n_i\} = \prod_i \frac{g_i^{n_i}}{n_i!}$$

7.6 Entropy of Various Statistics

To get the entropy, we have to find

$$
\begin{aligned}
S &= k_B \ln \Omega(E) \\
&= k_B \ln \sum_{\{n_i\}} W\{n_i\} \\
&\approx k_B \ln W\{\overline{n_i}\}
\end{aligned}
\tag{7.49}
$$

Using Sterling approximation, $W\{\overline{n_i}\}$ for Bosons is

$$W(\overline{n_i}) = \prod_i \frac{(\overline{n_i} + g_i - 1)!}{\overline{n_i}!(g_i - 1)!} \quad \text{(B-E)}$$

Then

$$\begin{aligned}
S &= k_B \ln W\{\overline{n_i}\} \\
&= k_B \sum_p \left[\overline{n}_p \ln\left(1 + \frac{g_p}{\overline{n}_p}\right) + g_p \ln\left(1 - \frac{\overline{n}_p}{g_p}\right) \right] \\
&= k_B \sum_p g_p \left[\frac{\ln e^{\beta(\epsilon_p - \mu)}}{e^{\beta(\epsilon_p - \mu)} - 1} - \ln\left(1 - e^{-\beta(\epsilon_p - \mu)}\right) \right] \quad \text{(B-E)} \quad (7.50)
\end{aligned}$$

For Fermi-Dirac statistics,

$$W\{\overline{n}_i\} = \prod_i \frac{g_i!}{\overline{n}_i!(g_i - \overline{n}_i)!} \quad \text{(F-D)}$$

Then

$$\begin{aligned}
S &= k_B \ln \Omega(E) \simeq k_B \ln W\{\bar{n}_i\} \\
&= k_B \sum_p \left[\overline{n}_p \ln\left(\frac{g_p}{\overline{n}_p} - 1\right) - g_p \ln\left(1 - \frac{\overline{n}_p}{g_p}\right) \right] \\
&= k_B \sum_p g_p \left[\frac{\beta(\epsilon_p - \mu)}{e^{\beta(\epsilon_p - \mu)} + 1} + \ln\left(1 + e^{-\beta(\epsilon_p - \mu)}\right) \right] \quad \text{(F-D)} \quad (7.51)
\end{aligned}$$

For Boltzmann statistics

$$\begin{aligned}
S &= k_B \ln \Omega(E) \\
&\cong k_B \sum_i \left(\overline{n}_i \ln g_i - \overline{n}_i \ln \overline{n}_i + \overline{n}_i \right) \\
&\approx k_B e^{\beta\mu} \sum_p g_p \, \beta \, e^{-\beta\epsilon_p}(\epsilon_p - \mu) \quad \text{(M-B)} \quad (7.52)
\end{aligned}$$

We thus see that, entropy expressions are immediate, if microcanonical ensemble is used.

Short Questions

1. Distinguish between pure state and a mixed state.

2. Define microcanonical ensemble for many particle quantum system. Define entropy in that.

3. Explain as to why there is a double averaging of a physical quantity in quantum statistics.

4. Show that $\rho^2 = \rho$ is necessary and sufficient condition for a pure state.

5. Obtain Boltzmann distribution as an asymptotic state or a classical limit of both Bose-Einstein ad Fermi-Dirac distribution.

6. Explain- 'For F-D statistics, the number of available states are always more than the number of particles of the system'.

Problems

Problem 1. Consider a gas of three particles with four available states. Find

a) Number of states available if the gas was (i) M-B (ii) F-D (iii) B-E.

Ans: (i) 64 (ii) 4 (iii) 20

b) Find the probability that there are three particles in a single state in case of (i) M-B (ii) F-D (iii) B-E.

Ans: (i) 4/64 (ii) 0/4 (iii) 4/20

c) Find the probability that there are two particles in a single state in case of (i) M-B (ii) F-D (iii) B-E.

Ans: (i) 36/64 (ii) 0/4 (iii) 12/20

d) Find the probability that there is a single particle in each state in case of (i) M-B (ii) F-D (iii) B-E.

Ans: (i) 24/64 (ii) 4/4 (iii) 4/20

(Hint: $g = 4, n = 3$ and B-E, F-D occupancy formula can be used. Make a table and count the states to get answer)

Problem 2. Consider a beam of light traveling in the z direction. Let $\begin{pmatrix} 1 \\ 0 \end{pmatrix}$ be a state of x polarized light and $\begin{pmatrix} 0 \\ 1 \end{pmatrix}$ be a y polarized light. Let the pure state be $\begin{pmatrix} a \\ b \end{pmatrix}$ with $|a|^2 + |b|^2 = 1$

a) Write density matrix of the pure state.

$$\text{Ans: } \rho = \begin{pmatrix} |a|^2 & ab^* \\ ba^* & |b|^2 \end{pmatrix}$$

b) Density matrix for 45° polarized state (Write $a = \frac{1}{\sqrt{2}}, b = \frac{1}{\sqrt{2}}$)

$$\text{Ans: } \rho = 1/2 \begin{pmatrix} 1 & 1 \\ 1 & 1 \end{pmatrix}$$

c) Mixture of 50% x polarized and 50% y polarized light

$$\text{Ans: } \rho = \left(\tfrac{1}{2}\right)\rho_x + \left(\tfrac{1}{2}\right)\rho_y = \begin{pmatrix} 1/2 & 0 \\ 0 & 1/2 \end{pmatrix}$$

d) Mixture of 50% 45° polarized and 50% 135° polarized light

$$\text{Ans: } \rho = \begin{pmatrix} 1/2 & 0 \\ 0 & 1/2 \end{pmatrix}$$

Problem 3. (i) Find $\langle x \rangle$ in terms of $\rho(x, x')$

$$\text{Ans: } \langle x \rangle = \int x \, dx \, \rho(x, x)$$

(ii) Find $\langle p \rangle$ in terms of $\rho(x, x')$

$$\text{Ans: } \langle p \rangle = i\hbar \int dx \left[\tfrac{\partial}{\partial x} \rho(xx') \right]_{x=x'}$$

Chapter 8

CANONICAL AND GRAND CANONICAL ENSEMBLES

In the previous two chapters, we saw how the thermodynamic quantity entropy S is related with microscopic states via microcanonical ensemble. Even though in principle we can calculate the thermodynamics using microcanonical ensemble, it is impractical because of difficulty in counting the states under constraints. Moreover, we never have an isolated system in practice. Consider a system under study. This system is embedded in a larger system. The larger system is called reservoir. Then there are three possibilities. (i) The system and reservoir are in contact such that the system and reservoir exchange energy only but not the matter. (ii) The system and reservoir exchange both, energy and matter. (iii) The system and reservoir interact so that the energy is exchanged and the volumes are adjusted but no exchange of matter. The two - namely reservoir and system form an isolated system with its total energy constant. For this combined system, we apply microcanonical ensemble. We then obtain the density matrix for the system of interest. Once this is obtained, we can get the properties of the system in a straight forward manner. We re-derive F-D and B-E distributions without factorials.

8.1   Canonical Ensemble

We consider a small system A in contact with a heat reservoir A' and weakly interacting with it. We are interested in the thermodynamic properties of A. $A << A'$ which means A' has many many more accessible states than those for A. Even though the system A is small, still it is macroscopic. Total system $A_t = A + A'$ is isolated with total energy E_t. Let T be the temperature of the reservoir. It is so big that even if a small amount of heat is removed or added in it, its temperature does not change. Since A_t is isolated we can apply ideas of microcanonical ensemble. Let the microstate of A be $|r\rangle$ with energy E_r. We are here picking up one particular independent microstate out of many microstates corresponding to the same energy. Since A and A' interact weakly, the energies are additive. Thus

$$E_t = E_r + E'$$

Since total number of states accessible to A' are very large, and the system A is in state E_r, total number of states accessible to A_t is $\Omega'(E') = \Omega'(E_t - E_r)$. All these states are spread around E_t in the small band of δE. Probability P_r that the system A is in state E_r is proportional to $\Omega'(E_t - E_r)$. Thus,

$$P_r = \frac{\Omega'(E_t - E_r)}{\sum_r \Omega'(E_t - E_r)} \tag{8.1}$$

Ω' is a huge number proportional to $E_t{}^f$. It is a very sensitive function. $\ln \Omega'$ is relatively slowly varying function. We expand

$$\ln \Omega'(E_t - E_r) = \ln \Omega'(E_t) - E_r \left(\frac{\partial \ln \Omega'}{\partial E'}\right)_0 \, ... \tag{8.2}$$

By our choice, A' is a heat reservoir at temperature T and $\dfrac{\partial \ln \Omega'}{\partial E'} = \dfrac{1}{k_B T} = \beta$. Unless otherwise stated we will use the notation $\beta = 1/k_B T$ from now onwards. Thus

$$\Omega'(E_t - E_r) = \Omega'(E_t) e^{-\beta E_r} \tag{8.3}$$

Then clearly,

$$P_r = \frac{e^{-\beta E_r}}{\sum_r e^{-\beta E_r}} \tag{8.4}$$

Equation(4) is called canonical distribution.

The result is the probability of the system to be in a particular state $|r\rangle$. But because of high degeneracy, we have many independent quantum states with same energy and are lying around E_t in the range δE. Let $\Omega(E)$ be such total number of states of the system A with energy E. Then, clearly, the probability that the system A has an energy E is

$$P(E) = C\Omega(E) e^{-\beta E} \tag{8.5}$$

This is a very general result. It is also very interesting. $\Omega(E)$ is a fast increasing function of energy E and $e^{-\beta E}$ is a fast decreasing function of E. The product of the two has a very sharp maximum as shown in figure (1). Let it be at $\tilde{E}$. The width of this maximum decreases as $\frac{1}{\sqrt{N}}$ when $N >> 1$. It turns out that $\tilde{E}$ is same as average energy $\langle E \rangle$ of system A.

The quantity $Z = \sum_r e^{-\beta E_r}$ is called the partition function, which is nothing but *sum over states*. Once the partition function is evaluated, equilibrium physical quantities of interest and related with our macroscopic system A can be calculated.

If we have a quantum system, then, we straightway take a result (4) for P_r. It is possible to do so because, in its derivation, there are no specific classical or quantum ideas used. Only a fact of conservation of energy, energy additivity (due to weak interaction

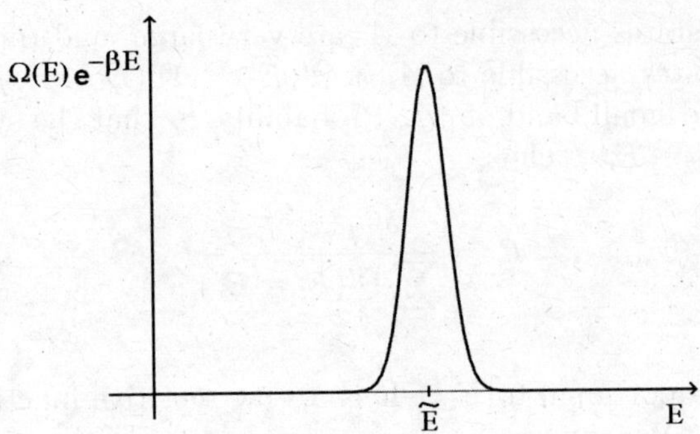

Figure 8.1: Probability density for energy at temperature T.

between A and A') and smoothness of $\ln \Omega(E)$ are used. Thus, if $\hat{O}$ is a physical quantity of A, then, using equation (4)

$$\langle O \rangle = \frac{\sum_k \left\langle k \left| \hat{O} \right| k \right\rangle e^{-\beta E_k}}{\sum_k e^{-\beta E_k}} \tag{8.6}$$

This result is so important that Feynman calls this expression as the *Summit of statistical mechanics*. He also writes "entire subject is either the slide down from the summit, as the principle is applied to various cases, or the climb-up to where the fundamental law is derived and the concepts of thermal equilibrium and temperature T are clarified."

If we compare equation (6) with equation (7.29) of density matrix, we can write

$$\begin{aligned} \hat{\rho} &= \sum_n |n\rangle\, e^{-\beta E_n}\, \langle n| \\ &= e^{-\beta H} \sum_n |n\rangle \langle n| \\ &= e^{-\beta H} \end{aligned} \tag{8.7}$$

Clearly, the density matrix in canonical ensemble is

$$\hat{\rho} = e^{-\beta H} \tag{8.8}$$

and

$$Z = \text{Tr}\,\hat{\rho} = \sum_n e^{-\beta E_n} \tag{8.9}$$

Thus, canonical ensemble for quantum statistics gives

$$\left\langle \hat{O} \right\rangle = \frac{\text{Tr}\,\hat{\rho}\hat{O}}{\text{Tr}\,\hat{\rho}} = \frac{\sum_k \left\langle k \left| \hat{O} \right| k \right\rangle e^{-\beta E_k}}{\sum_k e^{-\beta E_k}} \tag{8.10}$$

If E is continuous, let $\rho(E)$ be the density of states. Then, we write partition function as,

$$Z(\beta) = \int_0^\infty \rho(E)\, dE\, e^{-\beta E}. \tag{8.11}$$

Also,

$$P(E) = \frac{\rho(E)e^{-\beta E}}{Z} \tag{8.12}$$

Equation (11) clearly shows that $Z(\beta)$ is a Laplace transform of density of states. Laplace transform can be inverted to get the density of states (Bromwitch integral)

$$\hat{\rho}(E) = \frac{1}{2\pi i}\int_{\beta'-i\infty}^{\beta'+i\infty} Z(\beta)e^{\beta E}\, d\beta \qquad (\beta' > 0) \tag{8.13}$$

$\hat{\rho}(\beta) = e^{-\beta H}$ is an unnormalized density matrix. We will obtain its equation of motion, useful in many body physics.

$$-\frac{\partial\hat{\rho}}{\partial\beta} = \hat{H}e^{-\beta H} = \hat{H}\hat{\rho} \tag{8.14}$$

This is known as Bloch's equation.

We can compare this with Schrödinger equation as

$$i\hbar\,\frac{\partial|\psi\rangle}{\partial t} = \hat{H}|\psi\rangle = -\frac{\partial|\psi\rangle}{\partial(i\,t/\hbar)} = H|\psi\rangle$$

We thus see that the part played by β is same as that by $(it/\hbar)$. This fact enables us to use the results of Euclidean field theory. Initial condition imposed on the solution of equation (14) is

$$\rho(0) = \rho(T = \infty) = 1 \tag{8.15}$$

8.2　Partition Function and Thermodynamics

(i) Average Energy - the Internal Energy

It is to be noted that the particles of a macroscopic system has a distribution of energy. The average energy of the macroscopic system is the internal energy of it. Thus

$$\langle E\rangle = \frac{\sum_r P_r E_r}{\sum_r P_r}$$

$$= \frac{\sum_r E_r e^{-\beta E_r}}{Z}$$

$$= -\frac{\partial}{\partial\beta}\ln Z \tag{8.16}$$

From now on we will write internal energy as E. Thus, average energy can be found and so also the specific heat with the knowledge of Z.

(ii) Pressure

Suppose that the system has external parameters $X_1, X_2, X_3....X_k$. Let the parameters change by an infinitesimal amount $X_i \rightarrow X_i + dX_i$. Corresponding change in energy is

$$dE = \sum \frac{\partial E}{\partial X_i} dX_i$$

This must be equal to the work done by the system. The generalized force in this case is $X_i = \partial E/\partial X_i$. When the system has volume as an external parameter, then, clearly, $\frac{\partial E}{\partial V} = -\bar{P}$ the average pressure or simply the pressure. Thus

$$\bar{P} = \frac{-\sum P_r \frac{\partial E_r}{\partial V}}{Z}$$
$$= \frac{1}{\beta} \frac{\partial \ln Z}{\partial V} \tag{8.17}$$

Obviously the generalized force is

$$\bar{X} = \frac{1}{\beta} \frac{\partial \ln Z}{\partial X}$$

(iii) Entropy

Clearly, $Z = Z(\beta; X_1, X_2, ...X_k)$ and for a standard gas, $Z = Z(\beta, V)$
Thus,

$$d(\ln Z) = \frac{\partial (\ln Z)}{\partial \beta}.d\beta + \frac{\partial \ln Z}{\partial V}.dV$$
$$= -E\,d(\frac{1}{k_B T}) + \beta\, P\, dV$$
$$= -d(E\beta) + \beta\, dE + \beta\, P\, dV$$

Or,

$$d(\beta E + \ln Z) = \frac{dE}{k_B T} + \frac{PdV}{k_B T}$$

We then immediately identify,

$$S = k_B(\beta E + \ln Z) \tag{8.18}$$

(iv) Free Energy $F = E - TS$

From the expression of entropy, we can easily see,

$$Z = e^{-\beta F} = \sum_r e^{-\beta E_r}$$

or

$$F = -k_B T \ln Z = -k_B T \ ln \ (\text{Tr}\hat{\rho}) \tag{8.19}$$

8.3 Grand Canonical Distribution

For this distribution, the system A and A' are such that A' is a reservoir of temperature T. A is a small system of our interest. A is weakly interacting with A' and can exchange both matter (i.e. number of its particles) and also the energy. The distribution function or the required probability now depends not only on energy state E_r but also on number N of particles of A. We then denote energy levels as E_{rN} to show their explicit dependence on N. The probability that A is in a state r and has N particle is indicated by P_{rN}. We recall the derivation of Canonical distribution. Entropy of A' is $S' = S(E', N')$. Let N_0 be total number of particles of A and A'. Then, clearly, $E' = E_t - E_{rN}$, $N' = N_0 - N$. The probability P_{rN} is then,

$$P_{rN} = \ \text{constant} \ \times e^{S'(E_t - E_{rN}, N_0 - N)/k_B}$$

Like in canonical case, we expand entropy as

$$S'(E_t - E_{rN}, N_0 - N) \ = \ S'(E_t, N_0) - E_{rN}\left(\frac{\partial S'}{\partial E'}\right)_N - N\left(\frac{\partial S'}{\partial N'}\right)_E + ...$$

$$= \ S'(E_t, N_0) - \frac{E_{rN}}{T} + \frac{\mu N}{T}$$

Thus,

$$P_{rN} = Ce^{-\dfrac{E_{rN} - \mu N}{k_B T}} \tag{8.20}$$

where C is constant. Here, we used a thermodynamic relation

$$TdS = dE + PdV - \mu \ dN = k_B T \ d(\ln \Omega)$$

Equation (20) is a principle result of Grand Canonical ensemble.

Clearly, the normalized probability is,

$$P_{rN} = \frac{e^{-\beta(E_{rN} - \mu N)}}{\displaystyle\sum_r \sum_N e^{-\beta(E_{rN} - \mu N)}} \tag{8.21}$$

We now connect P_{rN} to entropy. Recall, the result of an illustrative problem (1) in chapter(6) as

$$S \ = \ -k_B \sum P_r \ln P_r$$

$$= \ -k_B \overline{\ln P_r} \tag{8.22}$$

where $\overline{\ln P_r}$ is average of $\ln P_r$. It will be generalized here that

$$S \ = \ -k_B \overline{\ln P_{rN}}$$

$$= \ -k_B \ln C - \frac{\mu \bar{N}}{T} + \frac{\bar{E}}{T}$$

or

$$k_B \ln C = \bar{E} - TS - \mu \bar{N} \tag{8.23}$$

But from thermodynamics, $\mu N = G$ the Gibbs free energy. Also $G = E - TS + PV = \mu N$.

Thus, writing $k_B T \ln C$ by Ω, the grand potential (Don't confuse this Ω with number of accessible states), we get

$$P_{rN} = e^{\beta(\Omega + \mu N - E_{rN})} \tag{8.24}$$

Equation (24) is a Gibbs distribution for variable number of particles and energy. Clearly, $\Omega = -PV$.

Thus the evaluation of grand potential directly leads to a pressure exerted by the system. Clearly, the normalization of P_{rN} requires sum over all quantum states for a given N and then over all values of N. Thus,

$$\sum_N \sum_r P_{rN} = e^{\beta \Omega} \sum_N \sum_r e^{-\beta(E_{rN} - \mu N)} = 1$$

or, the thermodynamic grand potential Ω is,

$$\Omega = -PV = -k_B T \ln \left[\sum_N e^{\beta \mu N} \sum_N e^{-\beta E_{rN}} \right] \tag{8.25}$$

8.4 Remarks: Microcanonical, Canonical & Grand Canonical Distributions

Following point are worth pondering.

(i) Canonical and grand canonical distributions are completely identical for all thermodynamic properties apart from total number of particles. Clearly, if we ignore fluctuations in N (i.e. hold N constant) then, $\Omega + \mu N = E - TS = F$ and distribution (24) coincides with canonical one given by equation (4)

(ii) Relation between canonical and microcanonical distribution is also similar. If we ignore fluctuations in energy and fix it in some very narrow range of energy, the two distributions are identical. This is seen in the figure (1) where probability density has a very sharp maximum which can be approximated as constant at energy $\bar{E}$.

The statement of equivalence of microcanonical and canonical ensemble, and of canonical and grand canonical ensemble, when we ignore fluctuations in energy and the number of particles respectively in true only in a restricted sense. This equivalence is possible if the energy is additive. Recall that in the derivation of canonical and grand canonical ensemble, we always have a system of interest to be very small as compared with the larger system (reservoir). The two are weakly interacting so that $E_t = E_A + E_B$. This happens in most of the cases of physical interest. There is a class of problems where the energy additivity is invalid. This happens in case of long range interactions like gravity. You may question, why not e.m. forces? It is true that the Coulomb interaction is long range. But there are two types of charges - positive and negative, with a strict electrical neutrality. This leads to the screening of Coulomb interaction making it effectively short ranged. With a short range interaction, energy additivity is possible. But gravity cannot be screened and remains always long range and attractive. Therefore we can not isolate two parts which are noninteracting. Then in that case the derivation of $P_r \propto e^{-\beta E_r}$ falls flat. We cannot then apply canonical as well as grand canonical probability distributions to the gravitational systems. As a matter of fact, if we apply the canonical and grand canonical ensembles to the systems with gravitational interactions between the particles, we get an unphysical result where $C_V < 0$. From the problem 2 of this chapter it is seen that the dispersion in energy $\overline{E^2} - \bar{E}^2 = K_B T^2 C_V$. This gives $C_V > 0$. To see how we get the negative specific heat for a gravitational system we use the Virial theorem[1]. It is an exact result valid for either periodic systems and/or bounded systems. If the potential energy of interaction between the particles is $V(r) = ar^{n+1}$, then we have a standard result

$$\overline{\text{K.E.}} = \left(\frac{n+1}{2}\right) \overline{\text{potential energy}} \tag{8.26}$$

For $n = -2$, we have a gravitational situation. Total energy$= E = \langle K.E. \rangle + \langle P.E. \rangle$.

Then, denoting the kinetic energy by K

$$E = \overline{K} + \overline{V} = \left(\frac{n+3}{n+1}\right) \overline{K}. \tag{8.27}$$

It is easy see that $n = -(3 + \epsilon)$ where ϵ is a small positive quantity, $E > 0$. For $n = -2$, a gravitational case, $E < 0$. If we assume E to be proportional to T, the absolute temperature, then we have $C_V = \left(\frac{\partial E}{\partial T}\right)_V < 0$. As a matter of fact, the concept of temperature in statistical mechanics is only in canonical and grand canonical ensemble. It is an attribute of a reservoir. It gets attached to the smaller systems of our interest through a contact. This implies that for the systems where there are long range forces, the only attribute is its conserved total energy. Therefore, one needs to use the microcanonical ensemble in case of gravitational systems.

We 'define' a specific heat in a microcanonical ensemble as follows. The entropy $S(E) = k_B \ln \Omega$. We construct a function $\beta(E) = \frac{1}{k_B}\frac{\partial}{\partial E}\ln \Omega$. This can in principle be inverted to give $E(\beta)$. Then the specific heat is $C_V = -\beta^2/k_B \left(\frac{\partial E}{\partial \beta}\right)$. Specific heat so defined may or may not be positive. So long as the canonical and the microcanonical ensembles are equivalent, $C_V > 0$. In phase transitions, the correlations build at T_C to be ∞. This makes definition of C_V at T_C difficult. In the presence of long range forces, we can not 'derive' a canonical distribution from a microcanonical one. One more aspect of gravitational interaction $(-1/r)$ is that the ground state energy of the N particle $(-1/r)$ potential system is $E \propto N^3$ and not $E \propto N$. To see these effects, an interesting toy model of a binary star is discussed by T. Padmanabhan. Binary stars has a hamiltonian

$$H(P, Q, p, r) = \frac{P^2}{2M} + \frac{p^2}{2\mu} - \frac{Gm^2}{r}$$

---

[1] For the detailed discussion of Virial theorem, see (a) Classical Mechanics by P.V.Panat p-135,(Narosa); (b) An Invitation to Astrophysics by T. Padmanabhan p-19 (World Scientific). See also his article "Statistical mechanics of gravitating system", Physics reports <u>188</u>,no 5, (1990) p-285

where $M = 2m$, the mass of the binary star with r as relative distance, p as relative momentum and P is a momentum of the center of mass of the binary. $\mu = m/2$ is the effective mass. Let a typical binary from a collection of binaries be confined in a sphere of radius R and smallest distance $r = a$. It is closest distance that the component stars of the binaries can come. Without (a, R) it is a standard Kepler problem. The number of accessible states $\Omega(E)$ of a constant energy surface in the microcanonical ensemble are

$$\Omega(E) = \int d^3P \; d^3Q \; d^3p \; d^3r \; \delta(E - H(P, p, Q, r)) \tag{8.28}$$

Its evaluation is tricky and we present final result as

$$\frac{\Omega(E)}{(Gm^2)^3} = \begin{cases} \frac{1}{3}R^3(-E)^{-1}(1 + aE/Gm^2)^3 & -\frac{Gm^2}{a} < E < -\frac{Gm^2}{R} \\ \frac{1}{3}R^3(-E)^{-1}\left[\left(1 + \frac{RE}{Gm^2}\right)^3 - \left(1 + \frac{aE}{Gm^2}\right)^3\right] & -\frac{Gm^2}{R} < E < \infty \end{cases} \tag{8.29}$$

Entropy is $S = k_B \ln \Omega$. We then 'define' a temperature in microcanonical ensemble as

$$\beta(E) = \frac{1}{k_B T(E)} = \frac{1}{k_B}\frac{\partial S}{\partial E}.$$

We then plot the absolute temperature T as function of E as shown in figure(2).

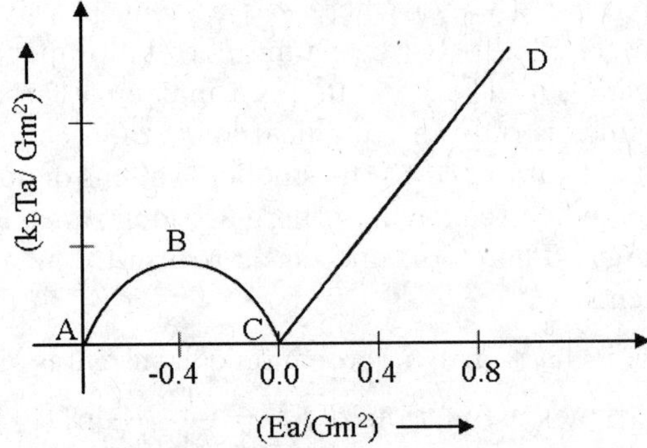

Figure 8.2: T versus E in microcanonical ensemble for the binary stars. Energy is measured in the unit of Gm^2/a.

From the figure, it is clear that the C_V is positive in the region AB, it is negative in the region BC and is positive beyond. If we solve the same problem using canonical ensemble - this is off course forcing it, then we evaluate

$$\begin{aligned} Z(\beta) &= \int d^3P d^3Q d^3p d^3r e^{-\beta H} \\ &= \frac{4}{3}\pi R^3 \int \exp[-\frac{\beta P^2}{2m}d^3P].\int \exp[-\frac{\beta p^2}{2\mu}d^3p.]4\pi \int_a^R r^2 dr \exp[\beta\frac{Gm^2}{r}] \\ &= t^3\left(\frac{R}{a}\right)^3 \int_1^{R/a} dx x^2 e^{1/xt} \end{aligned} \tag{8.30}$$

where $t = ak_BT/Gm^2$, a dimensionless temperature. Various limiting forms of this expression can be found and average energy $E(\beta) = -\partial \ln Z/\partial \beta$ gives C_V upon differentiation. We leave it as an exercise to complete the problem. T versus E graph of figure(3) shows that the region BC of microcanonical ensemble graph is flattened so that $C_V \geq 0$

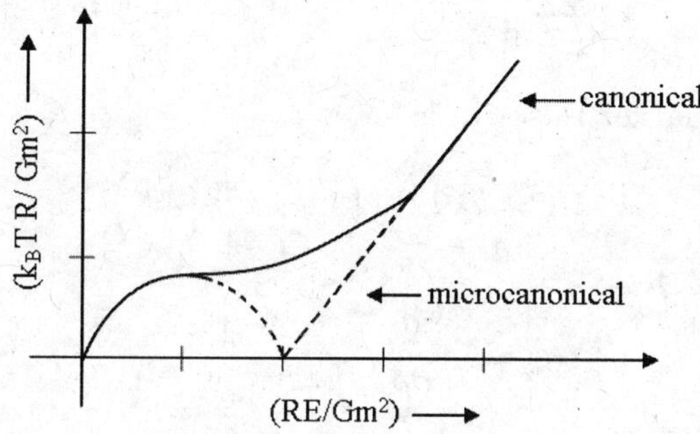

Figure 8.3: T versus E in canonical and microcanonical ensemble for a binary star. Energies are measured in the units of R/Gm^2.

8.5 Fermi and Bose Distributions

We rederive these distributions using grand potential instead of counting number of states in a narrow energy range of microcanonical ensemble.

Let there be n_1 particles with energy ϵ_1, n_2 with energy ϵ_2 etc. Then

$$\sum n_k = N$$

and
$$\sum n_k \epsilon_k = E$$

With these constraints, we write grand potential of equation (25) as

$$
\begin{aligned}
\Omega &= -k_B T \, \ln \sum e^{\beta\mu(n_1+n_2+\ldots)-\beta(\epsilon_1 n_1+\epsilon_2 n_2+\ldots)} \\
&= -k_B T \, \ln \sum_{\{n_k\}} e^{\beta(\mu-\epsilon_1)n_1+\beta(\mu-\epsilon_2)n_2+\ldots} \\
&= -k_B T \, \ln \sum_{\{n_k\}} \left(e^{\beta(\mu-\epsilon_k)}\right)^{n_k}
\end{aligned}
\tag{8.31}
$$

Bose Distribution:

n_k can take values from 0 to ∞. Thus for a state k

$$\Omega_k = -k_B T \ln \left[\sum_{n_k=0}^{\infty} \left(e^{\beta(\mu-\epsilon_k)}\right)^{n_k} \right] \tag{8.32}$$

This is a geometric series and must converge. This requires $e^{\beta(\mu - \epsilon_k)} < 1 \ \forall \ \epsilon_k$. This is possible only if $\mu < 0$. Thus, we get an important result that for a Bose gas, chemical potential is always negative. Thus,

$$\sum_{n_k = 0}^{\infty} e^{\beta(\mu - \epsilon_k)n_k} = \frac{1}{1 - e^{\beta(\mu - \epsilon_k)}}$$

This gives grand potential for Bose-gas as

$$\Omega_k = k_B T \ \ln\left(1 - e^{-\beta(\epsilon_k - \mu)}\right) \tag{8.33}$$

It is easy to see that

$$\bar{n}_k = -\frac{\partial \Omega_k}{\partial \mu} = \frac{1}{e^{\beta(\epsilon_k - \mu)} - 1} \tag{8.34}$$

which is Bose distribution.

Fermi Distribution:

Because of Pauli principle, $n_k = 0$ and 1. As a result, the grand potentials for the Fermi gas is,

$$\begin{aligned}
\Omega_k &= -k_B T \ \ln\left[\sum_{n_k}\left(e^{\beta(\mu - \epsilon_k)}\right)^{n_k}\right] \\
&= -k_B T \ \ln\left[1 + e^{-\beta(\epsilon_k - \mu)}\right] \tag{8.35}
\end{aligned}$$

and

$$\bar{n}_k = -\frac{\partial \Omega_k}{\partial \mu} = \frac{1}{e^{\beta(\epsilon_k - \mu)} + 1} \tag{8.36}$$

We thus see how easy it is to get F-D and B-E distributions from grand canonical distribution than the counting of states and relating them to entropy.

From equation (33) and (35) we immediately get pressure for two gases because $\Omega = -PV$. Thus, summing over all states k,

$$\begin{aligned}
\frac{PV}{k_B T} &= -\sum_k \ln\left(1 - e^{-\beta(\epsilon_k - \mu)}\right) \qquad \text{(Bose)} \\
&= \sum_k \ln\left(1 + e^{-\beta(\epsilon_k - \mu)}\right) \qquad \text{(Fermi)}
\end{aligned}$$

Illustrative Problems

Problem 1: An atom in atomic system has two levels ϵ_1 and ϵ_2 close to each other. It is excited by Laser of frequency ω such that $\hbar\omega = \epsilon_2 - \epsilon_1$. The system is at temperature

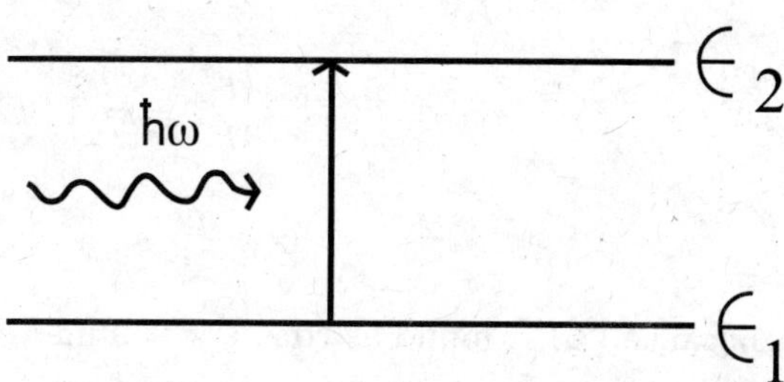

Figure 8.4:

T. Power of the laser absorbed is proportional to the population difference in two levels. Assume that $\epsilon_2 - \epsilon_1 << k_B T$, find the behaviour of power as a function of T.

Solution: Let there be N atoms. Number of atoms in state ϵ_1 are $\qquad N_1 = \dfrac{Ne^{-\beta\epsilon_1}}{e^{-\beta\epsilon_1} + e^{-\beta\epsilon_2}}$

and those in ϵ_2 are $N_2 = \dfrac{Ne^{-\beta\epsilon_2}}{e^{-\beta\epsilon_1} + e^{-\beta\epsilon_2}}$. The difference is

$$\frac{N_1 - N_2}{N} = \frac{\Delta N}{N} = \frac{e^{-\beta\epsilon_1} - e^{-\beta\epsilon_2}}{e^{-\beta\epsilon_1} + e^{-\beta\epsilon_2}} = \frac{e^{-\beta\epsilon_1}\left[1 - e^{\beta(\epsilon_1-\epsilon_2)}\right]}{e^{-\beta\epsilon_1}\left[1 + e^{\beta(\epsilon_1-\epsilon_2)}\right]}$$

Thus

$$\frac{\Delta N}{N} \cong \frac{1 - 1 + \beta(\epsilon_2 - \epsilon_1)}{1 + 1 + \beta(\epsilon_2 - \epsilon_1)} \approx \frac{\epsilon_2 - \epsilon_1}{2k_B T}$$

in the limit $\Delta\epsilon << k_B T$.

Therefore, the power absorbed is $\propto \dfrac{\hbar\omega}{2k_B T} \propto \dfrac{1}{T}$

Problem 2: Find Fluctuations in energy i.e. $\langle E^2 \rangle - \langle E \rangle^2$ (sometimes called dispersion in Energy) using canonical ensemble and show that $\langle E^2 \rangle - \langle E \rangle^2 = k_B T^2\, C_V$.

solution: From equation (16)

$$\langle E \rangle = -\frac{\partial \ln Z}{\partial \beta} = -\frac{1}{Z}\frac{\partial Z}{\partial \beta}$$

Similarly,

$$\langle E^2 \rangle = \frac{\sum E_r^2 e^{-\beta E_r}}{\sum e^{-\beta E_r}} = \frac{1}{Z}\frac{\partial^2 Z}{\partial \beta^2}.$$

Then,

$$\langle E^2 \rangle - \langle E \rangle^2 = \frac{1}{Z}\frac{\partial^2 Z}{\partial \beta^2} - \frac{1}{Z^2}\left(\frac{\partial Z}{\partial \beta}\right)^2 = -\frac{\partial \langle E \rangle}{\partial \beta}$$

But

$$\frac{\partial}{\partial \beta} = -k_B T^2 \frac{\partial}{\partial T}$$

Thus,

$$(\Delta E)^2 = \langle E^2 \rangle - \langle E \rangle^2 = k_B T^2 \frac{\partial \langle E \rangle}{\partial T} = k_B T^2 \, C_V$$

q.e.d

Here $\dfrac{\partial \langle E \rangle}{\partial T}$ is called C_V because $\langle E \rangle$ is found at constant volume.

This result is valid for classical as well as quantum statistics because (a) It is a thermodynamic result (b) In its derivation, nowhere the nature of statistics is invoked.

Problem 3: In this problem, we rederive F-D distribution by a principle of detailed balance. It relates to the collision equation. We assume that the gas of fermions is such that there are binary collision only. Also we assume that any two collisions are independent.

Let f_k be average number of particles in $|k\rangle$ state.

Let $1 + 2 \leftrightarrow 3 + 4$.

This means state $|1\rangle$ has occupancy f_1, $|2\rangle$ has f_2. It will go to states $|3\rangle$ and $|4\rangle$, if both has a probability $(1 - f_3)$ and $(1 - f_4)$. This is because of Pauli principle. If f_3 probability that a state $|3\rangle$ is occupied, then the probability that the state $|3\rangle$ has vacancy, is $(1 - f_3)$. Thus rate at which the reaction moves forward is $f_1 f_2 (1 - f_3)(1 - f_4) |M|^2$. Here $|M|^2$ is square of the matrix elements causing the reaction. Reverse reaction is equally probable. This is because of time reversal invariance at microscopic level. Rate of reverse reaction is $f_3 f_4 (1 - f_1)(1 - f_2) |M|^2$. Here matrix elements remain same. Principle of detailed balance states that these two rates are identical.

Thus,

$$f_1 f_2 (1 - f_3)(1 - f_4) = f_3 f_4 (1 - f_1)(1 - f_2)$$

Or

$$\frac{f_1}{1 - f_1} \cdot \frac{f_2}{1 - f_2} = \frac{f_3}{1 - f_3} \cdot \frac{f_4}{1 - f_4} \tag{8.37}$$

If there are many particles involved in collision, then,

$$\frac{f_1}{1 - f_1} \cdot \frac{f_2}{1 - f_2} \cdots \frac{f_k}{1 - f_k} = \frac{f_1'}{1 - f_1'} \cdot \frac{f_2'}{1 - f_2'} \cdots \frac{f_k'}{1 - f_k'}$$

In all these collisions, energy and number of particles are conserved. i.e.

$$\epsilon_1 + \epsilon_2 = \epsilon_3 + \epsilon_4$$

or,

$$\epsilon_1 + \epsilon_2 + \epsilon_3 + \dots \epsilon_k + \dots = \epsilon_1' + \epsilon_2' + \epsilon_3' + \dots \epsilon_k' + \dots$$

Solution to equation (37) can be f_k=constant. But this is trivial. we have seen that the density function in equilibrium is a function of constants of motion. There is another solution

$$\frac{f_k}{1 - f_k} = e^{-\alpha - \beta \epsilon_k}$$

or,

$$f_k = \frac{1}{e^{\beta \epsilon_k - \alpha} + 1}$$

which is Fermi-Dirac distribution.

Short Questions

1. If a system has average energy $\bar{E}$, can it have an energy $\bar{E} \pm \alpha \bar{E}$ when in contact with a reservoir? And with what probability? Assume temperature of reservoir is T.

2. Explain why $P(E)$ has a maximum at average E?

3. Explain how $e^{-\beta \hat{H}}$ is a density matrix in the canonical distribution.

4. Compare Bloch equation with Schrödinger equation.

5. What is the difference between microcanonical and canonical ensemble? When they are identical?

6. When canonical and grand canonical ensemble are identical?

7. Derive Bose distribution and Fermi distribution using grand Canonical ensemble.

8. Explain how various thermodynamic quantities can be found if the partition function is evaluated.

9. Can you guess what is a density matrix in grand canonical ensemble? Why?

Ans: $\rho = e^{-\beta(H - \mu N)}$

Problems

Problem 1. In case of Bose statistics, if f_k is occupancy of state $|k\rangle$, then probability of another Bose particle going to that state enhances and occupancy is $(1 + f_k)$. Use this result and the principle of detailed balance to get Bose distribution. (Hint: Use arguments similar to those used in illustrative problem (3))

Problem 2. A quantum harmonic oscillator of frequency ω is in contact with heat bath of low temperature T. The quantum of an energy of the oscillator is such that its energy is much larger than thermal energy $k_B T$. The oscillator can be in any of the quantum numbers $n = 0, 1, 2, ...$
a) Find ratio of probability that the oscillator is in $n = 1$ state to probability that it is in $n = 0$ state.

Ans: $p_1/p_0 = e^{-\hbar\omega/k_B T}$

b) Find average energy if only $n = 0$ and $n = 1$ states are populated.

$$\text{Ans: } \bar{E} = \frac{\hbar\omega}{2}\left(\frac{1 + 3e^{-\beta\hbar\omega}}{1 + e^{-\beta\hbar\omega}}\right)$$

Problem 3. Consider a single electron with spin $\vec{S} = \frac{1}{2}\hbar\hat{\sigma}$ in an external uniform magnetic field $\vec{B} = B_0\hat{k}$. Let μ_B be Bohr magneton for electron. Find density matrix of electron in canonical ensemble. Find average value of $S_Z \equiv \sigma_Z$ (Hint: $\hat{H} = -\mu_B\hat{\sigma}.\vec{B} = -\mu_B\sigma_Z B$.

But $\sigma_Z = \begin{pmatrix} 1 & 0 \\ 0 & -1 \end{pmatrix}$. Thus,

$$\hat{\rho} = \frac{e^{-\beta H}}{Tr e^{-\beta H}} = \begin{pmatrix} e^{\mu_B\beta B} & 0 \\ 0 & e^{-\beta\mu_B B} \end{pmatrix} \times \frac{1}{e^{\beta\mu_B B} + e^{-\beta\mu_B B}}$$

Clearly $\langle\sigma_Z\rangle = Tr(\hat{\rho}\,\sigma_Z) = \tanh(\beta\mu_B B))$

Problem 4. An ultra centrifuge contain a polymer solution at temperature T and is rotating with angular frequency ω. Find

a) How the relative density $n(r)$ of the polymer solution varies with radial distance r?

b) How will you find the molecular weight of the solution by measuring densities at r_1 and r_2?

(Hint: Equivalent potential is $-\frac{1}{2}m\omega^2 r^2$. By canonical distribution,

$$n(r) = n(0)\, e^{\frac{m\omega^2 r^2}{2k_B T}}$$

$$\frac{n(r_1)}{n(r_2)} = e^{\frac{m\omega^2(r_1^2 - r_2^2)}{2k_B T}}.$$

With N_a as Avogadro number,

the molecular weight $= \dfrac{2\,N_a k_B T}{\omega^2(r_1^2 - r_2^2)} \ln\left(\dfrac{n(r_1)}{n(r_2)}\right))$

Problem 5. At a given low temperature a gas of hydrogen molecule is a mixture of ortho and para hydrogen. At that low temperatures, only excitations possible are rotational one. In ortho hydrogen molecule, the proton nuclear spins are parallel, whereas they are antiparallel in para hydrogen. Rotational energies of para and ortho hydrogen are given by

$$E_{ortho} = \frac{\hbar^2}{2I}J(J+1) \quad (J = 1, 3, 5, ...)$$

$$E_{para} = \frac{\hbar^2}{2I}J(J+1) \quad (J = 0, 2, 4, ...)$$

Each state has $(2J+1)$ fold degeneracy. I is moment of inertia of H_2 molecule. Given $T = 20°K$, and bond length of H_2 as 0.74Å, find N_{ortho}/N_{para} at that temperature.

$$\frac{N_{ortho}}{N_{para}} = \frac{3 \sum\limits_{J(odd)} (2J+1)e^{\dfrac{-\beta\hbar^2}{2I}J(J+1)}}{\sum\limits_{J(even)} (2J+1)e^{\dfrac{-\beta\hbar^2}{2I}J(J+1)}}$$

Problem 6. Assume that an average energy of a gas molecule is $\frac{3}{2}k_BT$. By comparing the mean inter atomic distance with thermal de-Broglie wavelength, argue that the classical approximation of M-B statistics is valid if,

$$\left(\frac{V}{N}\right)^{1/3} >> \frac{h}{\sqrt{3mk_BT}}$$

Take $P = 76\ cm$ of Hg $\sim 10^6 dynes/cm^2$. $T = 27^oC$, and a gas of oxygen$\approx \frac{32}{6\times 10^{23}} \sim 5 \times 10^{-23} gm = m_{O_2}$ and show that oxygen gas at above conditions is classical.

Chapter 9

MAXWELL-BOLTZMANN VELOCITY DISTRIBUTION OF IDEAL GAS

Formulation of statistical mechanics which we discussed pertains to equilibrium situation and is mainly due to Willard Gibbs. Before this work, kinetic theory of gases was developed by Boltzmann and Maxwell. Main assumption was that gas is made up of large number of molecules confined in a volume and are in continuous random motion. The velocity of the molecules could vary and have a distribution. It was also proved by Boltzmann that in equilibrium, there is a uniform density of the gas and it is achieved due to collisions. This molecular model called kinetic theory of gases, was used to find equilibrium as well as non-equilibrium properties (Such as viscosity, thermal conductivity etc.). Boltzmann also discovered a transport equation and an H- theorem. Historically all these molecular ideas were not accepted by leading physicists of that time such as Lord Kelvin. It was only after Einstein's explanation of Brownian motion, the molecular theory of matter was accepted. In this chapter, we are going to discuss the velocity distribution of molecules of a gas in equilibrium. Non equilibrium aspects of the kinetic theory will be discussed in subsequent chapters.

9.1   Maxwell-Boltzmann Velocity Distribution

Consider as classical dilute gas confined in a volume V. It is considered classical because the de-Broglie wavelength associated with a molecule is much smaller than an average inter-particle separation. We also assume that the intermolecular distance is larger than the range of interaction between any two molecules. This means the molecules have kinetic energy only and thus can be treated as ideal. Real gases are discussed in subsequent chapter.

Our aim here is to find the number of molecules in a phase space volume $d^3r\ d^3p$ around $\vec{r}$ and $\vec{p}$. This number, on average will not change with time in equilibrium situation. Total energy of the gas of N molecules in volume V is

$$E = \sum_{k=1}^{N} \frac{p_k^2}{2m} \tag{9.1}$$

Here we are considering a single specie of the gas. Historically the question posed was not in terms of momenta but in terms of velocity. Then, probability that a molecule has a velocity lying between $\vec{v}$ and $\vec{v}+d\vec{v}$, and between the position $\vec{r}$ and $\vec{r}+d\vec{r}$, by canonical distribution is,

$$P(\vec{r},\vec{v})\ d^3r\ d^3v = \text{ constant } \times e^{-\dfrac{mv^2}{2k_BT}}\ d^3r\ d^3v \tag{9.2}$$

For a non ideal gas, this probability cannot be written so simply and we have to incorporate the effect of other molecules. We convert this probability into Maxwell Boltzmann form quiet easily. Let $f(\vec{r},\vec{v}) \times d^3r\ d^3v$ be mean number of particles in $d^3r\ d^3v$. Clearly, integral of it over entire volume and velocity will be total number of particles. Thus,

$$\text{constant} \int_{\{r\}} \int_{\{v\}} d^3r\ d^3v\ e^{-\dfrac{\beta m v^2}{2}} = N \tag{9.3}$$

The constant can be immediately obtained as,

$$\text{constant} \quad = \quad \dfrac{N}{V \int_0^\infty 4\pi\ v^2\ dv\ e^{-\beta v^2 m/2}}$$

$$= \quad n \left(\dfrac{m}{2\pi k_B T}\right)^{3/2}$$

where $n = \text{ density } = N/V$

Thus, we get,

$$f(\vec{v}) = n \left(\dfrac{m}{2\pi k_B T}\right)^{3/2} e^{-\dfrac{mv^2}{2k_BT}} \tag{9.4}$$

This is known as Maxwell-Boltzmann distribution of velocities. This is independent of position $\vec{r}$ indicating uniform density. This is because there is no non-equilibrium situation. Moreover, $f(\vec{v}) = f(|\vec{v}|) \equiv f(v)$. This means, the distribution is isotropic. If we want to find number of molecules with their speed between v and $v + dv$ then it is simply

$$f(v) = 4\pi n \left(\dfrac{m}{2\pi k_B T}\right)^{3/2} e^{-\dfrac{mv^2}{2k_BT}}\ v^2 \tag{9.5}$$

Clearly

$$\int_0^\infty f(v)dv = n$$

The most probable speed is at the maximum of (5) and is obtained by

$$\dfrac{df(v)}{dv} = 0.$$

This gives

$$v_{max} = \sqrt{\dfrac{2k_BT}{m}} \tag{9.6}$$

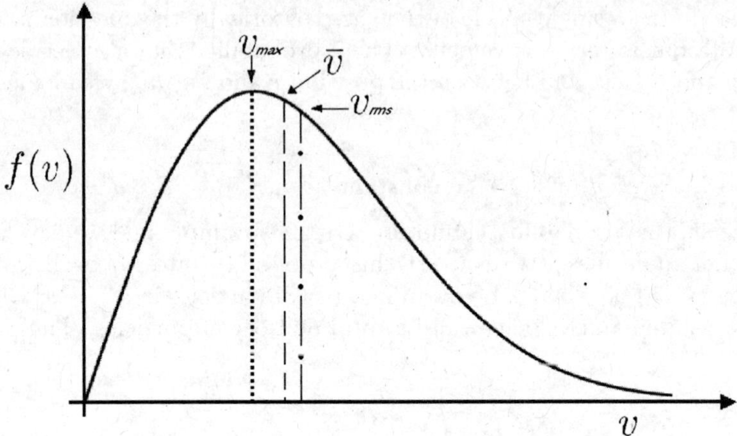

Figure 9.1: Maxwell distribution for distribution of molecular speed.

Mean velocity of the gas is zero because there are as many left movers as right movers in all the three dimensions. However, we can find average speed $\bar{v}$. Thus,

$$
\begin{aligned}
\bar{v} &= \frac{1}{n} \int_0^\infty v f(v)\, dv \\[2mm]
&= 4\pi \left(\frac{m}{2\pi k_B T} \right)^{3/2} \int_0^\infty v^3\, e^{-\frac{mv^2}{2k_B T}}\, dv \\[2mm]
&= 4\pi \left(\frac{m}{2\pi k_B T} \right)^{3/2} \left(\frac{2k_B T}{m} \right)^2 \int_0^\infty \xi^3 e^{-\xi^2}\, d\xi \\[2mm]
&= 2\pi \left(\frac{m}{2\pi k_B T} \right)^{3/2} \left(\frac{2k_B T}{m} \right)^2 = \sqrt{\frac{8}{\pi}\frac{k_B T}{m}}
\end{aligned}
\tag{9.7}
$$

We can also find mean square speed $\bar{v^2}$ as,

$$
\begin{aligned}
\bar{v^2} &= \frac{4\pi n}{n} \left(\frac{m}{2\pi k_B T} \right)^{3/2} \int_0^\infty v^4 e^{-\frac{mv^2}{2k_B T}}\, dv \\[2mm]
&= 4\pi \left(\frac{m}{2\pi k_B T} \right)^{3/2} \cdot \left(\frac{2k_B T}{m} \right)^{5/2} \int_0^\infty \xi^4 e^{-\xi^2}\, d\xi \\[2mm]
&= 4\pi \cdot \frac{3}{8}\sqrt{\pi} \left(\frac{m}{2\pi k_B T} \right)^{3/2} \cdot \left(\frac{2k_B T}{m} \right)^{5/2} = \frac{3k_B T}{m}
\end{aligned}
$$

The root mean square speed is then

$$
v_{rms} = \sqrt{\bar{v^2}} = \sqrt{\frac{3k_B T}{m}}
\tag{9.8}
$$

Note - all these various speeds are - some numerical factor times $\sqrt{k_BT/m}$ indicating that all these speeds increase with increase of temperature. The ratio is,

$$v_{rms} : \bar{v} : v_{max} = \sqrt{3} : \sqrt{\frac{8}{\pi}} : \sqrt{2} \qquad (9.9)$$

To get the numbers, find m =molecular wt/Avogadro number (gms). Typical values are in the range of about 500 m/s at room temperature which is around the speed of sound in the gas at that temperature.

9.2   Distribution of Component of Velocity

If we are interested in finding average number of molecules with velocity component between v_x and $v_x + dv_x$ irrespective of values of other components, then we have to integrate Maxwell distribution over v_y and v_z. Let $g(v_x)dv_x$ be such number. Then,

$$
\begin{aligned}
g(v_x)\, dv_x &= n\left(\frac{m}{2\pi k_B T}\right)^{3/2} dv_x\ \exp\left[-\frac{mv_x^2}{2k_B T}\right] \\
&\times \int_{-\infty}^{\infty} exp\left[-mv_y^2/2k_B T\right] dv_y \\
&\times \int_{-\infty}^{\infty} exp\left[-mv_z^2/2k_B T\right] dv_z \qquad (9.10)
\end{aligned}
$$

But,

$$\int_{-\infty}^{\infty} exp\left[-mv_y^2/2k_B T\right] dv_y = \sqrt{\frac{2k_B T}{m}} \int_0^{\infty} e^{-\xi^2}\, d\xi = \sqrt{\frac{2\pi k_B T}{m}}\ \text{etc.}$$

Thus

$$g(v_x)dv_x = n\left(\frac{m}{2\pi k_B T}\right)^{1/2} e^{-\frac{mv_x^2}{2k_B T}}\, dv_x \qquad (9.11)$$

This distribution is Gaussian with a half width $\sqrt{\frac{k_B T}{m}}$. Also, it is normalized and integrates to n. Clearly, $\bar{v}_x = 0$, since the distribution is symmetric around the origin. It is easy to see that

$$\overline{v_x^2} = \frac{k_B T}{m}.$$

Then,

$$\overline{v^2} = \overline{v_x^2} + \overline{v_y^2} + \overline{v_z^2} = \frac{k_B T}{m} + \frac{k_B T}{m} + \frac{k_B T}{m} = 3\frac{k_B T}{m}$$

which is same as v_{rms} calculated earlier.

Illustrative Problems

Problem 1. Atomic spectral lines are a result of its transitions from one atomic level to the other and its frequency ν_0 is given by Bohr quantum condition. But the emitting atom is not stationary and moves with certain velocity. As a result there is a Doppler shift in the frequency. Suppose the arm of a spectrometer is in X direction. Then the observed frequency ν is

$$\nu = \nu_0 \left(1 + \frac{v_x}{c}\right).$$

This leads to a spectral distribution of intensity $I(\nu)d\nu$. Find

(a) mean frequency observed in the spectroscope. (b) RMS frequency shift $(\Delta\nu)_{rms} = \left[\overline{(\nu - \bar{\nu})^2}\right]^{1/2}$ and (c) $I(\nu)d\nu$.

Solution: (a)

$$\bar{\nu} = \nu_0 \left(1 + \frac{\overline{v_x}}{c}\right) = \nu_0 \quad \text{since} \quad \overline{v}_x = 0$$

(b)

$$\nu^2 = \nu_0^2 \left(1 + \frac{2v_x}{c} + \frac{v_x^2}{c^2}\right)$$

$$\overline{\nu^2} = \nu_0^2 \left(1 + \frac{2\overline{v}_x}{c} + \frac{\overline{v_x^2}}{c^2}\right) = \nu_0^2 \left(1 + \frac{k_B T}{mc^2}\right)$$

$$\therefore (\Delta\nu)_{rms} = \left[\overline{\nu^2} - (\bar{\nu})^2\right]^{1/2}$$

$$= \left[\overline{\nu^2} - \nu_0^2\right]^{1/2} = \nu_0 \left(\frac{k_B T}{mc^2}\right)^{1/2}$$

(c)

$$v_x = \frac{c\,(\nu - \nu_0)}{\nu_0}$$

The distribution of v_x is

$$g(v_x)dv_x = n \left(\frac{m}{2\pi k_B T}\right)^{1/2} \times exp\left[-\frac{mc^2\,(\nu - \nu_0)^2}{2\nu_0^2 k_B T}\right]$$

Intensity distribution is proportional to $g(v_x)dv_x = g(\nu)cd\nu/\nu_0$

Thus, $I(\nu) = I_0 exp\left[-\frac{mc^2\,(\nu - \nu_0)^2}{2\nu_0^2 k_B T}\right]$

This is Gaussian with width $\nu_0 \left(\frac{k_B T}{mc^2}\right)^{1/2}$ as found in (b).
Experiments indicate excellent agreement with this line shape.

Problem 2. Find $\overline{v_x^2\, v_y}$, $\overline{(v_x + bv_y)^2}$, $\overline{v_x^2\, v_y^2}$, $\overline{v^2 v_x}$.

Solution: Note v_x, v_y and v_z are independent random variables,
Therefore $\overline{v_x\,v_y} = \bar{v}_x\bar{v}_y$ etc. Thus

$$
\begin{aligned}
\overline{v_x^2\,v_y} &= \overline{v_x^2}\,\overline{v_y} = 0 \\[2mm]
\overline{(v_x + b\,v_y)^2} &= \overline{v_x^2} + b^2\,\overline{v_y^2} + 2b\,\overline{v_x\,v_y} \\[2mm]
&= \frac{k_B T}{m} + b^2\frac{k_B T}{m} + 0 = \frac{k_B T}{m}\left(1 + b^2\right) \\[2mm]
\overline{v_x^2\,v_y^2} &= \overline{v_x^2}\,\overline{v_y^2} = \left(\frac{k_B T}{m}\right)^2 \\[2mm]
\overline{v^2\,v_x} &= \overline{v^2}\,\overline{v_x} = 0.
\end{aligned}
$$

9.3   Number of Molecules Hitting a Unit Area of a Container

Here, we are interested in the number of collisions per unit area per second received by the wall. This is important because, if there is a hole, we would like to know the leakage rate. Or, we can design an oven where we would want to get a beam of atoms.

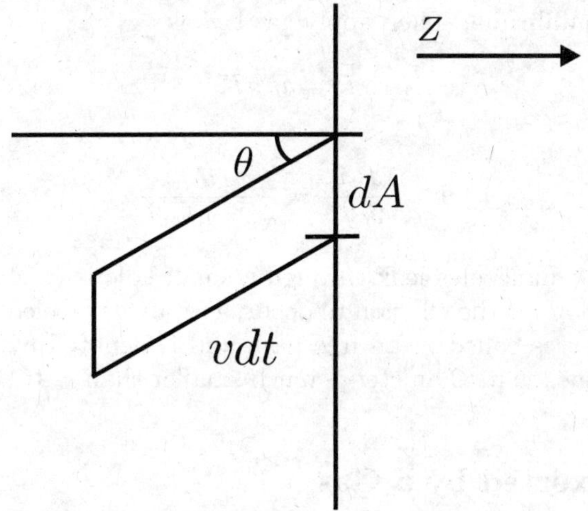

Figure 9.2: All molecules with a finite component of velocity v_z hit the area dA in time dt

Consider a wall in the Z direction as shown in figure(2). Let dA be small area. Construct a cylinder of length vdt and base of area dA inclined at an angle θ. Volume of this cylinder is $dA\,dt\,v\,\cos\theta = dA\,dt\,v_z$. We have, only the molecules of different values of v_z that are of interest. This is so because, they alone hit dA of the wall. Any velocity $\vec{v}$ which does not have v_z as a component, will not hit dA of the wall. Let $g(v_z)dv_z$ be mean number of molecules whose z component of velocity has range between v_z and $v_z + dv_z$. Then, Number of particles hitting dA in dt is,

$$
g(v_z)\,dv_z\,.\,dA\,.\,dt\,v\cos\theta = v_z\,g(v_z)\,dv_z\,dA\,dt. \tag{9.12}
$$

Thus,

$$\frac{\text{Total Number of particles}}{dA\ dt} = \int_0^\infty g(v_z)\ dv_z . v_z$$

$$= n\left(\frac{m}{2\pi k_B T}\right)^{1/2} \int_0^\infty v_z . exp\left[-\frac{mv_z^2}{2k_B T}\right]\ dv_z$$

$$= n\left(\frac{m}{2\pi k_B T}\right)^{1/2} \left(\frac{2k_B T}{m}\right) \int_0^\infty \xi\ e^{-\xi^2} d\xi$$

$$= n\left(\frac{2_B T}{\pi m}\right)^{1/2} \tag{9.13}$$

But the average speed $\bar{v} = \sqrt{\dfrac{8}{\pi}\dfrac{k_B T}{m}}$. Then, the flux Φ, which is total number of particles hitting unit area in one second is,

$$\Phi = \frac{1}{4} n\bar{v}. \tag{9.14}$$

If a sufficiently small hole to the container is present in the wall, the gas inside the container is approximately in equilibrium. Then apply gas law as

$$P = nk_B T$$

The leakage flux is

$$\Phi = \frac{P\ \bar{v}}{4k_B T} = \frac{P}{\sqrt{2\pi m k_B T}}. \tag{9.15}$$

The process that molecules leak through a small hole is called 'effusion'. There is a simple physical condition for the effusion to occur. Distance traveled by a molecule between two successive collisions is called mean free path and is denoted by l. "Sufficiently small" hole to the container means its diameter is much smaller than l. If this is the case, effusion takes place.

9.4 Pressure Exerted by a Gas

When ever a molecule hits the wall of a container, it is reflected back. We assume that the walls of the container are perfectly elastic and smooth. Momentum imparted to the wall by a molecule with velocity $\vec{v}$ hitting it is, $2m\vec{v}$ provided the wall is rigid and the collision is co-linear. We have calculated number of particles hitting area dA in z direction in time dt is

$$g(v_z)\ dv_z\ dA\ dt\ v_z.$$

Total momentum imparted to dA in time dt is

$$g(v_z)\ dv_z\ dA\ dt\ v_z\ 2m\ v_z. \tag{9.16}$$

Of $\vec{v}$, only v_z component will contribute to the total momentum. This is so because all other components of $\vec{v}$ are tangential to dA.

Pressure is Force / area = momentum/s/area

Thus pressure on the wall is

$$P = \int 2m{v_z}^2 \, g(v_z) \, dv_z$$

$$= 2mn \left(\frac{m}{2\pi k_B T}\right)^{1/2} \int_0^\infty {v_z}^2 . \, exp\left[-\frac{m{v_z}^2}{2k_B T}\right] \, dv_z$$

$$= 2mn \left(\frac{m}{2\pi k_B T}\right)^{1/2} \left(\frac{2k_B T}{m}\right)^{3/2} . \frac{\sqrt{\pi}}{4}$$

$$= nk_B T \tag{9.17}$$

This is an equation of state. But $v_{rms}^2 = \overline{v^2} = \frac{3k_B T}{m}$.
Thus

$$P = \frac{1}{3}mn\overline{v^2} \tag{9.18}$$

which is well known result of kinetic theory of gases.

9.5 Equipartition Theorem

If the hamiltonian of the system has a form,

$$H = ap^2 + bx^2$$

i.e. it is biquadratic then the equipartition theorem states that average value of each term is $k_B T/2$ provided the system of particles is classical.
Proof: The form of Hamiltonian for N particle system is

$$H = a\sum_{i=1}^{N} {p_i}^2 + b\sum_{i=1}^{N} {x_i}^2$$

Probability that the system has this energy is,

$$P = \frac{exp\left[-\beta \sum_{i=1}^{N}\left(ap_i^2 + bx_i{}^2\right)\right]}{\int dp_1 ..... \int dp_N \int dx_1 .... \int dx_N \; exp\left[-\beta \sum_{i=1}^{N}\left(ap_i^2 + bx_i^2\right)\right]}$$

$$= \left[\frac{e^{-\beta\left(ap^2+bx^2\right)}}{\int\int e^{-\beta(ap^2+bx^2)}dp \, dx}\right]^N$$

Thus for each term, which represent each degree of freedom, $a\langle p^2\rangle$ and $b\langle x^2\rangle$ can be found. For N particle system, we simply add these values. Thus,

$$\langle p^2\rangle = \frac{\int_{-\infty}^{\infty} p^2 e^{-\beta ap^2} dp \int_{-\infty}^{\infty} e^{-\beta bx^2} dx}{\int_{-\infty}^{\infty} e^{-\beta ap^2} dp \; \int_{-\infty}^{\infty} e^{-\beta bx^2} dx}$$

Denominator integral is $\left[\sqrt{ab}/\left(\pi k_B T\right)\right]^{-1}$. Thus, for each member of the N particle system,

$$\langle p^2 \rangle = \frac{\sqrt{ab}}{\pi k_B T} \left(\frac{1}{\beta a}\right)^{3/2} \left(\frac{1}{\beta b}\right)^{1/2} \cdot \frac{\sqrt{\pi}}{2} \cdot \sqrt{\pi}$$

$$= \frac{k_B T}{2a}. \tag{9.19}$$

Similarly,

$$\langle x^2 \rangle = \frac{k_B T}{2b}. \tag{9.20}$$

Then

$$a\langle p^2 \rangle = \frac{k_B T}{2} \text{ and } b\langle x^2 \rangle = \frac{k_B T}{2}.$$

and the total energy is,

$$E = \langle H \rangle = \frac{k_B T}{2} + \frac{k_B T}{2} = k_B T. \tag{9.21}$$

Thus average value of each term of the biquadratic Hamiltonian is $k_B T/2$

Short Questions

1. Can you have a gas of molecules such that all its molecules have same velocity? What is its entropy?

2. Maxwell-Boltzmann distribution does not depend upon the the position vector $\vec{x}$. Why?

3. Apply equipartition theorem to the ideal gas of N atoms to find its total energy.

$$\text{Ans:} E = \tfrac{3}{2} N k_B T$$

4. Under what condition of the size of a hole of container gives effusion and why?

5. Explain why $\overline{v_x{}^n} = \overline{v_y{}^n} = \overline{v_y{}^n} = 0$ for odd values of n.

6. A given box is separated by a membrane with a small hole of size smaller than mean free path. Left half of the container has a gas at pressure p_1 and at temperature T_1 and right half has pressure p_2 at T_2. The system is in equilibrium. Experimentally, it is found that $p_1/p_2 = \sqrt{T_1/T_2}$. Explain this result.

7. A mixture of Oxygen and Nitrogen is in equilibrium at a temperature T. Show

$$\left(v_{rms}^2\right)_{O_2} m_{O_2} = \left(v_{rms}^2\right)_{N_2} m_{N_2}$$

Problems

Problem 1. Find $\overline{v^n}$ for a gas obeying Maxwell velocity distribution

$$\text{Ans: } \overline{v^n} = \frac{2}{\sqrt{\pi}} \left(\frac{2k_B T}{m} \right)^{n/2} \Gamma \left(\frac{n+3}{2} \right)$$

(Hint: Use $\int_0^\infty e^{-ax^2} x^n \, dx = \frac{1}{2} \Gamma \left(\frac{n+1}{2} \right) \frac{1}{a^{(n+1)/2}}$)

Problem 2. Find r.m.s. value of a kinetic energy of a molecule.

$$\text{Ans: } \left(\overline{(\Delta \varepsilon)^2} \right)^{1/2} = \sqrt{\frac{7}{2}} \, (k_B T)$$

(Hint: $\bar{\varepsilon} = \frac{k_B T}{2}, \bar{\varepsilon^2} = \frac{m^2}{4} \overline{v^4} = \frac{m^2}{4} . 15 \left(\frac{k_B T}{m} \right)^2$)

Problem 3. Find number of molecules having speed larger than u_0

$$\text{Ans: } N_{u>u_0} = 1 + \left(\frac{2m}{\pi k_B T} \right)^{1/2} u_0 \, exp[-\frac{mu_0^2}{2k_B T}] - \frac{2}{\sqrt{\pi}} \int_0^{\sqrt{mu_0^2/2k_B T}} e^{-x^2} dx$$

(Hint: $N_{u>u_0} = 4\pi n \left(\frac{m}{2\pi k_B T} \right)^{3/2} \int_{u_0}^\infty v^2 exp[-\frac{mv^2}{2k_B T}] \, dv.$

Simplify $\int_0^\infty f(v) dv = \int_0^{u_0} f(v) dv + \int_{u_0}^\infty f(v) \, dv)$

Problem 4. Find Maxwell's energy distribution function from velocity distribution function.

$$\text{Ans: } g\left(\varepsilon \right) d\varepsilon = \frac{n\sqrt{\varepsilon}}{2\pi \sqrt{\pi}} \, exp[-\varepsilon/k_B T] \, d\varepsilon$$

(Hint: $f(v) dv = g(\varepsilon) d\varepsilon$. Find $\frac{dv}{d\varepsilon}$)

Problem 5. A vessel has volume V and contains a gas at temperature T and pressure P. Its wall has a small hole. Outside pressure is low enough so that there is no backflow. Find the time taken for a pressure to fall to its $(1/e)^{th}$ value.

$$\text{Ans: } N = N_0 exp \left[-\frac{A}{V} \left(\frac{k_B T}{2\pi m} \right)^{1/2} t \right] = N_0 e^{-t/\tau}$$

$$\text{Required time} = \frac{V}{A} \left(\frac{2\pi m}{k_B T} \right)^{1/2} = \frac{4V}{A\bar{v}} = \tau$$

(Hint: From equation (13) it is clear $\dfrac{dN}{dt} = -\displaystyle\int_0^A dA \int_0^\infty g(v_z)\, v_z\, dv_z$)

Problem 6. Consider rotating molecule. It has three moments of inertia (principle values) I_x, I_y and I_z and let ω_x, ω_y and ω_z be components of angular velocity $\vec{\omega}$. Let L_x, L_y and L_z be corresponding angular momenta. Then rotational energy[1] is $E_{rot} = \frac{1}{2}\left(I_x\omega_x^2 + I_y\omega_y^2 + I_z\omega_z^2\right) = \frac{1}{2}\left(\dfrac{L_x^2}{I_x} + \dfrac{L_y^2}{I_y} + \dfrac{L_z^2}{I_z}\right)$

Find (i) Probability distribution of angular momentum

(ii) Find $\overline{\omega^2}$ and $\overline{L^2} = \overline{L_x^2} + \overline{L_y^2} + \overline{L_z^2}$

(Hint: Probability that the system of rotating molecules in contact with a heat bath at temperature T is proportional to

$$\exp\left[-\frac{\beta}{2}\left(\frac{L_x^2}{I_x} + \frac{L_y^2}{I_y} + \frac{L_z^2}{I_z}\right)\right] dL_x\, dL_y\, dL_z$$

Constant of proportionality can be gotten by normalization.
We get the constant as

$$\text{Constant} = \frac{1}{(2\pi k_B T)^{3/2}\,\sqrt{I_x I_y I_z}}.$$

Thus probability distribution is,

$$\frac{1}{(2\pi k_B T)^{3/2}\,\sqrt{I_x I_y I_z}}\,\exp\left[-\frac{\beta}{2}\left(\frac{L_x^2}{I_x} + \frac{L_y^2}{I_y} + \frac{L_z^2}{I_z}\right)\right] dL_x\, dL_y\, dL_z.$$

$$\overline{L^2} = \overline{L_x^2} + \overline{L_y^2} + \overline{L_z^2} = k_B T\left(I_x + I_y + I_z\right)$$

and $\overline{\omega^2} = k_B T\left(\dfrac{1}{I_x} + \dfrac{1}{I_y} + \dfrac{1}{I_z}\right)$.)

---

[1]See Classical Mechanics, P-189 by P. V. Panat (Narosa Publishers)

Chapter 10

PARAMAGNETISM AND FERROMAGNETISM

In this chapter, we are going to study, how the localized atoms / ions with magnetic moment produce paramagnetism and in some cases ferromagnetism. For the sake of simplicity, we assume that the localized atoms / ions (hereafter called localized magnetic ions) are arranged in some crystalline form. If the atoms / ions have no net spin and therefore no net magnetic moment, then, there is no question of para or ferromagnetism. In this situation, we get a diamagnetism. We will not consider here diamagnetism. We will also not consider para or ferromagnetism in metals where itinerant (moving) electrons are involved.

When the spins are arranged, there are two types of interactions among them.

a) Spins cause dipole moment. The dipole moments interact among themselves as a dipole-dipole interaction as,

$$W = \left[\frac{\vec{m}_1.\vec{m}_2}{r^3} - \frac{3\,(\vec{m}_1.\vec{r})\,(\vec{m}_2.\vec{r})}{r^5} \right] \tag{10.1}$$

where $\vec{m}_1$ and $\vec{m}_2$ are the dipole moments of particles 1 and 2 separated by a distance r. This is a classical interaction.

b) There is another interaction called exchange interaction, discovered by Heisenberg. It is quantum mechanical in nature. Thus, if we have two spins $\vec{S}_1$ and $\vec{S}_2$ separated by a distance r, this interaction has a Hamiltonian

$$\mathcal{H}_{exch} = -2J\vec{S}_1.\vec{S}_2 \tag{10.2}$$

Clearly, if $\vec{S}_1$ and $\vec{S}_2$ are parallel, this interaction reduces the energy. If they are antiparallel, then energy is enhanced. J is known as coupling constant. $J > 0$ for ferromagnetic situation and $J < 0$ for anti ferromagnetic situation. J is normally a function of r but is usually taken to be constant unless $\vec{S}_1$ and $\vec{S}_2$ are very far apart than the nearest neighbor distance. In that case $H_{exch} \approx 0$. Hamiltonian given by equation (2) can be generalized to N particle system

$$\mathcal{H}_{exch} = -J\sum_{i=1}^{N}\sum_{k=1}^{N}\vec{S}_i.\vec{S}_j \tag{10.3}$$

Factor $\frac{1}{2}$ cancels the factor 2 for double counting. Hamiltonian of equation (3) is known as Heisenberg Hamiltonian. Magnetism is caused because of Heisenberg's exchange and not because of dipole - dipole interaction. We estimate this interaction to realize this important point. Let $m_1 = m_2 = \mu_B$ the Bohr magneton $\sim 9.2 \times 10^{-21}$ erg/Gauss. Take $r \cong 5\mathring{A} = 5 \times 10^{-8}$ cms. Then, W is of the order of

$$W \sim \frac{\mu_B^2}{r^3} \sim \frac{82 \times 10^{-42}}{125 \times 10^{-24}} \sim 0.7 \times 10^{-18} ergs.$$

Corresponding equivalent temperature $T = W/k_B$.

$$\therefore T \sim \frac{0.7 \times 10^{-18}}{1.4 \times 10^{-16}} \sim 10^{-2} K.$$

This means, for temperatures less than about $0.1K$, the ferromagnetic alignment with a magnetic dipole - dipole interaction is possible. However since the ferromagnetism is observed at room temperature and above, we rule out magnetic dipole - dipole interaction as a cause of ferromagnetism.

Let $\mu_B e\hbar/2mc$ be Bohr magneton for electron. Then, magnetic moment $\overline{\mu}$ associated with the electron is

$$\mu = g\mu_B \vec{J} \tag{10.4}$$

where g is Lande's g factor. $\vec{J}$ is a total angular momentum. For a discussion of 'g' see any standard text on modern physics like that by Eisberg and Resnick.

When the magnetic ion is kept in an external magnetic field, it is acted upon by a local magnetic field[1] consisting of external magnetic field plus the magnetic field due to surrounding environment. For paramagnetism, the local field is almost same as that of external magnetic field. For consideration of ferromagnetism, we need to include field of surrounding environment via Heisenberg exchange interaction.

10.1   Paramagnetism

Let $\vec{H}$ be external magnetic field. We are considering a situation where field produced by surrounding spins is negligible. As a result the local field is same as that of external field. Hamiltonian in this case is,

$$\begin{aligned} \mathcal{H} &= -\vec{\mu}.\vec{H} \\ &= -g\mu_B \vec{J}.\vec{H} \end{aligned} \tag{10.5}$$

Taking external field in z-direction, $\vec{J}.\vec{H} = J_z H$. But J_z is quantized ranging from $-J$ to J; i.e.

$$J_z = -J, -J+1, -J+2..., 0, 1, 2, ...J-1, J.$$

Allowed values of J_z are $(2J+1)$ which are projections of $\vec{J}$ on $\vec{H}$.

---

[1] See introduction to Electrodynamic, Ch. 6 by P.V.Panat(Narosa Publishers(2002))

There are no kinetic energy terms for the system because the magnetic ions are localized. The dimensionless parameter entering in a problem is $g\mu_B H/k_B T \equiv \xi$. Energy of the magnetic ion is

$$\varepsilon = -\frac{g\mu_B H J_z}{k_B T} = -\xi J_z \tag{10.6}$$

Partition function for the system is

$$\begin{aligned}
Z &= \sum_{J_z=-J}^{J} e^{\xi J_z} \\
&= e^{-\xi J} + e^{-\xi(J-1)} + e^{-\xi(J-2)} + \ldots \\
&\quad + 1 + e^{\xi} + e^{2\xi} + \ldots + e^{\xi(J-1)} + e^{\xi J}
\end{aligned} \tag{10.7}$$

This is a geometric series with a common ratio e^{ξ}. Answer of the sum is

$$\begin{aligned}
Z &= \frac{e^{-\xi J} - e^{\xi(J+1)}}{1 - e^{\xi}} \\
&= \frac{\sinh(J + 1/2)\xi}{\sinh(\xi/2)}
\end{aligned} \tag{10.8}$$

Average value of the magnetic moment is $\overline{\mu}_z$ as

$$\begin{aligned}
\overline{\mu}_z &= \frac{\sum_{J_z=-J}^{J} g\mu_B J_z \exp[g\mu_B J_z H/k_B T]}{\sum_{J_z=-J}^{J} \exp[g\mu_B J_z H/k_B T]} \\[2mm]
&= \frac{1}{\beta}\frac{\partial \ln Z}{\partial H}
\end{aligned} \tag{10.9} \tag{10.10}$$

Equation(8) is substituted in equation(10) to get

$$\overline{\mu}_z = g\mu_B J B_J(\xi) \tag{10.11}$$

where $B_J(\xi)$ is Brillouin function as given by

$$B_J(\xi) = \frac{1}{J}\left[\left(J + \frac{1}{2}\right)\coth\left[\left(J + \frac{1}{2}\right)\xi\right] - \frac{1}{2}\coth\frac{\xi}{2}\right] \tag{10.12}$$

Equation(11) is not physically very transparent. We consider its low and high temperature limits.

(i) High temperature limit: Here $\xi = \dfrac{g\mu_B H}{k_B T} \ll 1$. In this limit, the thermal energy far exceeds the energy level difference of the magnetic ion. Then,

$$\lim_{0<\xi\ll 1} B_J(\xi) = \frac{(J+1)\xi}{3} = \frac{(J+1)g\mu_B H}{3k_B T}.$$

Let N be number of magnetic ions/volume. Then M, the magnetic moment per unit volume is

$$M = N\overline{\mu}_z = N\frac{g^2\mu_B^2 J(J+1)}{3k_B T}H. \tag{10.13}$$

Paramagnetic susceptibility is, $\chi_T = (\partial M/\partial H)_T$.

$$\chi_T = N\frac{g^2\mu_B^2 J(J+1)}{3k_B T} \propto \frac{1}{T}$$

Thus,

$$\chi = \frac{C}{T} \tag{10.14}$$

where $C = \dfrac{Ng^2\mu_B^2 J(J+1)}{3k_B}$. Equation (14) is known as Curie's law.

Curie's law is verified for many systems where magnetic ions like Cr^{+++}, Fe^{++}, Gd^{+++} etc. are doped in crystal of chemical compounds. The agreement with Brillouin function is excellent.

(ii) Low temperature limit: Here $\xi = \dfrac{g\mu_B H}{k_B T} >> 1$. It means, the energy of the level separation of the magnetic ion is much larger than the thermal energy. As a result, most of the ions are in ground state

$$\lim_{\xi >> 1} B_J(\xi) = 1$$

Then,

$$M = N\overline{\mu_z} = Ng\mu_B J. \tag{10.15}$$

It means, all the ions are aligned along H, there is a saturation (maximum) magnetization. Susceptibility is zero.

10.2   Ferromagnetism: Weiss Theory

Consider a magnetic ion kept at j^{th} site. We show a typical spin arrangement on a square lattice in figure (1).

It is to be noted that the arrangement need not be a square lattice. We showed it for purpose of a diagram. We make a further approximation that $\vec{S_1}.\vec{S_2} = S_{1z}S_{2z}$. This ignores x and y components of spin. Strictly we should have

$$\vec{S_1}.\vec{S_2} = S_{1x}S_{2x} + S_{1y}S_{2y} + S_{1z}S_{2z} \tag{10.16}$$

If the external field is in z-direction, according to a vector atom model, $\langle S_{1x}\rangle = \langle S_{2x}\rangle = \langle S_{1y}\rangle = \langle S_{2y}\rangle = 0$. Only nonzero term is $S_{1z}S_{2z}$. This is known as Ising model. With this approximation, we also assume that only the nearest neighbor spins interact. In the figure(1), number of nearest neighbors of jth atom are four. For cubic lattice, this number is six. In general, let p be the nearest neighbors of jth magnetic ion. This leads to the Hamiltonian of jth atom is

$$\mathcal{H}_j = -g\mu_B S_{jz}H - 2JS_{jz}\sum_{k=1}^{p} S_{kz} \tag{10.17}$$

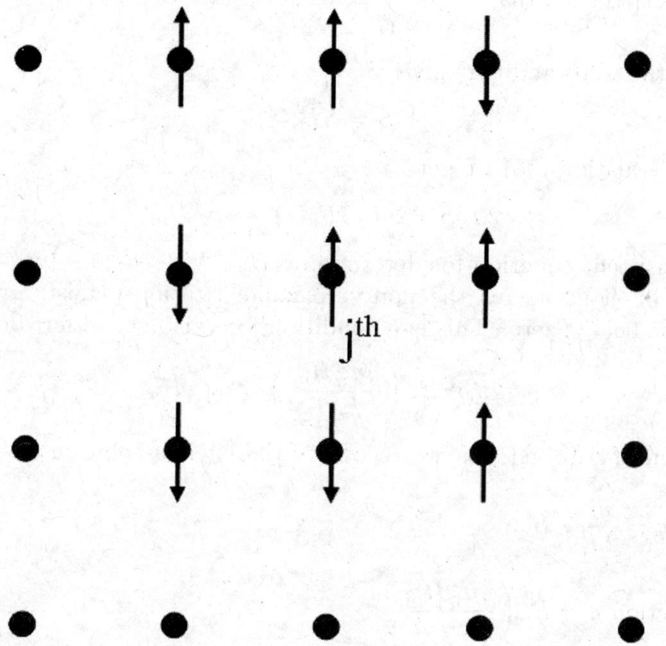

Figure 10.1: Spins at lattice site

In the spirit of mean field theory, we average the contribution of nearest neighbor spins in the form of local field H_m. It is,

$$g\mu_B H_m = 2J \overline{\sum_{k=1}^{p} S_{kz}} \qquad (10.18)$$

Then,

$$\begin{aligned}
\mathcal{H}_j &= -g\mu_B(H + H_m)S_{jz} \\
&= -g\mu_B H_{eff} S_{jz} \qquad (10.19)
\end{aligned}$$

where $H_{eff} = H + H_m$ is effective magnetic field. Compare equation (19) with equation (5). Then, going through identical steps from equations (5) to (11) except replacing H by H_{eff}, we get

$$M = Ng\mu_B S B_s(\xi) \qquad (10.20)$$

where

$$\xi = \frac{g\mu_B(H + H_m)}{k_B T}$$

Let $\bar{S}$ be average spin such that

$$M = Ng\mu_B \bar{S} \tag{10.21}$$

Comparing equation (20) and (21) gives,

$$\bar{S} = SB_s(\xi) \tag{10.22}$$

Comparing with equation (18) we get,

$$2Jp\bar{S} = 2JpSB_s(\xi) = g\mu_B H_m \tag{10.23}$$

This is a self consistent equation for determining H_m. We seek a solution of equation (23) such that $H = 0$. This means the non zero magnetization persists even when external aligning magnetic field is zero. This is a condition for existence of ferromagnetism. Thus,

$$2JpSB_s\left(\frac{g\mu_B H_m}{k_B T}\right) = g\mu_B H_m \tag{10.24}$$

To solve equation (24) to get H_m we recognize that it is a solution of intersection of two curves. They are,

a. straight line $y = \dfrac{g\mu_B}{2JpS}H_m$

and

b. Brillouin function $y = B_s\left(\dfrac{g\mu_B H_m}{k_B T}\right)$

Solution of equation (24) gives T_c. Following are simple mathematical limits

$$\lim_{\xi<<1} \coth\xi \cong \frac{1}{\xi} + \frac{\xi}{3} + ... \tag{10.25}$$

and

$$\lim_{\xi>>1} \coth\xi = 1 \tag{10.26}$$

Brillouin function appearing in equation(22) is same as given in equation (12) except J is replaced by S. Thus,

$$B_s(\xi) = \frac{1}{S}\left[(S+1/2)\coth(S+1/2)\xi - \frac{1}{2}\coth\frac{\xi}{2}\right] \tag{10.27}$$

and is shown in figure(2).

A moment's reflection will reveal that when the slope $\left(\dfrac{g\mu_B}{2JpS}\right)$ of straight line

$$y = \frac{g\mu_B}{2JpS}H_m = \frac{k_B T}{2JpS}\xi$$

is less than the slope $\dfrac{dB_s\left(\dfrac{g\mu_B}{k_B T}H_m\right)}{dH_m}$ at origin, a solution to equation (24) will exist. The transition temperature T_c is then that temperature where the two slopes match. Taking account the limiting forms of $\coth x$, we see that

$$\lim_{\xi<<1} B_S(\xi) = \frac{(S+1)}{3}\xi = \frac{(S+1)}{3}\frac{g\mu_B}{k_B T}H_m \tag{10.28}$$

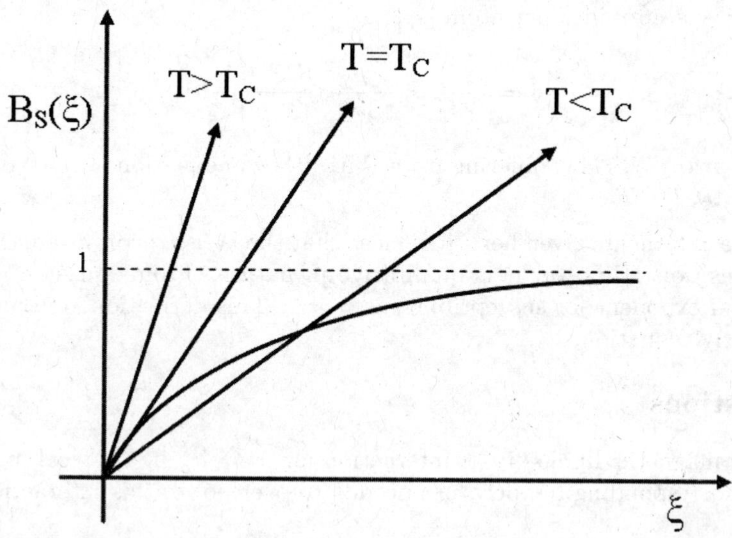

Figure 10.2: $y = B_S(\xi)$ and $y = \dfrac{k_B T}{2JpS}\xi$. Various straight lines show the temperature values relative to T_C

Using equation (24), we immediately find

$$T_c = \frac{2JpS(S+1)}{3k_B}. \tag{10.29}$$

This is mean field approximation. This is so because we are adding a local field to H. It is not negligible. Its effect is to align the spins even if the external field H is zero. Thermal agitations can not destroy the spin polarization for $T < T_c$. To find the susceptibility of ferromagnetic materials above T_c, we retain H in ξ as

$$\xi = \frac{g\mu_B(H + H_m)}{k_B T} \tag{10.30}$$

Using equations (22) and (23) we get,

$$B_S(\xi) = \frac{k_B T}{2JpS}\left(\xi - \frac{Hg\mu_B}{k_B T}\right). \tag{10.31}$$

Using $\xi << 1$ limit of equation (28) and the value of T_c given in equation (29), we get

$$\xi = \frac{g\mu_B}{k_B}\frac{H}{T - T_c}. \tag{10.32}$$

The magnetic moment given by equation (20) is,

$$\begin{aligned}
M &= Ng\mu_B S B_S\xi \\
&= \frac{Ng^2\mu_B^2}{3k_B}\frac{S(S+1)}{T - T_c}H.
\end{aligned} \tag{10.33}$$

Then, magnetic susceptibility per atom is,

$$\chi = \frac{M}{NH} = \frac{g^2 \mu_B{}^2}{3k_B} \frac{S(S+1)}{T - T_c}. \tag{10.34}$$

This is called Curie-Weiss law, differing from Curie's law of equation (14) for paramagnetism by replacing T by $T - T_c$.

From the arguments given here, it is clear that the Weiss theory is a mean field theory of a phase transition. However it is qualitative in nature. From equation (34) it is clear that, the critical exponent for susceptibility is $\gamma = 1$. This derivation is very similar to the derivation of vdW equation.

Short Questions

1. Estimate magnetic dipole-dipole interaction for $M = \mu_B$ in a typical magnetic crystal and the corresponding temperature needed to over power this interaction.

2. Differentiate between the Heisenberg model and the Ising model.

3. Show that the average magnetic moment is given by $\dfrac{1}{\beta} \dfrac{\partial \ln Z}{\partial H}$.

4. Using personal computer, plot $B_J(\xi)$ for different values of J Find the value of ξ when exact value of $B_J(\xi)$ and approximate value of $B_J(\xi) = \dfrac{(J+1)}{3}\xi$ differ from each other by about 5%.

5. Explain qualitatively, how Weiss theory leads to $T_c \neq 0$ and thus the ferromagnetism.

Problems

Problem 1. Consider a paramagnetism of spin 1/2 system. Show

(a) partition function $Z = \left(e^{\beta \mu_B H} + e^{-\beta \mu_B H}\right)$.

(b) Average magnetic moment at absolute temperature T is

$$M = N\mu_B \tanh\left(\frac{\mu_B H}{k_B T}\right)$$

(c) Corresponding Curie constant is

$$C = N\mu_B{}^2 / k_B$$

Problem 2. Consider a spin $1\hbar$ system with $m = 1, 0, -1$. It is degenerate so that $E = \epsilon$ for $m = \pm 1$ and $E = 0$ for $m = 0$. Find average energy of the system and its entropy.

$$\text{Ans:} \quad \bar{E} = \frac{2\epsilon e^{-\beta\epsilon}}{1 + 2e^{-\beta\epsilon}}$$

$$S = \frac{2k_B\epsilon}{T}\frac{e^{-\beta\epsilon}}{1 + 2e^{-\beta\epsilon}} + k_B \ln\left(1 + e^{-2\beta\epsilon}\right)$$

(Hint: Find $Z = 1 + 2e^{-\beta\epsilon}, \quad S = k_B\left[\ln Z + \beta\bar{E}\right]$ evaluate)

Problem 3. Repeat Weiss theory, in the absence of external magnetic field. Obtain expression for $\bar{E}$ and analyze its behavior for $T << T_c$, $T \approx T_c$ and $T >> T_c$ where T_c is curie temperature. Find specific heat in the three temperature ranges given.

$$\text{Ans:} \quad \bar{E} = -2JNpS^2 \qquad\qquad\qquad\qquad\qquad \text{for} \quad T << T_c$$

$$\bar{E} = -\frac{5N^2 k_B T S(S+1)}{S^2 + S + 1}\left[1 - \frac{3k_B T}{2pJS(S+1)}\right] \quad \text{for} \quad T \approx T_c$$

$$\bar{E} = 0 \qquad\qquad\qquad\qquad\qquad\qquad\qquad \text{for} \quad T >> T_c$$

(Hint: Find $Z = \dfrac{\sinh(S+1/2)\xi_m}{\sinh(\xi_m/2)}, \ \xi_m = \dfrac{g\mu_B}{k_B T}H_m$

$H_m = \dfrac{2JpS}{g\mu_B}B_S(\xi_m); \ \bar{E} = -\dfrac{\partial}{\partial\beta}\ln Z$.

Complete the calculations in the limit $\xi << 1, \xi \approx 1, \xi >> 1$.)

Chapter 11

IDEAL FERMI SYSTEM

Electrons are the building block of matter and also the earliest discovered elementary particle. It has a spin $(1/2)\ \hbar$. Electrons obey Pauli principle and hence we have a periodic table. They are called Fermions and have antisymmetric wave function. Metals contain free electrons. Study of a theory of metals was initiated by Lorentz and Drude. They calculated physical quantities like specific heat and conductivity. Even though their physics was almost correct, they went wrong in using Boltzmann statistics for the electrons in metals. They got wrong results. Sommerfeld applied Fermi statistics to the electrons in metal for the first time. He got answers agreeing with the experiments. This was also perhaps the first application of the Fermi statistics. Dirac clarified the relationship between the occupancy of the state and the antisymmetric nature of its wave function. That is why it is also called Fermi-Dirac statistics. Its second major application was due to Fowler and next by Chandrasekhar in astrophysics pertaining to the evolution of star. Subsequently, Landau, Mott, Wilson, Fröhlich, Peierls, Seitz etc. applied it to solid state physics problems.

11.1   Energy and Pressure of an Ideal Fermi Gas at $T = 0$

Consider collection of N noninteracting fermions in volume V at absolute temperature T. The kinetic energy of the fermion of mass m is $\epsilon_k = \hbar^2 k^2 / 2m$. Its momentum is $\vec{p} = \hbar \vec{k}$. There are different values of momentum and therefore different values of wave vector k. We have derived that the mean number of Fermi particles in a state k as

$$n_k = \frac{1}{e^{\beta(\epsilon_k - \mu)} + 1}. \tag{11.1}$$

Here, $\beta = \dfrac{1}{k_B T}$ and μ is a chemical potential. This is a distribution function of a perfect gas obeying Fermi statistics. It goes over to the Boltzmann distribution when μ is negative and its absolute value is very large. Thus, in that case $\exp\left[\beta\left(\epsilon_k - \mu\right)\right] >> 1$. Then equation (1) transforms to

$$n_k = e^{-\beta(\epsilon_k - \mu)} \tag{11.2}$$

which is a Boltzmann distribution. The plot of the Fermi function for $T > 0$ is shown in figure (1). Total number of the Fermi particles is

$$N = \sum_k n_k = \sum_k \frac{1}{e^{\beta(\epsilon_k - \mu)} + 1} \ . \tag{11.3}$$

At $\beta\mu = \dfrac{\mu}{k_B T} >> 1$ and $\epsilon_k << \mu$, $n_k = 1$. At $\epsilon_k = \mu$, $n_k = \dfrac{1}{2}$. The transition region of

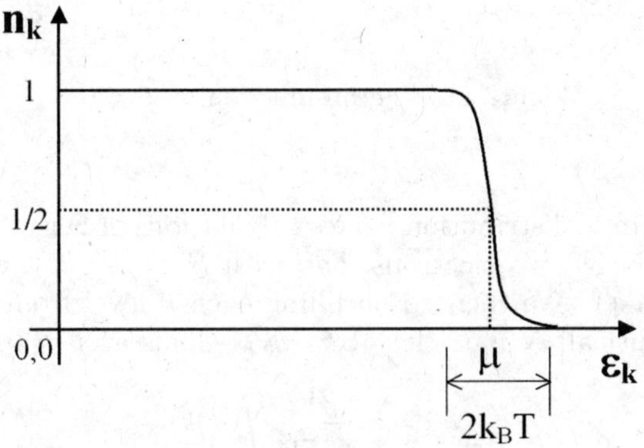

Figure 11.1: Fermi distribution at $T \neq 0$.

n_k is when n_k changes its value from 1 to 0. This energy interval is of the order of $2k_B T$ around μ. When $T \to 0$, $\beta\mu \to \infty$ and the transition region is very sharp. We can take Fermi distribution at $T = 0$ as shown in figure (2) as

$$n_k = \begin{cases} 1 & \epsilon_k < \mu_0 \\[2mm] 0 & \epsilon_k > \mu_0 \end{cases} \tag{11.4}$$

Here μ_0 is taken as a chemical potential of the Fermi gas at $T = 0$. It is also called as Fermi energy ϵ_F. Thus, $\mu_0 = \epsilon_F$ at $T = 0$. The corresponding wave vector k_F is called Fermi wave vector. The energy μ_0 is then $\epsilon_F = \hbar^2 k_F^2/2m$ called Fermi energy. Similarly a Fermi momentum is also defined as $\vec{p_F} = \hbar \vec{k_F}$.

We recall that (see appendix 1 for the detailed derivation) sum over k can be converted into an integral over k when various k levels are closely spaced. The formulae are

$$\begin{aligned} \sum_k f(k) &= \frac{V}{(2\pi)^3} \int f(k) d^3k & \text{in 3 dimensions} \\[2mm] &= \frac{A}{(2\pi)^2} \int f(k) d^2k & \text{in 2 dimensions} \\[2mm] &= \frac{L}{(2\pi)} \int f(k) dk & \text{in 1 dimension .} \end{aligned} \tag{11.5}$$

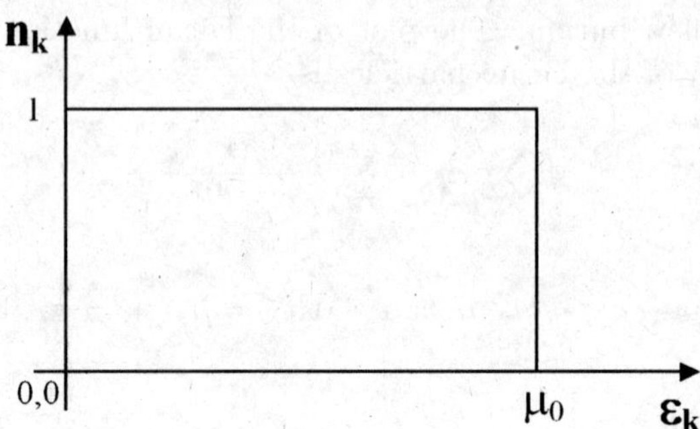

Figure 11.2: Fermi function at $T = 0$.

At $T = 0$, the form of distribution makes calculations of Fermi gas particularly simple. Each k level is occupied by two electrons, one with $S_z = +1/2$ (call it spin $\uparrow$) and other $S_z = -1/2$ (call it spin $\downarrow$). We then go on filling each k level from $k = 0$ onwards by two fermions until we exhaust all N particles at $k = k_F$. Thus at $T = 0$,

$$
\begin{aligned}
N &= \frac{2V}{(2\pi)^3} \int d^3k \\
&= \frac{V}{\pi^2} \int_0^{k_F} k^2 dk \\
&= \frac{V k_F^3}{3\pi^2}
\end{aligned}
$$

Thus, the Fermi wave vector k_F is given by

$$
k_F = \left(\frac{3\pi^2 N}{V} \right)^{1/3} \tag{11.6}
$$

in three dimensions.

The corresponding energy of the highest occupied k level i.e. of k_F is

$$
E_F = \frac{\hbar^2 k_F^2}{2m} \tag{11.7}
$$

This gives rise to Fermi momentum $p_F = \hbar k_F$.

Total energy of Fermions at $T = 0$ is

$$
\begin{aligned}
E_0 &= \sum_k \frac{\hbar^2 k^2}{2m} = \frac{V 4\pi}{(2\pi)^3} \int_0^{k_F} k^4 dk \\
&= \frac{V}{(\pi)^2} \frac{\hbar^2}{2m} \frac{k_F^5}{5} = \frac{V}{(\pi)^2} \frac{\hbar^2}{2m} \frac{3\pi^2 N}{V} \frac{k_F^2}{5} \\
&= \frac{3}{5} N \frac{\hbar^2 k_F^2}{2m} = \frac{3}{5} N \epsilon_F \tag{11.8}
\end{aligned}
$$

Here, a factor of 2 comes because of spin degeneracy $(2S + 1)$ for $S = \dfrac{1}{2}$. Pressure of the Fermi gas at $T = 0$ is nonzero since the entire gas is not standstill but has a kinetic energy even at $T = 0$. The pressure P of a Fermi gas is

$$P = -\frac{\partial E}{\partial V} = \frac{2}{5}\frac{N}{V}E_f .\tag{11.9}$$

The formulae (6) to (9) for energy E and pressure P can be used as an excellent approximation when the temperature T is 'close' to zero. We define an equivalent temperature of E_F by

$$E_F = \frac{\hbar^2 k_F^2}{2m} = k_B T_F .\tag{11.10}$$

T_F is called Fermi temperature. We can now define what is meant by T "close" to zero. This limit is actually achieved when $T/T_F << 1$. T_F is sometimes called as a degeneracy temperature. Electron gas or for that matter, any fermion gas at $T = 0$ is called a completely degenerate gas. Electron gas at a temperature T s.t. $T << T_F$ is called strongly degenerate electron gas.

Illustrative Problems

Problem 1: Consider a plasma of N electron and ions of charge Ze each. Show that, as the plasma density increases, it tends to become an ideal gas and obtain the condition on its density to realize it.

Solution: When the kinetic energy of electrons in a plasma dominates over the Coulombic potential energy between the electrons and nuclei of charge Ze then plasma tends to become an ideal gas. To arrive at the restriction on the density the argument goes as follows. Consider N electrons in a plasma in volume V . Let corresponding number of ions be N' such that $N = ZN'$. The volume per ion is $V/N' = \frac{ZV}{N} \approx a^3$. Here, a is the average distance between the electrons and the nuclei. Coulombic potential energy is of the order of $\frac{Ze^2}{a}$. Corresponding electron kinetic energy of the plasma is $\approx E_F$. The desired condition is then $E_F >> Ze^2/a$
or

$$\frac{N}{V} >> Z^2 \left(\frac{me^2}{\hbar^2}\right)^3$$

Clearly, the corresponding Fermi temperature is of the order of $Z^{4/3}$.

Problem 2: Find k_F, E_0, E_F and P for extremely relativistic electron gas in 3 dimension

Solution: Extreme relativistic limit is obtained when rest energy is $mc^2 << cp$. Then in that case

$$E_p = \sqrt{(c^2 p^2) + m^2 c^4} \approx cp .$$

or

$$E_k = \hbar c k$$

Clearly,

$$k_F = \left(3\pi^2 N/V\right)^{1/3}$$

$$
\begin{aligned}
E_0 &= \frac{V}{(2\pi)^3}.2.4\pi^2\hbar c \int_0^{k_F} k^3 dk \\
&= \frac{3}{4}Nc(\hbar k_F)
\end{aligned}
\tag{11.11}
$$

and corresponding pressure $= P = -\partial E/\partial V$ is obtained to be

$$P = \frac{E_0}{3V} = \frac{1}{4}\frac{N}{V}c\hbar k_F \tag{11.12}$$

11.2   Thermodynamic Quantities of an Ideal Fermi Gas at Finite Temperature

(a) Chemical Potential $\mu(T)$

Chemical potential at $T \neq 0$ is determined by the implicit relation

$$N = \sum_k n_k = \frac{V}{(2\pi)^3}.2.4\pi \int_0^{\infty} \frac{k^2 dk}{e^{\beta(\epsilon_k - \mu)} + 1} \tag{11.13}$$

Here, the upper limit of integration is extended to ∞ because the maximum momentum is not restricted to $p_F = \hbar(3\pi^2 N/V)^{1/3}$, but excitation can occur to any level with appropriate probability. With $\frac{\lambda}{2\pi} = \dfrac{\hbar}{\sqrt{2mk_BT}}$, and $(\hbar^2 k^2/2mk_BT) = \xi$, $n = N/V$ and $\beta\mu = \alpha$, we transform equation(13) to

$$\left(\frac{n\lambda^3}{4\pi}\right) = \int_0^{\infty} \frac{\xi^{1/2}d\xi}{e^{\xi - \alpha} + 1} \tag{11.14}$$

As shown in Appendix(4), equation (7)

$$\left(\frac{n\lambda^3}{4\pi}\right) = \frac{2}{3}\alpha^{3/2}\left(1 + \frac{\pi^2}{8\alpha^2} + ...\right) \tag{11.15}$$

Taking the approximate value of $\alpha \approx \frac{E_F}{k_BT} = \frac{T_F}{T} >> 1$, equation (15) simplified as,

$$\mu = E_F\left(1 - \frac{\pi^2}{12}\left(\frac{T}{T_F}\right)^2 + ...\right) \tag{11.16}$$

(b) Similarly, energy of the ideal Fermi gas at $T \neq 0$ is given by

$$E = \frac{V}{(2\pi)^3}.2.4\pi\frac{\hbar^2}{2m}\int_0^{\infty} \frac{k^4 dk}{e^{\beta(\epsilon_k - \mu)} + 1}. \tag{11.17}$$

With the previous substitutions made for evaluation of equation (13), we get

$$\frac{E}{V} = \frac{8\hbar^2\pi^3}{m\lambda^5}\left[\frac{2}{5}\alpha^{5/2} + \frac{\pi^2}{4}\alpha^{1/2}\right]$$

(11.18)

We can substitute correct finite temperature value of α using equation (16) as,

$$\sqrt{\alpha} = \sqrt{\beta\mu(T)} = \left(\frac{T_F}{T}\right)^{1/2}\left[1 - \frac{\pi^2}{24}\left(\frac{T}{T_F}\right)^2 + \ldots\right]$$

(11.19)

Then, with $\left(3\pi^2 n\right)^{2/3}\frac{\lambda^2}{4\pi^2} \cong \alpha \cong (T_F/T)$,

$$\frac{E}{N} = \frac{3}{5}E_F\left(1 + \frac{5}{12}\pi^2\left(\frac{T}{T_F}\right)^2 + \ldots\right)$$

(11.20)

(c) Pressure of the Fermi gas is obtained by evaluating the grand partition function

$$\frac{PV}{k_BT} = \ln Q(\mu,V,T) = \sum_k \ln\left(1 + e^{\alpha-\beta\epsilon_k}\right)$$

(11.21)

$$= \frac{V}{(2\pi)^3}.4\pi.2.\int_0^\infty k^2 dk \ln\left(1 + \alpha e^{-\beta\epsilon_k}\right)$$

Integrating by parts and after some simplification, we get,

$$\frac{PV}{k_BT} = \frac{2}{3}\frac{1}{k_BT}\frac{V}{(\pi)^2}\frac{\hbar^2}{2m}\int_0^\infty \frac{k^4 dk}{e^{\beta\epsilon_k-\alpha}+1} = \frac{2}{3}\frac{E}{k_BT}$$

(11.22)

Clearly then, equation (22) simplifies to

$$PV = \frac{2}{5}NE_F\left[1 + \frac{5}{12}\pi^2\left(\frac{k_BT}{E_F}\right)^2 + \ldots\right]$$

(11.23)

Clearly, we see from equation (23) that $P \neq 0$ even at very low temperatures because only two particles are at zero momentum state and all others have finite momentum due to Pauli principle.

(d) Specific heat of Fermi gas: This is historically a very important problem. Crudely speaking, in the metal, the ions are regularly placed, forming a lattice and the electrons are freely moving inside the metal. Lattice vibrations are the lattice excitations. As we will see when we discuss the applications of Bose statistics, lattice contributes to a specific heat $\approx T^3$. Electrons of the metal contribute very little to the specific heat, because, for a given temperature T, only few electrons (in the neighborhood of k_BT around E_F) are excited. This is obvious from the figure (1) of the Fermi distribution. Thus, the specific heat of metal at low temperature must behave as

$$C_V = AT + BT^3$$

(11.24)

Here the constants A and B are such that, for metals, at very low temperatures

$$C_V = AT \tag{11.25}$$

where

$$A = \frac{\pi^2}{2} \frac{N k_B^2 T}{E_F}. \tag{11.26}$$

It means, the electronic contribution to specific heat is linear in T.

If the electrons obeyed classical statistics, we have an equipartition theorem. In that case, we will get a constant value of C_V. This contradicts experimental result. Experimentally, equation (25) is correct at low temperatures for metal. Equation (25) and (26) are immediate, when we differentiate equation (20).

(e) Entropy of a Fermi gas: We use the second law of thermodynamics to find entropy as

$$S = \int_0^T \frac{C_V \, dT}{T} \tag{11.27}$$

We use expression (26) for C_V as

$$C_V = \frac{\pi^2}{2} \frac{N k_B^2 T}{E_F}$$

$$\therefore S = \frac{\pi^2}{2} \frac{N k_B}{T_F} \int_0^T dT = \frac{\pi^2}{2} \frac{N k_B T}{T_F} \tag{11.28}$$

We clearly see that as $T \to 0$, $S \to 0$. This is true because, at $T = 0$, the system has only one accessible state and hence $S = k_B \ln 1 = 0$

(f) Free energy of the Fermi gas: Free energy is

$$F = E - TS$$

using equations (20) and (25), we get

$$F(V,T) = \frac{3}{5} N E_F \left[1 - \frac{5}{12} \pi^2 \left(\frac{k_B T}{E_F} \right)^2 + \dots \right] \tag{11.29}$$

(g) Enthalpy and Gibbs free energy of Fermi gas:
Enthalpy

$$H = E + PV$$

Using equation (20) and (23), we get

$$H = N E_F \left(1 + \frac{5}{12} \pi^2 \left(\frac{T}{T_F} \right)^2 + \dots \right) \tag{11.30}$$

Gibbs free energy is

$$G = E - TS + PV = F + PV \tag{11.31}$$

Using equation (23) and (30), we get,

$$G = NE_F \left(1 - \frac{\pi^2}{12}\left(\frac{T}{T_F}\right)^2 + ...\right) \tag{11.32}$$

From equation (16)

$$G = N\mu \tag{11.33}$$

Illustrative Problem

Problem 3: Estimate Fermi energy E_F and Fermi temperature T_F of (i) electrons in typical metal (ii) nucleon in heavy nucleus (iii) He^3 atom in liquid He^3; atomic volume being 46 $A^{\circ 3}$/atom.

Solution: We have to find $\epsilon_F = \hbar^2 k_F^2 / 2mk_B T_F$ with $k_F = \left(\dfrac{3\pi^2 N}{V}\right)^{1/3}$. Thus,

(i) Take Sodium as a typical metal.

Electron concentration is $\dfrac{N}{V} = n = 2.65 \times 10^{22}/c.c..$

$$\therefore k_F = \left(3\pi^2 \times 2.65 \times 10^{22}\right)^{1/3} \approx 0.92 \times 10^8 / cm$$

$$E_F = \frac{\hbar^2}{2m}.k_F^2 \cong \frac{10^{-54} \times 10^{16}}{2 \times \times 9.1 \times 10^{-28}} \, ergs$$

But, $1.6 \times 10^{-12} erg = 1eV$

$$\therefore E_F \cong \frac{10^{-54} \times 10^{16}}{18.2 \times \times 10^{-28} \times 1.6 \times 10^{-12}} \cong 3.23 eV$$

Corresponding Fermi temperature is easy to obtain by $k_B T(in\ ergs) = E_F(in\ ergs)$. Or, everyone should remember a following rule of thumb. Room temperature$\approx 300^\circ K = 1/40 eV$

$$\therefore 3.2\ eV = 3.2 \times 40 \times 300 = 38400\ K.$$

This clearly shows that even at $1000^\circ C$, a typical metal can be treated almost at $T = 0\ K$. The temperature corrections to the thermodynamic quantities that we applied, being of the order of (T^2/T_F^2), formulae like (16 to 30) are excellent in accuracy.

(ii) Nucleons are Fermions. We do not distinguish between protons and neutrons for simplification. Let A be atomic mass number. We also assume that mass of each nucleon is about $1000 MeV/c^2$ in this approximate calculation. Here c is speed of light $c \approx 3 \times 10^8 m/s$ From nuclear physics, radius of nucleus is $R = R_0 A^{1/3}$ and its volume is $(4/3)\pi R^3 \cong R_0^3\ A$. Here, $R_0 = 1.4 \times 10^{-13} cms$. Consider a nucleus of $A = 200$

$$\therefore V = 4 \times 200 \times 10^{-39} \times 2.744 = 21.952 \times 10^{-37} cm^3$$

$$\therefore k_F = \left(\frac{3\pi^2 \times 200}{22 \times 10^{-37}} \right)^{1/3} \approx 31 \times 10^{12} cm^{-1}$$

$$E_F = \frac{\hbar^2 k_F^2}{2m} \cong \frac{10^{-54} \times 96 \times 10^{24}}{2 \times 200 \times 1.67 \times 10^{-24}} \approx 10^{-7} \times 1.7 ergs \ .$$

$$\therefore E_F \cong \frac{1.7 \times 10^{-7}}{1.6 \times 10^{-12}} \cong 10^5 eV = 0.1 MeV$$

and

$$T_F = \frac{1.7 \times 10^{-7}}{1.38 \times 10^{-16}} \approx 10^9 K \ .$$

(iii) For He^3, Atomic volume$= (46 \times 10^{-8})^3 /atom \cong 8.7 \times 10^{-20} cm^3$

$$k_F = \left[\frac{3\pi^2 \times 1}{(46 \times 10^{-8})^3} \right]^{1/3} \cong \frac{3}{46 \times 10^{-8}} \cong 6.5 \times 10^6 cm^{-1}$$

$$E_F = \frac{10^{-54} \times 42 \times 10^{10}}{3 \times 1.67 \times 10^{-24}} \cong 8 \times 10^{-20} ergs, \ T_F = \frac{8 \times 10^{-20}}{1.38 \times 10^{-16}} \approx 6 \times 10^{-4} K$$

Thus, He^3 has very low degeneracy temperature.

Superconductivity of metals is due to formation of bound pairs of electrons. He^3 also exhibits superfluidity but is vastly different from He^4 superfluidity (Boson condensation). He^3-He^3 form bound pairs at millidegree kelvin, because, to form a pair, one needs a small attractive interaction and a degenerate Fermi gas background. This is possible if $T < T_F$. Thus He^3 superfluidity, with similar mechanism of metallic superconductivity (BCS type), is possible at very low temperature of the order of milli kelvin.

11.3   Applications of F-D Statistics

(a)<u>Thermionic emission:</u> Metals have free electrons but they are not emitted spontaneously. This means, near its surface, some kind of potential exist. This forms a potential well. This potential well is not infinitely deep, because, otherwise we will not be able to extract electrons from the metal. Thus we will have to model the electrons in the metal to be confined in a potential well of finite depth. The model is such that there are levels of energy that start from bottom of the well, each level has two electrons. The well is deep enough so that there are enough levels to accommodate all the valance electrons. Actually, what happens in the model is following. Consider a following jellium model.

Here we assume, as if the positive charge is uniformly distributed up to surface, which terminates at $z = z_0$. Deep in the metal, density of the positive charge ρ_0 is uniform. Negative charge due to electrons are mobile. Near the surface at z_0, the electrons ooze out for $z > z_0$ and locally the charge imbalance is created.

As seen from figure (11.3), a dipole layer is created near the metallic surface. Width of this layer is couple of angstroms. This dipole layer creates a finite potential well where in the electrons in various levels are stacked. The same figure also shows the potential created

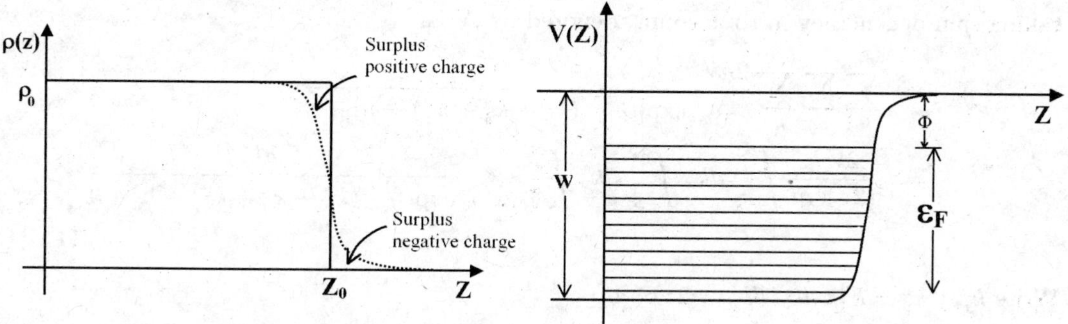

Figure 11.3: Left panel: Charge distribution near metallic surface. Right panel: Work function of the metal.

by the dipole layer. Potential at $z = \infty$ is then taken to be zero. Inside the metal, electrons move freely and more or less, we know that, at least in simple metals, they can be treated as noninteracting. The well is filled up upto $E = E_F < W$. The difference $W - E_F = \phi$ is called work function of the metal and is its an important parameter. If somehow, we can provide a minimum energy ϕ to the electron, it comes out of the surface. The various properties of surfaces of simple metal such as surface energy, work function, dipole layers and surface rearrangement (details of structure of positive and negative charge densities) can be calculated by using what is known of Density functional theory[1]. Even though these calculations are not difficult they are highly computational.

Consider an electron at the bottom of the potential well in figure (11.3). Velocity of electron v_z greater than $\sqrt{2W/m}$ will be necessary for its escape. At the same time, electrons near Fermi level will require $v_z > \sqrt{2\phi/m}$ for escape. This necessary increase of energy of electrons can be achieved by supplying a thermal energy i.e. heating. Resulting emission was studied in details by Richardson and is called thermionic emission. In the earlier days, thermionic emission was important because most of radio technology was based upon it. The important nonlinear element of radio was thermionic valve where the heated cathodes were electron emitters.

In such emission, important experimental quantity is the current. We will assume that any electron with z component of velocity $v_z > \sqrt{2W/m}$, when hit the surface, escapes. Consider now a cylindrical surface of unit cross-section area and of length v_z. All the electrons in this cylinder will hit the surface within 1 second and will escape. Thus number of electrons hitting a unit area of the metal in 1 second and having $v_z > \sqrt{2W/m}$ are needed to be found. Corresponding wave vector k_z is equal to $\sqrt{2mW}/\hbar$. The required number

[1] Angelos Michaelides and Matthias Scheffler, 'An introduction to theory of metal surfaces', $http://www.fhi-berlin.mpg.de/th/publications/Michaelides\_Scheffler\_Chapter\_Submission.pdf$

taking spin degeneracy in to account, denoted ny N is,

$$
\begin{aligned}
N &= \sum_{k_x}\sum_{k_y}\sum_{k_z} \frac{2\hbar k_z}{m}\frac{1}{\exp[\beta\frac{\hbar^2}{2m}\left(k_x^2+k_y^2+k_z^2\right)-\mu]+1} \\
&= \frac{2V.\hbar}{(2\pi)^3 m}\int_{-\infty}^{\infty}dk_x\int_{-\infty}^{\infty}dk_y\int_{\sqrt{2mW}/\hbar}^{\infty}\frac{k_z dk_z}{\exp[\frac{\beta\hbar^2}{2m}\left(k_x^2+k_y^2+k_z^2\right)-\alpha]+1}
\end{aligned}
$$

$$\text{(11.34)}$$

Write $k_x^2+k_y^2=k_\perp^2$, $dk_x\,dk_y=2\pi k_\perp dk_\perp$.

$$
\begin{aligned}
\therefore\quad N &= \frac{2V}{(2\pi)^2}\frac{\hbar}{m}\int_{\sqrt{2mW}/\hbar}^{\infty}k_z dk_z\int_0^{\infty}\frac{k_\perp dk_\perp}{\exp[\frac{\beta\hbar^2}{2m}\left(k_\perp^2++k_z^2\right)-\alpha]+1} \\
&= \frac{V}{2\pi^2}\frac{mk_BT}{\hbar^2}\frac{\hbar}{m}\int_{\sqrt{2mW}/\hbar}^{\infty}k_z dk_z\ln\left[1+\exp\left[\alpha-\frac{\beta\hbar^2 k_z^2}{2m}\right]\right]
\end{aligned}
$$

$$\text{(11.35)}$$

We now make approximation, $\alpha-\dfrac{\beta\hbar^2 k_z^2}{2m}=\dfrac{\mu-\varepsilon_z}{2m}$ is close to $\dfrac{\varepsilon_F-W}{k_BT}=-\dfrac{\phi}{k_BT}$. Characteristic work functions are of the order of few eV. Even at 1500^oC $(\phi/k_BT)$is around $(1/8)$ eV. This means $e^{-\phi/k_BT}<<1$. Then we use $\ln(1+x)\approx x$ for $0<x<1$.

    Thus,

$$
\begin{aligned}
N &= \frac{V}{2\pi^2}\frac{mk_BT}{\hbar^2}\int_{\sqrt{2mW}/\hbar}^{\infty}k_z dk_z\exp\left[\alpha-\frac{\beta\hbar^2 k_z^2}{2m}\right] \\
&= \frac{V}{2\pi^2}\frac{mk_BT}{\hbar^2}.\frac{\hbar}{m}\exp\left[\mu/k_BT\right]\frac{mk_BT}{\hbar^2}\exp\left[-W/k_BT\right] \\
&= V\frac{4\pi m}{h^3}\left(k_BT\right)^2\exp\left[(\mu-W)/k_BT\right]
\end{aligned}
$$

$$\text{(11.36)}$$

Thermionic current density is $J=Ne/V$. Thus

$$
J = \frac{4\pi me}{h^3}\left(k_BT\right)^2\exp\left[(\mu-W)/k_BT\right]
$$

$$\text{(11.37)}$$

Now really comes the question. To find $\alpha=e^{\mu/k_BT}$, what statistics one should use. If we use the classical value of $\alpha=e^{\mu/k_BT}$, we get current as,

$$
J_{class} = ne\left(\frac{k_B}{2\pi m}\right)^{1/2}T^{1/2}\exp\left[-W/k_BT\right]
$$

$$\text{(11.38)}$$

    Here, we need to interpret $W=\phi$ =work function. If we use F-D statistics, then, $\mu\approx\varepsilon_F$ even when the temperature $T\sim 1500^oC$ since $(T/T_F)^2\sim 10^{-2}$. Then, $\mu-W\approx\varepsilon_F-W=\phi$. The thermionic current is

$$
J = \frac{4\pi me}{h^3}k_B^2 T^2\exp\left[(-\phi)/k_BT\right]
$$

$$\text{(11.39)}$$

Detailed experimental analysis in conjunction with electron diffraction shows that equation (39) describes the reality better than the equation (38).

These current equations are still incorrect since we have left some physics - so called Schottky effect.

What happens is that the electric field F is applied to extract the electron from cathode. We assume a uniform field in the region and is perpendicular to the metallic surface. Then the potential well becomes shallower by an amount eFz where z is distance of thermionic electron from the surface. then, if the electron is at a distance z, its image is at $-z$. The two will attract with a force $-\dfrac{e^2}{(2z)^2}$ leading to a potential $-e^2/4z$. Then effective value of W due to these electric effects is

$$W_{eff} = W - eFz - \frac{e^2}{4z} \tag{11.40}$$

and has a maximum value $W_{eff}^{max} = W - e^{3/2}F^{1/2}$. Writing J_0 as value of the current as given by the equation (39), we see that the observed current due to Schottky effect is

$$J_F = J_0 \exp\left[(e^{3/2}F^{1/2})/k_BT\right] \tag{11.41}$$

The theory presented here is valid for fields of the order of Million volts/cm. For higher fields the cold emission results and have to consider quantum mechanical tunneling.

This cold emission of electrons from metals and semiconductors upon the application of high electric fields of the order of 10^9 V/m is called field emission. It is governed by Fowler-Nordheim equation

$$I = aV^2 \exp\left(-\frac{b\phi^{\frac{3}{2}}}{V}\right), \tag{11.42}$$

where a and b are constants, ϕ is the work function, and V is the applied voltage, and I is the field emission current. The Fowler-Nordheim plot is a graph of $\log(I/V^2)$ vs. $1/V$, as shown in Figure 11.4.

Short Questions

1. Draw the Fermi distribution at $T \neq 0$. Explain why in the energy width $2k_BT$ around μ, the distribution goes to zero.

2. Why can we apply the $T =)$ result to the electronic properties of copper metal even at $1000^\circ C$?

3. Under what conditions F-D distribution becomes M-B distribution?

4. Explain why the Fermi gas exerts the pressure even at $T = 0$?

5. Explain quantitatively as to why the specific heat of electron gas is linear with T where $T << T_F$.

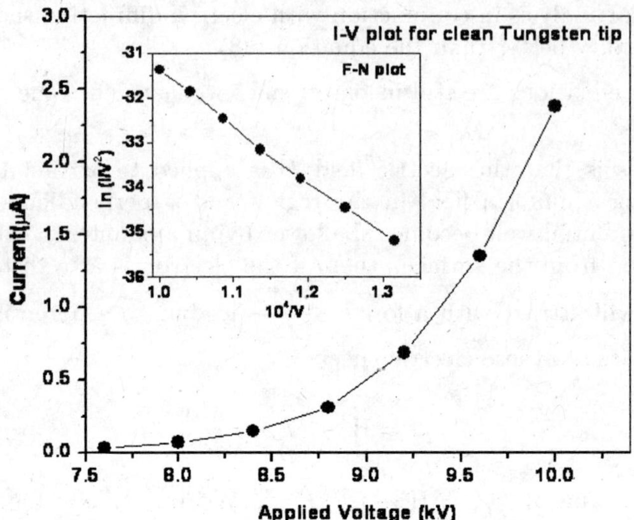

Figure 11.4: Fowler-Nordheim plot for Tungsten. The inset shows $\log(I/V^2)$ as a function of $10^4/V$. Courtesy: D. S. Joag, Department of Physics, University of Pune.

6. Explain the behavior of thermionic current when the electron obey classical statistics.

7. Explain microscopically as to what is meant by the work function?

8. Explain what is shottky effect.

Problems

Problem 1: (a) Estimate ε_F for Na. Atomic weight of $Na = 23$, and its density $\rho_{Na} = 0.95\ gm/c.c.$

(b) Experiment needs a $100\ gm$ sample of Na to be cooled from $1\ K$ to $0.3\ K$ by keeping it in contact with liquid Helium 3 at its temperature $0.3K$. Lattice specific heat at these temperatures can be ignored compared with electronic part. If 0.8 Joule heat is necessary to evaporate $1\ c.c.$ of He_3, find how many $c.c.$ of He_3 are vaporized when the sodium sample was cooled.

$$\left(\text{Hint: (b) Heat removed} = \int_{0.3}^{1} C_V\,dT = \frac{\pi^2}{2}\cdot\frac{\nu R}{\varepsilon_F}\cdot\frac{T^2}{2}k_B\bigg|_{0.3}^{1}\quad \text{where } \nu \text{ is number of moles.}\right.$$

The volume of He_3 evaporated is

$$\frac{\pi^2}{4}\cdot\frac{95}{23}\cdot\frac{8.3143\times10^7}{0.8\times10^7}\cdot\frac{18.2\times1.38\times10^{-16}}{10^{-10}}\cdot\left(1-\frac{9}{100}\right)\bigg)$$

Problem 2: Show that the classical value of thermionic current is

$$J_{class} = ne \left(\frac{k_B T}{2\pi m}\right)^{1/2} \exp\left[-\phi/k_B T\right]$$

Problem 3: Consider a metal surface in equilibrium with its own electrons at a temperature T. Density of the electrons outside the metal surface is so small that the classical statistics apply, and also the temperature T is such that $k_B T << \phi$. By equating chemical potential of electrons for outside and inside the surface, show that the equilibrium electron density outside metal surface is

$$n = \frac{2}{h^3}\left(2\pi m k_B T\right)^{3/2} \exp\left[-\phi/k_B T\right]$$

(Hint: $\mu_{classical} = -k_B T \ln\left[\sum \exp\left[-\beta p^2/2m\right]/N\right] = -k_B T \ln\left[\frac{8\pi^3}{4\pi^{3/2} n \lambda^3}\right]$

and $\dfrac{\lambda}{2\pi} = \dfrac{\hbar}{\sqrt{2 m k_B T}}$. Equate $\mu_{classical}$ with $-\phi$ and get the desired result.)

Problem 4: Fill the missing steps and obtain all thermionic potentials, viz, E, S, H, G and F, as given in the section 3 of this chapter.

Chapter 12

ELECTRONS IN MAGNETIC FIELD

Aim of this chapter is to discuss statistical mechanics of electrons (Fermions) in an external magnetic field $\vec{B}$. The resulting effects lead to para and diamagnetism. This theory is applicable to metals only, where the electrons are modeled as a free, degenerate electron gas. We will not discuss ferromagnetism in metals. We discussed magnetism in insulators in chapter (10) where in, we assume a tiny magnetic moment localized at each lattice point and interacting among themselves, either by Ising or Heisenberg Hamiltonians or their variants.

Thus, our system is itinerant (moving) free electrons of mass m - each having spin 1/2, intrinsic magnetic moment $\mu_B = \dfrac{e\hbar}{2mc} =$ Bohr magneton and a wave vector k $(p = \hbar k)$. In the absence of external magnetic field, one has a standard electron gas. By application of the magnetic field, two things happen, (i) magnetic moment of electrons, on average is aligned along the magnetic field resulting in paramagnetism. This paramagnetism is called Pauli paramagnetism (ii) electrons follow a helical path. This makes the current in such a manner that the magnetic field produced by it is in opposite direction to the external magnetic field. This is diamagnetism - first discussed by Landau and is called Landau diamagnetism.

When the spin - orbit interaction is weak, Pauli paramagnetic susceptibility and the Landau diamagnetic susceptibility are additive. However, paramagnetic contribution dominates over diamagnetic contribution to the susceptibility.

To discuss the diamagnetic part, we have to discuss the motion of electron in external magnetic field[1]. Quantum mechanical treatment of free electron in external magnetic field, where its eigenfunctions and eigen values are obtained by solving Schrödinger equation, is to be found in Quantum Mechanics - Vol 3 Landau and Lifshitz (Pergamon Press(1975)). In this chapter, we will only summarize the result.

---

[1]See Classical Mechanics by P. V. Panat page 70, (Narosa Publishers(2004)). See also Introduction to Electrodynamics by Capri and Panat page 207, (Narosa Publishers(2002))

12.1   Electron in Magnetic Field

a) Vector potential $\vec{A}(\vec{r})$ for external magnetic field $\vec{B}$ is such that

$$\vec{B} = \nabla \times \vec{A} \tag{12.1}$$

$\vec{A}$ is not unique and there are many gauge equivalent vector potentials that give same $\vec{B}$.

b) With $\vec{v}$ as velocity, the canonical momentum $\vec{p}$ is related with the mechanical momentum $m\vec{v}$ as,

$$
\begin{aligned}
\vec{p} &= m\vec{v} + e\vec{A} && \text{(MKSI)} \\
&= m\vec{v} + \frac{e}{c}\vec{A} && \text{(Gaussian)}
\end{aligned}
\tag{12.2}
$$

Corresponding Hamiltonian is,

$$
\begin{aligned}
H(\vec{p}, \vec{r}) &= \frac{1}{2m}\left(\vec{p} - \frac{e}{c}\vec{A}\right)^2 && \text{(Gaussian)} \\
&= \frac{1}{2m}\left(\vec{p} - e\vec{A}\right)^2 && \text{(MKSI)}
\end{aligned}
\tag{12.3}
$$

The orbit is obtained by solving

$$\frac{\partial H}{\partial p_i} = \dot{q}_i, \quad \frac{\partial H}{\partial q_i} = -\dot{p}_i \tag{12.4}$$

c) The trajectory so obtained is helical path for the electron. The $x - y$ motion is a circle with angular frequency $\omega = \dfrac{e|\vec{B}|}{mc}$ (gaussian) and a free motion along z axis. Here we assume that $\vec{B}$ is uniform and is along z axis.

d) The $x - y$ motion can be thought of as a simple harmonic motion with frequency $\omega = \dfrac{eB}{m[c]}$ which, when quantized, leads to the energy $\left(n + \frac{1}{2}\right)\hbar\omega$. The z motion is free, contributing $\dfrac{\hbar^2 k_z^2}{2m}$ to energy. Thus schrödinger equation, when solved for the situation, is expected to give energy eigen values as

$$\epsilon(k_z, n) = \frac{\hbar^2 k_z^2}{2m} + \left(n + \frac{1}{2}\right)\hbar\omega \tag{12.5}$$

Exactly same result is obtained by Landau. He also obtains eigenfunction as Harmonic oscillator wave functions multiplied by $e^{ik_z z}$. Only difference is that, the center of oscillator is shifted. Various harmonic oscillator levels denoted by n are called Landau levels.

12.2   Degeneracy of Levels

We will use gaussian system of units.

Quantum mechanically, energy associated with the circular orbit is quantized in the units of $\dfrac{eB}{m[c]}\hbar$. The circular orbit motion is labeled by n and this energy is discrete.

It must be noted that motion in magnetic field is such that, the energy of the charged particle does not change; only its momentum changes. Also, since Lorentz force does not affect a motion parallel to B (since it is zero in this case), p_z is constant and therefore $\hbar^2 k_z^2/2m = p_z^2/2m$ is constant. If $\vec{B}$ was absent, contribution to energy due to $x-y$ motion would have been $\left(p_x^2 + p_y^2\right)/2m$. This means all the $x - y$ motion is transformed in to discrete harmonic oscillator levels. Thus, there is a high level of degeneracy in the levels, which we calculate. Clearly, all those levels for which $\left(p_x^2 + p_y^2\right)/2m$ lie between $\dfrac{eB}{m[c]}n\hbar$ and $\dfrac{eB}{m[c]}(n+1)\hbar$, coalesce together, in a single level and $p_z^2/2m$ is unaffected. This single level is characterized by the quantum number n.

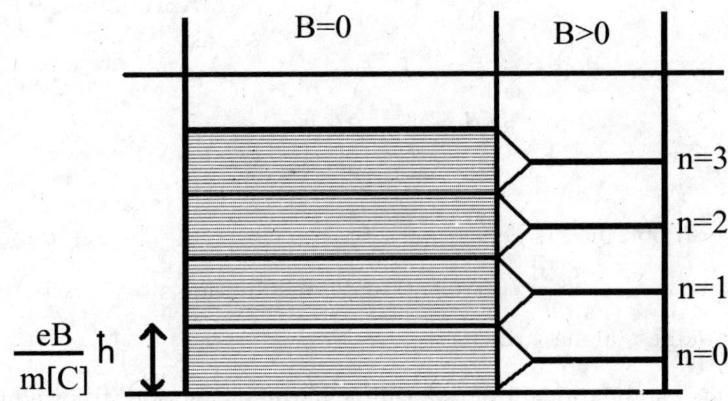

Figure 12.1: Single particle energy levels of p_x and p_y coalesce into discrete harmonic oscillator levels

Clearly, the degeneracy of the level n is equal to total levels

$$\frac{1}{(2\pi)^2}\int dx\ dy\ \ dk_x\ dk_y.$$

Thus, g, the degeneracy is,

$$
\begin{aligned}
g &= \frac{L_x L_y}{(2\pi)^2} . 2\pi \frac{k^2}{2} \\[2mm]
&= \frac{L_x L_y}{4\pi} k^2 \\[2mm]
&= \frac{L_x L_y}{4\pi} . \frac{1}{\hbar^2} \frac{\hbar^2 k^2}{2m} 2m \\[2mm]
&= \frac{L_x L_y}{4\pi\hbar^2} . 2m \left[\frac{eB}{m[c]} \hbar \left\{ (n+1) - n \right\} \right]
\end{aligned}
$$

$$
\begin{aligned}
\text{or } \; g &= \frac{L_x L_y}{2\pi} \frac{eB}{\hbar c}. \quad \text{(in Gaussian units.)} \\[2mm]
&= \frac{L_x L_y}{2\pi} \frac{eB}{\hbar}. \quad \text{(in MKSI units.)}
\end{aligned}
\tag{12.6}
$$

Here $k^2 = k_x^2 + k_y^2$. Degeneracy g is also sometimes called as multiplicity factor. Note that the multiplicity factor is independent of n. Observe that higher the value of magnetic field, higher is the degeneracy.

Filling factor is an important concept that is useful to discuss quantum Hall effect. The degeneracy factor g gives total number of states in a given Landau Level (LL) in a given area.

Thus, number of states/area in a given LL is $\dfrac{eB}{2\pi\hbar} = \dfrac{eB}{h}$. This is also same as the number of flux quantas threading unit area. Basic unit of flux - fluxon is $\dfrac{h}{e} = \phi_0$.

As the degeneracy depends upon B, we want to know as to how many LL are filled. This factor is known as filling factor ν. For partial filling of LL, ν is a fraction. Thus,

$$
\begin{aligned}
\nu &= \frac{\text{Area density of particles}}{\text{Number of states/area in LL}} \\[2mm]
&= \frac{nh}{eB} \qquad \text{in MKSI} \\[2mm]
&= \frac{nhc}{eB} \qquad \text{in gaussian units}
\end{aligned}
\tag{12.7}
$$

Clearly, as the magnetic induction increases, filling factor decreases. Quantum limit is reacted when only one Landau Level is reached

Short Questions

1. Explain, how various free electron levels degenerate into various Landau Levels. Explain dependence of multiplicity factor on B.

2. Explain why an electron motion perpendicular to B is described by harmonic oscillator energy values where as, those along the magnetic field are free particle motion.

3. Show that electronic motion perpendicular to B is circular with angular frequency $\omega = \dfrac{eB}{m[c]}$.

4. Explain a difference between mechanical momentum $m\vec{v}$ and canonical momentum $\vec{p}$.

5. Show how filling factor decreases with increase in B.

Problems

Problem 1. Solve Schrödinder equation for the Hamiltonian

$$H(\vec{p}, r) = \frac{1}{2m}\left(\vec{p} \quad e\vec{A}(\vec{r})\right)^2$$

where $\vec{B}$ is uniform and is along z axis. (Hint: See Landau- Lifshitz Quantum Mechanics - Non Relativistic theory, Chapter 16 (Pergamon Press(1958)))

Chapter 13

MAGNETISM OF ELECTRON GAS

13.1 Landau Diamagnetism

We already have calculated the degeneracies of various Landau levels and are now in a position to find the susceptibility of an electron gas arising out of its orbit.

The energy levels in c.g.s. units of an electron in a good metal in presence of magnetic induction $\vec{B} = B\hat{k}$ as,

$$\epsilon(n, k_z) = \left(n + \frac{1}{2}\right) \frac{\hbar e B}{m[c]} + \frac{\hbar^2 k_z^2}{2m}. \tag{13.1}$$

Here m is an effective mass (band mass) of the electron

We write $\omega_c = \dfrac{eB}{m[c]} =$ the cyclotron frequency. Grand partition function is given by

$$\ln Z = \sum_{\text{all states}} \ln\left[1 + \exp \beta \left(\mu - \epsilon(n, k_z)\right)\right]$$

$$= \frac{L_z}{2\pi} \cdot \frac{L_x L_y}{2\pi} \cdot \frac{eB}{\hbar c} \int_{-\infty}^{\infty} dk_z \sum_n \ln\left\{1 + \exp \beta \left[\mu - \left(n + \frac{1}{2}\right)\hbar\omega_c - \frac{\hbar^2 k_z^2}{2m}\right]\right\}$$

$$= \frac{V}{(2\pi)^2} \cdot \frac{m\omega_c}{\hbar} \int_{-\infty}^{\infty} dk_z \sum_n \ln\left\{1 + \exp \beta \left[\mu - \left(n + \frac{1}{2}\right)\hbar\omega_c - \frac{\hbar^2 k_z^2}{2m}\right]\right\} \tag{13.2}$$

Here we used the degeneracy factor of the Landau levels for $x - y$ motion. Evaluation of this integral is not possible in a closed form. We therefore study this problem in the low and the high temperature limit.

High Temperature Limit of $\ln Z$

We retain for high T , $\ln(1 + x) \approx x$ where $x << 1$. Thus, at high T

$$\ln Z \cong \frac{V e^{\beta\mu}}{(2\pi)^2} \cdot \frac{m\omega_c}{\hbar} \int_{-\infty}^{\infty} dk_z \exp\left[-\frac{\beta\hbar^2 k_z^2}{2m}\right] \sum_{n=0}^{\infty} \exp\left[-\beta(n + \frac{1}{2})\hbar\omega_c\right] \tag{13.3}$$

The value of a k_z integral is $\sqrt{\dfrac{2\pi m k_B T}{\hbar^2}}$. We also use

$$\sum_{n=0}^{\infty} \exp[-nx] = \frac{1}{1-e^{-x}} = \frac{e^{x/2}}{2\sinh x/2}.$$

Then, after some simplification,

$$\ln Z = \frac{V e^{\beta\mu}}{(2\pi)^2} \cdot \frac{m\omega_c}{\hbar} \sqrt{\frac{2\pi m k_B T}{\hbar^2}} \cdot \frac{1}{2\sinh\left(\frac{\hbar\omega_c}{2k_B T}\right)}. \qquad (13.4)$$

Equilibrium number $\bar{N}$ of electrons is obtained from the grand partition function by taking its derivative w.r.t $e^{\beta\mu}$. We also write $\lambda = \dfrac{h}{\sqrt{2\pi m k_B T}}$ and the Bohr magneton $\mu_B = \dfrac{e\hbar}{2m[c]}$. Then,

$$\frac{\bar{N}}{V} = \frac{1}{\lambda^3} \cdot \left[\mu_B . B . \frac{1}{k_B T}\right] \frac{e^{\beta\mu}}{\sinh\left[\dfrac{\mu_B B}{k_B T}\right]} \qquad (13.5)$$

Writing $x = \dfrac{\mu_B B}{k_B T}$, $n = \dfrac{\bar{N}}{V}$ and the magnetic moment

$$M = \frac{1}{\beta}\left(\frac{\partial}{\partial B}\ln Z\right)_{\mu,V,T} \qquad (13.6)$$

Differentiating equation(3) w.r.t. B, we get

$$M = \frac{V}{\lambda^3}\, exp[\beta\mu]\, \mu_B\, \frac{1}{\sinh x}[1 - x\coth x]$$

Eliminating $\exp[\beta\mu]$ using equation (5) we get,

$$M = -\bar{N}\mu_B\left\{\coth x - \frac{1}{x}\right\} \equiv -N\mu_B L(x). \qquad (13.7)$$

Here, we used $\bar{N} = N$ and $L(x)$ is a Langevin function we encountered in the study of paramagnetism. The negative sign in equation (7) is indicative of a diamagnetism since $L(x) > 0$ for $x > 0$. A further simplification is possible if we have weak field i.e. $x << 1$.

Then, $\left(\coth x - \dfrac{1}{x}\right)$ can be expanded for $x << 1$, to get

$$M = -N\mu_B^2 B/3k_B T \qquad (13.8)$$

And the corresponding diamagnetic susceptibility is,

$$\chi = -\frac{1}{3}\frac{n\,\mu_B^2}{k_B T} \qquad (13.9)$$

This is similar to Curie Law.

13.2 Orbital Magnetism at $T = 0K$

This is an interesting subject in that the magnetization M and the susceptibility χ discontinuously vary as a function of B at very low temperatures. This effect is known as de Hass-van Alphen effect[1].

To understand the physics of the problem, we will ignore the motion of the charged particles along the direction of the uniform magnetic induction. We assume $\vec{B} = B\hat{k}$ and consider a motion of the carriers in the $X - Y$ plane only. For simplicity, we consider $T = 0$ situation. For $T/T_F << 1$, the sharp discontinuities that we calculate will be little bit rounded off.

Normally, without the magnetic fields, density of states in 2 dimensions is constant and is given by

$$\frac{d\rho}{d\epsilon} = A\frac{m}{\pi \hbar^2}. \tag{13.10}$$

Also from the equation (12.6) degeneracy of the Landau level (LL) is $g = A\dfrac{eB}{2\pi\hbar[c]}$. Spin degeneracy will increase it by factor 2. The energies of LL are characterized by the quantum number n and,

$$\epsilon_n = (n + 1/2)\,\hbar\omega_c \quad ; \quad \omega_c = \frac{eB}{m[c]} \tag{13.11}$$

Clearly, the external $\vec{B}$ decides how many charge carriers are accommodated in a LL. If there are N particles, we can have an extreme limit of very high B such that $g > N$. Then this extreme situation, known as quantum limit, will be such that all the carriers will be in the ground state $n = 0$. Then,

$$E_0 = \frac{\hbar\omega_c}{2}.N = \frac{e\hbar}{2m[c]}BN \tag{13.12}$$

If the magnitude of B is such that, $g < N$, then the lower LLs will be full and the remaining one will be partially occupied and still higher levels will be empty. The energy spectrum describing occupancies is shown in fig.(1). Our aim here is to find the ground state energy E_0 as a function of B. Then, from the definition,

$$M = -\frac{\partial E_0}{\partial B}$$

and

$$\chi = \frac{\partial M}{\partial B} = -\frac{\partial^2 E_0}{\partial B^2} \tag{13.13}$$

for a given volume V (or here, given area A), M and χ can be found.

Since up to a level n, we have a complete filling, therefore $N > g(n + 1)$. We write (n+1) here because n starts from zero. The $(n + 1)^{th}$ level accommodates $[N - g(n + 1)]$ carriers. Clearly then, following inequality holds:

---

[1]Discussion of this section heavily leans on the treatment by K.Huang, Statistical Mechanics (John Wiley p. 243)

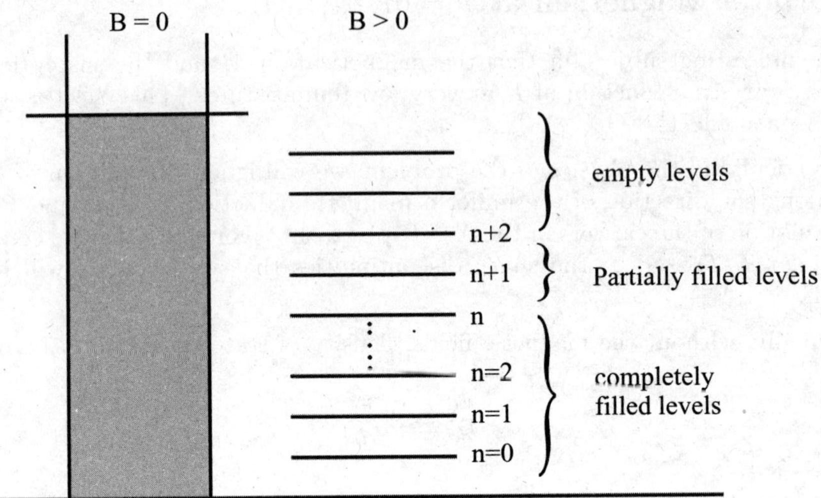

Figure 13.1: Landau Levels up to n are totally filled, $n+1$ is partially occupied and levels $n+2$ and up are empty.

$$(n+1)g < N < (n+2)g \tag{13.14}$$

or

$$\frac{1}{n+1} > \frac{g}{N} > \frac{1}{n+2} \tag{13.15}$$

But

$$\frac{g}{N} = \frac{AeB}{2\pi\hbar[c]} \cdot \frac{2}{N} \qquad \text{(Gaussian units)} \tag{13.16}$$

Here factor 2 is due to spin degeneracy, for MKSI units, factor in the square bracket, namely c is absent.

Denote, $B_0 = \pi\hbar[c]N/eA$. Then,

$$\frac{1}{n+1} > \frac{B}{B_0} > \frac{1}{n+2} \tag{13.17}$$

Clearly, if the value of B is such that it satisfies the equation (17), then E_0 is given by

$$
\begin{aligned}
E_0(B) &= g\sum_{k=0}^{n}\left(k+\frac{1}{2}\right)\hbar\omega_c + [N-(n+1)g]\,\hbar\omega_c\left(n+\frac{3}{2}\right) \\
&= NB_0\frac{e\hbar}{m[c]}\left(\frac{B}{B_0}\right)\left[\left(n+\frac{3}{2}\right)-\frac{1}{2}(n+1)(n+2)\frac{B}{B_0}\right] \tag{13.18}
\end{aligned}
$$

While evaluating $E_0(B)$ in equation (18) we have used

$$\sum_{k=0}^{n} k = \frac{n(n+1)}{2} \quad \text{and} \quad \sum_{k=0}^{n} \frac{1}{2} = \frac{1}{2}(n+1)$$

For extreme quantum case, $E_0(B) = N\left(\dfrac{e\hbar}{2mc}\right)B$. Write $x = B/B_0$.

Then,

$$\frac{1}{N}E_0(B) = \begin{cases} \dfrac{e\hbar}{2m[c]}B_0 x & x > 1 \\[2mm] \dfrac{e\hbar}{2m[c]}B_0 x\left[(2n+3)-(n+1)(n+2)x\right] & \text{when} \\[2mm] & \dfrac{1}{n+1} > x > \dfrac{1}{n+2} \end{cases} \tag{13.19}$$

Here the range of n is $(0,1,2,...)$.

Then, magnetic moment per particle is,

$$M = \begin{cases} -\mu_B & x > 1 \\[2mm] \mu_B\left[2(n+1)(n+2)x - (2n+3)\right] & \text{when} \\[2mm] & \dfrac{1}{n+1} > x > \dfrac{1}{n+2} \end{cases} \tag{13.20}$$

Here, $\mu_B = \dfrac{e\hbar}{2m[c]}$ is Bohr magneton.

Corresponding susceptibility is

$$\chi = \begin{cases} 0 & x > 1 \\[2mm] \dfrac{2\mu_B(n+1)(n+2)}{B_0} & \text{when} \\[2mm] & \dfrac{1}{n+1} > x > \dfrac{1}{n+2} \end{cases} \tag{13.21}$$

Equation(20) and (21) are the principle results. We plot M/μ_B versus $x = B/B_0$ and also, $\chi/\left(\frac{2\mu_B}{B_0}\right)$ versus x.

Here, for $x = 1, n = 0$, $M/\mu_B = 1$; for $x = 1/2, n = 0$; $M/\mu_B = -1$. Thus if $n = 0$ and higher levels are filled, and B changes from $B_0/2$ to B_0; the magnetic moment flips its sign. Similarly, if levels with $n > 1$ are also occupied, then (M/μ_B) will flip its sign as the magnetic induction B changes its value from $B_0/3$ to $B_0/2$; and so on. From figure(2) we can infer that χ is proportional to slope of AB when $\dfrac{1}{2} < x < 1$ and is 2. Similarly the slope of CD gives value of χ when $\dfrac{1}{3} < x < \dfrac{1}{2}$ and is 6; and so on.

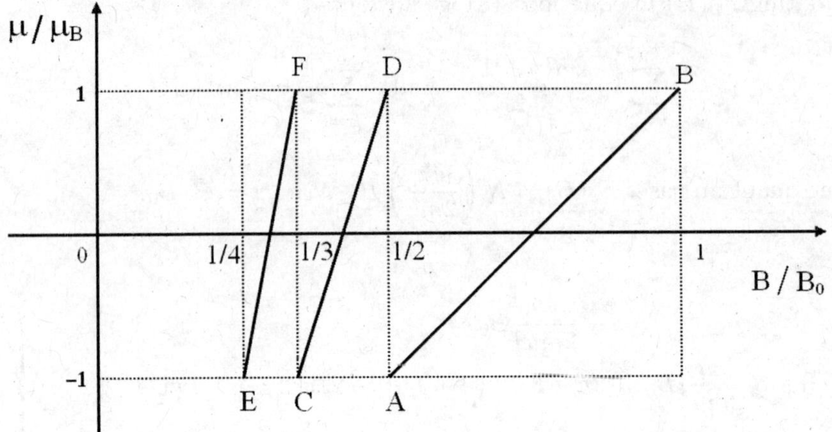

Figure 13.2: Magnetic moment in units of μ_B as a function of B/B_0

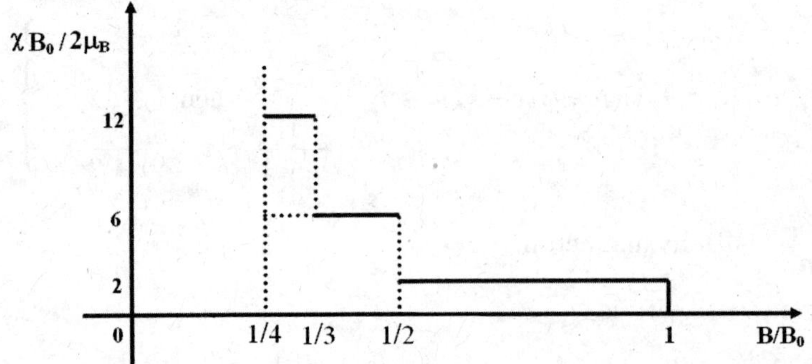

Figure 13.3: Behaviour of χ with B/B_0.

Taking Z motion and $T \neq 0$ will smoothen out discontinuities in the graphs of M and χ; but the oscillatory behavior of M with B still persists. This effect is known as de-Haas-Van Alphen -Shubunikov effect.

In periodic solids, we have the arrangement of lattice points according to some point group symmetry. Then, the Fermi level is different in different directions. When we connect all the Fermi level points, we get what is known as the Fermi surface. For free electrons this surface is a sphere of radius $k = k_F$ in k space. Realistic Fermi surface has necks, non spherical shapes and so on. Famous picture of Fermi surface of copper due to Shoenberg is exemplary. It is given in any standard solid state physics book such as Ziman. de Haas van Alphen effect along with technique of cyclotron resonance are useful tools to experimentally determine the realistic metallic Fermi surfaces.

13.3 Diamagnetic Susceptibility at All Temperatures in Weak Magnetic Field

Here we consider a weak magnetic field at all temperatures which means $\mu_B B << k_B T$. We calculate the diamagnetic susceptibility of an electron gas. For this we use Euler-Mclaurin summation formula.

We make a little digression here. Our functions used here are all nicely behaving (all requisite number of derivatives exist and are continuous). Then Euler summation formula states

$$\frac{f(0) + f(n)}{2} + \sum_{k=1}^{n-1} f(k) = \int_0^n f(x)dx + \sum_{k=1}^{p} \frac{B_{2k}}{(2k)!}\left[f^{(2k-1)}(n) - f^{(2k-1)}(0)\right] \qquad (13.22)$$

A closed look at the l.h.s. of this formula shows that it is nothing but a trapezoid rule type sum. B_{2k} are Bernoulli numbers. $B_1 = -\frac{1}{2}$, $B_2 = \frac{1}{6}$, $B_4 = -\frac{1}{30}$, $B_3 = B_5 = B_7... = 0$. Here we force n to ∞. The sum to be evaluated is given in equation(2) as,

$$f(n) = \sum_{n=0}^{\infty} \ln\left(1 + \exp\left\{\beta\left[\mu - \left(n + \frac{1}{2}\right)\hbar\omega_c - \frac{\hbar^2 k_z^2}{2m}\right]\right\}\right)$$

As $n \to \infty$, $f(n) \to 0$ and $f'(n) \to 0$.
Then retaining only $k = 1$ term, in $f^{(2k-1)}$ type terms in equation (22), we approximately get,

$$\sum_{k=0}^{\infty} f\left(k + \frac{1}{2}\right) = \int_0^{\infty} f(x)dx - \frac{1}{12}f'(0)$$

Using Euler formula,

$$f(n) = \int_0^{\infty} \ln\left(1 + \exp\left[\beta\left(\mu - \frac{\hbar^2 k_z^2}{2m}\right) - \beta\hbar\omega_c x\right]\right) dx$$
$$+ \frac{1}{12}\left\{\frac{\left(-\beta\hbar\omega_c \exp\left[\beta\left(\mu - \hbar^2 k_z^2/2m\right)\right]\right)}{1 + \exp\left[\beta\left(\mu - \hbar^2 k_z^2/2m\right)\right]}\right\} \qquad (13.23)$$

Thus,

$$\ln Z = \frac{V}{(2\pi)^2}\cdot\frac{m\omega_c}{\hbar}\left\{\int_{-\infty}^{\infty} dk_z \int_0^{\infty} dx \ln\left[1 + \exp\left[\beta\left(\mu - \frac{\hbar^2 k_z^2}{2m}\right)\right]\exp\left[-\beta\hbar\omega_c x\right]\right]\right.$$
$$\left. - \frac{1}{12}\beta\hbar\omega_c \int_{-\infty}^{\infty} dk_z \cdot\frac{\exp\left[\beta\left(\mu - \frac{\hbar^2 k_z^2}{2m}\right)\right]}{\exp\left[\beta\left(\mu - \frac{\hbar^2 k_z^2}{2m}\right)\right] + 1}\right\} \qquad (13.24)$$

First term of equation (23) is independent of B. This is true because, substitute $\beta\hbar\omega_c x = y$, and you will realize that the ω_c cancels.
The first term is proportional to

$$\frac{m k_B T}{\hbar^2}\int_{-\infty}^{\infty} dk_z \int_0^{\infty} dy \ln\left\{1 + \exp\left[\beta\left(\mu - \frac{\hbar^2 k_z^2}{2m}\right) - y\right]\right\}$$

which clearly is independent of B. We ignore it because ultimately we have to differentiate $\ln Z$ with B.

Then,

$$\ln Z \cong -\frac{1}{6}\frac{\mu_B B}{k_B T}\cdot\frac{V}{(2\pi)^2}\frac{m}{\hbar}\frac{eB}{mc}\int_{-\infty}^{\infty}\frac{dk}{\exp\left[\beta\left(\frac{\hbar^2 k^2}{2m}-\mu\right)\right]+1}$$

$$= -\frac{1}{6}\frac{V\left(\mu_B B\right)^2}{h^3}\left(2m\right)^{3/2}\beta^{1/2}\,2\int_0^{\infty}\frac{y^{-1/2}dy}{e^{y-\alpha}+1}$$

Definite integral $\displaystyle\int_0^{\infty}\frac{y^{-1/2}dy}{e^{y-\alpha}+1}$ can be evaluated by the methods of Appendix(4). It is $\Gamma\left(\frac{1}{2}\right)f_{1/2}\left(e^{\alpha}\right)$.

Clearly the integral at $T << T_F$ is, to a leading order,

$$2\sqrt{\alpha}=2\left(\frac{\epsilon_F}{k_B T}\right)^{1/2}\left(1-\frac{\pi^2}{24}\left(\frac{k_B T}{\epsilon_F}\right)^2+...\right).$$

Susceptibility

$$\chi=\frac{1}{\beta V B}\left(\frac{\partial}{\partial B}\ln Z\right)_{\mu,V,T} \tag{13.25}$$

With all these intermediate steps, being used, we get

$$\chi=\frac{(2\pi m)^{3/2}\,\mu_B^2}{3h^3}\left(k_B T\right)^{1/2}f_{1/2}\left(e^{\alpha}\right) \tag{13.26}$$

This expression at low temperature reduces to

$$\chi\cong -\frac{1}{2}\frac{n\mu_B^2}{\epsilon_F}\left(1-\frac{\pi^2}{24}\left(\frac{T}{T_F}\right)^2+...\right) \tag{13.27}$$

This indicates that the diamagnetic susceptibility at low temperatures and in a weak field is practically constant. We will show that contribution of paramagnetic susceptibility due to Pauli is three times more in magnitude than its diamagnetic part. Results of diamagnetic susceptibility of an electron gas are due to Landau.

13.4   Pauli's Paramagnetic Susceptibility

Sommerfeld was first to recognize that the electrons in metal obey F-D statistics and not Boltzmann statistics. Earlier, Drude had worked out a free electron theory of metal with Boltzmann statistics for electrons. This gave wrong results. Pauli, being one of the graduate students of Sommerfeld, was influenced by him and applied F.D. statistics to find a susceptibility of collection of electrons. He had himself worked on the problem of spin of electrons. Bohr had suggested a possibility of intrinsic magnetic moment of electron due to its orbital and intrinsic spin motion. Pauli calculated the contribution to magnetic moment due to spin alignment of electrons with external magnetic induction. This gives rise to paramagnetism, now known as Pauli's paramagnetism. Subsequently, Landau discovered diamagnetism due to orbital motion.

Consider N electrons confined in volume V of metal. It will be a random mixture of itinerant electrons. Each k level has up or down electron spin component S_z and all the levels are filled up to E_F. Finite temperature smear is very small when $T << T_F$. When the uniform external magnetic field is applied, N_+ electron spins are aligned parallel to B, whereas N_- electrons are aligned antiparallel to B. A typical electron of $\{N_+\}$ group has its energy lowered as $\epsilon_+ = p^2/2m - \mu_B B$, whereas, a typical electron of $\{N_-\}$ group will have a higher energy $\epsilon_- = p^2/2m + \mu_B B$. Here μ_B is Bohr magneton.

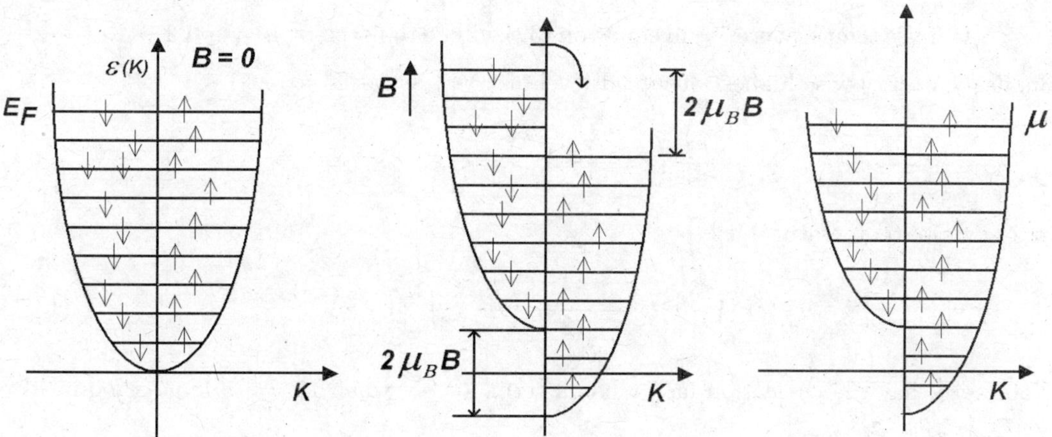

Figure 13.4: Free electrons without and with magnetic induction B. $\{N_-\}$ set of electrons move up by energy $\mu_B B$ whereas, $\{N_+\}$ set of electrons are lowered in energy by $\mu_B B$. Transition take place from $\{N_-\}$ to $\{N_+\}$ set resulting in same chemical potential.

Clearly, if $N_+ = N_- = N/2$, then there is no magnetism. However, if $N_+ > N_-$; then there is overall magnetization. Thus, we write

$$N = N_+ + N_- \tag{13.28}$$

and magnetic moment M as

$$M = \mu_B (N_+ - N_-) \tag{13.29}$$

We write spin polarization r as

$$r = \frac{N_+ - N_-}{N_+ + N_-} = \frac{N_+ - N_-}{N} \tag{13.30}$$

Then

$$N_+ = \frac{N}{2}(1 + r) \quad and \quad N_- = \frac{N}{2}(1 - r) \tag{13.31}$$

Clearly, highest occupied level of $\{N_+\}$ set at $T = 0$ is at

$$
\begin{aligned}
\frac{\hbar^2}{2m}\left(6\pi^2\frac{N_+}{V}\right)^{2/3} &= \frac{\hbar^2}{2m}\left(3\pi^2\frac{N}{V}\right)^{2/3}(1+r)^{2/3} \\
&= \frac{\hbar^2}{2m}k_F^2(1+r)^{2/3} \\
&= \epsilon_F(1+r)^{2/3}
\end{aligned}
\tag{13.32}
$$

Note here the factor $\left(6\pi^2\frac{N_+}{V}\right)^{2/3}$. Can you reason why the factor 6 and not 3 ?

At finite temperature, ϵ_F in equation (32) will be replaced by $\mu = \epsilon_F\left(1 - \frac{\pi^2}{12}\left(\frac{T}{T_F}\right)^2\right)$. Similarly, energy wise highest occupied level of $\{N_-\}$ set at $T = 0$ is

$$
\frac{\hbar^2}{2m}\left(6\pi^2\frac{N_-}{V}\right)^{2/3} = \epsilon_F(1-r)^{2/3}
\tag{13.33}
$$

and at finite temperature, it is

$$
\epsilon_F(1-r)^{2/3} - \frac{\pi^2}{12}\epsilon_F\left(\frac{k_BT}{\epsilon_F}\right)^2(1-r)^{-2/3}
\tag{13.34}
$$

Note here that the correction factor has a term $\frac{(k_BT)^2}{\epsilon_F}$ when $B = 0$. It gets modified to $\frac{(k_BT)^2}{\epsilon_F}(1-r)^{-2/3}$.

Thus, the energy difference between $\{N_+\}$ set and $\{N_-\}$ set is

$$
\epsilon_F\left\{\left[(1+r)^{2/3} - (1-r)^{2/3}\right] - \frac{\pi^2}{12}\frac{(k_BT)^2}{\epsilon_F^2}\left[(1+r)^{-2/3} - (1-r)^{-2/3}\right]\right\} = 2\mu_B B
\tag{13.35}
$$

For small $r << 1$, this equation reduces to

$$
\frac{4}{3}r\epsilon_F\left(1 + \frac{\pi^2}{12}\left(\frac{k_BT}{\epsilon_F}\right)^2 + ...\right) = 2\mu_B B
\tag{13.36}
$$

Equation (36) immediately gives spin polarization

$$
r = \frac{3}{2}\frac{\mu_B B}{\epsilon_F}\left(1 - \frac{\pi^2}{12}\left(\frac{k_BT}{\epsilon_F}\right)^2\right)
\tag{13.37}
$$

Then,

$$
M = \mu_B r N
$$

Or, magnetic moment per electron is

$$
\frac{M}{N} = \frac{3}{2}\frac{\mu_B^2 B}{\epsilon_F}\left(1 - \frac{\pi^2}{12}\left(\frac{k_BT}{\epsilon_F}\right)^2\right)
\tag{13.38}
$$

Corresponding susceptibility per electron is

$$\chi = \frac{M}{NB} = \frac{3}{2}\frac{\mu_B^2}{\epsilon_F}\left(1 - \frac{\pi^2}{12}\left(\frac{k_B T}{\epsilon_F}\right)^2\right) \tag{13.39}$$

Clearly, this contribution is almost independent of temperature and numerically it is three times that of diamagnetism.

13.5 Quantum Hall Effect

Conventional Hall effect is a consequence of Lorentz force. Consider a conducting strip sample as shown in figure (5) and kept in a uniform magnetic induction $\vec{B} = B\hat{k}$. Let the voltage be applied along x axis. Because a current moves in x direction, its charges experience $q(\vec{v} \times \vec{B})$ force. As a result, the positive and negative charges accumulate at the edges of the sample as shown in figure (5).

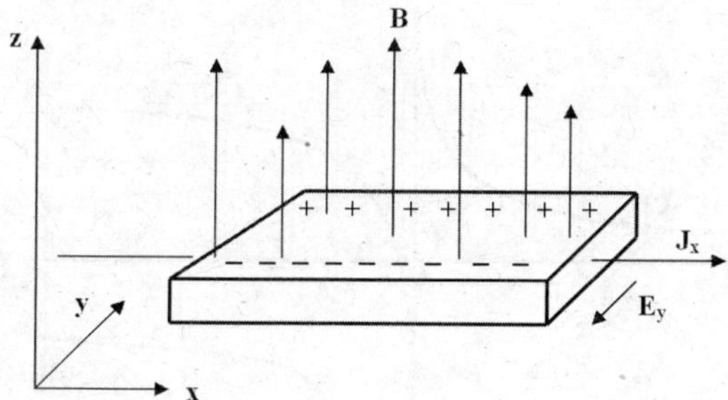

Figure 13.5: Hall effect. Because of a presence of $\vec{B}$, charges accumulate on sides.

This creates electric field E_y. Magnetic field is so applied that its effect is canceled by the effect of E_y on the carriers. Thus, $qv_x B_z = qE_y$. In Hall effect experiment, a transverse voltage is measured for a fixed current and Hall resistance is recorded. It is seen that the Hall resistance ρ_{xx} increase linearly with B. Hall coefficient $R_H = \dfrac{E_y}{B_z J_x} = \dfrac{1}{nq}$ is measured. Its sign gives the nature of the charge carries. If it is negative, then the carriers are holes in a semiconducting sample. For further details see a standard book on solid state physics like that of Kittel.

Quantum Hall effect (QHE) is of two kinds. One type is an integer QHE and the other is a fractional Quantum Hall effect (FQHE). We will discuss integer QHE in this book. It is observed in two dimensional electron gas (2-DEG) samples. MOSFET junction or superlattices have 2-DEG.

At low temperatures it is observed that a series of plateau or steps appear in the Hall resistance as a function of B. These steps occur at precise value of the resistance at integral multiple of basic unit of resistance h/e^2. It is also called the quantized resistance. To observe QHE one requires a very high magnetic fields in the range of several teslas.

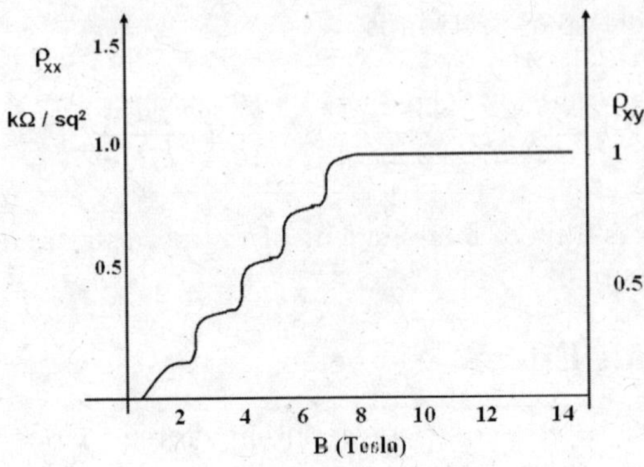

Figure 13.6: QHE in GaAs-GaAlAs heterostructure.

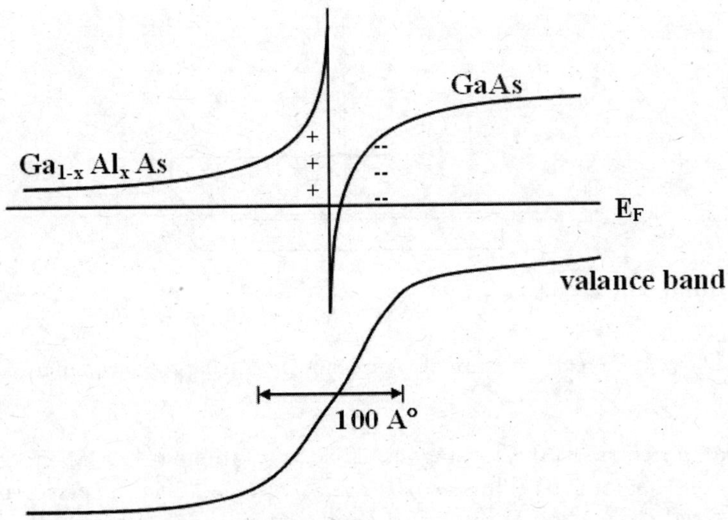

Figure 13.7: Heterostructure of GaAs-GaAlAs.

We show QHE observed in GaAs-GaAlAs heterostructure in the figure (6). The pattern is observed at about $T = 30 \ mK$ with carrier concentration of $10^{11}/cm^2$. GaAs is a direct band gap semiconductor. It means, the bottom of the conduction band is directly above the top of its its valance band. GaAlAs is doped n-type. First, with the help of MBE a film of GaAs is deposited and then another film of the n doped GaAlAs is deposited over it. Junction region of the two materials is around $100\overset{\circ}{A}$. Its band diagram is shown in figure (7). To reach an equilibrium, Fermi level across the junction should be same. This becomes possible because electrons from n-type doped region of GaAlAs are transferred in GaAs region near the junction. This results in a dipole layer of about $100\overset{\circ}{A}$ thickness at the junction. The excess electrons are confined in a junction region and can move freely in the plane of a junction but can not move across the junction. Thus they form 2-DEG.

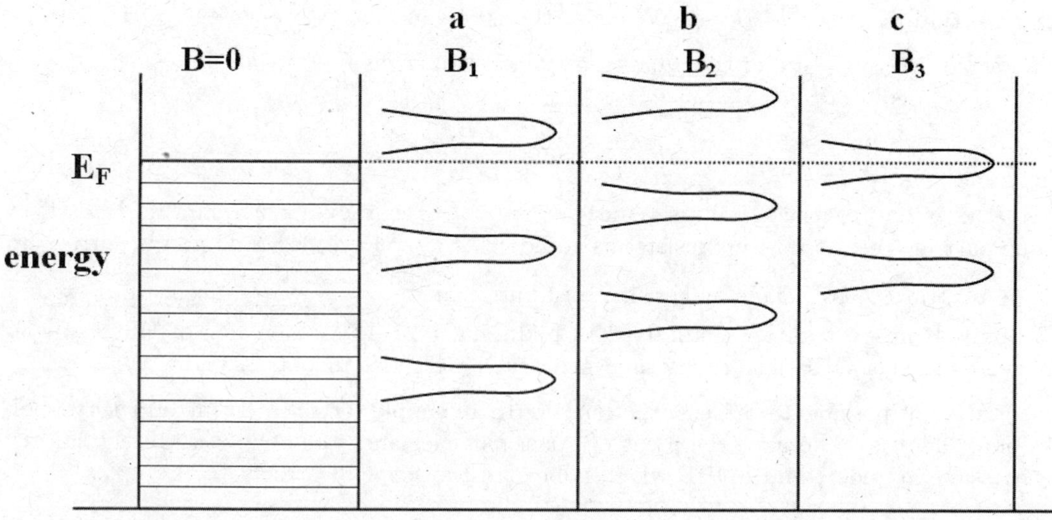

Figure 13.8: 2DEG has constant density of states when $B = 0$. As B is applied, the momentum levels transform it to LL. Note the LL for $B_1 < B_2 < B_3$. Note also the crossing of LL at E_F

When a magnetic field is applied, highly degenerate Landau Levels (LL) are formed. Separation between the levels increases with increasing B and so also their degeneracy. This leads to the movement of LL. When ϵ_F is between two Landau Levels or , completely above all the Landau Levels, the electrons in it can not be scattered into other states. The transport is then without resistance. We then say that the electrons are localized in the region.

Classical Hall effect has the Hall resistance as B/nq. The number of states in a Landau Level is eB/h. Let there be i LL below ϵ_F Suppose they are completely filled. Then the possible number of carriers are ieB/h. Corresponding Hall resistance is h/ie^2. Figure(8) depicts the scenario of QHE. It is seen that as B increase, LL start moving towards ϵ_F. Unless the LL crosses ϵ_F, current is zero. Only when the situation shown in (c) part of figure (8) occurs, i.e ϵ_F is in LL, voltage changes and non zero resistance occurs. Hall conductance is given by

$$G_H = e[c]\frac{\partial n(\epsilon_F)}{\partial B} \tag{13.40}$$

With ν as a filling factor, $n = \dfrac{\nu e B}{h[c]}$. Thus,

$$G_H = \frac{\nu e^2}{h[c]} \tag{13.41}$$

and Hall resistance is

$$R_H(\nu) = \frac{1}{G_H} = \frac{h}{e^2}\left(\frac{1}{\nu}\right) \tag{13.42}$$

Classical Hall resistance is B/nq. We claim that the equation (42) of integer QHE is same as above. With $\nu = \dfrac{nh}{eB}$ and the Quantum Hall resistance as $\dfrac{h}{ie^2}$, then,

$$\frac{h}{ie^2} = \frac{B\nu}{ien} = \frac{B}{en}.$$

It is same as that of usual Hall resistance since $\nu = i$. Accuracy of the quantum Hall effect is one part in 10^9. As a result resistance standard is $h/e^2 = 25812806\ \Omega$ and is correct up to one part in 2×10^8. This makes fine structure value $\dfrac{e^2}{\hbar c} \approx \dfrac{1}{137.036}$, accurate to within 0.3 ppm. Primary resistance standard is 1 klitzing= $25813\ \Omega$ in honor of Klitzing who discovered integer QHE. Klitzing was awarded a Nobel Prize in year 1985.

It is not possible to discuss fractional QHE in simple terms in an introductory book like ours. This is so because the ideas of quasi particles and some Quantum Field Theory is necessary to understand FQHE which is beyond the scope of present book.

Chapter 14

PROBLEMS OF PHYSICAL INTEREST

Now that the formalism of equilibrium of statistical mechanics is complete, we solve many problems of physical interest in this chapter.

Problem 1: Obtain partition function for ideal gas and calculate its equation of state and the specific heat and entropy.

Solution: For ideal gas, $E = \sum_{i=1}^{N} p_i^2/2m$.

$$
\begin{aligned}
Z &= \frac{1}{h^{3N}} \int \exp\left[-\beta \sum_{i=1}^{N} p_i^2/2m\right] d^3p_1 \, d^3p_2 ... d^3P_N \, d^3r_1 \, d^3r_2 ... d^3r_N \\
&= \frac{V^N}{h^{3N}} \int \exp\left[-\beta p_1^2/2m\right] d^3p_1 \int \exp\left[-\beta p_2^2/2m\right] d^3p_2 ..... \int \exp\left[-\beta p_N^2/2m\right] d^3p_N
\end{aligned}
$$

But

$$
\begin{aligned}
\int \exp\left[-\beta p^2/2m\right] d^3p &= 4\pi \int_0^\infty \exp\left[-\beta p^2/2m\right] p^2 \, dp \\
&= 4\pi \left(\frac{2m}{\beta}\right)^{3/2} \int_0^\infty \xi^2 \exp\left[-\xi^2\right] d\xi \\
&= 4\pi \left(\frac{2m}{\beta}\right)^{3/2} \frac{\sqrt{\pi}}{4} \\
&= (2\pi m k_B T)^{3/2}
\end{aligned}
$$

Thus $Z = \dfrac{V^N (2\pi m k_B T)^{3N/2}}{h^{3N}}$

Write $\dfrac{h}{\sqrt{2\pi m k_B T}} = \lambda_{th}$ the thermal de-Broglie wavelength.

Then,

$$Z = \left(\frac{V}{\lambda_{th}^3}\right)^N \tag{14.1}$$

(i) Free energy

$$F = -k_B T \ln Z = -N k_B T \left[\ln V - 3 \ln \lambda_{th}\right]$$

(ii)

$$\langle E \rangle = \bar{E} = -\frac{\partial \ln Z}{\partial \beta} = \frac{3}{2} N k_B T$$

(iii)

$$C_v = \left.\frac{\partial \bar{E}}{\partial T}\right)_V = \frac{3}{2} N k_B.$$

For one mole of a gas, $N = N_A$, the Avogadro number and $N_A k_B = R$ the gas constant.
Thus $C_V = \frac{3}{2} R \ /mole$

(iv)

$$\begin{aligned}
S &= k_B \left[\ln Z + \beta \bar{E}\right] \\
&= k_B \left[N \ln V + \frac{3N}{2} \ln T + \frac{3N}{2} \ln\left(\frac{2\pi m k_B}{h^2}\right) + \frac{3}{2} N\right]
\end{aligned}$$

We write constant $\sigma = \frac{3}{2} + \frac{3}{2} \ln\left(\frac{2\pi m k_B}{h^2}\right)$
Thus

$$S = N k_B \left[\ln V + \frac{3}{2} \ln T + \sigma\right]$$

Problem 2: <u>Gibbs Paradox</u>: Consider an ideal gas contained in a box at pressure P and temperature T. The box is divided into two parts of volume V_1 and V_2 by a partition. Let S be the entropy of the gas before partition was inserted. If S_1 and S_2 are entropies of the gas in V_1 and V_2, then find

$$S - S_1 - S_2$$

Solution:

$$V = V_1 + V_2, \quad N = N_1 + N_2$$

$$S = N k_B \left[\ln V + \frac{3}{2} \ln T + \sigma\right]$$

$$S_1 = N_1 k_B \left[\ln V_1 + \frac{3}{2} \ln T + \sigma\right]$$

$$S_2 = N_2 k_B \left[\ln V_2 + \frac{3}{2} \ln T + \sigma\right]$$

Clearly $\Delta S = S - (S_1 + S_2) = Nk_B \left[\ln V \right] - N_1 k_B \ln V_1 - N_2 k_B \ln V_2$
Then,

$$\Delta S = k_B \ln \left(\frac{V^N}{V_1^{N_1} V_2^{N_2}} \right)$$

This answer is wrong !! This is a paradox. We have made a mistake in calculating the partition function. Correct answer is zero, because putting a partition or removing it from the box does not alter number of accessible states to the system. Thus, correctly, we should get,

$$S = S_1 + S_2.$$

Problem 3: Resolve this paradox

Solution: Here the difficulty is that we have treated the gas molecules of a same kind to be distinguishable. This is strictly not correct. Since the molecules are identical, exchange among them should not give a physically distinct state. If there are N particles, there are $N!$ different ways of exchange and we have over counted states by the same amount. In quantum mechanics, correct counting is done automatically. On the other hand, in classical statistics, this correction of indistinguishability need to be done explicitly. Thus, the correct partition function is,

$$Z = \frac{1}{N!} \frac{V^N (2\pi m k_B T)^{3N/2}}{h^{3N}}$$

Then,

$$S = k_B \left[\ln Z + \beta \bar{E} \right].$$

$\bar{E}$ will not change by $1/N!$ factor. Using Sterling approximation we write,

$$N! \simeq N \ln N - N \text{ for } N >> 1$$

Then,

$$S = k_B \left[N \ln \left(\frac{V}{N} \right) + \frac{3}{2} N \ln T + N(\sigma + 1) \right].$$

By adding a partition or removing it in the given situation leaves $\left(\frac{V}{N} \right)$ unaltered. Thus,

$$\frac{V}{N} = \frac{V_1}{N_1} = \frac{V_2}{N_2} = \frac{1}{\rho}$$

where ρ is a density.

$$\therefore S_1 = k_B \left[N_1 \ln \left(\frac{V_1}{N_1} \right) + \frac{3}{2} N_1 \ln T + N_1 (1 + \sigma) \right]$$

$$S_2 = k_B \left[N_2 \ln \left(\frac{V_2}{N_2} \right) + \frac{3}{2} N_2 \ln T + N_2 (1 + \sigma) \right]$$

Then, by simple calculation and using the above expressions for S, S_1 and S_2 it is easy to see that

$$S - S_1 - S_2 = 0$$

We see then that, by correct counting, we can resolve the Gibb's paradox.

The correct classical partition function is

$$Z = \frac{1}{N!} \frac{V^N}{\lambda_{th}^{3N}}.$$

With this choice of Z, we get the correct entropy as

$$S = N k_B \left[\ln \frac{V}{N} + \frac{3}{2} \ln T + 1 + \sigma \right]$$

where

$$\sigma = \frac{3}{2} + \frac{3}{2} \ln \left(\frac{2\pi m}{h^2} \right).$$

Problem 4: Consider a collection of a non interacting quantum harmonic oscillators of frequency ω each. Find the partition function, its average energy, free energy, specific heat and entropy.

Solution: For a harmonic oscillator $E_n = (n + \frac{1}{2})\hbar\omega$, where $n = 0, 1, 2....$

(1) Partition function: A partition function for a single oscillator is

$$\begin{aligned}
Z &= \sum_n \exp\left[-\beta(n + \frac{1}{2})\hbar\omega\right] \\
&= \exp\left[-\beta\hbar\omega/2\right]\left[1 + \exp\left[-\beta\hbar\omega\right] + \exp\left[-2\beta\hbar\omega\right] + \exp\left[-3\beta\hbar\omega\right] + +...\right].
\end{aligned}$$

This is a geometric series with common ratio $\exp\left[-\beta\hbar\omega\right] < 1$.
Its sum is

$$Z = \frac{\exp\left[-\beta\hbar\omega/2\right]}{1 - \exp\left[-\beta\hbar\omega\right]}.$$

For N oscillators it is Z^N.

(2) Free energy:

$$\begin{aligned}
F &= -N k_B T \ln Z \\
&= \frac{N\hbar\omega}{2} + N k_B T \ln(1 - \exp\left[-\beta\hbar\omega\right]).
\end{aligned}$$

(3) Internal energy:

$$\bar{E} = -\frac{\partial}{\partial\beta} \ln Z = \frac{\hbar\omega}{2} + \frac{\hbar\omega}{\exp\left[\hbar\omega/k_B T\right] - 1}$$

From this expression, we immediately see the Bose factor with zero chemical potential.

(4) Specific heat at constant volume:

$$C_V = \frac{\partial \bar{E}}{\partial T} = k_B \left(\frac{\hbar\omega}{k_B T}\right)^2 \frac{\exp\left[\frac{\hbar\omega}{k_B T}\right]}{\left(\exp\left[\frac{\hbar\omega}{k_B T}\right] - 1\right)^2}.$$

We see then that, $C_V \to 0$ as $T \to 0$.

(5) Entropy/particle:

$$\begin{aligned} S &= k_B(\ln Z + \beta\bar{E}) \\ &= k_B\left[\frac{x}{e^x - 1} - \ln\left(1 - e^{-x}\right)\right] \end{aligned}$$

Where $x = \left(\dfrac{\hbar\omega}{k_B T}\right)$.

Problem 5: Obtain the classical limit of the physical quantities calculated in problem 4.

Solution: In classical limit, the thermal energy is much larger than the energy level difference in quantum case. This means the system is excited to very high levels. Mathematically, it means $x = \hbar\omega/k_B T << 1$. The low temperature value is given by $x >> 1$ and gives extreme quantum limit. Thus, classically

(1) For the high temperature limit, $x << 1$. Then,

$$Z \approx \frac{1 + x/2 + x^2/8 + \ldots}{1 + x + x^2/2! + \ldots - 1} \approx \frac{1}{2} + \frac{k_B T}{\hbar\omega}.$$

(2)

$$\bar{E} = \frac{\hbar\omega}{2} + \hbar\omega\frac{1}{e^x - 1} \approx \frac{\hbar\omega}{2} + \hbar\omega.\frac{k_B T}{\hbar\omega} \approx k_B T$$

Here we see the same result as given by the equipartition theorem. You will get same result by taking $\left(-\dfrac{\partial}{\partial\beta}\ln Z\right)$ in (1).

(3) Free energy: Free energy in classical limit is $F = k_B T \ln\left(\dfrac{\hbar\omega}{k_B T}\right) - \dfrac{\hbar\omega}{2}$

(4) $C_V/particle = k_B$

(5) Entropy

$$S \approx k_B\left[1 + 1 - x\right] = 2k_B - \frac{\hbar\omega}{k_B T} \approx 2k_B$$

For N oscillators, E, F, C_V and S will simply scale by a factor N. How will the Z of (1) get modified?

Problem 6: Find the chemical potential of an ideal classical gas.

Solution:
Method 1: Gibb's free energy $G = F + PV = E - TS + PV$.
Also, $G = n\mu$. Thus, with $PV = Nk_BT$, using the Sterling approximation, and results of the problem (1),

$$
\begin{aligned}
\mu &= \frac{F + Nk_BT}{N} \\
&= k_BT\left[\ln\left(\frac{N}{V}\right) - \frac{3}{2}\ln T - \frac{3}{2}\ln\left(\frac{2\pi mk_B}{h^2}\right)\right].
\end{aligned}
$$

Method 2: We could invoke a discussion of the classical limit of quantum statistics, where we have seen that in classical limit,

$$
n_k = \exp\left[\beta(\mu - \epsilon_k)\right].
$$

Summing over all states, $\sum n_k = N$ and $\sum_p \to \frac{V}{h^3}\int d^3p$ we get,

$$
N = \frac{V}{h^3}\exp\left[\beta\mu\right]\int \exp\left[-\beta p^2/2m\right]d^3p.
$$

Carrying out the integration and simplifying,

$$
\mu = k_BT\ln\left(\frac{N}{V}\right) - \frac{3}{2}k_BT\ln\left(\frac{2\pi mk_B}{h^2}\right) - \frac{3}{2}k_BT\ln T
$$

which is same as above result.

Problem 7: Consider a sensitive spring balance of a spring constant α. The balance is in the environment whose temperature is T. A small object of mass m is suspended to the spring. Find

    a Mean elongation of the spring.

    b Thermal fluctuation in its position i.e. $\left[\overline{(x - \overline{x})^2}\right]$

    c Obviously the balance will fail to measure the mass when the mean elongation equals r.m.s. thermal fluctuation. Find this limit on mass.

Solution
(a) $mg = \alpha x$ or $x = mg/\alpha$

(b) By equipartition theorem,

$$
\overline{V} = \frac{1}{2}\alpha\overline{x^2} = k_BT/2
$$

or ,

$$\overline{x^2} = \frac{k_B T}{\alpha}$$

But $\overline{x} = 0$, thus, $\overline{(x - \overline{x})^2} = k_B T/\alpha$

(c) $\sqrt{k_B T/\alpha} = \dfrac{Mg}{\alpha}$ or $M = \sqrt{\dfrac{\alpha k_B T}{g^2}}$

Problem 8: Consider an ideal gas of N molecules enclosed in a volume V. In addition to translational motion, the molecules have rotational and vibrational motion and electronic motion. To a first approximation, energy of the molecules can be written in an additive manner as,

$$E = E_{tr} + E_{vib} + E_{rot} + E_{el}.$$

Show that the partition function Z is

$$Z = \prod_i Z_i$$

where i runs from translational motion to the electronic motion.

Solution:

$$
\begin{aligned}
Z &= \sum_{allstates} \exp\left[-\beta(E_{tr} + E_{vib} + E_{rot} + E_{el})\right] \\
&= \left(\sum \exp\left[-\beta E_{tr}\right]\right) \left(\sum \exp\left[-\beta E_{rot}\right]\right) \left(\sum \exp\left[-\beta E_{vib}\right]\right) \left(\sum \exp\left[-\beta E_{elec}\right]\right) \\
&= Z_{tr} Z_{rot} Z_{vib} Z_{elec}
\end{aligned}
$$

Problem 9: Find the value of C_V and C_P due to a vibrational contribution at high temperatures for a gas of molecules with n atoms.

Solution For a molecule of n atoms, it has $6n$ degrees of freedom. The molecule has a center of mass which has three degrees of freedom due to the translational motion in X, Y, Z directions. This motion corresponds to a motion of the molecule as a whole. There are three rotations about three axes of molecule. Thus, vibrational degrees of freedom for a non-co-linear molecule are $l = 6n - 6$. For a co-linear molecule, there is no rotational motion along its axis. Therefore for the co-linear molecule, the vibrational degrees of freedom are $l = 6n - 5$. Consider a gas of N molecules. By the equipartition theorem, its internal energy due to vibrational motion is $N\dfrac{l k_B T}{2}$. This means, $C_V = Nl k_B/2$. C_P can be estimated as $C_P - C_V = N k_B$. Thus,

$$C_P = N \frac{l+2}{2} k_B.$$

Problem 10: Consider typical diatomic molecules such as H_2, N_2, NO. Find their dissociation energies and corresponding equivalent temperatures. The molecules are excited in electronic level where the excitation energies are comparable to the dissociation energies. Compare their energies with the vibrational energies and rotational energies. Spectra of these molecules are obtained by striking a discharge in the gas tubes. Argue which motion will contribute to the specific heat at high temperatures.

Solution: A standard result that is useful to be remembered by all is ,

$$\frac{1}{40}\, eV \equiv 300K = 27^{o}C \cong \text{room temperature}$$

Consider a following data collected from standard tables.

No.	Molecule	Dissociation energy	Vibrational energy	Rotational energy
1	H_2	$4.33\ eV \equiv 52000K$	$0.5\ eV \equiv 6000K$	$0.007\ eV \equiv 85K$
2	N_2	$7.08\ eV \equiv 85000K$	$0.28\ eV \equiv 3360K$	$0.00025\ eV \equiv 3K$
3	NO	$5.08\ eV \equiv 61000K$	$0.224\ eV \equiv 2688K$	$0.0002\ eV \equiv 2.4K$

It is clear from the table that

$$\epsilon_{elec} \gg \epsilon_{vib} \gg \epsilon_{rot}.$$

The temperature in a discharge tube is around $1200K$. Thus at this temperature, there are practically no electronic excitations. Only the vibrational and rotational excitations take place and that to at a very high rotational quantum number.

Problem 11: Find a partition function for a diatomic molecule for

  (i) translational motion

 (ii) Electronic motion

(iii) vibrational motion

(iv) Rotational motion

Solution: (i) Translational motion is a motion of center of mass of the molecule. Therefore putting $m = m_1 + m_2$ we get,

$$Z_{tr} = \frac{1}{N!}\frac{V^N}{h^{3N}}\left[2\pi(m_1 + m_2)k_B T\right]^{3N/2}.$$

(ii) As argued in problem 10, electronic state remains the ground state ϵ_0 of the molecule. Then,

$$Z_{ele} \cong e^{-\beta\epsilon_0}$$

(iii) Vibrational energy is given by $E_n = \left(n + \frac{1}{2}\right)\hbar\omega$. As given in problem 4 ,

$$Z_{vib} = \frac{\exp\left(\beta\hbar\omega/2\right)}{\exp\left(\beta\hbar\omega\right) - 1}$$

(iv) Rotational motion: We consider the diatomic molecule of two different atoms (heteronuclear molecule) in this problem. The rotational energy of a heteronuclear diatomic molecule is given by

$$E_{rot} = \frac{\hbar^2}{2I}K(K+1)$$

where K is a rotational quantum number. I is moment of inertia, $I = \mu r_0^2$. Here, r_0 is an equilibrium distance between the two atoms of the molecule. $\mu = \dfrac{m_1 m_2}{m_1 + m_2}$ is a reduced mass of the molecule. As clear from the table of problem 10 the rotational energy is the smallest energy among all the energies of the molecule. Taking $(2K + 1)$ as a rotational degeneracy we get

$$Z_{rot} = \sum_{K=0}^{\infty} (2K + 1)\exp\left[-\frac{\beta\hbar^2}{2I}K(K+1)\right].$$

Low temperature limit: $\dfrac{\hbar^2}{2Ik_BT} \gtrsim 1$ then

$$Z_{rot} \approx 1 + 3e^{-\beta\hbar^2/I}$$

$$F_{rot} = -Nk_BT \ln\left(1 + 3e^{-\beta\hbar^2/I}\right)$$

High temperature limit: $\dfrac{\hbar^2}{2Ik_BT} << 1.$
Here, large values of K are excited. Then,

$$Z_{rot} \approx \int_0^{\infty} 2K \exp\left[-\frac{\beta\hbar^2 K^2}{2I}\right] dK = \frac{2k_BTI}{\hbar^2}$$

Problem 12: Under what conditions the various contributions to the specific heat due to different degrees of freedom of a diatomic molecule is additive? Find the separate contributions of different motions and discuss their behavior.

Solution: If we can split the energy of a diatomic molecule additively i.e. $E = E_{tr} + E_{elec} + E_{vob} + E_{rot}$, then as seen from the problem 8 we can write

$$Z = Z_{tr}\, Z_{elec}\, Z_{vib}\, Z_{rot}.$$

Consequently we can write ,

$$F = -k_BT \left[\ln Z_{tr} + \ln Z_{elec} + \ln Z_{vib} + \ln Z_{rot}\right]$$

Then the average energy is

$$\overline{E} = -\frac{\partial}{\partial \beta}\left(\ln Z_{tr} + \ln Z_{elec} + \ln Z_{vib} + \ln Z_{rot}\right)$$

which is additive.

$$C_V = \left.\frac{\partial \overline{E}}{\partial T}\right)_V = C_V|_{tr} + C_V|_{elec} + C_V|_{vib} + C_V|_{rot} .$$

However nature is not that simple. Whenever there is a rotational motion, the rotator does not remain rigid. As a result, the bond length r_0 between the two atoms of the molecules increases. It means, the moment of inertia I increases as compared to what it was without a rotation. This then induces a coupling between the vibrational and the rotational motion. There is another important effect, namely, the vibrational potential is no more strictly quadratic. Its anharmonic potential part starts operating. Then the energy depends upon the amplitude in classical motion. When these effects become predominant we can no longer write the energy additively and therefore we can not have the additive C_V.

Thus consider an electronic motion. Then

$$F_{elec} = -Nk_B T \ln(1 + \exp\left(-\epsilon_0/k_B T\right))$$

$$\overline{E} = \frac{\epsilon_0}{e^{\beta \epsilon_0} + 1}$$

and

$$C_V|_{elec} = \frac{k_B \left(\epsilon_0/k_B T\right)^2}{\left(1 + \exp\left[\epsilon_0/k_B T\right]\right)\left(1 + \exp\left[-\epsilon_0/k_B T\right]\right)}$$

$C_V|_{elec} \to 0$ as $T \to 0$ or $T \to \infty$.

From problem 4

$$C_V|_{vib} = k_B \left(\frac{\hbar\omega}{k_B T}\right)^2 \frac{\exp\left[\frac{\hbar\omega}{k_B T}\right]}{\left(\exp\left[\frac{\hbar\omega}{k_B T}\right] - 1\right)^2}$$

For a rotational part only the low and the high temperature limit of the partition function can be found. Thus, using result of problem 11, we see that at low temperatures

$$C_{rot} = 3Nk_B \left(\frac{\hbar^2}{I k_B T}\right)^2 \exp\left(-\frac{\hbar^2}{I k_B T}\right)$$

On the other hand, for high temperature, $C_V = k_B/molecule$. It is seen from the figure 2 that the rotational contribution to the specific heat rises to a maximum value up to about $1.1k_B$ at around $T = 0.8\hbar^2/(2I k_B)$ and then asymptotically tends to its high temperature value at k_B.

Problem 13: Obtain rotational partition function for a gas of homo nuclear molecules in low and high temperature limit. Why do you need to consider a case of homo nuclear molecules separate from a hetero nuclear molecular case?

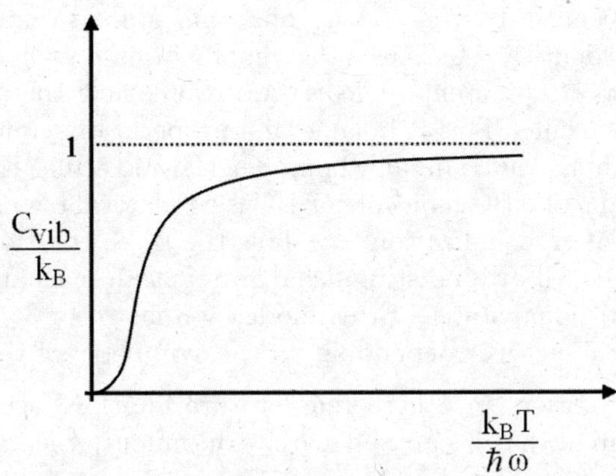

Figure 14.1: Vibrational contribution to the diatomic molecule specific heat as a function of the absolute temperature.

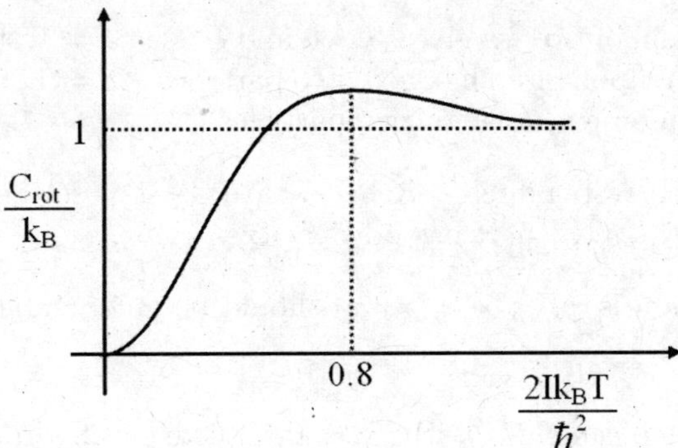

Figure 14.2: Rotational contribution to the specific heat of a diatomic molecule.

Solution Homo nuclear diatomic molecule is made up of two identical atoms. e.g. H_2, N_2, O_2 etc. On the other hand, molecules like NO, CO are hetero nuclear molecules.

The exchange properties of the two types of molecules are different. In case of H_2, if there is a rotation from ϕ to $\phi + \pi$, the quantum state is same since two hydrogen atoms forming H_2 are identical. On the other hand, in case of hetero nuclear molecules, the states ϕ and $\phi + \pi$ are different. Clearly we over count the states by a factor 2 in case of homo nuclear molecule. This is in general a property of symmetry. It can be seen in general. Consider an equilateral triangular homo nuclear molecule. Here, we see the rotation perpendicular to the plane of ABC molecule. ABC is equilateral triangle with vertices having same atom. For the sake of visualization, we show the three rotations through 120° where $ACB \rightarrow BAC \rightarrow CBA$. All are indistinguishable quantum mechanically. As a result, we will be counting the rotational states of this mode by a factor of 3. In general, we say that we over count states by a factor σ depending on the symmetry of the molecule.

For a rotational motion, in a high temperature limit, K becomes large and energy difference becomes so small that it can be treated continuous. Clearly, expression of Z_{rot} in problem 11 (iv), becomes,

$$Z_{rot} = \frac{2k_B T I}{\sigma \hbar^2}.$$

For a homonuclear diatomic molecule, $\sigma = 2$ and

$$Z_{rot} = \frac{k_B T I}{\hbar^2} \quad \text{and } C_V = N k_B.$$

At low temperatures, quantum effects become predominant. Consider a case of H_2 where in the effect of homo nuclear nature is predominant. This is so because its factor $\hbar^2/2Ik_B$ has a highest value among the homo nuclear diatomic molecules. Following quantum mechanical facts are worth recalling.

(a) Combination of spin of two $\frac{1}{2}\hbar$ gives $S = 0$ and $S = 1$. $S = 0$ state is an antisymmetric state (changes sign upon interchange of two particles.) $S = 1$ state is symmetric. (Does not change sign upon exchange of two particles).

(b) Rotational state corresponding to K has a parity $(-1)^K$ so that $K_{even}(0, 2, 4, 6, ..)$ has even parity and $K_{odd}(1, 3, 5...)$ has odd parity.

(c) Total wave function is $\psi_{spin} \times \psi_{rot}$ and should be antisymmetric. This is so because nuclear spin is $i = \frac{1}{2}\hbar$.

Thus hydrogen molecule H_2 with $S = 0$, called parahydrogen will have K values $0, 2, 4, ...even, ...$ etc. to make total wave function antisymmetric.

Similarly for H_2, $S = 1$ called ortho hydrogen will have its rotational quantum numbers $K = 1, 3, 5, ...$odd integers.
This give

$$Z_{rot}\big|_{para} = \sum_{K=0,2,4...} (2K + 1) \exp\left(\frac{-\beta \hbar^2 K(K + 1)}{2I}\right)$$

$$Z_{rot}|_{ortho} = \sum_{K=1,3,5\ldots} (2K+1) \exp\left(\frac{-\beta\hbar^2 K(K+1)}{2I}\right)$$

Ortho hydrogen has $(2i+1) = 3$ as a statistical weight whereas, for parahydrogen has $(2i+1) = 1$ as a statistical weight. Then,

$$Z_{rot} = \frac{3}{4} Z_{rot}|_{ortho} + \frac{1}{4} Z_{rot}|_{para}$$

Rotational contribution to the specific heat is

$$C_V = \frac{3}{4} C_{ortho} + \frac{1}{4} C_{para}$$

Problem 14: Consider a following interesting 2 dimensional situation. A quiet surface of superfluid liquid He^4 has a surface tension σ_0. Free energy of a surface area A is $\sigma_0 A$. Surface of the He^4 is excited so that waves (called Rayleigh's capillary waves) are created whose dispersion relation is $\omega_k^2 = \frac{\sigma_0}{\rho} k^3$. Here ρ is density of liquid He^4. Because of these excitations (waves) surface tension changes to σ and this free energy σA is given by

$$\sigma A = \sigma_0 A + k_B T \sum_k \ln(1 - e^{-\beta\hbar\omega_k})$$

Find $\sigma(T)$.

Solution: In two dimensions

$$\sum_k \to \frac{A}{(2\pi)^2} \int .$$

Thus,

$$\sum_k \ln(1 - \exp(-\beta\hbar\omega_k)) = \frac{A}{(2\pi)^2} . 2\pi \int_0^\infty k \, dk \ln(1 - \exp(-\beta\hbar\omega_k))$$

$$= \frac{A}{2\pi} \left(\frac{\rho}{\sigma_0}\right)^{2/3} . \frac{2}{3} \int_0^\infty \omega^{1/3} \ln(1 - \exp(-\beta\hbar\omega)) d\omega$$

Integrating by parts and simplifying we get

$$= -A \frac{\hbar}{4\pi} \left(\frac{\rho}{\sigma_0}\right)^{2/3} \left(\frac{k_B T}{\hbar}\right)^{7/3} \int_0^\infty \frac{\xi^{4/3}}{e^\xi - 1} .$$

Value of the above definite integral is approximately 0.13.
Thus,

$$\sigma = \sigma_0 - 0.13 \frac{(k_B T)^{7/3}}{\hbar^{4/3}} \left(\frac{\rho}{\sigma_0}\right)^{2/3}$$

This behavior of decreasing surface tension $\sim T^{7/3}$ is well verified by the experiments.

Problem 15: It is well known that due to high temperatures, some of the atoms, from otherwise a perfectly crystalline solid, are displaced from lattice position to an interstitial position. Let n be the number of interstitial atoms and ε_0 be the energy necessary to dislodge the atom from lattice to interstitial position. With N normal lattice points in volume V, $(N-n)$ lattice positions are arranged so also the n interstitial atoms. Find the equilibrium distribution $n(T)$.

Solution: N lattice points have N interstitial positions. Thus n atoms can be arranged in N positions, $\dfrac{N!}{n!(N-n)!}$ ways. Number of ways in which $(N-n)$ atoms can be arranged in N positions is $\dfrac{N!}{(N-n)!n!}$. The total weight thus is

$$W = \left[\frac{N!}{(N-n)!n!} \right]^2 .$$

Entropy is, by using Sterling formula,

$$S = k_B \ln W \cong 2k_B \left[N \ln N - n \ln n - (N-n) \ln(N-n) \right]$$

The internal energy of the lattice is $n\varepsilon_0$.
The free energy is

$$F = \overline{E} - TS = n\varepsilon_0 - 2k_B T \left[N \ln N - n \ln n - (N-n) \ln(N-n) \right] .$$

For equilibrium, $\partial F/\partial n = 0$. Taking the derivative and simplifying we get

$$n = \frac{N}{1 + e^{\varepsilon_0/2k_B T}}$$

Problem 16: Given there states of energy 0, ε and 3ε and two particles at an absolute temperature T, find the partition function when
(i) Particles obey Maxwell- Boltzmann distribution
(ii) Particles obey Bose-Einstein distribution
(iii) Particles obey Fermi -Dirac distribution.

Solution: Let us show occupancy in various cases

	I			II			III	
$\varepsilon = 0$	ε	3ε	$\varepsilon = 0$	ε	3ε	$\varepsilon = 0$	ε	3ε
AB	-	-	AA	-	-	A	A	-
-	AB	-	-	AA	-	-	A	A
-	-	AB	-	-	AA	A	-	A
A	B	-	A	A	-			
B	A	-	-	A	A			
-	A	B	A	-	A			
-	B	A						
A	-	B						
B	-	A						

Here I corresponds to Maxwell -Boltzmann occupancy, II corresponds to Bose -Einstein occupancy and III corresponds to Fermi-Dirac occupancy.

Partition function in case of
(i) Maxwell-Boltzmann statistics

$$Z_{M-B} = 1 + e^{-2\beta\varepsilon} + e^{-6\beta\varepsilon} + 2e^{-\beta\varepsilon} + 2e^{-4\beta\varepsilon} + 2e^{-3\beta\varepsilon}$$

(ii) Bose- Einstein statistics

$$Z_{B-E} = 1 + e^{-2\beta\varepsilon} + e^{-6\beta\varepsilon} + e^{-\beta\varepsilon} + e^{-4\beta\varepsilon} + e^{-3\beta\varepsilon}$$

(iii) Fermi-Dirac statistics

$$Z_{F-D} = e^{-\beta\varepsilon} + e^{-4\beta\varepsilon} + e^{-3\beta\varepsilon}$$

Problem 17: Consider N' adsorbed atoms on the surface of area A. These atoms are held on the surface and have a binding energy ε_0. Then, as these adsorbed atoms can move freely on the surface, the energy of the atom is $p^2/2m - \varepsilon_0$. Find chemical potential μ' of this 2 dimensional gas of adsorbed atoms in classical approximation. This surface is exposed to the air and the air and the adsorbed gas are in equilibrium. If P is pressure of air find equilibrium density $n' = N'/A$ at temperature T.

Solution: This problem is of practical importance to estimate the equilibrium concentration n'. First we calculate μ' as follows.

$$n_k = \exp\left[\beta\left(\mu' - \varepsilon_k\right)\right]$$

$$
\begin{aligned}
\therefore \quad N' &= \sum_P \exp\left[\beta\left(\mu' - p^2/2m + \varepsilon_0\right)\right] \\[2mm]
&= \frac{A}{h^2} 2\pi \exp\left[\beta\mu' + \beta\varepsilon_0\right] \int_0^\infty p\, dp \exp\left[\left(-\beta p^2/2m\right)\right] \\[2mm]
&= \frac{\pi A}{h^2} \exp\left[\beta\mu' + \beta\varepsilon_0\right]\left(2mk_B T\right)
\end{aligned}
$$

Thus, simplifying ,

$$\frac{\mu'}{k_B T} = -\frac{\varepsilon_0}{k_B T} + \ln n' - \ln\left(\frac{2\pi m k_B}{h^2}\right) - \ln T$$

For equilibrium, (from thermodynamics) with surrounding air of chemical potential μ we must have $\mu = \mu'$. From problem (4)

$$\frac{\mu}{k_B T} = \left[\ln\left(\frac{N}{V}\right) - \frac{3}{2}\ln T - \frac{3}{2}\ln\left(\frac{2\pi m k_B}{h^2}\right)\right]$$

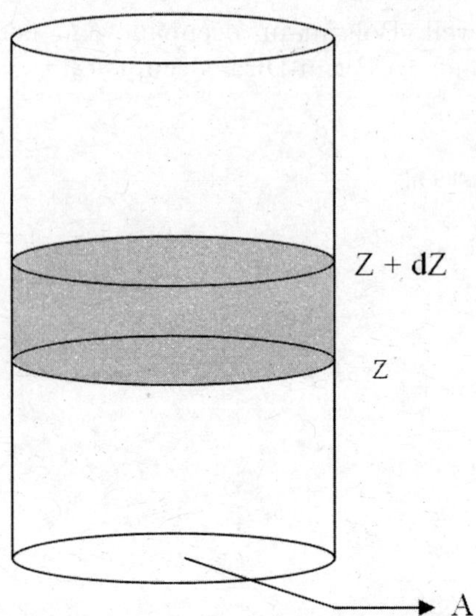

Figure 14.3: Pressure difference between two ends of the slice balanced by weight of the gas

But the air surrounding the surface is almost ideal. We use $P = \dfrac{N}{V} k_B T$ for surrounding air pressure.

Comparing

$$\frac{\mu}{k_B T} = \frac{\mu'}{k_B T}.$$

After simplifying, we get desired result as,

$$n' = \frac{Ph}{(2\pi m)^{1/2}} \left(k_B T\right)^{-3/2} \exp\left(\beta \varepsilon_0\right).$$

Problem 18: Earth's atmosphere can be very crudely thought of as an ideal gas in the field of uniform gravity g and at a constant temperature T. (Actually condition of constant T is not accurately satisfied). By examining the condition of a hydrostatic equilibrium for a slice of a gas confined within z and $z + dz$, derive the expression for local number density $n(z)$ of the air as a function of z.

Solution:

Consider a column of air in the cylinder as shown in figure 14.3. The pressure difference between z and $z + dz$ is equal to pressure exerted by the weight of the air in the slice . Thus

$$[P(z) - P(z + dz)]\, A = \rho\, A\, g\, dz$$

where ρ is mass density. Taylor expanding, we get,

$$-\frac{\partial P}{\partial z} = nmg$$

where $\rho = nm$, n is number density of air molecule of mass m. Write $P = nk_BT$
Thus,

$$\frac{\partial n}{\partial z} = -\frac{nmg}{k_BT}$$

Or, integrating, with n_0 as a number density of air at the ground, we get,

$$n(z) = n_0 \exp\left(-\frac{mgz}{k_BT}\right).$$

This can be obtained very easily using canonical distribution. Here probability that the molecule is at a height z, irrespective of its momentum is

$$P(z) \propto \exp\left(-\frac{V}{k_BT}\right) = C \exp\left(-\frac{mgz}{k_BT}\right)$$

The number density then is $n(z) = n_0 \exp\left(-\dfrac{mgz}{k_BT}\right)$
This is also called as barometric distribution.

Problem 19: Evaluate a density matrix for a free particle in coordinate representation i.e. find $\left\langle \vec{r} \,|\rho|\, \vec{r'} \right\rangle$. Evaluate the partition function also.

Solution: Free particle wave-function is

$$\psi_{\vec{k}} = \frac{1}{\sqrt{V}} e^{i\vec{k}.\vec{r}} = \left\langle \vec{r}\,|\,\vec{k} \right\rangle$$

where $\vec{k}$ is a wave vector $= \vec{p}/\hbar$ and V is confining volume. Then, with $\hat{H} = \hbar^2 k^2/2m$,

$$\left\langle \vec{r}\left|e^{-\beta\hat{H}}\right|\vec{r'}\right\rangle = \sum_k \left\langle \vec{r}|\vec{k}\right\rangle \exp\left(-\frac{\beta\hbar^2 k^2}{2m}\right)\left\langle \vec{k}|\vec{r'}\right\rangle$$

$$= \frac{1}{V}\sum_k \exp\left(-\beta\frac{\hbar^2 k^2}{2m} + i\vec{k}.(\vec{r}-\vec{r'})\right)$$

$$= \rho_{\vec{r}-\vec{r'}}$$

But,

$$\sum_k = \frac{L^d}{(2\pi)^d}\int d^d k \qquad \text{for } d \text{ dimensions}$$

For 3 dimensions, then,

$$\left\langle \vec{r}\left|e^{-\beta\hat{H}}\right|\vec{r'}\right\rangle = \frac{2\pi}{(2\pi)^3}\int_0^\infty k^2 dk \exp\left(-\beta\frac{\hbar^2 k^2}{2m}\right)\int_0^\pi \exp\left(ik|\vec{r}-\vec{r'}|\cos\theta\right)\sin\theta\, d\theta$$

$$= \frac{1}{(2\pi)^2}2\int_0^\infty k^2 dk \exp\left(-\beta\frac{\hbar^2 k^2}{2m}\right)\frac{\sin\left(k|\vec{r}-\vec{r'}|\right)}{k|\vec{r}-\vec{r'}|}$$

$$= \left(\frac{m}{2\pi\beta\hbar^2}\right)^{3/2}\exp\left(-\frac{mk_BT}{2\hbar^2}|\vec{r}-\vec{r'}|^2\right)$$

$$\begin{aligned}
Z &= Tr\, e^{-\beta H}\\
&= \int \left\langle \vec{r}\,\middle|\,e^{-\beta \hat{H}}\,\middle|\,\vec{r}\,\right\rangle d^3 r\\
&= V\left(\frac{m}{2\pi\beta\hbar^2}\right)^{3/2}\\
&= V\left(\frac{m2\pi k_B T}{h^2}\right)^{3/2}
\end{aligned}$$

which is same as what we got earlier. For N particle system, $Z = \dfrac{V^N}{\lambda^{3N}}$ but without $\dfrac{1}{N!}$ factor. What is the reason for the wrong answer? See problem (24).

Problem 20: With the density matrix calculated in problem (19), find average E per particle.

Solution:

$$E = Tr\left(\rho\hat{H}\right) = -\frac{\hbar^2}{2m}\left(\frac{2\pi m k_B T}{h^2}\right)^{3/2}\int\left[\nabla^2 \exp\left(-\frac{2\pi m k_B T}{h^2}r\right)\right]_{r=0} d^3 r.$$

Use $\nabla^2 = \left(\dfrac{\partial^2}{\partial r^2} + \dfrac{2}{r}\dfrac{\partial}{\partial r}\right)$. After completing the steps, one gets,

$$E = \frac{3}{2}k_B T$$

Problem 21: To separate U_{235} from U_{238} from naturally occurring uranium, a gas of fluorides $U_{235}F^{19}$ and $U_{238}F^{19}$ is passed through a porous partition. Abundance of U_{235} and U_{238} is 0.7 and 99.3 % in natural mixture. Thus the concentration ratio in a natural mixture is $C(U_{235})/C(U_{238}) = \dfrac{0.7}{99.3}$. The gas is confined in a vessel of volume V and A is total area of pores in the wall. Find ratio $C'(U_{235})/C'(U_{238})$ after a time t in a collecting chamber attached to the vessel at its porous wall. Express your answer in terms of molecular weights $M(235)$ and $M(238)$. The ratio C_1/C_2 is maintained constant by approximately replenishing the supply in the vessel.

Solution: From problem (5) of chapter (??)

$$C_1' = C_1 \exp\left(-t/\tau_1\right), \quad C_2' = C_2 \exp\left(-t/\tau_2\right)$$

where $\tau_{1,2} = \dfrac{V}{A}\left(\dfrac{2\pi}{k_B T}\right)^{1/2}\sqrt{m_{1,2}} \equiv \alpha(T,V,A)\sqrt{m_{1,2}}$

Thus,

$$\begin{aligned}
\frac{C_1'}{C_2'} &= \left(\frac{C_1}{C_2}\right)\exp\left(-\frac{t}{\alpha(T,V,A)}\left(\frac{1}{\sqrt{m_1}} - \frac{1}{\sqrt{m_2}}\right)\right)\\
&= \left(\frac{C_1}{C_2}\right)\exp\left(-\frac{tA}{V}\left(\frac{RT}{2\pi M(235)}\right)^{1/2}\left(1 - \sqrt{\frac{M(235)}{M(238)}}\right)\right)
\end{aligned}$$

This method of an isotope separation was used for separating fissile U_{235} isotope from non fissile U_{238} in Manhattan Project. This method was developed by Neddermyre.

Problem 22: Consider a small system of volume V_1 under consideration in contact with a reservoirs of volume $V_2 >> V_1$. The contact is established by a mobile wall so that the small system and the reservoirs can exchange energy, the volumes V_1 and V_2 change around their mean, keeping $V = V_1 + V_2$ fixed, but there is no exchange of particles. In the discussion, ignore a kinetic energy $P_W^2/2M_W$ of the wall of mass M_W. Find corresponding partition function. Find mean square fluctuation in volume.

Solution: We follow the steps exactly similar to those we followed when we discussed canonical and grand canonical ensemble cases. Here,

$$V = V_1 + V_2, \quad E = E_1 + E_2$$

and, number of states accessible to the reservoir are

$$\Omega\left(E_2, V_2\right) = \Omega\left(E - E_1, V - V_1\right).$$

Clearly, probability P_r that the system has energy state E_1 and volume V_1 is

$$P_r \equiv P_{E_1, V_1} = \frac{\Omega\left(E - E_1, V - V_1\right)}{\sum \Omega} = \text{constant } \Omega\left(E - E_1, V - V_1\right)$$

$$\therefore \quad \ln P_r = \ln(\text{ constant }) + \ln \Omega\left(E - E_1, V - V_1\right)$$

But $\ln \Omega\left(E - E_1, V - V_1\right) = \ln \Omega(E, V) - E_1 \dfrac{\partial \ln \Omega}{\partial E} - V_1 \dfrac{\partial \ln \Omega}{\partial V}$. Thus, with $(\partial \ln \Omega / \partial E) = (k_B T)^{-1}$ etc,

$$P_r = C \exp\left(-\frac{E + PV}{k_B T}\right)$$

Clearly, the partition function in this case is $Z(P, T)$ such that

$$Z(P, T) = \sum \exp\left(-\frac{E + PV}{k_B T}\right)$$

Thus, the enthalpy function appears in case of volume and energy fluctuating ensemble.

With the procedures already discussed, we can write

$$\exp\left(-G/k_B T\right) = Z(P, T)$$

where G is Gibb's free energy. (Prove this assertion).
From the definition

$$dG = -SdT + \overline{V}dP$$

$$\overline{V} = \left(\frac{\partial G}{\partial P}\right)_T = \frac{\sum V \exp\left(-\beta(E + PV)\right)}{\sum \exp\left(-\beta(E + PV)\right)}.$$

$$\frac{\partial \overline{V}}{\partial P} = \frac{Z\left(-\beta \sum V^2 \exp\left[-\beta(E+PV)\right]\right) + \beta\left(\sum V \exp\left[-\beta(E+PV)\right]\right)}{Z^2}$$

$$= -\frac{1}{k_B T}\left(\overline{V^2} - \left(\overline{V}\right)^2\right)$$

But isothermal compressibility is $\kappa = -\dfrac{1}{\overline{V}}\dfrac{\partial \overline{V}}{\partial P}$.

Thus,

$$\left(\frac{\overline{V^2} - \overline{V}^2}{\overline{V}^2}\right) = \frac{(\Delta V)^2}{\left(\overline{V}^2\right)} = \frac{k_B T}{\overline{V}}\kappa.$$

For an ideal gas, $\overline{V} = N k_B T/P$, $\kappa = \overline{V}/N k_B T$

Thus,

$$\frac{\sqrt{\overline{V^2} - \overline{V}^2}}{\overline{V}} = \frac{1}{\sqrt{N}} \to 0 \text{ as } N >> 1$$

This ensemble is isothermal-isobaric ensemble.

Problem 23: Find mean square deviation in pressure,

$$\overline{P^2} - \overline{P}^2$$

using canonical ensemble.

Solution:

$$\overline{P} = \frac{\sum\limits_{r}\left(-\dfrac{\partial E_r}{\partial V}\right)\exp\left(-\beta E_r\right)}{\sum\limits_{r}\exp\left(-\beta E_r\right)} = \frac{\sum\limits_{r}\left(-\dfrac{\partial E_r}{\partial V}\right)\exp\left(-\beta E_r\right)}{Z}$$

Thus,

$$\left(\frac{\partial \overline{P}}{\partial V}\right)_T Z + (\beta)\overline{P}\sum\limits_{r}\left(-\frac{\partial E_r}{\partial V}\right)\exp\left(-\beta E_r\right)$$

$$= \sum\limits_{r}\left(-\frac{\partial^2 E_r}{\partial V^2}\right)\exp\left(-\beta E_r\right) + \frac{1}{k_B T}\sum\limits_{r}\left(-\frac{\partial E_r}{\partial V}\right)^2\exp\left(-\beta E_r\right)$$

$$\therefore \quad \left(\frac{\partial \overline{P}}{\partial V}\right)_T + \frac{\overline{P}^2}{k_B T} = \overline{\left(-\frac{\partial^2 E}{\partial V^2}\right)} + \frac{\overline{P^2}}{k_B T}$$

or

$$\frac{\overline{P^2} - \overline{P}^2}{k_B T} = \left(\frac{\partial \overline{P}}{\partial V}\right)_T + \overline{\left(\frac{\partial^2 E}{\partial V^2}\right)_T}$$

Or,

$$\overline{P^2} - \overline{P}^2 = k_B T\left[\left(\frac{\partial \overline{P}}{\partial V}\right)_T - \overline{\frac{\partial P}{\partial V}}\right]$$

Problem 24: Find the mean square fluctuation in enthalpy.

Solution:

$$Z(P,T) = \sum \exp\left(-\beta H_{enth}\right)$$

where H_{enth} is enthalpy $= E + PV \equiv H$
Clearly,

$$\frac{\partial \ln Z}{\partial \beta} = -\frac{\sum H_{enth} \exp\left(-\beta H_{enth}\right)}{Z} = -\overline{H}$$

$$\frac{\partial^2 \ln Z}{\partial \beta^2} = \overline{H^2} - \overline{H}^2 = \left(\frac{\partial \overline{H}}{\partial \beta}\right)_P = k_B T^2 \left(\frac{\partial \overline{H}}{\partial T}\right)_P$$

Also, from the definition of enthalpy, $dH = TdS + VdP$

$$\therefore \left(\frac{\partial \overline{H}}{\partial T}\right)_P = T\frac{\partial S}{\partial T} = C_P$$

Then,

$$\frac{\overline{H^2} - \overline{H}^2}{\overline{H}^2} = \frac{k_B T^2 C_P}{\overline{H}^2}$$

Problem 25: If quantum mechanics is the correct description of nature, then we should get partition function for free particle as $\dfrac{V^N}{N!\lambda^{3N}}$. However, problem (19) gives the answer without $(1/N!)$ factor. Explain !

Solution: In problem (19), we calculated $\rho_{\vec{r},\vec{r}'}$ as,

$$\rho_{\vec{r},\vec{r}'} = \frac{1}{\lambda^3} \exp\left(-\frac{|\vec{r} - \vec{r}'|^2}{\lambda^2}\right)$$

Taking trace of $\rho_{\vec{r},\vec{r}'}$ gives correct partition function for one particle. Raising it to the power N for N particle system is <u>wrong</u>. In effect, then , we are writing N particle state as

$$|\vec{r_1}, \vec{r_2}, ...\vec{r_N}\rangle = |\vec{r_1}\rangle |\vec{r_2}\rangle ... |\vec{r_N}\rangle$$

This is not correct. We must properly symmetrize the state for N identical particles. If N particles are Fermions, then the N particle state is Slater determinant and is given by

$$\psi_{p_1...p_N}^{Fermi}(\vec{r_1}, \vec{r_2}...\vec{r_N}) = \frac{1}{\sqrt{N!}} \begin{vmatrix} \psi_{p_1}(\vec{r_1}) & \psi_{p_1}(\vec{r_2}) & ... & \psi_{p_1}(\vec{r_N}) \\ \psi_{p_2}(\vec{r_1}) & \psi_{p_2}(\vec{r_2}) & ... & \psi_{p_2}(\vec{r_N}) \\ \cdot & & & \\ \cdot & & & \\ \cdot & & & \\ \psi_{p_N}(\vec{r_1}) & \psi_{p_N}(\vec{r_2}) & ... & \psi_{p_N}(\vec{r_N}) \end{vmatrix} \equiv \frac{1}{\sqrt{N!}} det\,(\vec{r_1}, \vec{r_2}, ...\vec{r_N})$$

If we write

$$\rho_{\vec{r_1},\vec{r_2},....\vec{r_N};\ \vec{r_1'},\vec{r_2'},....\vec{r_N'}} = \frac{1}{\lambda^{3N}} \cdot \exp\left[-\frac{|\vec{r_1}-\vec{r_1'}|^2 + |\vec{r_2}-\vec{r_2'}|^2 + ... + |\vec{r_N}-\vec{r_N'}|^2}{\lambda^2}\right].$$

Then we find,

$$
\begin{aligned}
Z &= Tr\ \rho \\
&= \lim_{\vec{r_i}\to\vec{r_i'}} \frac{1}{N!}\ \frac{1}{\lambda^{3N}} \int \left[det\ (\vec{r_1},\vec{r_2},....\vec{r_N}) \exp\left[-\frac{|\vec{r_1}-\vec{r_1'}|^2 + |\vec{r_2}-\vec{r_2'}|^2 + ... + |\vec{r_N}-\vec{r_N'}|^2}{\lambda^2}\right] \right. \\
&\qquad \left. \times\ det\ (\vec{r_1'},\vec{r_2'},....\vec{r_N'}) \right] d^3r_1 d^3r_2 ... d^3r_N \\
&= \frac{V^N}{N!}\ \frac{1}{\lambda^{3N}}
\end{aligned}
$$

for every i running from 1 to N.

Clearly, we see that the correct counting is made in quantum mechanics and the Gibb's paradox is avoided.

Problem 26: You are given a three level system with ground state $E_0 = 0$, first excited state has energy $E_1 = k_B T$ and second excited state is $E_2 = 2k_B T$. N particles are distributed in these three states at temperature T in equilibrium. Equilibrium energy of the system is $10000 k_B T$. Find N.

Solution:

N_0 : Number of particles in ground state.

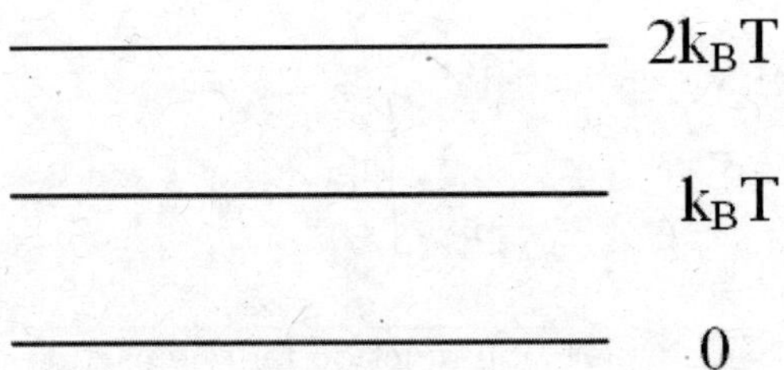

Figure 14.4: Energy levels for Problem 26.

N_1 : Number of particle in first excited state.

N_2 : Number of particle in second excited state.

Clearly, $\dfrac{N_1}{N_0} = \exp\left(-\beta k_B T\right) = \dfrac{1}{e}$

Similarly, $\dfrac{N_2}{N_0} = \dfrac{1}{e^2}$.

$$N = N_0\left(1 + \frac{1}{e} + \frac{1}{e^2}\right); \quad E = N_1 E_1 + N_2 E_2 = \frac{N k_B T\left(\frac{1}{e} + \frac{2}{e^2}\right)}{\left(1 + \frac{1}{e} + \frac{2}{e^2}\right)}$$

Put $E = 10000 k_B T$. Then find $N = \dfrac{10^4 \left(1 + e^{-1} + e^{-2}\right)}{(e^{-1} + 2e^{-2})} \approx 24000$.

Problem 27: N spin half particles are located at N lattice points. They may or may not be interacting. It is known that, at absolute zero, the system is ferromagnetic. Spin are totally randomized at high temperature. If $C(T)$ is a specific heat of the system at T, show that

$$\int_0^\infty \frac{C(T)dT}{T} = N k_B \ln 2$$

irrespective of mechanism of ferromagnetism.

Solution: $S = k_B \ln \Omega$ where Ω is number of accessible states. For high temperatures, $\Omega = 2^N \Rightarrow S_\infty = N k_B \ln 2$.
At $T = 0$, $\Omega = 1$, the ferromagnetic state. Thus $S_0 = N k_B \ln 1 = 0$.
Since entropy is a thermodynamic state,

$$S_\infty - S_0 = S_\infty = N k_B \ln 2 = \int_0^\infty \frac{C(T)dT}{T}.$$

In this counting, no mechanism of ferromagnetism is invoked

Problem 28: You are given N non interacting equilateral triangles at the corners of which spin $1/2$ particles is fixed. Typical equilateral triangles has Hamiltonian

$$H = \frac{\lambda}{3}(\vec{\sigma_1}.\vec{\sigma_2} + \vec{\sigma_2}.\vec{\sigma_3} + \vec{\sigma_3}.\vec{\sigma_1})$$

with $\vec{S} = \frac{1}{2}\vec{\sigma}$ and $\vec{\sigma_1}$ is kept at 1st corner etc. Find partition function.

Solution: Here $\vec{S_1}, \vec{S_2}, \vec{S_3}$ and $\vec{S} = \vec{S_1} + \vec{S_2} + \vec{S_3}$ are constants of motion. Then,

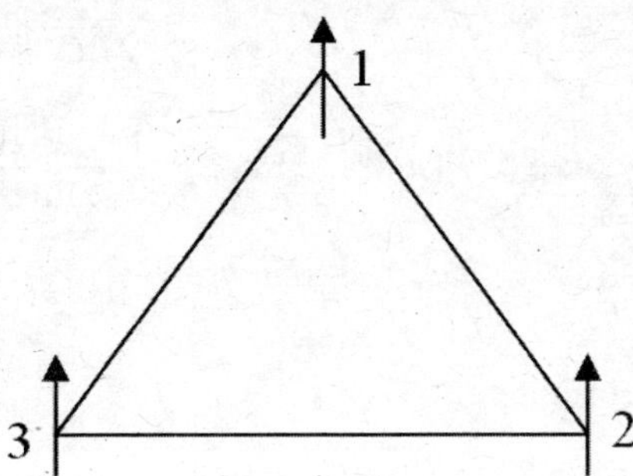

Figure 14.5: Spin-1/2 particles sitting at the vertices of an equilateral triangle.

$$\left(\vec{S_1}.\vec{S_2} + \vec{S_2}.\vec{S_3} + \vec{S_3}\vec{S_1}\right) = \frac{S^2}{2} - \frac{S_1^2 + S_2^2 + S_3^2}{2}.$$

$$\left\langle \vec{S_1}.\vec{S_2} + \vec{S_2}.\vec{S_3} + \vec{S_3}\vec{S_1} \right\rangle = \frac{1}{2}S(S+1)\hbar^2 - \frac{3}{2}.\frac{1}{2}.\frac{3}{2}\hbar^2$$

Thus, eigen values of H given as,

$$H = \frac{\lambda}{6}(4S(S+1) - 9)$$

Possible values of S are $\frac{3}{2}$ and $\frac{1}{2}$. Putting these,
we get, $E_1 = \lambda$ and $E_2 = -\lambda$.
$S = 3/2$ has a degeneracy $2S + 1 = 4$ where as $S = 1/2$ has degeneracy 2. But there are
two independent combinations giving degeneracy 4. Partition function of the one triangle
is,

$$Z = 4 \times 2\frac{\left(e^{\beta\lambda} + e^{-\beta\lambda}\right)}{2} = 8\cosh(\beta\lambda)$$

$\therefore$ For N triangles,

$$Z = 8^N \cosh^N(\beta\lambda).$$

Problem 29: $L - C$ circuit is an electrical analogue of mechanical vibrations (See Capri
and Panat 'Introduction to electrodynamics' P-260 Narosa). The energy of the undamped
$L - C$ circuit is

$$H = \frac{1}{2}L\left(\frac{dQ}{dt}\right)^2 + \frac{Q^2}{2C}$$

where Q is the charge in the circuit and L and C are inductor and capacitor joined in
parallel. Find $r.m.s.$ noise voltage as a function of temperature.

Solution: $L - C$ circuit has oscillating frequency $\omega = 1/\sqrt{LC}$.
The energy of the oscillator can be quantized to give

$$E_n = (n + \frac{1}{2})\hbar\omega \ ; \ \omega = \frac{1}{\sqrt{LC}}.$$

$$Z = \sum_{n=0}^{\infty} \exp\left(-\beta(n + 1/2)\hbar\omega\right) = \frac{1}{2}\frac{1}{\sinh(\beta\hbar\omega/2)}$$

$$E = -\frac{\partial}{\partial\beta}\ln Z = \frac{\hbar\omega}{2}\coth(\beta\hbar\omega/2)$$

Clearly, for harmonic oscillator,

$$\left\langle \frac{1}{2}LI^2 \right\rangle = \left\langle \frac{1}{2}\frac{Q^2}{C} \right\rangle = \left\langle \frac{1}{2}CV^2 \right\rangle = \frac{E}{2}$$

$$\therefore \qquad \langle V^2 \rangle \ = \ \frac{\hbar\omega}{2C}\coth(\beta\hbar\omega/2)$$

$$\langle I^2 \rangle \ = \ \frac{\hbar\omega}{2L}\coth(\beta\hbar\omega/2)$$

At high temperature limit, $\hbar\omega << k_B T$, we get

$$\langle V^2 \rangle = \frac{k_B T}{C} \quad ; \quad \langle I^2 \rangle = \frac{k_B T}{L}.$$

At low temperature limit, $\hbar\omega >> k_B T$,

$$\langle V^2 \rangle = \frac{\hbar\omega}{2C} \quad ; \quad \langle I^2 \rangle = \frac{\hbar\omega}{2L}.$$

Note, at low temperature, the noise is independent of temperature.

Problem 30: Consider a liquid in contact with its vapor. The vapor has a small droplet of the liquid and is of radius r. P_∞ is a vapor pressure of the vapor and P_r is a vapor pressure for the droplet. Find P_r in terms of appropriate parameters.

Solution: P_r and P_∞ differ because of surface tension σ. As vapor and the drops are in equilibrium, Gibb's free energy has attained minimum. Thus, if g_v and g_d are Gibb's energies per unit mass for vapor and the drop, then,

$$G = M_d g_d + M_r g_r + 4\pi\sigma r^2$$

where M_d is mass of the drop, M_v is mass of the vapor. With mass conservation, $\delta M_d = -\delta M_v$.

$$\delta G = \delta M_d \left(g_d - g_v\right) + 8\pi\sigma r \frac{\partial r}{\partial M_d}\delta M_d = 0.$$

But $\rho\frac{4}{3}\pi r^3 = M_d$ or $\dfrac{\partial r}{\partial M_d} = \dfrac{1}{4\pi r^2 \rho}$
Thus,

$$g_v - g_d = \frac{2\sigma}{\rho r}.$$

But, $\dfrac{G}{N} = g$, $(\partial g/\partial P)_T = \dfrac{1}{\rho}$

$$\therefore \left(\frac{\partial g_v}{\partial P}\right)_T - \left(\frac{\partial g_d}{\partial P}\right)_T = -\frac{2\sigma}{\rho r^2}\frac{\partial r}{\partial P}$$

$$\left(\frac{\partial g_v}{\partial P}\right)_T = \frac{1}{\rho_v} >> \left(\frac{\partial g_d}{\partial P}\right)_T = \frac{1}{\rho_d}$$

Moreover, assume vapor to be ideal $\Rightarrow P = \rho_v k_B T$.
Using these relations and integrating and using $N_v k_B = R$, we get,

$$P_r = P_\infty \exp\left(\frac{2\sigma M}{\rho R T r}\right)$$

where M is molecular weight of the vapor. This result is derived by Lord Kelvin and is used in cloud physics.

Chapter 15

CHEMICAL REACTIONS

The topic of chemical reactions is vast and of practical importance. This is so because a vast chemical industry relies on it. There are many issues that are studied

a. When two or more chemicals are in contact, then, will the reaction take place and what are reaction products?

b. If the reaction takes place, then what is the rate of reaction and can it be accelerated? It is well known that most of the organic reactions are snail paced.

   It is found that a mixture of H_2 and O_2 can remain without a formation of water if undisturbed . But a single spark products a reaction immediately. Spark acts as catalyst.

c. What is the role of catalysis in the reaction?

d. What are the concentrations of the reaction products after the equilibrium is established?

e. What is the effect of external parameters such as pressure and temperature on the rate of reaction as well as on concentration of reaction products.

Since there are many excellent books written by the physical chemists on this topic, we will not address most of the above questions. Discussion of the rate of reaction involves non equilibrium statistical mechanics and hence will not be discussed.

We will discuss only the equilibrium stage of the reactions.

There are two kinds of chemical reactions, homogeneous reactions and heterogeneous reactions. In homogeneous reactions, the reactants and the products are in same phase, i.e. they are all either gases, or liquids or solids.

In heterogeneous reactions, some products and some reactants may have different phases during the reaction.

We will restricts here only to homogeneous chemical reactions.

We will assume the temperature to be same throughout the reaction space. In addition, if the reaction takes place such that its volume does not change, then $\delta F = 0$. On the other hand, if besides temperature, the pressure is same throughout the reaction space, then $\delta G = 0$. The discussion with these assumption is ideal. Actual reactions may have temperature and pressure gradients. None the less, the idealized discussion is useful. To give the example, around the first world war, ammonia was synthesized by Haber with N_2 from air and H_2. The yield of ammonia in the reaction

$$N_2 + 3H_2 \rightleftharpoons 2NH_3$$

increases with increase of pressure. Haber arrived at this conclusion from Nernst heat theorem and succeeded in manufacturing ammonia.

15.1   Law of Mass Action

Let A_i be the chemical symbols for reagents. Then the chemical reaction is written as

$$\sum \nu_i A_i = 0 \tag{15.1}$$

where ν_i are positive or negative integers.
Examples:

(i) $H_2 + S \rightleftharpoons H_2S$ Or, $H_2 + S - H_2S = 0$.
Then $\nu_{H_2} = 1$, $\nu_S = 1$, $\nu_{H_2S} = -1$

(ii) $2H_2 + O_2 \rightleftharpoons 2H_2O$ or $2H_2 + O_2 - 2H_2O = 0$
$\nu_{H_2} = 2$, $\nu_{O_2} = 1$, $\nu_{H_2O} = -2$

(iii) $H_2 + CO_2 \rightleftharpoons H_2O + CO$
$\therefore \nu_{H_2} = 1, \nu_{CO_2} = 1, \nu_{H_2O} = -1$ and $\nu_{CO} = -1$

Let us discuss the case where pressure and temperature remain same in the reaction space. Then

$$\delta G \equiv \delta G(P, T, N_1, N_2...) = 0 \tag{15.2}$$

Here $N_1, N_2...$ are the number of particles of different substances that participate in chemical reaction. Under the condition of constant pressure and temperature, equation (2) imply

$$\sum_i \frac{\partial G}{\partial N_i} dN_i = 0 \tag{15.3}$$

The change dN_i is not arbitrary but it is consistent with balance of the reaction. Clearly, dN_i is proportional to ν_i to be consistent with (1). Using the fact that $\partial G/\partial N_i = \mu_i$, we write equation (3) as

$$\sum_i \nu_i \mu_i = 0 \tag{15.4}$$

Equation (4) is valid for homogeneous and heterogeneous reactions because, in equilibrium, μ is same for the gaseous and liquid phase.

We have found chemical potential of an ideal gas (prob. 6 ch. 14) as,

$$\mu = k_B T \ln\left(\frac{N}{V}\right) + k_B T \ln\left(\lambda^3\right) \tag{15.5}$$

We want to express this in terms of pressure by using $\dfrac{N}{V} = \dfrac{P}{k_B T}$. Thus,

$$\mu = k_B T \ln P + k_B T \ln\left(\frac{\lambda^3}{k_B T}\right) \tag{15.6}$$

Writing $\chi(T) = k_B T \ln(\lambda^3/k_B T)$ as a pure function of temperature, we get for i^{th} specie

$$\mu_i = k_B T \ln(P_i) + \chi_i(T) \tag{15.7}$$

We make the approximation that gases, participating in chemical reaction are treated as ideal. Then, the pressure in the reaction chamber is sum of the partial pressures $P_i = (N_i k_B T)/V$ of the participants.

Let C_i be concentration of ith specie and is $C_i = N_i/N = N_i/\left(\sum_i N_i\right)$. Partial pressure $P_i = C_i P$. Dalton's law of partial pressure is $P = \sum P_i$. Then, equation (4) becomes

$$\sum \nu_i \mu_i = 0 = k_B T \sum_i \nu_i \ln P_i + \sum \nu_i \chi_i \tag{15.8}$$

This implies

$$\prod_i P_i^{\nu_i} = \exp\left[-\beta \sum \nu_i \chi_i\right] \tag{15.9}$$

We introduce a notation

$$K_P(T) = \exp\left[-\beta \sum \nu_i \chi_i\right] \tag{15.10}$$

$K_P(T)$ is called reaction constant. Then,

$$\prod_i P_i^{\nu_i} = K_P(T) \tag{15.11}$$

Equation (11) can be written in terms of concentration C_i by noting that $P_i = C_i P$ as

$$\prod C_i^{\nu_i} = P^{-\sum \nu_i} K_P(T) \equiv K_c(P,T) \tag{15.12}$$

It must be noted that $K_c(P,T)$ is a function of pressure and temperature and is constant. The product $\prod_i C_i^{\nu_i}$ is constant and at constant pressure and temperature is known as "law of mass action".

$K_c(P,T)$ is called chemical equilibrium constant.

Pressure dependence of equilibrium at a fixed T is totally determined by a factor $P^{-\sum \nu_i}$.

15.2 Heat of Reaction

Let us denote reactants by symbol A and the reaction products by B. We will write the chemical reaction as

$$a_1 A_1 + a_2 A_2 + ... + a_k A_k = b_1 B_1 + b_2 B_2 + ...b_s B_s \qquad (15.13)$$

Let $\lambda_{A_1}, \lambda_{A_2}...\lambda_{A_k}$ be number of moles of $A_1, A_2, ...A_k$ respectively. We also write $\lambda_{B_1} \lambda_{B_2}...$ as number of moles of $B_1, B_2, ...$ respectively.

Due to reaction, initial values of λ change. But they cannot change arbitrarily. They must change consistent with the reaction of equation (13).

Let E_{A_i} be the internal energy of ith reactant and E_{B_k} be the internal energy if k_{th} product B_k.

Then, change in internal energy as a result of the reaction is

$$\Delta E = \sum_{k=1}^{s} (b_k E_{B_k}) - \sum_{i=1}^{k} (a_i E_{A_i}) \qquad (15.14)$$

Let Q_V be heat evolved when a reaction

$$a_1 A_1 + a_2 A_2 + ... + a_k A_k = b_1 B_1 + b_2 B_2 + ...b_s B_s + Q_V \qquad (15.15)$$

takes place at a constant volume. Then,

$$Q_V = -\Delta E \qquad (15.16)$$

Q_V is called heat of reaction performed at constant volume.

If the reaction (13) takes place at a constant pressure, then, a volume of the system changes. Additional energy $P\Delta V$ need to be subtracted from Q_V as given by equation (16). Thus, for a chemical reaction, performed at constant pressure,

$$Q_P = -\Delta E - P\Delta V = -\Delta(E + PV) = -\Delta H \qquad (15.17)$$

where H is enthalpy. Thus heat of reaction in an isobaric case is a change in enthalpy.

15.3 Discussion of Some Chemical Reactions

(i) Reactions without change of molar number

$$H_2 + I_2 \rightleftharpoons 2HI$$

$$N_2 + O_2 \rightleftharpoons 2NO$$

Let n moles of HI be present of which a fraction x dissociates. Then, $P_{HI} = n(1 - x)k_B T$; $P_{H_2} = P_{I_2} = nx k_B T$. Then putting it in the law of mass action (12), we get

$$K_c = \frac{P_{H_2} P_{I_2}}{P_{HI}^2} = \frac{x^2}{(1 - x)^2} \qquad (15.18)$$

This is independent of P and therefore, change in pressure will not change the amount of product of the reaction.

(ii) Dissociation in to two identical molecules

Well known example is that of N_2O_4 which almost completely dissociates at low pressures.

$$N_2O_4 \rightleftharpoons 2NO_2$$

Let n_0 be number of moles of undissociated N_2O_4. Let fraction x dissociate. Then, $P_{N_2O_4} \propto n_0(1-x)$, $P_{NO_2} \propto 2n_0x$. Total mole of the gas is $n_0(1+x)$. Then,

$$K_p(T) = \frac{(2n_0x)^2}{n_0(1-x)} \cdot \frac{P}{n_0(1+x)} = \frac{4x^2}{1-x^2}P$$

Clearly, to keep $K_p(T)$ fixed, decrease in P will cause increase in x leading to a large dissociation of N_2O_4 at low pressure.

(iii) Haber's synthesis of Ammonia

Consider

$$N_2 + 3H_2 \rightleftharpoons 2NH_3$$

This reaction can be enhanced by catalyst such as Fe or Mn or Os.

Let three volumes of hydrogen and one volume of nitrogen be kept in a reaction space at pressure P and temperature T. Let x be the NH_3 fraction product. Then, $P_{NH_3} = xP$, $P_{N_2} = \frac{1}{4}(1-x)P$ and $P_{H_2} = \frac{3}{4}(1-x)P$.

By the law of mass action,

$$\frac{P_{NH_3}^2}{P_{N_2}P_{H_2}^3} = K_P(T) = \frac{x^2}{(1-x)^4}\frac{1}{P^2}\frac{256}{27}$$

Clearly, at any fixed T, as P increases, x increases. Thus, compression of a mixture of N_2 and H_2 is needed to get a good yield of NH_3.

Haber got a Noble prize of chemistry in 1918 for the synthesis if ammonia.

15.4   Saha's Ionization Formula

When a gas is heated, it ionizes. This can be looked upon as a chemical reaction. In a stellar structure, the temperature at the core can be as high as $10^8 K$ and at its surface it is in the range of $10^4 K$. At such high temperatures, all the elements are in an ionized gaseous state. These elements are in various degrees of ionization. This depends upon ionization potential and temperature. These elements have an emission spectrum characteristic of the element. From intensity of these lines (which depends upon the concentration of the species) one can find the temperature of the star. Prof. M.N. Saha discovered ionization formula in 1922. After that, flood of activity started in a stellar spectroscopy as well as corona spectroscopy of the Sun.

Thermodynamically, ionization reaction, say,

$$Na \rightleftharpoons Na^+ + e^-$$

is a chemical reaction. Simultaneously, one may have

$$Na^+ \rightleftharpoons Na^{++} + e^-$$

and similar ionization are possible.

In general these simultaneous reactions can be expressed as

$$\left. \begin{array}{l} A_0 \rightleftharpoons A_1 + e^- \\ A_1 \rightleftharpoons A_2 + e^- \\ ------ \end{array} \right\} \tag{15.19}$$

etc. A_0 is a neutral atom; $A_1, A_2, A_3...$ etc are singly, doubly, triply ionized atom A.

Let C_0 be the concentration of neutral atoms and $C_1, C_2, C_3...$ be concentration of $A_1, A_2, A_3...$ etc. There is overall charge neutrality. Let C_e be the concentration of electrons. Then, by the overall charge neutrality,

$$C_e = C_1 + 2C_2 + 3C_3 + ... \tag{15.20}$$

Then, the law of mass action [equation(12)] gives

$$\frac{C_n}{C_{n-1}C_e} = P K_P^{(n)}(T) \tag{15.21}$$

Here $-\nu_{n-1} - \nu_e + \nu_n = -1 - 1 + 1 = -1$ and $P^{-\sum \nu_i} = P$. Then $K_P^{(n)}$ is evaluated for $A_{n-1} \leftrightarrow A_n + e^-$. Also,

$$K_P^{(n)} = \exp\left[-\beta \sum \nu_i \chi_i\right]$$

where $\sum_i \nu_i \chi_i = -\nu_{n-1}\chi_{n-1} - \nu_e\chi_e + \nu_n\chi_n$. One has to be careful in evaluating $\chi_n(T)$. This situation is different from earlier chemical reactions in that, we have to consider, not only the translational motion, but also internal energy levels of the atom. This modification is easy and we have to add the energy level ϵ_i of the ith atom in χ_i of equation (7). We can then write

$$\chi_i(T) = \epsilon_i - C_{p_i} \ln(k_B T) - k_B T \xi_i \tag{15.22}$$

Here $C_p = \frac{5}{3} k_B$ and ξ is

$$\xi = \ln\left[g \left(\frac{2\pi m}{h^2}\right)^{3/2}\right] \tag{15.23}$$

where g is degeneracy (statistical weight) of ground state of atom. Let L and S be angular and spin momentum of atom. The degeneracy for electron is 2, whereas for the ground state it is $(2L+1)(2S+1)$. Then, using equations (10), (12), (20), (21) and (22), we obtain

$$K_P^{(n)}(T) = \frac{g_{n-1}}{g_n} \frac{\lambda^3}{k_B T} e^{\beta I_n}$$

where $I_n = \epsilon_n - \epsilon_{n-1}$ is energy of nth ionization or n^{th} ionization potential of the atom.

When temperature increases, equilibrium constant $PK_P{}^{(n)}(T)$ becomes nearly one. In fact if I is smallest ionization potential, then $n\lambda^3$ for $k_B T \sim I$, small. This compensates the exponential factor. It turns out that even at $k_B T < I$, there is a considerable ionization of the gas.

Let x be the fraction of ionization of a gas. Clearly x is a ratio of number of ionized atoms to the total number of atoms. We consider a singly ionized system.

Hence,

$$C_e = C_1 = \frac{x}{1+x} \text{ and } C_0 = \frac{1-x}{1+x}$$

Equation (20) then becomes

$$\frac{1-x^2}{x^2} = PK_P{}^{(1)} = \frac{Pg_0}{g_1}\frac{\lambda^3}{k_B T}\exp\left[I_1/k_B T\right] \tag{15.24}$$

Equation (24) is called Saha's ionization formula. Thus,

$$x = \frac{1}{\sqrt{1 + PK_P{}^{(1)}}}$$

Short Questions

1. State, what are the issues that are studied relating the chemical reaction.

2. We obtained the expression $\sum \mu_i \nu_i = 0$ by looking at minimum of Gibb's potential. Prove that same result is obtained even by looking at minimum of Helmholtz free energy F.

3. Explain the difference between heat of reaction when V and T constant and when P and T are constant.

4. Following reaction is carried out at a constant temperature T.

$$CO_2 + H_2 \rightleftharpoons CO + H_2O$$

There is chemical equilibrium and reaction is carried out in a vessel of volume V. If the volume of this vessel is increased, does the relative concentration of CO_2 increase, decrease or stays same? Why?

5. Show that the concentration contents of reactions, where no molar number changes (i.e. $\sum \nu_i = 0$ are unchanged with pressure variation.)

Problems

Problem 1: For a fixed V and T, show that we have to analyze F and show that the reaction constant in that case is

$$K_N(V,T) = V^{\sum \nu_i} K_n(T)$$

where $K_n(T) = \exp\left[-\beta \sum \nu_i \chi_i\right]$
Also show

$$\frac{d}{dT} \ln K_n = -\frac{Q_V}{k_B T^2}$$

(Hint: Refer Reif chapter 8 where he analyzes F and not G)

Problem 2: Show that

$$\frac{d}{dT} \ln K_p = -\frac{Q_P}{k_B T^2}$$

Problem 3: Consider a matter at such a high temperature when $k_B T \gtrsim mc^2$. Here m is mass of electron and $mc^2 \simeq 0.5 MeV$. At these temperatures, electron-positron pairs are formed. Pair annihilation leads to emission of γ ray. Thus,

$$e^+ + e^- \rightarrow \gamma$$

is a chemical reaction. It is given that chemical potential of photon is zero and $N_{e^+} = N_{e^-}$. Then show
(i) $\mu^+ = \mu^- = 0$

(ii) For $k_B T \gg mc^2$, take $\epsilon = cp$ and show

$$N^+ = N^- = \frac{V}{\pi^2} \left(\frac{k_B T}{c\hbar}\right)^3 \int_0^\infty \frac{x^2 dx}{e^x + 1}$$

and

$$\text{(iii)} E^+ = E^- = \frac{7\pi^2}{120} k_B T \left(\frac{k_B T}{c\hbar}\right)^3 V$$

Chapter 16

White Dwarf Stars

Story of the life cycle of starts is very interesting and is a major topic of study in Astrophysics. This involves Quantum Mechanics, Statistical mechanics, and Special and General Theory of Relativity. We present a brief account of stellar evolution. First let us consider a very brief account of fusion process. We know that a given nucleus is characterized by Z protons and N neutrons with its mass number $A = Z + N$. Usually, $N > Z$, but for many nuclei with low A ($N \approx Z \approx A/2$). The nuclei are bound together by an attractive strong force which is charge independent. The strong force has a short range $\leq 10^{-13}$cm. It is zero beyond this distance. It is so strong that Coulomb repulsion between two protons, separated by the distances $< 10^{-13}$cm (range of strong force)is overpowered by the strong force. The strong force between two neutrons, between the neutron and proton and between proton and proton is same when they are separated at distance less than its range. These nuclear systems are bound. It is found that the mass m_A of nucleus is such that

$$m_A < [Z\, m_p + (A - Z)m_N],\qquad(16.1)$$

where m_p is mass of proton and m_N is mass of neutron. This binding comes from mass-energy relation of Einstein. Masses of nuclei are very accurately known by mass spectroscopic measurements. Binding energy B of the nucleus is defined as,

$$B = (Z\, m_p + (A - Z)m_N)\, c^2.\qquad(16.2)$$

Here c is the speed of light.

It is found that binding energy per nucleon, i.e. B/A is almost constant except for very light or very heavy nuclei. From this diagram it is clear that lighter elements like H, He are such that they have a tendency to fuse if they are influenced by a strong interactions.

Thus if a protons and a neutron come closer than the range of the strong force, they will tend to produce deuteron. Similarly, under similar conditions, a deuteron and a proton will tend to produce He^3 nucleus. However, the biggest question is, how to bring a deuteron and a proton so close when there is a Coulomb repulsion? Here some kind of a

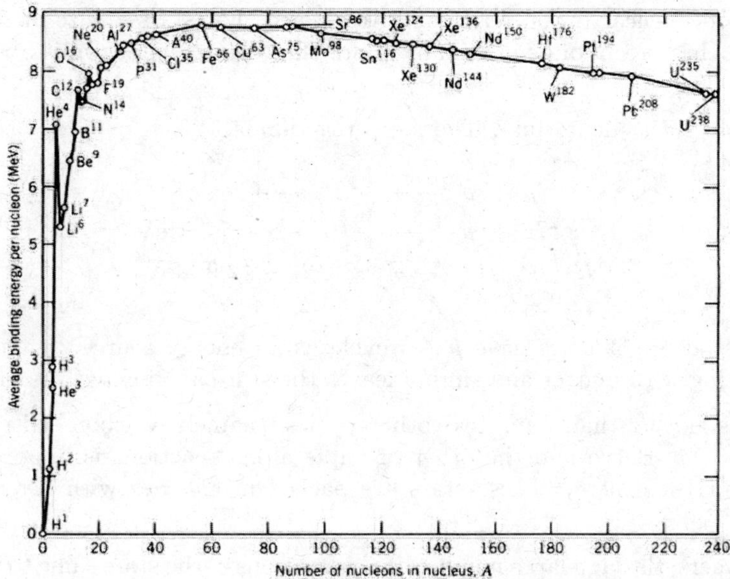

Figure 16.1: binding energy per nucleon in MeV versus nucleon number.

barrier—Coulomb barrier is formed. To overcome it, the particle must have a high energy, i.e. a gas of proton and Helium must be at high temperatures. The temperature equivalent in energy to the barrier height is more than about 10 million K. This is precisely the core temperature of most stars on main sequence. Fusion is a main source of energy of a star. Thus, in short, a core of a star is slowly burning hydrogen bomb.

On the other hand, elements with large A like Uranium, have a tendency of fission rather than fusion. As far as stellar evolution is concerned, fission has no role to play.

It is believed that a big bang explosion took around 13 billion years ago. A lot of gas of Hydrogen and Helium was produced soon. Many intermediate steps from explosion to hydrogen atom are speculative and are of no concern to us. With this the nebulae are formed. Nebula is a Greek work meaning cloud. Main constituent of the cloud is hydrogen (about 97%) and Helium as rest.

Within a nebula, there are variations of density of the clouds of dust and gas. Clearly denser regions have more atoms than the regions of comparable size with lower density. Somehow, (and we still do not clearly understand the mechanism) a center for gravitational attraction is created. Probably, this is similar to homogeneous nucleation. Thus a star begins to form and is known as Proto star. With increase in density of a matter at a core of the Proto Star; kinetic energy of hydrogen atoms increase, making the core hot. This increases outward pressure. The two competing effects one due to gravitational attraction and the other due to kinetic outward pressure balance each other in equilibrium. Energy provider to the star is fusion reaction of hydrogen. Depending upon the size of star, one of the three types of nuclear fusion reactions takes place. For stars of size of sun or smaller, proton−proton chain reaction is dominant. Here the core temperature is about 15 million degree Kelvin.

We denote ^2H as deuterium. The $p - p$ reaction is,

$$p + p \longrightarrow {}^2H + e + \bar{\nu}_e$$
$$^2H + p \longrightarrow {}^3He + \gamma + 5.5 \text{ MeV}$$
$$^3He + {}^3He \longrightarrow \alpha + 2p + 12.9 \text{ MeV} \tag{16.3}$$

Most of the bright stars have $p - p$ cycle as an energy source. $\bar{\nu}_e$ is antineutrino, weakly interacting with matter and simply leaves the star carrying with it some energy.

For stars hotter than sun, two other cycles namely Carbon−Nitrogen−Oxygen (CNO) proposed by Bethe and the other of triple alpha reactions is important. We will not worry about them here. These stars are packed in galaxies with beautiful spiral or elliptic shapes.

Astronomers, through large number of observations of the stars came to the conclusion that the luminosity of the star (proportional to its brightness) is directly proportional to its color the predominant emitted wavelength. Moreover, the proportionality constant is almost same for most of the stars.

Predominantly emitted wavelength is related with surface temperature of the star. A plot of Luminosity versus surface temperature is known as Hertzprung Russel (HR) diagram shown in figure (.2)[1]

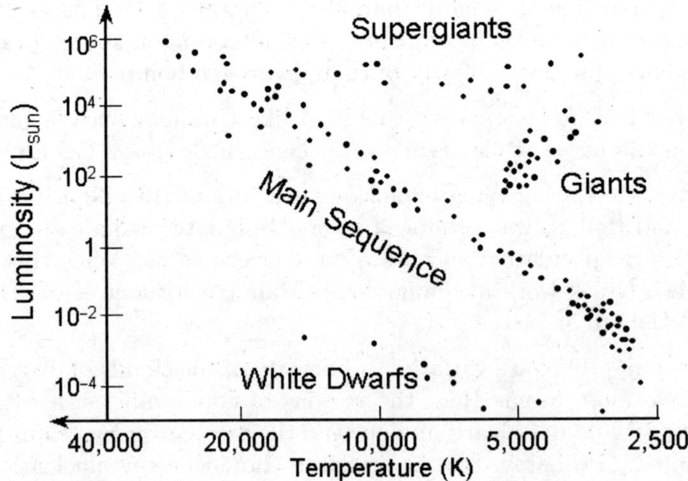

Figure 16.2: Hertzprung-Russel diagram.

---

[1] By courtesy of Prof. Richard Pogge of Astronomy Department, Ohio State University.

Note the direction of temperature in the graph. Star in upper left corner is bright and hot. Sun is in the middle of H−R diagram. Stars in lower right parts are cold and dim. Most of the stars lie in a region called main sequence as shown. As one goes from right to left on main sequence, stars become hotter and brighter. On the upper right corner of H−R diagram and not lying on main sequence are red giants where as, those not lying on the main sequence and on the lower left corner are white dwarfs, which is the topic of study of this chapter.

The stars on main sequence burn for billions of years. During this time, their fuel supply of hydrogen fuel is getting exhausted or burnt off. The core is now mostly Helium and has become cool. Outer layer of this star has still some hydrogen but the layer is not hot enough to $p-p$ cycle. Gravitational attraction still continues. The outer layer is pulled in, becoming hot. The layer then swells and the star becomes giant. It is no more on the main sequence. Meanwhile, the core helium nuclei are hot and undergo other fusion cycles like CND, and heavier elements in the core are formed.

When all the hydrogen fuel supply is over, and in the last phase of the evolution of stars, we have white dwarfs. By what exact process, they are formed, is still a topic of research. We give some chronology in the cosmic evolutions:

1. Big bang $\sim$ 15 billion years ago.

2. Nebulae and galaxies $\sim$ 13 billion years ago.

3. Black holes, neutron stars, red giant and white dwarfs $\sim$ 11 billion years ago.

4. Planets and solar system $\sim$ 4.7 billion years ago.

5. Origin of life, formation of biomolecules $\sim$ 4 billion years ago.

6. Human beings $\sim$ 100 thousand years ago.

7. Chinese and Hindu culture $\sim$ 4000 years ago

16.1   Problem Posed by a White Dwarf Star

Problem posed by the white dwarf star is, why it is away from the main sequence of Hertzprung-Russel (HR) diagram, and that, why they are dull (faint) brightness? From the trend in HR diagram, it is clear that brighter the star, more is its colour towards white or blue. Following are the facts about white dwarf stars.

1. Observations of these stars indicated that its predominant constituent is Helium. This is so because these stars are relatively old − means, by thermonuclear reaction at their center, most of the hydrogen is burnt up. This means most of the hydrogen is converted into helium. Thermonuclear reactions involving Helium have a much slower time scale. This is main reason for their faintness. Gravitational attraction pulls a mass of the star inwards and gravitational energy is released. This energy is responsible for the faint white color of white dwarf star.

2. Its core temperature is roughly same as that of core temperature of sun. Numerically, it is of the order of 10^7 °K $= T_\odot$. Here, the subscript $\odot$ denotes to quantities referring to sun. Roughly 12000 °K is 1eV. Thus $T_\odot$ is about 1000eV which is much higher than ionization potentials of Helium atom. This means the Helium gas in the core of white dwarf star is totally ionized, which means we have a release of two electrons and one α particle from each helium ionization. Thus,

$$^4_2He \longrightarrow {}^4_2He^{++} + 2e^- \tag{16.4}$$

Thus, if there are N electrons in a white dwarf , then there are $(N/2)$ alpha particles. Each alpha has a mass approximately equal to 4 proton masses. Then, the mass of white dwarf is

$$M = N(m_e + 2m_p) \approx 2Nm_p, \tag{16.5}$$

where m_e is mass of electron and m_p is mass of proton.

3. White dwarf star is spherical, with mass density $\rho \approx 10^7$g/cc. Thus, the electron density in the white dwarf star is,

$$n = \frac{N}{V} = \frac{M/2m_p}{M/\rho} = \frac{\rho}{2m_p} \tag{16.6}$$

Clearly, with $m_p \approx 9.1 \times 10^{-24}$gm, electron density n is of the order of $10^{30}/\text{cm}^3$

4. Fermi momentum corresponding to the electron density $N = 10^{30}/\text{cm}$ is

$$p_F = \hbar(3\pi^2 n)^{(1/3)} \simeq 10^{-27} \times 3 \times 10^{10} = 3 \times 10^{-17}\text{gm cm/sec} \tag{16.7}$$

and the Fermi energy is,

$$E_F = \frac{p_F^2}{2m_e} = \frac{9 \times 10^{-34}}{2 \times 9.1 \times 10^{-28}} \cong 0.5 \times 10^{-6}ergs \cong 3 \times 10^5 e.V. \tag{16.8}$$

5. Fermi temperature is $T_F = 12000 \times 3 \times 10^5 = 3.6 \times 10^9 \ K$

Core temperature of white dwarf star is $T = 10^7 \ K$ Thus,

$$\frac{T}{T_F} = \frac{10^7}{3.6 \times 10^9} \cong 3 \times 10^{-3} \tag{16.9}$$

This means the electron gas in white dwarf star is highly degenerated. We therefore make $T = 0$ calculations.

6. Rest mass of electron in energy units is $m_e \simeq 0.5 Mev = 5.0 \times 10^5 e.V.$ This is quiet comparable with Fermi energy $\sim$ kinetic energy. Since these two energies are comparable, we must apply special relativity to attack the problem of white dwarf star.

16.2 Model of a White Dwarf Star

The model of white dwarf star is,

(a)that the white dwarf star is spherical, and composed of Helium nuclei and a highly degenerate electron gas which is relativistic.

(b)Because the electron gas obeys Pauli exclusion principle, even at T=0, there is a pressure exerted by the electron gas and, for its confinement in the spherical geometry, this pressure is radial and outwards. For equilibrium of the star, this must be balanced by the attractive force towards the center of the star. This attractive force is gravity.

(c) The effects that we have neglected are,

(i) Radiation pressure that is present due to Helium burning.

(ii) Electron-electron interaction energy is ignored.

(iii) Some highly energetic electrons may scatter to produce electron-position pairs. These pairs may annihilate, radiating energy.

(iv) In a weak Leptonic processes, neutrinos may be produced which leave the system, almost without interacting with it. It is thus a constant drain of energy.

(d) The problem is reduced barely to the balance between Pauli repulsive pressure and gravitational attraction towards the centre of white dwarf star.

(e) The motion of alpha particles is neglected because they are heavy as compared with electrons. Their presence is necessary for the stability of the star.

16.3 Model Calculations

1. Fermi momentum at $T = 0$ is

   $$p_F = \hbar(3\pi^2 N/V)^{1/3}.$$

2. Kinetic energy of relativistic electrons of white dwarf is

   $$E_p = \sqrt{c^2 p^2 + m^2 c^4} - mc^2$$

   Thus,

   $$E_p = mc^2 \left[\left(1 + \frac{p^2}{m^2 c^2} \right)^{1/2} - 1 \right]$$

   Where m is rest mass of electron.

3. Ground state energy of electrons in the white dwarf star is,

$$E_0 = 8\pi \frac{V}{(2\pi\hbar)^3} \int_0^{p_F} p^2 \sqrt{c^2 p^2 + m^2 c^4}\, dp \qquad (16.10)$$

Pressure exerted is $-\dfrac{\partial E_0}{\partial V}$.

We make a transformation $p/mc = x$ and $p_F/mc = x_F$.

Also, the upper limit of integration becomes x_F. We consider extreme relativistic limit so that $x_F = P_F/mc \gg 1$. Then it is easy to see,

$$P \cong \frac{2\pi\,(mc)^5}{3mh^3}\left(x_F^4 - x_F^2\right) \qquad (16.11)$$

4. For a white dwarf star,

$$M = 2m_p\,N \text{ and } R = \left(\frac{3V}{4\pi}\right)^{1/3}$$

We take $\dfrac{\hbar}{mc}$ as a Compton wavelength of electron. Then,

$$x_F = \frac{p_F}{mc} = \frac{\hbar k_F}{mc} = \frac{\hbar}{mc}\left[\frac{3\pi^2 N}{V}\right]^{1/3}.$$

Thus in terms of R and M we write,

$$x_F = \frac{\hbar}{mc}\frac{1}{R}\left[\frac{9\pi M}{8m_p}\right]^{1/3} \qquad (16.12)$$

Define a dimensionless quantities, as,

$\overline{R} = \dfrac{R}{\hbar/mc}$, $\overline{M} = \left[\dfrac{9\pi M}{8m_p}\right]$. Then $x_F = \overline{M}^{\frac{1}{3}}/\overline{R}$.

Pressure exerted by the electron gas is,

$$P = K\left[\frac{\overline{M}^{4/3}}{\overline{R}^4} - \frac{\overline{M}^{2/3}}{\overline{R}^2}\right] \qquad (16.13)$$

where $K = \dfrac{2\pi\,(mc)^5}{3mh^3}$.

5. We apply now equilibrium condition by equating this Pauli pressure with pressure due to gravitational interaction. Gravitational force $\sim (1/r^2)$. Then, we know that if we have a sphere of uniform mass M and of radius R, then its self energy is $3/5\, GM^2/R$. Here, the factor $3/5$ is not important but has come there because we have assumed uniform mass density. Generally, there will be a factor $\alpha \approx 1$ in general. Thus the

gravitational self energy can be written as $-\alpha G\, M^2/R$. Expressing $R = \left(\dfrac{3V}{4\pi}\right)^{1/3}$ and differentiating it w.r.t. V we get gravitational pressure, as,

$$P_g = \frac{\alpha}{4\pi}\frac{GM^2}{R^4} \tag{16.14}$$

Measuring R in the units of $\dfrac{\hbar}{mc}$, M in the units of $\dfrac{8m_p}{9\pi}$ we get the gravitational pressure as

$$P_g = \frac{K'\overline{M}^{\,2}}{\overline{R}^{\,4}} \tag{16.15}$$

where,

$$K' = \frac{\alpha\, G}{4\pi}\left(\frac{8m_p}{9\pi}\right)^2\left(\frac{mc}{\hbar}\right)^4$$

6. Equating Pauli pressure with pressure due to gravity,

$$\frac{K'\overline{M}^{\,2}}{\overline{R}^{\,4}} = K\left(\frac{\overline{M}^{\,4/3}}{\overline{R}^{\,4}} - \frac{\overline{M}^{\,2/3}}{\overline{R}^{\,2}}\right) \tag{16.16}$$

Or,

$$\overline{R} = \overline{M}^{\,1/3}\left[1 - \left(\frac{\overline{M}}{\overline{M_0}}\right)^{\frac{2}{3}}\right]^{1/2} \tag{16.17}$$

where

$$\overline{M_0} = \left(\frac{K}{K'}\right)^{\frac{3}{2}} = \left(\frac{27\pi\hbar c}{64\alpha G m_p^{\,2}}\right)^{\frac{3}{2}} \tag{16.18}$$

From this relation, it is clear that $\overline{R}$ is imaginary if $\overline{M} > \overline{M_0}$. This is impossible. Therefore, there is a maximum limit on the mass of white dwarf. Putting constants $\hbar$, c, G and m_p and taking $\alpha \approx 1$ we get,

$$M_0 = \frac{8}{9\pi}m_p\,\overline{M_0} \approx 10^{33}\,gms \simeq M_\odot \tag{16.19}$$

Thus, according to this analysis, white dwarf cannot have a mass larger than solar mass. Detailed calculations can be made by taking correct density distribution and hence correct value of α. The limiting value of the mass of white dwarf star turns out to be about $1.44\,M_\odot$. This is celebrated Chandrasekhar limit on a size of white dwarf star.

Short Questions

1. Why one has to consider fusion as a source of energy of star and not the fission?

2. Which fusion energy cycle is responsible for stars on main sequence of H-R diagram? How much energy is released in the above single fusion cycle?

3. Describe Herzprung - Russel diagram.

4. Explain why the electron gas in the white dwarf star is highly degenerate?

5. Why we need to use special relativity to estimate Pauli pressure?

6. What physical processes are ignored while the considering the model of white dwarf star?

7. Show that the gravitational self energy of a sphere of mass M and of radius R, with uniform density is $(3/5)\, G\, M^2/R$.

Problems

Problem 1: Show that the ground state energy of electrons in the white dwarf star is,

$$\frac{E_0}{V} = \frac{\pi(mc)^5}{3mh^3}\, F(x_F)$$

where,

$$[F(x_F) = [8x_F^3\, \{\sqrt{(1+x_F^2)} - 1\} - x_F\, \sqrt{(1+x_F^2)}\, (2x_F^2 - 3)$$

$$+3Sinh^{-1}x_F]$$

and, $x_F = p_F/mc$.

Problem 2: If in the star, $T \gg T_F$, electrons can be treated as a Boltzmann gas. Thus, $P = Nk_B\,T/V$. Show that radius of the star in this situation is

$$R = \frac{2}{3}\frac{\alpha M G m_p}{k_B T} \tag{16.20}$$

Problem 3: Show that $x_F = p_F/mc \ll 1$ is a non relativistic limit. In this limit, show that the Pauli pressure is

$$(4/5)\frac{K\,\overline{M}^{5/3}}{\overline{R}^5}$$

Problem 4: In a non relativistic limit of an electron motion in the star, show that the radius of the star decreases as mass of the star increases. Obtain explicit relation

$$\text{Ans: } \overline{R}\,\overline{M}^{1/3} = (4K/5K')$$

Chapter 17

IDEAL BOSE SYSTEM I

The nature is made up of two types of particles, those having half integral spin (obeying the Pauli principle) and those having integral spin. The first one (with half integral spin) have anti symmetric wave function upon the exchange of the two particles. The second one (with integral spin) have a symmetric wave function upon the exchange of two particles. They are called as Fermions and Bosons respectively. We have already studied the statistical mechanics of the Fermions. We will be studying a statistical mechanics of the assembly of non-interacting Bosons in this chapter. We first consider a system of Bosons with zero chemical potential in this chapter. In the next chapter we will discuss a Bose system with non-zero chemical potential.

There are many systems in nature where $\mu = 0$. Most significant example of such a system is a collection of photons. Others are phonons (quantized lattice vibrations), magnons (quantized spin wave excitations), plasmons (quantized plasma excitations) and so on. All these are the collective excitations of appropriate systems. Thus, the photons are quantized modes of the electromagnetic field. It can be shown that the electromagnetic energy due to $\vec{E}$ and $\vec{B}$ field

$$\mathcal{E} = \frac{1}{8\pi} \int (E^2 + B^2) d\tau \tag{17.1}$$

can be written as an energy of a collection of harmonic oscillators with frequency ω such that[1]

$$\mathcal{E} = \sum (n + 1/2)\, \hbar\omega_k \tag{17.2}$$

where

$$\vec{E}(\vec{r}, t) = \vec{E}_0(\vec{r}, t) \exp\left(i(\vec{k} \cdot \vec{x} - \omega t)\right). \tag{17.3}$$

Similarly

$$\vec{B} = \vec{k} \times \vec{E}/|\vec{k}|.$$

A similar analysis can be made of phonons where the lattice vibration energy can be represented as a collection of normal modes of vibration. In general, a wave of frequency ω

---

[1] 'Quantum Mechanics' by L. I. Schiff, 3rd edition, McGraw-Hill, ch. 14 pp. 516

is set in as a collective oscillation with a wave vector $k = \dfrac{2\pi}{\lambda}$. The functional relationship between ω and k need to be supplied to study the statistical mechanics of such excitations. In all these excitations $\mu = 0$. We give some examples of these excitations with a dispersion relation $\omega = f(k)$ as

$$
\begin{array}{ll}
\omega = ck & \text{light wave} \\
\omega = v_s k & \text{phonons, } v_s = \text{sound velocity} \\
\omega = Ak^2 & \text{ferromagnetic magnons} \\
\omega = Bk & \text{Anti ferromagnetic magnons}
\end{array}
\tag{17.4}
$$

All of these are the elementary excitations of solids where $\mu = 0$. All these mentioned are Bosons. They are stable only at very low temperature.

17.1 Partition Function

Since the chemical potential is zero for all the elementary excitations, we use canonical partition function. All these elementary excitations have the energy spectrum of the form

$$
E_n = (n + 1/2)\,\hbar\omega
\tag{17.5}
$$

and a known dispersion relation $\omega = f(k)$.

$$
\begin{aligned}
Z &= \sum_n \exp\left(-\beta E_n\right) \\
&= \sum_n \exp\left(-\beta n\hbar\omega\right) \cdot \exp\left(-\beta\hbar\omega/2\right) \\
&= \frac{\exp\left(-\beta\hbar\omega/2\right)}{1 - \exp\left(-\beta\hbar\omega\right)}
\end{aligned}
\tag{17.6}
$$

From Problem (4) of chapter (14) we get the average energy for a mode ω_k as,

$$
\overline{E}_k = \frac{\hbar\omega_k}{2} + \frac{\hbar\omega_k}{\exp\left(\beta\hbar\omega_k\right) - 1}
\tag{17.7}
$$

Here, we clearly see that $\dfrac{1}{\exp\left(\beta\hbar\omega_k\right) - 1}$ is a famous Bose factor, giving average occupancy of the state $|k\rangle$. This can also be gotten by putting $\mu = 0$ in the expression of grand canonical ensembles. Total energy of the system can be found by adding contribution of all the modes i.e. over all the states $|k\rangle$. This is possible because the systems that give rise to above mentioned elementary excitations are linear. Thus, ignoring the infinity due to $\sum \hbar\omega_k/2$, which anyway is a constant, to be subtracted,

$$
E = \sum_k \overline{E}_k = \sum \frac{\hbar\omega_k}{\exp\left(\beta\hbar\omega_k\right) - 1}
\tag{17.8}
$$

The sum over k can be converted into an integral in d dimensions,

$$
\sum_k \rightarrow \frac{L^d}{(2\pi)^d} \int d^d k
\tag{17.9}
$$

we specialize in three dimensions. Then,

$$E = \frac{V}{(2\pi)^3} \int \frac{\hbar\omega_k}{\exp(\beta\hbar\omega_k) - 1} d^3k \tag{17.10}$$

This summation is system dependent i.e. on the type of the elementary excitations. If we assume isotropy then $d^3k = 4\pi k^2 dk$. Then, the equation (10) becomes

$$E = \frac{4\pi V}{(2\pi)^3} \int \frac{k^2 \dfrac{dk}{d\omega_k} \hbar\omega_k \, d\omega_k}{\exp(\beta\hbar\omega_k) - 1} \tag{17.11}$$

The relationship $\omega = f(k)$ is invertible, and therefore $\dfrac{dk}{d\omega_k}$ can be easily found. Then,

$$E = \frac{V}{(2\pi^2)} \int_0^{\omega_{max}} \frac{\left(f^{-1}(\omega_k)\right)^2 . \hbar\omega_k \dfrac{dk}{d\omega_k} d\omega_k}{\exp(\beta\hbar\omega_k) - 1} \tag{17.12}$$

Here the upper limit of frequency ω_{max} could be ∞ as in case of photons or it will depend upon N the number of ions in a solid. N will decide the maximum phonon frequency. We now specialize in two important cases. One is that of a Black - Body radiation and the other is the calculation of specific heat of solids due to lattice vibrations. Debye, among others made the calculation of specific heat of crystalline solids.

17.2   Problem of Specific Heat of Solids

Specific heat measurements is an important experimental activity which sheds light upon phonon spectra as well as on phase transition. Experimentally, one always measures C_P and not C_V. This is so because after giving a certain amount of heat to solid, to keep its volume fixed would require the application of enormous amount of pressure. This is impractical. The C_V is inferred from C_P by thermodynamic formula

$$C_P - C_V = \frac{VT\alpha^2}{\kappa} \tag{17.13}$$

This difference in value is few percent. However, for gases this difference is significant.

According to the third law of thermodynamics (Nernst Theorem) , entropy must go to its constant limiting value (usually zero) as the absolute temperature $T \to 0$. This means the slope $\left(\dfrac{\partial S}{\partial T}\right)_V$ and $\left(\dfrac{\partial S}{\partial T}\right)_P$ is finite. Then, $C_P = T\left(\dfrac{\partial S}{\partial T}\right)_P$ and $C_V = T\left(\dfrac{\partial S}{\partial T}\right)_V$ must go to zero as T goes to zero.

Here comes a difficulty. We apply the equipartition theorem for the solid of N atoms situated at N lattice sites. This situation can be thought of as a collection of N oscillators oscillating in three dimensions. By equipartition theorem, we find the energy of this solid to be

$$\begin{aligned}
E &= N\left(\frac{3}{2}k_B T\right) + N\left(\frac{3}{2}k_B T\right) \\
&= \text{K.E. contribution} + \text{P.E. contribution}
\end{aligned}$$

The specific heat of the solid is then $C_V = 3Nk_B$ which does not go to zero as T goes to zero. For one gram mole, $N = N_A$ the Avogadro number, and $R = N_A k_B$. The molar specific heat at constant volume is $C_V = 3R$. This is 'Dulong and Petit law'. The value $3R$ is found to be true for almost all solids at high temperatures. However, experimentally it is also found that the heat capacity behaves as T^3 at low temperatures. This implies that the classical theory fails to explain the complete behavior of the heat capacity on the entire temperature range.

Einstein recognized this as a failure of classical theory and applied the newly discovered quantum ideas to the problem. His results could give a correct high temperature value and also $C_V \to 0$ as $T \to 0$ but not the experimental T^3 law. Einstein's arguments regarding the lattice dynamics of the solids were over simplistic. Einstein assumed that all the atoms in solid vibrate with one frequency ω. He treated the solid as assembly of $3N$ independent one-dimensional harmonic oscillators. Evidently

$$\overline{E} = \frac{3N\hbar\omega}{\exp\left(\beta\hbar\omega\right) - 1}$$

and

$$C_V = 3Nk_B \left(\frac{\hbar\omega}{k_BT}\right)^2 \frac{\exp\left(\hbar\omega/k_BT\right)}{\left(\exp\left(\dfrac{\hbar\omega}{k_BT}\right) - 1\right)^2}$$

We define Einstein temperature $\Theta_E = (\hbar\omega/k_B)$. Then, the molar specific heat is

$$C_V = 3R \left(\frac{\Theta_E}{T}\right)^2 \frac{\exp\left(\Theta_E/T\right)}{\left(\exp\left(\dfrac{\Theta_E}{T}\right) - 1\right)^2} \tag{17.14}$$

This has a correct high temperature limit of $3R$ when $\Theta_E/T << 1$.

At low temperatures, $\Theta_E/T >> 1$ and

$$C_V \to 3R \left(\frac{\Theta_E}{T}\right)^2 \exp\left[-\left(\Theta_E/T\right)\right]$$

This clearly goes to zero as T goes to zero.

The behavior of heat capacity in Einstein's model is exponential rather than T^3.

Einstein's assumption of the solid vibrating with one frequency is wrong. We have to correctly enumerate various modes of oscillations of the lattice. This was done by Debye. Even though the Einstein's model is wrong it emphasized the need to incorporate quantum theory ideas for correct calculations of the specific heat.

17.3   Specific Heat Due to Lattice Vibrations: Debye's Theory

We have calculated the specific heat of electrons in metal. There are free electrons in the metal and the corresponding ions arranged in a regular lattice points. The lattice vibrations require energy. Thus they contribute to the specific heat. For a metal, if there are no magnetic or other interactions,the contribution to the specific heat comes from both, the electronic motion and the lattice vibrations. On the other hand, for non magnetic insulators, the lattice contribution is the only one to its specific heat since there are no free electrons. The vibrational spectrum of a lattice can be calculated if a crystal structure and its appropriate spring constants are known. It can be determined experimentally by the neutron scattering and/or by x-ray scattering. Details of these calculations and experiments can be in Solid State Physics volumes edited by Seitz and Turnbull (see e.g. Maraduddin, Montroll, Weiss and Ipatowa Academic Press 1971). We show a ω versus k spectrum of a typical solid in Figure (1). It is seen that there are many modes of oscillation. Some modes, called optical

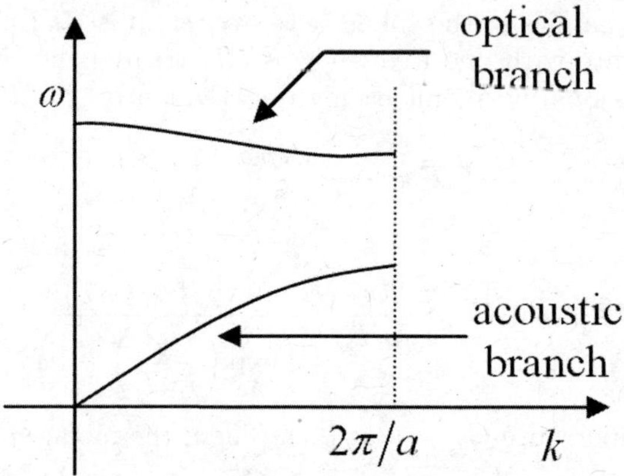

Figure 17.1: Typical lattice dispersion.

modes, are of high frequency and are almost dispersionless. They have a very narrow band width. We can treat these oscillations of one frequency only. For these modes, $\omega \neq 0$ at $k = 0$. Einstein model suits these modes.There are other types of modes called Acoustic modes of the solid. They have a dispersion $\omega = v_s k$ where v_s is a sound velocity. These modes are of low frequency. They have a low energy. Therefore they are easily excitable and contribute to the lattice specific heat in a major way. A quantum of lattice vibration is known as phonon. The figure (2) shows density $\sigma(\omega)$of phonon modes as a function of their frequency. Thus, $\sigma(\omega)d\omega$ gives number of modes of lattice oscillations that lie between ω and $\omega + d\omega$.

If there are N atoms, then there are $3N$ degrees of freedom and we have a common sense sum rule as

$$\int_0^{\omega_{max}} \sigma(\omega)d\omega = 3N \tag{17.15}$$

Since $\sigma(\omega)$ is a very complicated function of ω, Debye approximated it with a quadratic form with a cut off frequency ω_D. This is shown as a dashed curve in the figure(2). Thus,

$$\int_0^{\omega_{max}} \sigma(\omega)d\omega = \int_0^{\omega_D} \sigma_D(\omega)d\omega = 3N \tag{17.16}$$

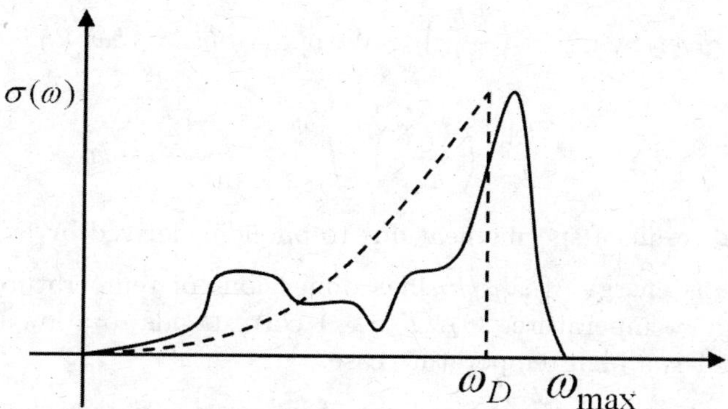

Figure 17.2: Typical spectrum of lattice vibrations - Dashed curves shows Debye approximation.

Also, $\sigma_D(\omega) = 0$ for $\omega > \omega_D$. We can now easily calculate σ_D. We know that the number of modes of oscillations between k and $k + dk$ are,

$$\sigma(k)dk = \frac{4\pi V k^2 dk}{(2\pi)^3}$$

$$= \sigma(\omega)d\omega$$

$\therefore \sigma(\omega) = \dfrac{Vk^2}{2\pi^2}\dfrac{dk}{d\omega}$. But $\omega = v_s k$. Thus,

$$\sigma_D(\omega) = \frac{V\omega^2}{2\pi^2 v_s^3}. \tag{17.17}$$

The Debye frequency ω_D can be found as,

$$3N = \frac{V}{2\pi^2 v_s^3}\int_0^{\omega_D} \omega^2 d\omega \tag{17.18}$$

In actual solid there are two transverse and one longitudinal modes of oscillations. We will go wrong in finding ω_D unless we account for this fact at least in an approximate way. For simplicity we take the speed of longitudinal and the transverse waves to be same. Then the above equation is corrected as,

$$3N = 3\frac{V}{2\pi^2 v_s^3}\int_0^{\omega_D} \omega^2 d\omega$$

Or,

$$\omega_D = v_s \left(\frac{6\pi^2 N}{V}\right)^{1/3} \tag{17.19}$$

Equation (12) can now be used to find the energy due to lattice vibrations in crystalline solids. We obviously take the maximum frequency of vibration to be $\omega_{max} = \omega_D$.

$$E = \frac{3V}{2\pi^2 v_s^3}\int_0^{\omega_D} \frac{\hbar\omega\omega^2 d\omega}{e^{\beta\hbar\omega} - 1} \tag{17.20}$$

The specific heat is given by $C_V = \left(\dfrac{\partial E}{\partial T}\right)_V$. With $x = \beta\hbar\omega = (\hbar\omega/k_BT)$, the equation can be written as,

$$C_V = k_B \frac{3V}{2\pi^2}\left(\frac{k_BT}{\hbar v_s}\right)^3 \int_0^{\beta\hbar\omega_D} \frac{x^4 e^x}{(e^x - 1)^2}\,dx \qquad (17.21)$$

This is the principal result of specific heat due to phonons derived by Debye.

Quantity $\hbar\omega_D$ is energy. $(\hbar\omega_D/k_B)$ has dimensions of temperature. We call $\Theta_D = (\hbar\omega_D/k_B)$ as a Debye temperature. $\Theta_D/T >> 1$ corresponds to a low temperature limit, whereas, $\Theta_D/T << 1$ is a high temperature case.

a) Low temperature case: Here the upper limit of integration of equation(21) can safely be put to ∞. Then the integral

$$\int_0^\infty \frac{x^4 e^x}{(e^x - 1)^2}\,dx$$

is a pure number with its value $4\pi^4/15$. Then in a low temperature limit, C_V per unit volume is,

$$\frac{C_V}{V} = \frac{2\pi^2 k_B}{5}\left(\frac{k_BT}{\hbar v_s}\right)^3 \qquad (17.22)$$

which is famous T^3 law of Debye.

b) High temperature case: In this case, the values of x are restricted in the range $0 < x < \Theta_D/T << 1$. The integrand can be expanded as, $\dfrac{x^4 e^x}{(e^x - 1)^2} \approx \dfrac{x^4}{x^2}$. Then,

$$\int_0^{\beta\hbar\omega_D} \frac{x^4 e^x}{(e^x - 1)^2}\,dx \approx \frac{1}{3}\left(\frac{\hbar\omega_D}{k_BT}\right)^3 \qquad (17.23)$$

Then after some simplification, we get

$$C_V = 3Nk_B \qquad (17.24)$$

which is famous Dulong and Petit value.

17.4   Black-Body Radiation

The problem of Black body radiation is a historic problem that was solved by Planck (1901) and changed our course of thinking in the twentieth century i.e. the birth of quantum ideas.

It is interesting that the study of radiation as a mode of transport of energy was started by Kirchhoff where he conceived many ideas, among them the idea of a Black-body. He was a professor of physics at Berlin. After his retirement, his chair was offered to H. Hertz who experimentally showed the existence of electromagnetic waves. Hertz declined the offer. Then the chair was offered to Boltzmann who also declined it. Finally, Planck was asked and he accepted the chair at Berlin where he completed the programme of Black-body radiation of Kirchhoff.

We know that when an e. m. wave falls on a material, fraction r of it is reflected, fraction t is transmitted and fraction a is absorbed. Clearly,

$$r + t + a = 1 \tag{17.25}$$

All these fractions are dependent on wavelength. Thus, $r = r_\lambda$, $t = t_\lambda$ and $a = a_\lambda$.

We now define substances of limiting properties vis a vis the radiation. They are

1. Perfectly Black-body: It is such that $r_\lambda = t_\lambda = 0$ and $a_\lambda = 1$. Carbon suit (lamp black) is almost a perfect Black-body where, for visible light, $a_\lambda = 0.96$. Platinum black has $a_\lambda = 0.98$ in the same wavelength range.

2. Perfectly white body: Here, $r_\lambda = 1$, $t_\lambda = a_\lambda = 0$. A piece of white chalk or white paper are the examples.

Prevost made an important contribution by realizing that all the substances emit radiation at any non-zero temperature. Moreover, the amount of radiation increases with increasing temperature of the substance. From this it is clear that if a body is in equilibrium with a radiation, then it radiates as much a energy a it receives.

We now define emissive power or emissivity e_λ of a wavelength λ as follows.

$e_\lambda d\lambda$ is an amount of power radiated normally by a unit area in a unit solid angle. Let $u_\lambda d\lambda$ be the amount of energy radiated by dA in a solid angle $d\Omega$ in time dt. Then

$$e_\lambda d\lambda = \frac{u_\lambda d\lambda}{dA \cdot d\Omega \cdot dt}.$$

If the energy Q_λ falls on the body and a fraction a_λ of it is absorbed, a_λ is called absorptivity or absorptive power. For a Black-body, $a_\lambda = 1$.

Black-body radiation enclosed in an enclosure has following properties.

a. The radiation has all the frequencies.

b. The radiation is homogeneous - meaning its density is uniform in an enclosure.

c. The radiation is isotropic. It means for a given frequency ω, the density of radiation is same in all directions. It is thus independent of $\vec{k}/|\vec{k}|$ but depends only on $|\vec{k}|$.

d. The radiation density in the enclosure does neither depend on shape of it nor on the material of the enclosure. It also also does not depend upon the volume of the enclosure. It depends only upon the temperature of the enclosure.

We now state and deduce the Kirchhoff's law. It states that if E_λ is an emissive power of a Black-body at a wavelength λ and at a temperature T then, the ratio of emissive power e_λ to the absorptive power a_λ of *any* material at a temperature T is equal to E_λ. Thus, by Kirchhoff's law,

$$E_\lambda = \frac{e_\lambda}{a_\lambda}$$

This can be proved as follows. Consider an enclosure filled with a radiation of wavelength between λ and $d\lambda$. Let a material body sit in the enclosure. In an equilibrium, by Prevost's theory, amount of energy radiated by the body $e_\lambda d\lambda$ must be equal to the energy absorbed by it. If dQ is the energy of incident radiation of the body, then $a_\lambda dQ$ is absorbed. Thus,

$$e_\lambda d\lambda = a_\lambda dQ \ .$$

For a perfectly Black-body $a_\lambda = 1$. Then, $E_\lambda d\lambda = dQ$. Thus,

$$\frac{e_\lambda}{a_\lambda} = \frac{dQ}{d\lambda} = E_\lambda \tag{17.26}$$

q.e.d.

Here we have used a crucial fact that whatever the material body - black or otherwise, placed inside an enclosure with a radiation, the nature of radiation does not change in an equilibrium. There are many ways of making black-bodies. We show here the simplest one

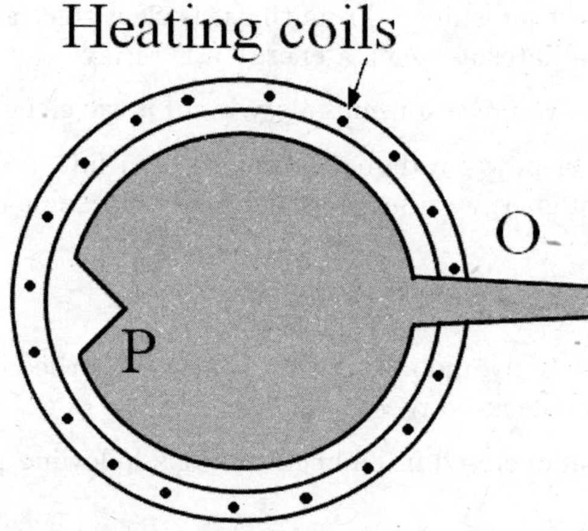

Figure 17.3: Fery's Black body.

due to Fery. It is a hollow metallic sphere blackened inside. It is heated by an electric current passing around its walls. It is blackened inside so that the equilibrium is reached quickly. It has a hole O from which the transport of radiation can take place. If any radiation goes inside, it equilibriates with inside radiation, reflected back and forth and is unlikely to come out. Thus it gets absorbed almost completely. A hump at P prevents direct reflection of radiation to O.

Qualitatively, Kirchhoff's law means that if a body is able emit certain radiation, it will absorb the same radiation when it falls on it. Microscopically, we understand this by the following facts.

1. Atoms and molecules have energy levels.

2. The process of absorption of light of frequency ω takes place when a transition from a lower level A to an upper level B is such that $\hbar\omega = E_B - E_A$. This is possible provided the selection rules are satisfied. Number of transitions per second are proportional to $|\langle B|H'|A\rangle|^2$ where H' is a transition causing part of the Hamiltonian.

3. In the emission process exactly the reverse process occurs. By the principle of detailed balance the square matrix elements $|\langle B|H'|A\rangle|^2$ remains same. The ratio of rate of emission to the rate of absorption of wavelength λ is e_λ/a_λ and it remains same.[2]

This has got interesting consequences. We know that a green glass looks green because it absorbs all the colors except green. The major color that is absorbed by the green glass is its complementary color - namely red. When a green glass is heated in a furnace and is taken out, it glows as red. Similarly, a hot piece of red glass glows as green.

These ideas of Kirchhoff opened up a branch of spectroscopy and also of astrophysics. This enabled the detailed study of solar spectrum (emission and absorption spectrum in a visible region).

17.5 Planck's Law

The photon states are represented by a solution of Maxwell's equation. Thus a photon of frequency ω and a wave-vector k can be represented as

$$\vec{E}(\vec{k},\omega) = \vec{E}_0 e^{i(\vec{k}\cdot\vec{r} - \omega t)}$$

Here $\omega = ck$. Photon is a massless particle and its energy is $\epsilon_k = \hbar\omega_k$. There are two polarization states of photon. Thus using equation (11), the energy density $u_{\omega_k} d\omega_k$ for a particular mode ω_k is

$$\begin{aligned}
u_{\omega_k} d\omega_k &= \frac{4\pi \times 2}{8\pi^3} \frac{k^2 \dfrac{dk}{d\omega_k} \hbar\omega_k d\omega_k}{\exp\left(\beta\hbar\omega_k\right) - 1} \\
&= \frac{(\hbar\omega_k)\omega_k^2}{\pi^2 c^3} \frac{1}{\exp\left(\hbar\omega_k/k_B T\right) - 1} d\omega_k
\end{aligned} \qquad (17.27)$$

Thus, the energy density for a frequency ω is

$$u_\omega = \frac{\hbar\omega^3}{\pi^2 c^3} \frac{1}{\exp\left(\hbar\omega/k_B T\right) - 1} \qquad (17.28)$$

This is the celebrated Planck's formula for distribution of energy of electromagnetic field in various frequency modes at a temperature T. We write the dimensionless parameter $\eta = \hbar\omega/k_B T$. Then, the Planck's law becomes

$$u_\omega d\omega = \frac{\hbar}{\pi^2 c^3} \left(\frac{k_B T}{\hbar}\right)^4 \frac{\eta^3 d\eta}{e^\eta - 1}. \qquad (17.29)$$

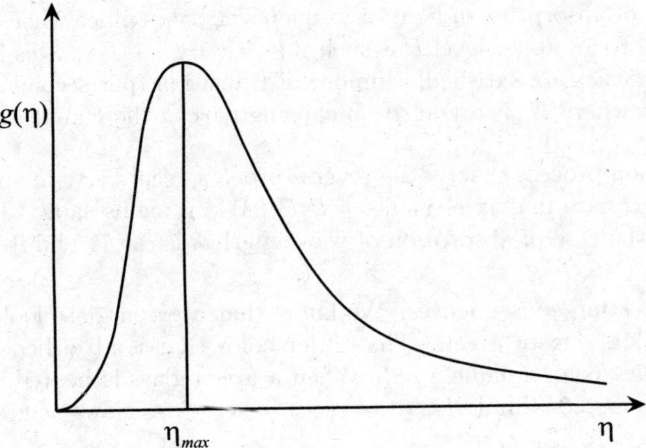

Figure 17.4: A function $g(\eta) = \dfrac{\eta^3}{e^\eta - 1}$.

Let $g(\eta) = \dfrac{\eta^3}{e^\eta - 1}$. Clearly, $g(\eta)$ has a maximum for some fixed value of η. Suppose for a temperature T_1 of a Black-body a maximum occurs at frequency ω_1. Similarly at absolute temperature T_2, it occurs at a frequency ω_2. Then,

$$\eta_{max} = \frac{\hbar\omega_1}{k_B T_1} = \frac{\hbar\omega_2}{k_B T_2} . \tag{17.30}$$

Or, if λ_{max} is a wavelength corresponding to η_{max}, then , from above scaling,

$$\frac{ch}{k_B \lambda_{max} T} = constant.$$

The constant is 4.9651. Thus

$$\lambda_{max} T = ch/(k_B \times 4.9651) = constant \tag{17.31}$$

This is famous Wein's displacement law. Thus the maxima of a Black-body curve shift with different temperatures as shown in figure (5). Here we plot the experiential parameter e_λ as a function of λ. The experimental curves shown here are similar to those obtained by Lummer and Pringsheim. It must be noted that, historically, the displacement law was obtained by Wein by purely thermodynamic arguments. From the graph it is clear that the value of λ_{max} decreases as the temperature T increases in accordance with Wein's law.

---

[2]'Quantum Mechanics' by L.I. Schiff, 3rd edition, sec 35

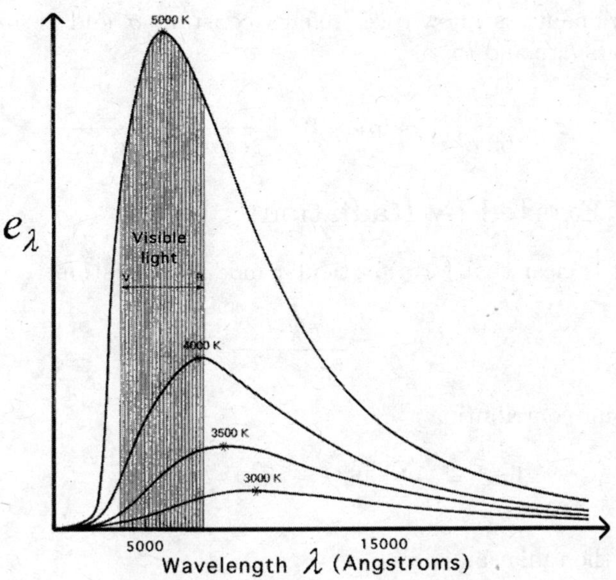

Figure 17.5: Emissivity e_λ versus λ at different black body temperatures.

17.6 Mean Total Energy

Mean (average) total Black-body energy density is sum of u_ω over all frequencies ω. We denote the mean total energy at absolute temperature T as $\overline{u}_0(T)$. Then,

$$\overline{u}_0(T) \;=\; \int_0^\infty u(\omega, T)\, d\omega$$

$$=\; \frac{\hbar}{\pi^2 c^3} \left(\frac{k_B T}{\hbar}\right)^4 \int_0^\infty \frac{\eta^3 d\eta}{e^\eta - 1}$$

Value of the integral $\displaystyle\int_0^\infty \frac{x^3\, dx}{e^x - 1} = \frac{\pi^4}{15}$. Thus,

$$\overline{u}_0(T) = \frac{\pi^2}{15} \frac{(k_B T)^4}{(\hbar c)^3}. \tag{17.32}$$

Here, $\overline{u}_0(T)$ is a density of radiation in an enclosure of temperature T. From the construction of a Black-body the amount of power loss from an aperture per unit area due to radiation is denoted by H and

$$H = \frac{c\overline{u}_0}{4} = \frac{\pi^2 \hbar c^2}{60} \left(\frac{k_B}{\hbar c}\right)^4 T^4 \equiv \sigma T^4 \,. \tag{17.33}$$

Constant of proportionality is known as Stefen's constant σ and is found here from the fundamental constants $\hbar$, c and k_B as

$$\sigma = \frac{\pi^2}{60} \frac{k_B^4}{c^2 \hbar^3} \sim 5.7 \times 10^{-5} \frac{erg}{sec\ cm^2\ degree^4} \qquad (17.34)$$

17.7 Pressure Exerted by Radiation

From equation (6) it is clear that for a particular mode of radiation

$$Z = \frac{e^{-\beta \hbar \omega / 2}}{1 - e^{-\beta \hbar \omega}}$$

Ignoring the zero point contribution,

$$\ln Z = -\sum_k \ln \left(1 - \exp\left[-\beta \hbar \omega_k\right]\right) \ .$$

Pressure exerted by the radiation is

$$\begin{aligned}
\overline{P} &= k_B T \frac{\partial \ln Z}{\partial V} \\
&= k_B T \left(\frac{-2}{(2\pi)^3}\right) \int_0^\infty \ln\left(1 - e^{-\beta \hbar c k}\right) 4\pi k^2 dk \qquad (17.35)
\end{aligned}$$

Here a factor 2 comes from two polarizations. This integral is easy and it is left as an exercise to show

$$\overline{P} = \frac{\overline{u}_0}{3}.$$

Thus, the radiation pressure depends upon mean energy density and is a function of T only.

17.8 Cosmic Radiation Background

Penzias and Wilson discovered a background radiation in the universe. They observed it to be isotropic. They conjectured that this radiation was created at the time of Big-bang. At the early times the matter was a high density of electrons, ionized hydrogen and ionized helium. The radiation and matter was strongly coupled under temperatures around 4000 to $5000K$ persisted.

As universe expanded adiabatically, the radiation cooled down. Because of expanding matter, photon frequencies red shifted. Today, the temperature of this radiation is $3K$. It is assumed here that this cosmic background radiation is a Black-body radiation and is confined in a universe which acts as a cavity. Thus, $3K$ radiation is a signature of the recombination events of a capture of electrons by H or He nuclei - which happened tens of billions of years ago.

This experiment of Penzias and Wilson is very difficult one, in that the elimination of background noise from the signal in a detector is difficult. Penzias and Wilson got Nobel prize in the year 1978 for the discovery of this microwave cosmic background radiation.

17.9   High and Low Frequency Limits of Planck's Law

In a Planck's law

$$u_\omega d\omega = \frac{\hbar\omega^3}{\pi^2 c^3}\,\frac{d\omega}{\exp\left(\hbar\omega/k_B T\right) - 1}$$

The relevant dimensionless parameter is $\hbar\omega/k_B T$. When $\dfrac{\hbar\omega}{k_B T} >> 1$, it is a high frequency limit. Here

$$u_\omega d\omega = \frac{\hbar\omega^3}{\pi^2 c^3}\,e^{-(\hbar\omega/k_B T)}\,d\omega \tag{17.36}$$

This is known as Wein's law. In its original form Wein's law was,

$$u_\lambda d\lambda = \frac{A}{\lambda^5}\exp\left(-c_2/\lambda T\right) d\lambda \tag{17.37}$$

Where A and λ are constant. Wein did not know quantum theory. He obtained this law by purely thermodynamic arguments. We know that $\omega = 2\pi c/\lambda$. $\therefore \omega^3 d\omega = (2\pi c)^4 d\lambda/\lambda^5$. Lumping $\hbar$, $\pi\,c$ and k_B in appropriate manner we get, A and c_2. Wein had almost come close to Planck's assumption, but not quiet. He argued that the oscillator emitting radiation has a kinetic energy $\dfrac{1}{2}mv^2$. He further assumed that the kinetic energy is proportional to the frequency of the oscillator giving $1/2mv^2 = a\nu$. The probability that the oscillator has this kinetic energy is proportional to $e^{-a\nu/k_B T}$.

The total number of states corresponding to the frequency ν are proportional to $\nu^2 d\nu = A\nu^2 d\nu = \dfrac{Ac^3}{\lambda^4}d\lambda$. Dumping all the constants together as the general constants, energy then is proportional to

$$\frac{A}{\lambda^5}\exp\left(-a/\lambda T\right) d\lambda$$

which is Wein's law.

In the low frequency limit, $\hbar\omega/k_B T << 1$. Then, $\left(e^{\hbar\omega/k_B T} - 1\right) \approx \hbar\omega/k_B T$. Planck's law then reduces to

$$u_\omega d\omega = \frac{\hbar\omega^3}{\pi^2 c^3}\cdot\frac{k_B T}{\hbar\omega}\cdot d\omega$$

Thus,

$$u_\omega = \frac{\omega^2 (k_B T)}{\pi^2 c^3} \tag{17.38}$$

This is known as Rayleigh - Jean's law. This law can be understood as follows. From equation (17) we have seen that the number of modes of oscillations per unit volume per unit angular frequency interval $\sigma(\omega)$ where

$$\frac{\sigma(\omega)}{V} = \frac{\omega^2}{2\pi^2 c^3}\,.$$

There are two polarizations. According to classical analysis each mode carries a energy $k_B T$. Thus energy density is

$$u_\omega = \frac{\omega^2}{\pi^2 c^3}k_B T$$

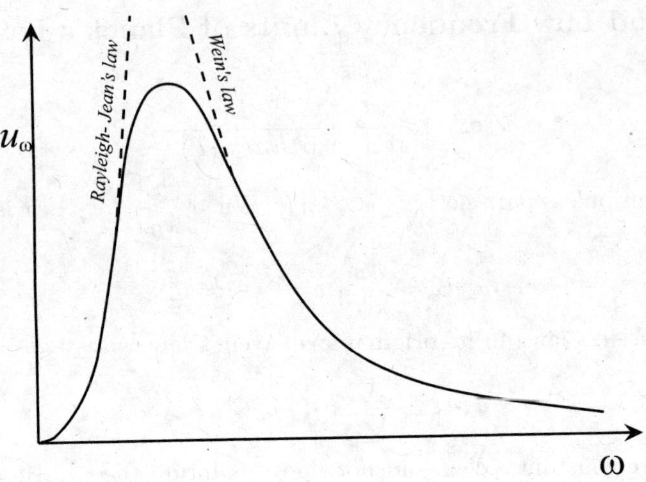

Figure 17.6: Limiting forms of Planck's law are shown by dotted lines.

which is the Rayleigh - Jean's law. We thus see that the complete Black-body spectrum can not be understood by a classical statistical mechanics.

17.10   Einstein's Derivation of Planck's Law

We present here Einstein's derivation because it shades the light on physical process of absorption and emission. He also skillfully uses the principles of detailed balance. Bohr model gives a knowledge of a process of absorption and also partially of emission process. In Bohr's model, only mechanism for emission transition was due to self interaction of electron at higher level resulting to jump of it to lower level. This transition leads to spontaneous emission. There is another process of emission called induced emission. Induced emission takes place only in presence of external e. m. field. These transitions are described by transition probabilities. They have maximum value when Bohr condition $\hbar\omega_{12} = E_2 - E_1$ is satisfied. Einstein discovered the process of induced emission and emphasized its need in deriving Planck's law. He correctly anticipated that the transition probability for induced emission is proportional to the intensity of external e. m. field. The process responsible for emission and absorption are shown below.

1. Absorption process: Let N_1 be number of atoms in a state $|1\rangle$. Let probability of transition from $|1\rangle \rightarrow |2\rangle$ i.e. of absorption be B_{12}. Let radiation density be $u(\omega_{12})$. Then number of absorptions are

$$N_1 u(\omega_{12}) B_{12}$$

2. Spontaneous emission process: This process does not require external e. m. field. Let probability of transition from state $|2\rangle \rightarrow |1\rangle$ i.e. emission be A_{21}. Let N_2 be number of atoms in state $|2\rangle$. Then number of spontaneous emission are $N_2 A_{21}$.

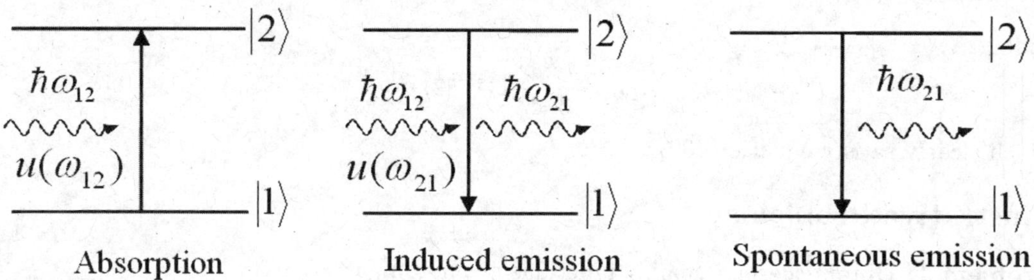

Figure 17.7: Radiation process.

3. Induced emission process: It is similar to absorption process depending upon intensity of incident radiation. Let B_{21} be probability of induced emission. Then, number of induced emission are

$$N_2 u\left(\omega_{21}\right) B_{21}$$

For a thermodynamical equilibrium, a detailed balance would imply the number of absorption should be equal to the number of emissions. It means,

$$N_1 u\left(\omega_{12}\right) B_{12} = N_2 u\left(\omega_{21}\right) B_{21} + N_2 A_{21}$$

Clearly, $|\omega_{21}| = |\omega_{12}| \equiv \omega$. $u(\omega_{12}) = u(\omega_{21})$. Moreover, at an absolute temperature T, $\dfrac{N_2}{N_1} = \exp\left(-\dfrac{\hbar\omega}{k_B T}\right)$. Then,

$$u\left(\omega\right) = \frac{A_{21}}{B_{21}\left(\dfrac{B_{12}}{B_{21}} \exp\left[\beta\hbar\omega\right] - 1\right)}$$

This expression must be identical with Planck's law. This is possible if $B_{12} = B_{21}$ and ,

$$A_{21} = \frac{\hbar\omega^3}{\pi^2 c^3} B_{21}. \tag{17.39}$$

Thus, we have a connection between A_{21}, B_{12} abd B_{21} which are known as Einstein coefficients. They are intrinsic atomic properties. These coefficients play an important role in Laser where the population inversion ($N_2 > N_1$) is necessary. This is not possible with two levels. We need at least three levels for population inversion, levels $|1\rangle$, $|2\rangle$ and $|3\rangle$ such that transitions from $|1\rangle$ to $|2\rangle$ and $|2\rangle$ to $|3\rangle$ are allowed but transitions from $|2\rangle$ to $|1\rangle$ are forbidden.

The Einstein coefficients were obtained in terms of microscopic parameters by Dirac[3] in his seminal papers on quantum electrodynamics as

$$B_{12} = \frac{4\pi^2 e^2}{3\hbar^2} |\langle 1|\vec{r}|2\rangle|^2 = B_{21}$$

---

[3]See for example 'Quantum Mechanics' by L. I. Schiff ch. 11 and 14, McGraw-Hill, third edition

and

$$A_{12} = \frac{4e^2\omega^3}{3\hbar c^3} |\langle 1|\vec{r}|2\rangle|^2$$

which clearly satisfy equation (39)

Illustrative Problems

Problem 1: Obtain thermodynamic functions of photon gas.

1. Internal energy $E = u(T)V = \sigma V T^4$.

2. $C_V = \left.\dfrac{\partial E}{\partial T}\right|_V = 4\sigma V T^3$.

3. Entropy $S = \displaystyle\int_0^T \frac{C_V dT}{T} = \frac{4}{3}\sigma V T^3$.

4. Free energy $F = E - TS = \sigma V T^4 - \dfrac{4}{3}\sigma V T^4 = -\dfrac{1}{3}\sigma V T^4$

5. Pressure $P = \left(-\dfrac{\partial F}{\partial V}\right)_T = \dfrac{1}{3}\sigma T^4 = \dfrac{ST}{4V} = \dfrac{u}{3}$

6. Equation of state is,

$$PV^{4/3} = S^{4/3} \left(\frac{3}{4\sigma}\right)^{1/3}$$

7. Isothermal compressibility $\kappa_T = -\dfrac{1}{V}\left(\dfrac{\partial V}{\partial P}\right)_T$

   But P is independent of V and is

$$P = u/3 = \frac{\sigma T^4}{3}.$$

Thus $\left(\dfrac{\partial V}{\partial T}\right)_T = \infty$ and thus $\kappa_T = \infty$.

Thus the photon gas is infinitely compressible in isothermal process.

8. Adiabatic compressibility :

   Here $\kappa_s = -\dfrac{1}{V}\left(\dfrac{\partial V}{\partial P}\right)_S$.

   By First law,

$$0 = dE + PdV$$

or,

$$\left(\frac{\partial E}{\partial V}\right)_S = -P$$

Using (6),

$$\left(\frac{\partial V}{\partial P}\right)_S = -\frac{3}{4}\frac{V}{P}$$

Then,

$$\kappa_S = \frac{3}{4P}$$

Problem 2: Run the Carnot cycle with a photon gas as a working substance and show that its efficiency is same as obtained by using ideal gas as the working substance[4].

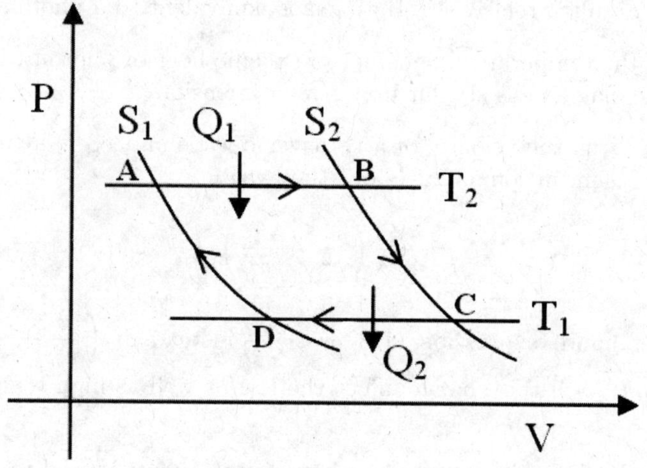

Figure 17.8: Carnot engine run by photon gas as a working substance. The two isotherms at T_1 and T_2 are horizontal because u is a function of T only.

Solution Consider two adiabats S_1 and S_2 ($S_1 < S_2$). As the pressure by the radiation does not depend upon volume, (it is a function of T alone) AB and CD are isotherms at temperatures T_2 and T_1 respectively. Here $Q = T\Delta S$. Thus heat intake during isothermal expansion along AB is $Q_1 = T_2(S_2 - S_1)$. Heat rejected is $Q_2 = T_1(S_2 - S_1)$. It must be noted here that Q_1 is not the work done during isothermal expansion. This is unlike the Carnot engine with ideal gas. The adiabats BC and DA obviously have $Q = 0$. This means $E_C - E_B = -W(B \to C)$.

Efficiency is

$$\eta = \frac{Q_1 - Q_2}{Q_1} = 1 - \frac{T_1}{T_2}$$

which is same as ideal gas Carnot engine.

---

[4]M. Howard Lee, Am. J. Phys. <u>69</u>, 874 (2001)

It is interesting to to note that $\dfrac{dP}{dT} = \dfrac{S}{V}$. Compare this with Clausius-Clapeyron equation,

$$\frac{dP}{dT} = \frac{S_2 - S_1}{V_2 - V_1} .$$

This is same if $S_1 = V_1 = 0$. Vanishing of entropy and volume of one phase means phase transition. We will show in the next chapter that the ideal Bose- Einstein gas with $\mu \neq 0$ undergoes a phase transition when $\mu(T_c) = 0$. Then it is like a photon gas. Here $L = TS$ the latent heat and is not zero.

Short Questions

1. Explain how an ultra relativistic Bose gas is equivalent to a photon gas.

2. Explain why the temperature behavior of specific heat of photons, phonons and anti-ferromagnetic magnons is similar in a given dimension?

3. For a solid if v_{st} is the velocity of a transverse wave and v_{sl} is a velocity for a longitudinal wave, then, in some text books they write

$$3N = \frac{V}{2\pi^2} \left(\frac{2}{v_{st}^2} + \frac{1}{v_{sl}^2} \right) \int_0^{\omega_D} \omega^2 d\omega$$

   On the other hand, expression (19) uses $\dfrac{3}{v_S^2}$ instead of $\left(\dfrac{2}{v_{st}^2} + \dfrac{1}{v_{sl}^2} \right)$. Justify the approximation. Will this change affect the low as well as high temperature behavior of C_V?

4. Can you use the Einstein model to account for the contribution to C_V due to optical modes? At low temperatures, the contribution to C_V will dominate from acoustic modes or from optical modes? why?

5. Show that if the elementary excitation has a form $\omega = Ak^s$ and $\mu = 0$, then energy is d dimensions behaves as $T^{\frac{d+s}{s}}$ and the specific heat behaves as $T^{d/s}$ with absolute temperature. (Use simple scaling arguments for the energy expression modifying it appropriately for d dimensions.)

6. Find λ_{max} for the microwave cosmic background radiation at $3K$.

7. Explain what is meant by low frequency and the high frequency limits of Planck's law. Approximate Planck's law in the above limits.

Problems

Problem 1: Solar constant is defines as the amount of solar power received by $1cm^2$ area of a black surface held at right angles to the Sun's rays and placed at mean distance of earth from sun.

Assume Sun to be a Black-body at absolute temperature T and of radius $R = 7 \times 10^{10}$ cm. Mean distance from Sun to Earth is $L = 1.5 \times 10^{13}$ cm. Value of a solar constant is $S = 1.937$ $cal/cm^2/min$. Find the temperature of the Sun.

$$\text{Ans: } T = 5732K$$

(Hint: Amount of power lost by Sun $= 4\pi R^2 \sigma T^4$. Draw a concentric sphere of radius L. Then by definition, $S = \dfrac{4\pi R^2}{4\pi L^2}\sigma T^4$. Put the numbers and get the answer.)

Problem 2: If L is means radius of Earth's orbit around Sun and it is found that the radius of Mercury's orbit is $0.39L$ and that of a Mars' orbit is $1.52L$. Find the solar constant at Mars and at Mercury if it is $S = 1.937$ for earth.

$$\text{Ans: S for Mercury} = \frac{1.937}{(0.39)^2} = 12.735$$
$$\text{S for Mars is} = \frac{1.937}{(1.52)^2} = 0.838.$$

Problem 3: It is known that an atom bomb is spherical and has a size of 10 cm radius. Two fissile half spheres are separated. When the explosion is desired they are brought together and imploded symmetrically by dynamic charges. A temperature of about $10^6 K$ is produced. Find,

A. Amount of power radiated from the surface of the bomb.

B. $Power/m^2$ at a distance of $1km$ from the point of explosion.

C. λ_{max}, the wavelength corresponding to the peak in the power spectrum.

Assume the exploded bomb to be a Black-body

$$\text{Ans: (A) } 71.26 \times 10^{14} \ W$$
$$\text{(B) } 5.67 \times 10^8 \ W/m^2$$
$$\text{(C) } 2.92 \times 10^{-11} \ cm$$

Problem 4: Use the parameters given in problem (1) and assume

1. Temperature of the Sun as $5732K$

2. Earth absorbs all the radiation all the radiations that falls on it.

3. Steady state

Notice that the Earth revolves round itself and therefore whole of its area is exposed to the radiation. Find the approximate temperature of Earth.

$$\text{Ans: } T_{earth} = \sqrt{\frac{R}{2L}}\, T_{sun} \approx 300K$$

Problem 5: Define a function $D(\xi) = \dfrac{3}{\xi^3} \displaystyle\int_0^\xi \dfrac{x^3}{e^x - 1}\, dx$. Define Debye temperature $k_B \Theta_D = \hbar \omega_D$.

Find, in this approximation

(a) $\ln Z$,   (b) average energy $\overline{E}$ (c) entropy S.

Ignore zero point energy contribution.

(Hint: $\ln Z = -\displaystyle\int_0^{\omega_D} \sigma(\omega) \ln(1 - e^{\beta \hbar \omega})\, d\omega$.

$\sigma(\omega) = \dfrac{3V}{2\pi^2}\dfrac{\omega^2}{v_S^3}$, $\dfrac{\omega_D^3}{v_s^3} = \dfrac{\sigma \pi^2 N}{V}$. Integrate by part and get result.

Similarly fine $\overline{E} = -\dfrac{\partial \ln Z}{\partial \beta}$ and $S = k_B(\beta \overline{E} + \ln Z)$)

$$\text{Ans: (a) } \ln Z = -3N \ln(1 - e^{-\Theta_D/T}) + ND\left(\frac{\Theta_D}{T}\right)$$

$$\text{(b) } \overline{E} = 3Nk_B T D\left(\frac{\Theta_D}{T}\right).$$

$$\text{(c)} S = -Nk_B \left[3\ln\left(1 - \exp\left(-\frac{\Theta_D}{T}\right)\right) - 4D\left(\frac{\Theta_D}{T}\right)\right].$$

Problem 6: Show that the mean square dispersion for the photon (or phonon), or for any elementary excitation (Bosonic) is

$$\frac{\overline{n_k^2} - \overline{n_k}^2}{\overline{n_k}^2} = \frac{1}{\overline{n_k}} + 1\ .$$

Chapter 18

IDEAL BOSE SYSTEM II

In this chapter, we will consider an ideal Bose gas system with $\mu > 0$. This means system is made up of Bose particles which have a nonzero mass. This gas has a phase transition. In the condensed phase, we have macroscopic number of particles with zero momentum. This transition occurs at a certain transition temperature T_C. For $T < T_C$ number of particles in zero momentum state go on increasing as T is lowered. At $T = 0$, all the particles of the system are in condensed phase. It must be noted that for an ideal Maxwell-Boltzmann gas and ideal Fermi gas, there is no phase transition. On the other hand the ideal Bose gas undergoes a phase transition. This is purely a quantum effect. This happens because of two reasons. (1) If a Bose particle occupies a certain quantum state say $|k\rangle$, then another identical Bose particle has an enhanced probability of occupying $|k\rangle$. (2) As the temperature is lowered, de-Broglie thermal wavelength $\lambda = \dfrac{\hbar}{\sqrt{2mk_BT}}$ increases. As a result, correlation between the identical Bosons increases. This happens for fermions too. However, for Fermions, if $|k\rangle$ is occupied, another identical Fermions have a zero probability of occupying it. Therefore no phase transition for fermions.

Gas of Helium 4 liquifies at very low temperature around $4K$. If the temperature is further lowered at $2.186K$, it becomes superfluid. This is thought to be due to Bose -Einstein condensation but of interacting system. Two Helium atoms interact with each other through a van der Waals interaction.

Almost truly non interacting Bose system was constructed by Cornell et.al. in 1995. The temperatures involved were about $10^{-9}K$. For this experimental verification of the Bose- Einstein condensation, Cornell, Weimann and Kitterleg got Nobel prize in 2001.

Actually the condensation was discovered by Einstein when he generalized Bose's derivation of Black body radiation for Bose particles of non zero mass in 1924. When K.Onnes discovered superfluidity of liquid Helium in 1911, theoretical explanation of identifying it as a case of Bose-Einstein condensation was suggested by London (1938). However, Landau in 1941 disputed this explanation and discovered his famous spectrum of phonons and rotons (elementary excitations of Helium II liquid) and explained superfluidity.

Finally, Feynman (1954) considered various excitations of Bosonic system and derived

Landau's phonon and roton spectrum.

18.1   Thermodynamic Functions of a Bose-Einstein Gas

We have deduced that the grand partition function for an ideal gas is

$$\frac{PV}{K_B T} = \ln \ \Omega(\mu, V, T) = -\sum_k \ln(1 - \exp\left[\beta(\mu - \epsilon_k)\right]) \tag{18.1}$$

where μ is chemical potential and is a function of temperature and $\epsilon_k = (\hbar^2 k^2/2m)$.

    Total number of particles are

$$N = \sum_k n_k = \sum_k \frac{1}{\exp\left[\beta(\epsilon_k - \mu)\right] - 1} \tag{18.2}$$

While converting the sums in equation (1) and (2), there come the problems. Normally, we would have written equation (2) as

$$N = \frac{V}{(2\pi)^3} \int_0^\infty \frac{d^3 k}{\exp\left[\beta(\frac{\hbar^2 k^2}{2m} - \mu)\right] - 1} \tag{18.3}$$

Here, $\epsilon_k = 0$ i.e. ground state and $\mu = 0$, if occur simultaneously then, there is a divergence at the lower limit of the integral. To avoid this situation, we separate $\vec{k} = 0$ state contribution and then convert the sum to an integral. For brevity, we write $z = e^{\beta\mu}$. z is called fugacity. Then equations (1) and (2) become,

$$\frac{P}{k_B T} = -\frac{1}{(2\pi)^3}.4\pi \int_0^\infty k^2 dk \ln\left(1 - ze^{-\beta\epsilon_k}\right) - \frac{1}{V}\ln(1 - z) \tag{18.4}$$

and

$$\frac{N}{V} = \frac{4\pi}{(2\pi)^3} \int_0^\infty \frac{k^2 dk}{z^{-1}e^{\beta\epsilon_k} - 1} + \frac{1}{V}\frac{z}{1 - z} \tag{18.5}$$

These integrals can be cast in to Bose-Einstein (B-E) integrals, as defined in an appendix. They are

$$\mathcal{J}_\nu(\alpha) = \frac{1}{\Gamma(\nu)} \int_0^\infty \frac{x^{\nu-1}dx}{e^{x+\alpha} - 1} = \sum_{n=1}^\infty \frac{e^{-n\alpha}}{n^\nu} \tag{18.6}$$

Then, after some simple algebra, equation (4) and (5) become

$$\frac{P}{k_B T} = \frac{1}{\lambda^3}\mathcal{J}_{5/2}(\alpha) - \frac{1}{V}\ln(1 - z) \tag{18.7}$$

and

$$\frac{N}{V} = \frac{1}{\lambda^3}\mathcal{J}_{3/2}(\alpha) + \frac{1}{V}\frac{z}{1 - z} \tag{18.8}$$

where $\lambda = \dfrac{h}{\sqrt{2\pi m k_B T}}$

In equations (7) and (8); $e^{-\alpha} = e^{-\beta\mu}$. N are total number of Bose particles confined in a volume V. When $\alpha = 0$ but $\vec{k} = 0 = \epsilon_k$, then, average occupancy of $k = 0$ state is,

$$N_0 = \frac{z}{1-z} \tag{18.9}$$

Thus, we may write

$$\frac{N}{V} = \frac{1}{\lambda^3}\mathcal{J}_{3/2}(\alpha) + \frac{N_0}{V} \tag{18.10}$$

Consider now following facts.

(i) As $z = e^{+\beta\mu}$ is such that $0 \leq z \leq 1$, $\mathcal{J}_{3/2}$ is bounded from above by $\mathcal{J}_{3/2}(\alpha = 0) = \mathcal{J}_{3/2}(z = 1)$ which is finite. Fugacity is range bound $(0 \leq z \leq 1)$. Otherwise equation (7) will not be satisfied. This means $-\infty \leq \mu \leq 0$. In contrast to this, for a Fermi gas μ ranges as $-\infty \leq \mu \leq E_F$. $\mu \to -\infty$ gives M-B limit. (ii) The number of particles in non zero momentum are given by $\frac{1}{\lambda^3}\mathcal{J}_{3/2}(\alpha)$. As the temperature goes on decreasing, lambda goes on increasing and $\left(1/\lambda^3\mathcal{J}_{3/2}(\alpha)\right)$ the number of particles in nonzero momentum state decrease.

(iii) A stage then comes when $N = N_0$

(iv) We reinterpret equation (9) as furnishing the density of Bosons in zero momentum state. Thus

$$\frac{N_0}{V} = \left(n\lambda^3 - \mathcal{J}_{3/2}(\alpha)\right)/\lambda^3. \tag{18.11}$$

Since $\mathcal{J}_{3/2}(\alpha = 0) = \sum_{n=1}^{\infty} \frac{1}{n^{3/2}} = \zeta(3/2) \cong 2.612$ is a pure number, we obtain T_c as

$$k_B T_c = \frac{h^2}{2\pi m}\left(\frac{n}{2.612}\right)^{2/3}. \tag{18.12}$$

T_c is known as a transition temperature. For $T < T_c$, the fraction of zero momentum density, N_0/V, increases and is equal to N/V at $T = 0$.

We say that in the momentum space, $\vec{k} = 0$ or $\epsilon_k = 0$ state is macroscopically occupied at temperatures less than T_c. This is termed as a Bose -Einstein condensation and the transition occurs at temperature T_c as given by equation (12).

To get $z = e^{\beta\mu}$ as a function of T and N/V, we solve equation (8). This needs to be done numerically. We give here the numerical result in the figure (1)

In the limit $V \to \infty$, the result is $z = 1$ or $\mu = 0$ for $n\lambda^3 \geq 2.612$ and z is a root of $n\lambda^3 = \mathcal{J}_{3/2}(z)$ when $n\lambda^3 < 2.612$. Incidentally if we put the parameters of liquid He^4 in to the expression of T_c, we get $T_c = 3.14K$. This is close to the experimental value $2.2K$. This is the temperature below which superfluidity in He^4 occurs. He$_I^4$ is called a normal phase of Helium and is above $T > T_c$. Superfluid phase is called He$_{II}^4$ phase and occurs at $T < T_c$

We now find the co-existence curve of normal and superfluid phase. For this we rewrite equation (8) as

$$N = \frac{1}{e^\alpha - 1} + \frac{V}{\lambda^3}\mathcal{J}_{3/2}(\alpha). \tag{18.13}$$

But,

$$n\lambda_c^3 = \mathcal{J}_{3/2}(\alpha = 0) .$$

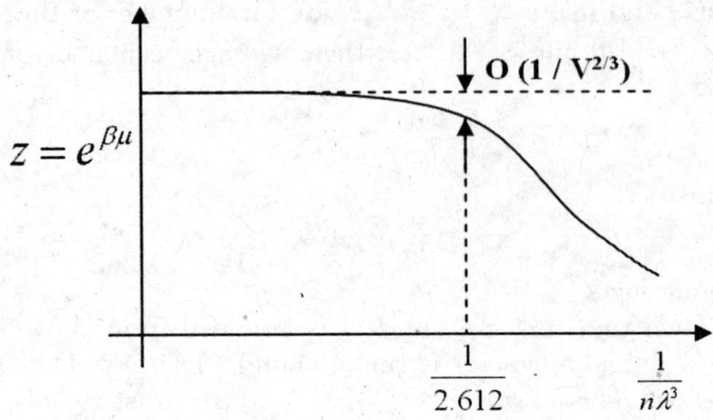

Figure 18.1: Fugacity of ideal Bose gas in volume V.

Then

$$N = \frac{1}{e^\alpha - 1} + N\left(\frac{T}{T_c}\right)^{3/2}\frac{\mathcal{J}_{3/2}(\alpha)}{\mathcal{J}_{3/2}(0)} \tag{18.14}$$

As seen from the figure(1), for $T < T_c$, $\alpha - 1 \approx \vartheta\left(\dfrac{1}{V}\right)$. Then, $N_0 = \dfrac{1}{e^\alpha - 1} \approx 1/\alpha$. Here we use ϑ to symbolize "of the order of".

We can approximate $\mathcal{J}_{3/2}(\alpha)$ by $\mathcal{J}_{3/2}(\alpha = 0)$ for the entire $0 < T < T_c$ range. We then get

$$\frac{N_0}{N} = \left(1 - \left(\frac{T}{T_c}\right)^{3/2}\right) \qquad T < T_c \tag{18.15}$$

For $T > T_c$, $N_0/N \to 0$ for macroscopic N. This means $\dfrac{1}{e^\alpha - 1} \approx 0$ and

$$\mathcal{J}_{3/2}(\alpha) = \left(\frac{T_c}{T}\right)^{3/2}\mathcal{J}_{3/2}(0) \tag{18.16}$$

From appendix, it is clear that

$$\mathcal{J}_{3/2}(\alpha) \cong -3.545\sqrt{\alpha} + 2.612 \qquad \text{for } 0 < \alpha < 1\ .$$

Write $C = 3.545/2.612 = 1.36$.

From equation (16), we find, $\alpha = \beta\mu$ for $T > T_c$ but close to T_c as,

$$\mu(T) \cong \frac{1}{k_B T}\left[\frac{1}{1.36N}\right]^{2/3}\left[1 + \frac{T - T_c}{T_c}\left(\frac{N\xi^2(3/2)}{4\pi}\right)^{1/3}\right] \tag{18.17}$$

Thus in the same limit,

$$\mu(T) \sim \vartheta\left(\frac{1}{N^{2/3}}\right) \sim \vartheta\left(\frac{1}{V^{2/3}}\right) \to 0$$

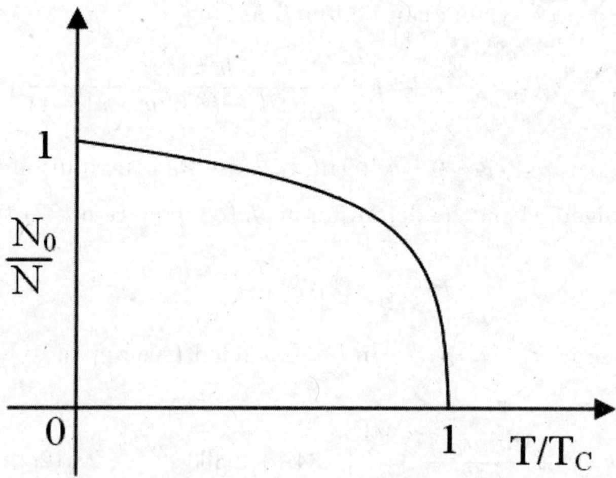

Figure 18.2: Normalized condensate density versus T/T_c.

in a thermodynamic limit $N \to \infty, V \to \infty$, $N/V = n$ (finite).

Pressure in condensed and non-condensed phase is found by analyzing the equation (7). For $T > T_c$

$$\frac{P}{k_B T} \approx \frac{(2\pi m k_B T)^{3/2}}{h^3} \mathcal{J}_{5/2}(\alpha) \equiv \frac{P_+}{k_B T}$$

$$\approx \frac{(2\pi m k_B T)^{3/2}}{h^3} \left[\xi(5/2) - 2.612\alpha + \vartheta(\alpha^{3/2}) \right]. \tag{18.18}$$

The value of $\alpha = \beta\mu$ is substituted using equation (16). For $T < T_c$, on the other hand, $\mu = 0 = \alpha$. Here

$$\frac{P}{k_B T} = \frac{(2\pi m k_B T)^{3/2}}{h^3} \mathcal{J}_{5/2}(\alpha = 0) = \frac{(2\pi m k_B T)^{3/2}}{h^3} \xi(5/2)$$

$$= \frac{(2\pi m k_B T)^{3/2}}{h^3} (1.342) \equiv \frac{P_-}{k_B T} \tag{18.19}$$

Clearly, P_- is a function of T only. A quantity of interest is a specific volume $\dfrac{V}{N}$ at $T = T_c$. This is given by equation (10) where $N_0 = 0$ at $T = T_c$. That means

$$v_c = \left(\frac{V}{N} \right)_{T=T_c} = \frac{\lambda_c^3}{\mathcal{J}_{3/2}(\alpha = 0)} = \frac{\lambda_c^3}{2.612} \tag{18.20}$$

Then, equation (15) can be written as

$$\left(\frac{N_0}{N} \right)_{T<T_c} = 1 - \left(\frac{T}{T_c} \right)^{3/2} = 1 - \frac{v}{v_c} \tag{18.21}$$

Internal energy of the Bose system can be found as

$$E = \sum_k n_k \epsilon_k = \frac{4\pi V}{(2\pi)^3} \int_0^\infty \frac{\hbar^2 k^2 / 2m}{\exp\left(\beta\left[\hbar^2 k^2 / 2m - \mu\right] - 1\right)} k^2 dk \tag{18.22}$$

Here, we have not separated $\epsilon_k = 0 = k$ term, because its contribution is always zero.

With simple algebra and the definition of $\mathcal{J}_\nu(\alpha)$, equation (22) transforms to

$$E = \frac{3}{2} V (k_B T) \frac{\mathcal{J}_{5/2}(\alpha)}{\lambda^3} . \tag{18.23}$$

For $T > T_c$, but close to T_c, $\mathcal{J}_{5/2}(\alpha)$ can be expanded (see appendix) and energy E_+ can be found as

$$E_+ = \frac{3}{2} V (k_B T) \left(\frac{2\pi m k_B T}{h^2}\right)^{3/2} \left[1.342 + 2.363\alpha^{3/2} - 2.612\alpha + ...\right] \tag{18.24}$$

with α as given by equation (17).

Internal energy below T_c is E_- where

$$E_- = \frac{3}{2} V (k_B T) \left(\frac{\xi(5/2)}{\lambda^3}\right) = \frac{3 \times 1.342}{2} V (k_B T)^{5/2} \left(\frac{2\pi m}{h^3}\right)^{3/2} . \tag{18.25}$$

18.2   Illustrative problem

Problem 1: Find C_V for Bose gas for $T < T_c$ and for $T > T_c$. Plot C_V versus T and discuss the nature of phase transition.

Solution: Equation (23) gives the value of energy in general. E_+ gives the value of internal energy for $T > T_c$ and E_- gives the value for $T < T_c$.

From equation (25) for E_-, for $T < T_c$ we get,

$$\begin{aligned}
\left.\frac{C_V}{N k_B}\right)_{T < T_c} &= \frac{15}{4} \times 1.342 \, T^{3/2} \left(\frac{2\pi m k_B}{h^3}\right)^{3/2} \frac{V}{N} \\
&= \frac{15}{4} \times \xi(5/2) \frac{1}{n\lambda^3} \propto T^{3/2} .
\end{aligned} \tag{18.26}$$

For $T > T_c$, we use equation (24) where the temperature enters through $\alpha = \beta\mu$ via equation (17). However, there is a simpler method. Consider an implicit relation between α and T as given by equation (9). For $T > T_c$, $N_0 \approx 0$. Thus,

$$n\lambda^3 = \mathcal{J}_{3/2}(\alpha) \tag{18.27}$$

gives $\alpha = \alpha(T)$. Equation (23) can be written as

$$\frac{E_+}{N} = \frac{3}{2} k_B T \frac{\mathcal{J}_{5/2}(\alpha)}{\mathcal{J}_{3/2}(\alpha)} . \tag{18.28}$$

Thus,

$$\left.\frac{C_V}{Nk_B}\right)_{T>T_c} = \frac{3}{2}\frac{d}{dT}\left(T\frac{\mathcal{J}_{5/2}(\alpha)}{\mathcal{J}_{3/2}(\alpha)}\right)$$

$$= \frac{3}{2}\left[\frac{\mathcal{J}_{5/2}(\alpha)}{\mathcal{J}_{3/2}(\alpha)} + T\left(1 - \frac{\mathcal{J}_{5/2}(\alpha)\mathcal{J}_{1/2}(\alpha)}{\mathcal{J}_{3/2}^2(\alpha)}\right)\frac{d\alpha}{dT}\right] \qquad (18.29)$$

He we have used the relation (see appendix)

$$\frac{d}{d\alpha}\mathcal{J}_\nu(\alpha) = \mathcal{J}_{\nu-1}(\alpha)$$

From equation (26), after differentiation we get

$$\frac{d\alpha}{dT} = -\frac{3}{2}\frac{1}{T}\frac{\mathcal{J}_{3/2}(\alpha)}{\mathcal{J}_{1/2}(\alpha)}\ .$$

With these equation and after some simplification we get

$$\left.\frac{C_V}{Nk_B}\right)_{T>T_c} = \frac{15}{4}\frac{\mathcal{J}_{5/2}(\alpha)}{\mathcal{J}_{3/2}(\alpha)} - \frac{9}{4}\frac{\mathcal{J}_{3/2}(\alpha)}{\mathcal{J}_{1/2}(\alpha)}\ .$$

When $T = T_c$, $\alpha = 0$ and

$$\left.\frac{C_V}{Nk_B}\right)_{T_c} = \frac{15}{4}\frac{\mathcal{J}_{5/2}(0)}{\mathcal{J}_{3/2}(0)} = 1.925$$

This is because $\mathcal{J}_{1/2}(0)$ diverges. Equation (26) also reduces to the same value at $T = T_c$ indicating that C_V continues at T_c for Bose- Einstein condensation. As $T >> T_c$ i.e. $\alpha >> 1$, all $\mathcal{J}_\nu(\alpha) \to e^{-\alpha}$. This means

$$\left.\frac{C_V}{Nk_B}\right)_{T>>T_c} \to \frac{15}{4} - \frac{9}{4} = \frac{3}{2}$$

which is a classical result.

In terms of Ehrenfest classification of phase transition, the temperature derivative of C_V is discontinuous for Bose- Einstein condensation. Therefore it is a third order phase transition.

It is important to note that the normal to superfluid transition of Helium is drastically different as compared with ideal Bose-gas. The specific heat diverges logarithmically with temperature at T_c in Helium.

18.3   Recent Experiments on BEC

Most important aspect of Bose- Einstein condensation is,

1. It takes place in k space. (i.e. $\vec{k} = 0$ is macroscopically occupied.) and

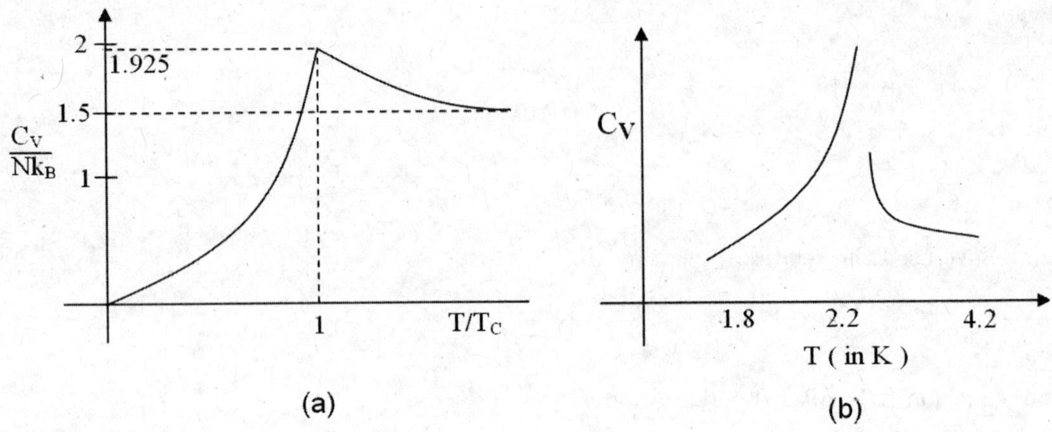

Figure 18.3: (a) Specific heat as a function of T/T_c for Bose-Einstein gas. (b) Experimental specific heat for superfluid transition in Helium.

2. The particles are noninteracting.

In reality the particles interact. Even He^4 atoms interact with each other strongly. Superfluidity in He^4 is now accepted to be a B.E. condensation. Earlier there was a controversy, if the superfluidity is a result of B.E. condensation. Landau explained superfluidity of He^4 due to his empirical phonon - roton spectrum of excitation. London argued in favor of B.E. condensation and Tisza gave a two fluid model based on these ideas. It was a seminal work of Feynman who obtained elementary excitation spectrum of Landau by looking at the excitations of Bose system.[1] Bogoliubov also obtained the Landau spectrum by studying interacting Bose gas.

To experimentally verify B.E. condensation one would need a gas of noninteracting Bosons. Such a gas is impossible to obtain. We therefore need a situation such that any two particles, on an average have a distance r_{int}, to a much larger than the range of particle-particle interaction r_d. Range of this interaction is approximately of the order of the s wave scattering length (a_s). Another important requirement is that the thermal de-Broglie wavelength $\lambda_{dB} = \dfrac{h}{\sqrt{2\pi m k_B T}}$ must be much larger than r_{int}. Thus, when $\lambda_{dB} >> r_{int} >> r_d \sim a_s$ is satisfied for Bosons, we have practically a system of non interacting Bosons. In this gas we have a possibility of observing Bose- Einstein condensation.

Clearly for liquid He^4, $r_{int} \sim 3.5\overset{\circ}{A}$ and $a_s \sim 2.5\overset{\circ}{A}$. Thus, the liquid Helium is a strongly interacting Bosonic system and is not suitable for observing the B.E. condensation.

Experimenters would need a collection of Bosons confined in a specific volume. This required confinements of atoms by traps. Such traps were designed by David Pritchard of MIT. It consists of what is known as anti Helmholtz coils. Helmholtz coils are two co-axial

---

[1] R. P. Feynman, Phys. Rev. __94__, 262 - 277 (1954)

parallel circular coils of radius a carrying a current I in the same direction and are separated by the distance a. The plane of the coils is at right angles to their axis. This arrangement has a property that the magnetic induction B is almost constant at a distance of about $a/5$ around the center of the line. (See figure (4))

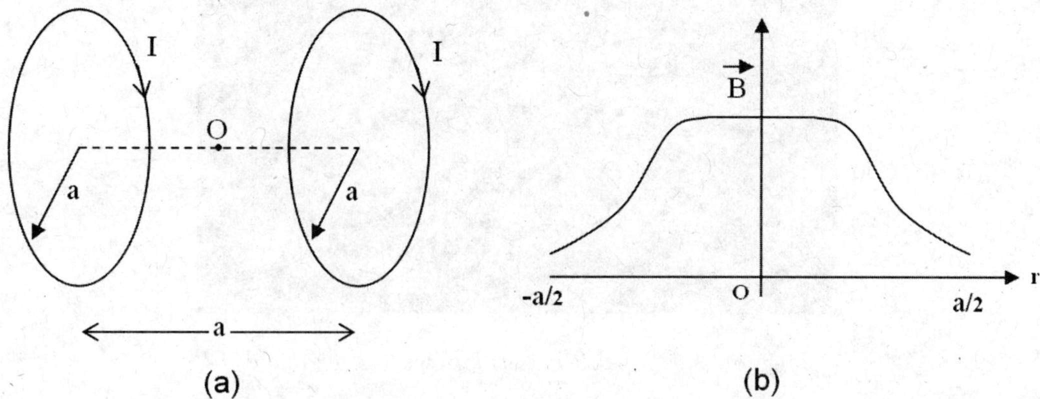

Figure 18.4: (a) Helmholtz coil and a field distribution across its axis.   (b) Region of almost constant magnetic field near the center of the Helmholtz coil.

Anti Helmholtz coil is Helmholtz coil but with current in two coils at opposite direction. This produces a quadrupole field. It is such that $\vec{B}$ is almost zero at the center. The condensed atoms are trapped at the center of the anti Helmholtz coil which acts as trap. This trap is shown in figure (5)

The requirements mentioned above can be satisfied by the alkali atoms whose electronic valance is one. But one needs integral spin for B.E. condensation. This requires a coupling of nuclear spin with the electronic spin. However this has a problem because the nuclear magnetic moment μ_N is smaller than the electronic magnetic moment by at least four orders of magnitude.

For the coupling of electronic spin $\vec{S}$ with the nuclear spin $\vec{I}$ to be significant, one needs a very low temperature such that $k_B T$ is much smaller than the hyperfine interactions. Experiments of Anderson[2] et.al. involved Rubidium isotope ^{87}Rb. One has to be correct in choosing the isotope of alkali item. Let $\vec{F}$ be total atomic spin, $\vec{I}$ the nuclear spin and $\vec{J}$ the spin due to electronic degree for the atom. Then

$$\vec{F} = \vec{I} + \vec{J}$$

and has a range of $I + J$ to $|I - J|$.

For alkalies, $S = J = 1/2\hbar$. When F is integral spin, then, it is a B.E. condensation candidate. So far, ^{87}Rb, ^{23}Na, ^{7}Li and ^{85}Rb have undergone BEC. All of them have $I =$

[2]MH Anderson, JR Ensher, MR Matthews, CE Weiman, and E. Cornell, Science <u>269</u>, 198 (1995)

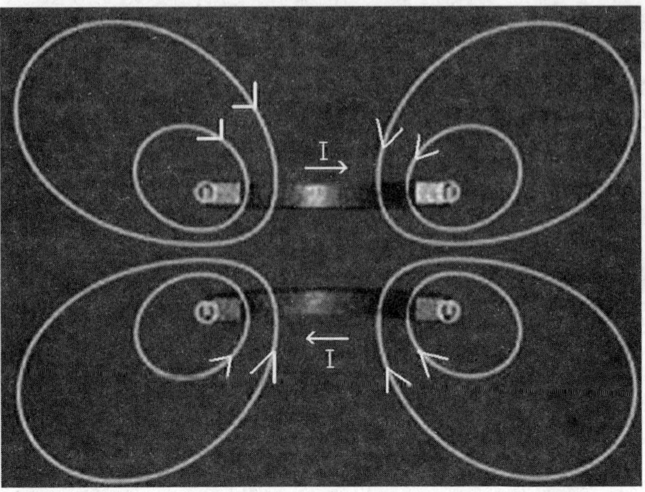

Figure 18.5: Anti Helmholtz coil with $\vec{B}$ field lines.

$3/2\hbar$. Thus for these atoms, F has value either 2 or 1. Let A be the mass number of the atom. For alkali atoms Z is odd. To get Bosons from them, we must have A also to be odd. Then $(Z + A)$ is even number of Fermions in alkali atoms giving integral value of F. Clearly, ^{6}Li or ^{86}Rb will not undergo BEC.

We estimate a typical value of T_c. The formula for T_c is,

$$T_c = \frac{3.31\hbar}{mk_B}n^{2/3} .$$
(18.30)

Typical values of the density of ^{87}Rb in experiment are of the order of $n \sim 10^{12}/cm^3$. Putting $m = (37m_P + 37m_e + 50m_N)$ we get $T_c \sim 114nK$. This is in kHz frequency range. Hyperfine interaction energy $\Delta E_{H.F.}$ for a typical alkali atom is about a fraction of a kelvin. (0.3 K for Rb). It means,. below about $0.1K$ the coupling between the electronic and nuclear spin remains intact. Thus, at the temperatures of around $10^{-7}K$ and below, the Bose-Einstein condensation takes place.

Bose-Einstein condensation of alkali atoms in the traps need to be interpreted properly. The gas of alkali atoms is dilute so that the atoms almost do not interact among themselves. But all the atoms in the central region of the trap are subjected to a harmonic potential of the form

$$V(\rho, \theta, z) = \frac{1}{2}m\omega_\perp^2\rho^2 + \frac{1}{2}m\omega_z^2Z^2.$$

This is a asymmetric potential. For simplification, we may take the average of angular frequency and take the average potential $V(r) = \frac{1}{2}m\omega^2r^2$. Here, characteristic radius 'b' of the potential is defined as

$$\hbar\omega = \frac{\hbar^2}{mb^2} .$$
(18.31)

Then, the ground state wave-function is

$$\psi_0(r) = \langle r|0\rangle = \frac{1}{(\pi b^2)^{3/4}} \exp\left(-\frac{r^2}{2b^2}\right) \tag{18.32}$$

Here, BEC means, macroscopic fraction of atoms go in the ground state $\psi_0(r)$. This is very much different from original free particle state $\langle r|k\rangle = \frac{1}{(2\pi)^{3/2}} e^{i\vec{k}\cdot\vec{r}}$, and hence $\vec{k} = 0$ as a ground state.

It can be shown that the condensation temperature is,

$$bT_c = (1.202)^{-1/3} \left(\frac{\hbar^2}{mk_B}\right) \left(\frac{N}{b^3}\right)^{1/3}. \tag{18.33}$$

However, if one takes a limit $N \to \infty$, $b \to \infty$, N/b^3 finite, then we immediately see that $T_c = 0$. This means strictly there is no finite critical temperature. Then how do we understand BEC? Actually, for a finite value of b and N, the above defined T_c characterizes B.E. condensation. It is observed that below the T_c of equation (32), $|\psi_0(r)|^2$ starts peaking at the center as T decreases. We thus see that more and more alkali atoms condense in to $\langle \vec{r}|0\rangle$ state. This was precisely observed by Anderson et.al. for ^{87}Rb atomic cloud. Since $|\psi_0(r)|^2$ is Gaussian, its Fourier transform is also Gaussian with a peak at $k = 0$, this allows us to say equivalently more and more number of particles occupy $k = 0$ state. Figure (6) shows the computer simulation of this effect. We will now see how the low temperatures in nano kelvin range are attained. We will discuss the Doppler cooling of atoms which can cool the atoms to the temperatures around $10^{-4}K$. To cool them below about $10^{-6}K$, we use Sysiphus cooling and evaporation cooling. We will not discuss Sysiphus cooling because it will take us in Quantum Optics too far.

Recall that whenever an e.m. radiation of frequency ω falls on an atom with an energy level difference $\hbar\omega_0$, then the probability amplitude for absorption is proportional to $\frac{1}{(\omega - \omega_0) + i\Gamma/2}$ where Γ is a natural line width[3]. In calculations of the probability amplitude, it is assumed that the atom is at rest. Actually this is not the case. The atom has a motion. Suppose that it has velocity $\vec{v}$ when a radiation of wave vector $\vec{k}$ and frequency ω falls on it. Then the Doppler effect comes in to play. If the atom is moving towards the light source, then the frequency ω is blue shifted (i.e. ω increases). On the other hand, if the atom is receding away from the source, the frequency ω appears to be red shifted to the atom (i.e. ω appears to be decreased). It turns out that this frequency shift is by an amount $-\vec{v}\cdot\vec{k}$ Thus the denominator in the probability amplitude for absorption changes to $\left[\left(\omega - \omega_0 - \vec{k}\cdot\vec{v}\right) + i\Gamma/2\right]^{-1}$. The quantity $\Delta = -\omega + \omega_0$ is called detuning. More the detuning, lesser is the absorption of the e.m. wave. With the Doppler effect, the effective detuning is $\Delta' = (\Delta + \vec{k}\cdot\vec{v})$. Consider now a one dimensional situation. Suppose that the atom is moving along two counter propagating laser beams with detuning. Then $\Delta'_1 = \Delta + |\vec{k}||\vec{v}|$ and $\Delta'_2 = \Delta - |\vec{k}||\vec{v}|$ are the frequencies observed by the atom. These means more energy will be absorbed from one beam than the other. This means, the atomic

---

[3] see Quantum Mechanics, L I Schiff Ch.11

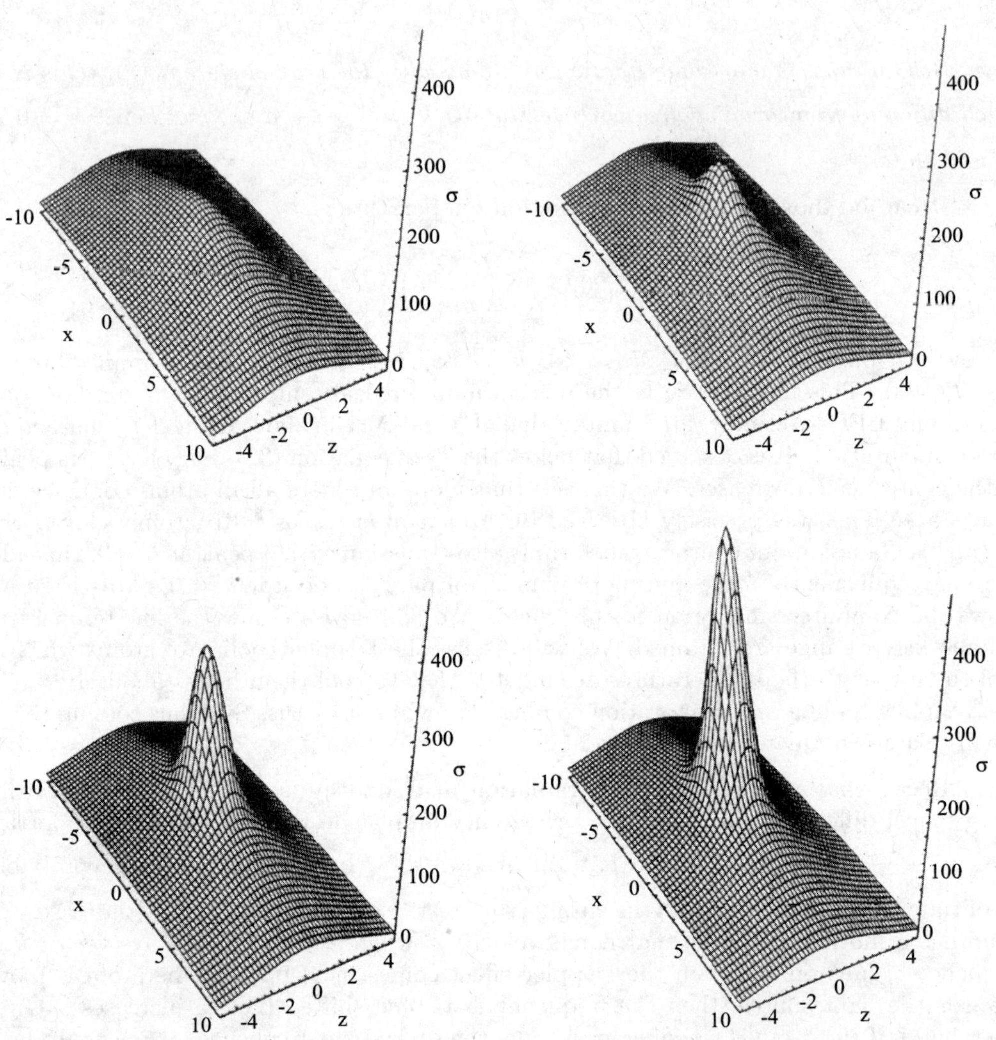

Figure 18.6: The integrated two dimensional density profile of 20,000 non-interacting bosons in a harmonic trap with $b_z = 0.9845 \ \mu m$ and $b\perp = 1.655 \ \mu m$. x and z are measured in units of μm. The temperatures are, from left to right and from top to bottom, 72, 71, 70 and 69 nK respectively. The condensation temperature is 71 nK. (See S. A. Chin, H.A. Forbert and E. Krotscheck in 'Condensed Matter Theories', Volume 12 page 27, Ed. by John Clark and P. V. Panat, Nova Science publishers, 1997).

momentum is reduced by $2|\vec{k}||\vec{v}|/c$. It is obviously due to the fact that one of the beams has a frequency closer to the resonance than the other. The forces[4] exerted by the two counter propagating beams on the atom are not balanced, resulting a net force opposite to $\vec{v}$. This reduction in velocity means reduction of kinetic temperature. Since the velocities of atom have three components, three counter propagating beams in x, y, and z are used in atomic cooling experiments. Atomic cloud is at the origin in the trap. Eventually the cloud is cooled. There is a limit to cooling by this method. When the Doppler cooling is balanced by the heating due to spontaneous emission, the limit is reached. For an atomic transition, spontaneous emission rate is proportional to the line width $\Gamma/2$ and also to the Einstein's A coefficient. When corresponding energy, $\hbar\Gamma/2$ becomes comparable with $k_B T$ we reach the limit of Doppler cooling. Thus,

$$k_B T_{max} = \frac{\hbar\Gamma}{2} \, . \tag{18.34}$$

For ^{87}Rb , $E_{ns} \to E_{np}$ transition, $\hbar\Gamma = 3 \times 10^{-4} K$. Thus the lowest temperature achieved is $1.5 \times 10^{-4} K$. For Na, $T_{max} \sim 2.4 \times 10^{-4} K$. It is interesting to note that, in a single photon absorption, due to atomic recoil, speed loss is about 3 cm/sec for Na. In one second, the number of absorptions = transition rate $\sim \Gamma = 10^9$ per second. Atom thus experiences enormous acceleration $\sim 10^7 m/s^2$.

We report the simulation results of B.E. condensation of 20,000 Bose particles as a function of temperature. T_c in this case is $71 nK$.

We discuss atoms trapped in an anisotropic trap of figure(6). The trap has frequencies $\omega_x = \omega_y = \omega_\perp \neq \omega_z$. The energy levels are

$$\epsilon_{ijk} = \hbar\omega_\perp(i + j) + \hbar\omega_z k$$

Here the zero point energy of an atom in the trap is

$$\epsilon_{0,0,0} = (\hbar\omega_\perp + \hbar\omega_z/2)$$

and is adjusted in μ. A little reflection will reveal that this zero point energy $=\mu$ at T_c. Fugacity $z = e^{\beta\mu}$ is implicitly determined as

$$N = \sum_{i,j,k} \frac{1}{e^{\beta(\epsilon_{ijk} - \mu)} - 1} \equiv \sum_{i,j,k} N_{i,j,k} \tag{18.35}$$

Once the chemical potential or the fugacity is known, we can find integrated two dimensional profile.

$$\rho(x, z) = \int dy \sum_{i,j,k} N_{i,j,k} |\psi_i(x)|^2 |\psi_j(y)|^2 |\psi_k(z)|^2 \tag{18.36}$$

---

[4]It can be shown that there are two kinds of forces on an atom in e.m. field. One is called dissipative force and has a largest value $\frac{1}{2}\hbar k\Gamma$ Here $\Gamma/2$ is a transition rate of an atom from lower level to the upper level. In each photon absorption momentum transferred is $\hbar k$. Thus total rate of momentum transferred is equal to the dissipative force having maximum value $\frac{1}{2}\hbar k\Gamma$. There is another force called dipolar force resulting from the shape of the beam profile of the laser. It is responsible for trapping of atom. For details see Gorden and Ashkin, Phys. Rev.21, 1606 (1980). See also, Elements of Quantum Optics by Meystre and Sargent Springer, 3rd edition (2003)

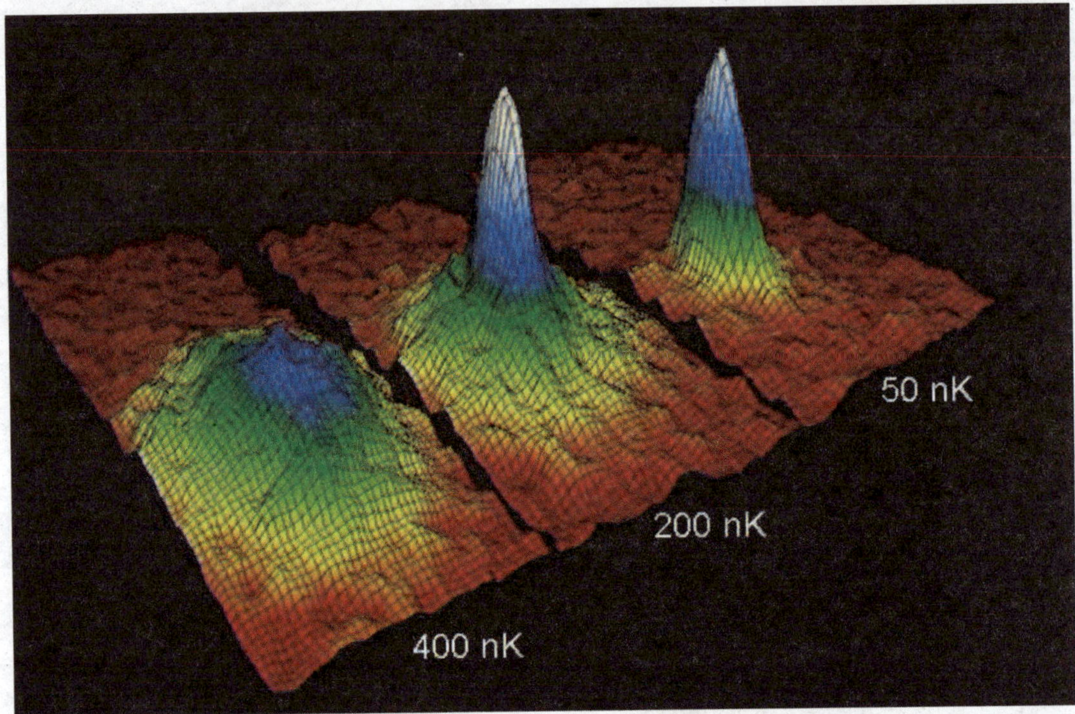

Figure 18.7: The first condensate figures obtained by Prof. Cornell's group at Jila Colorado

The sums were carried out with indices i, j and k ranging 400 states each. Resulting $\rho(x, z)$ for temperatures T=72nK, $T_c = 71nK$, 70nK and 69nK is shown in figure (6). The simulations[5] were carried out by Siu A Chin et. al. We also show the condensate picture due to Anderson et. al. in figure (7).

Short Questions

1. Explain physical reasons for Bose-Einstein condensation.

2. Why a condensation does not occur in Fermi gases?

3. It is required to separate $\epsilon_k = 0$ state in the equation for N where as such a separation is unnecessary in the energy expression. Why?

4. Explain the behaviour of the chemical potential of the ideal Bose gas when $T < T_c$, $T = T_c$ and $T >> T_c$.

5. Obtain the behaviour of $C_V(T)$ for $T < T_c$.

---

[5] Condensed Matter Theories Vol 12 Edited by John W Clark and P V Panat Nova Science Publishers (1996)

6. Recent experiments of BEC use alkali metals as Bosonic candidates. They have a valency one with effective spin angular momentum $\hbar/2$. Explain this apparent contradiction.

7. Find the de-Broglie wavelength of Cs, Rb, and Na atoms used in recent BEC experiments at $T = 10^{-9}K$.

Problems

Problem 1: Find the jump of $\left(\dfrac{dC_V}{dt}\right)$ at $T = T_c$ for an ideal gas.

(Hint: Use $\mathcal{J}_{-1/2}(\alpha) = \dfrac{1}{2}\dfrac{\sqrt{\pi}}{\alpha^{3/2}}$ as $\alpha \to 0$, $\mathcal{J}_{1/2}(\alpha) \cong 1.773/\sqrt{\alpha}$, and asymptotic properties of $\mathcal{J}_\nu(\alpha)$ from appendix. Get

$$\frac{dC_V^+}{dT} = -\frac{3}{2}\frac{Nk_B}{T}\frac{\mathcal{J}_{3/2}}{\mathcal{J}_{1/2}}\left[\frac{3}{2} - \frac{15}{4}\frac{\mathcal{J}_{5/2}\mathcal{J}_{1/2}}{\mathcal{J}_{3/2}^2} + \frac{9}{4}\frac{\mathcal{J}_{3/2}\mathcal{J}_{-1/2}}{\mathcal{J}_{1/2}^2}\right]$$

and close to T_c,

$$\frac{dC_V^-}{dT} = \frac{45}{8}\frac{Nk_B}{T_c}\frac{\xi(5/2)}{\xi(3/2)}.$$

Get the difference at T_c.)

$$\text{Ans: } \Delta\left(\frac{dC_V}{dT}\right) = \frac{3.66Nk_B}{T_c}$$

Problem 2: Calculate the grand partition function for ideal two dimensional Bose-Einstein gas. Find the equation that relates number of particles with μ and T. Show that there is no Bose- Einstein condensation at a temperature T other than $T = 0$.

(Hint: $\displaystyle\sum_k = \frac{A}{(2\pi)^2}\int d^2k$ in $2 - D$. Grand partition function is

$$\frac{PA}{k_BT} = \ln\,\Omega(N, A, T) = -\ln(1 - z) - \frac{A}{\lambda^2}\mathcal{J}_2(\alpha); \quad z = e^{-\alpha}, \alpha = \beta\mu$$

Then,

$$N = \frac{z}{1 - z} + \frac{A}{\lambda^2}\mathcal{J}_1(\alpha).$$

Here we used $N = z\dfrac{\partial}{\partial z}\ln\Omega$. From appendix $\mathcal{J}_1(z = 1) \to \infty$.

$\therefore N = N_0 + \dfrac{A}{\lambda^2}\mathcal{J}_1(\alpha = 0)$ is satisfied only at $T = 0$. Thus no B-E condensation for $T > 0$ in $2 - D$.)

Problem 3: Show that an equation for an adiabat for ideal Bose gas is $PV^{5/3} =$ constant.

Problem 4: Show that the transition line in $P - T$ plane for B.E. condensation is

$$P = \left(\frac{2\pi m}{h^2}\right)^{3/2} (k_B T)^{5/2} \mathcal{J}_{5/2}(\alpha = 0).$$

Show that the isotherms of ideal Bose gas are given as shown in figure(8). Show also

$$P\left(\frac{V}{N}\right)^{5/3} = \frac{h^2}{2\pi m} \frac{\mathcal{J}_{5/2}(\alpha = 0)}{\left[\mathcal{J}_{3/2}(\alpha = 0)\right]^{5/3}}.$$

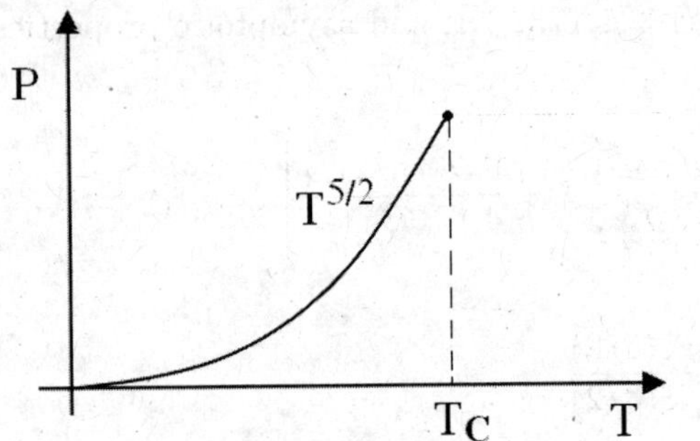

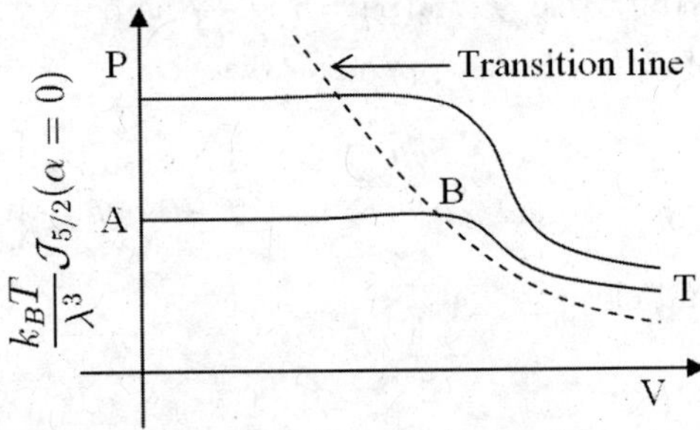

Figure 18.8: Transition line of B.E. condensation in P-T plane (Problem 4)

Chapter 19

Transport Phenomena: Elementary Exposé

Study of transport phenomena in materials involve (a) Transport of mass (Diffusion) (b) Transport of charges (Electrical conductivity) (c) Transport of heat (Thermal conductivity) and their combinations such as Thermoelectric effects. In all these processes there is no equilibrium. The system may have attained a steady state but still it is a non-equilibrium process. Study of the processes involves use of Boltzmann transport equation and/or fluctuation-dissipation theorem. Since this study is as vast as that of equilibrium processes, we will restrict ourselves to its elementary exposition. Emphasis is on the physical processes. Maxwell and independently Boltzmann initiated this study. Boltzmann asked a pertinent question as to how a system of large number of particles attains equilibrium? In the process he discovered a transport equation and his famous H-theorem.

We take a kinetic theory approach, give a hand waving arguments emphasize the approximations involved and give one detailed calculation.

19.1 Hand-waving (Elementary) Derivation of Transport Coefficients

In these calculations following simplifications are made.

1. We ignore the velocity distribution of the molecules. We assume that the molecules move with the average speed $\langle u \rangle = \sqrt{\dfrac{8k_B T}{m\pi}} = \overline{u}$

2. The gas is dilute so that for a large fraction of observation time the molecules do not interact and thus have a free motion. It means the average time between two collisions is much larger than the time involved in the collision.

3. Collision are described by a two body scattering process and the triple collisions are very infrequent. We ignore them.

19.2 Collision Time and Mean Free Path

Since the molecules are in the incessant random motion and undergoing collisions, the average distance traveled between two collisions called as mean free path (mfp) is of physical interest. Let $P(x)$ be a probability that a representative molecule travels a distance x without collision. Let $w\,dx$ be a probability that the molecule suffers a collision between x and $x + dx$.

Then, the probability that the molecule <u>does not</u> suffer a collision when it travels a distance $x + dx$ is

$$P(x + dx) \cong P(x) + \delta x \frac{dP}{dx} = P(x)(1 - w\,dx) \tag{19.1}$$

Then

$$\frac{dP}{dx} = -P(x)w\,dx$$

Or,

$$P(x) = e^{-wx}. \tag{19.2}$$

Here we use the boundary condition $P(0) = 1$.
The probability is normalized as

$$C \int_0^\infty e^{-wx} dx = 1, \text{ giving } C = w. \tag{19.3}$$

The quantity $\dfrac{1}{w} = l$ is called mean free path(mfp). Thus we write probability that the particle travels a distance x without suffering a collision as

$$P(x) = \frac{1}{l} e^{-x/l} \tag{19.4}$$

Average distance traveled without collision is

$$\overline{x} = \int_0^\infty x P(x) dx = \frac{1}{l}.l^2 \int_0^\infty \xi e^{-\xi} d\xi = l. \tag{19.5}$$

This justifies $w^{-1} = l = \overline{x}$ as mean free path.

It is easy to see that if the collisions are considered as a function of time (rather than a distance) then

$$P(t)dt = e^{-t/\tau} \frac{dt}{\tau}. \tag{19.6}$$

Here τ is a mean time between collisions and is also called as 'relaxation time'. $P(t)dt$ gives a probability that the molecule does not suffer a collision for a time t and then suffers a collision between t and $t + \delta t$.

Relation between collision time τ, mean free path l and collision cross section[1] can be found as follows.

---

[1]For detailed discussion of differential cross section, hard sphere scattering etc. see Classical Mechanics by P V Panat, Ch. 12, Narosa (2005)

Scattering cross section $\sigma(E)$ is a function of the energy of incident particle. It can be calculated classically or quantum mechanically depending upon the situation.

Let n be average number of molecules with an average speed $\bar{u}$. Consider $\overline{V}$ as the mean relative speed of these molecules which will range between 0 and $2\bar{u}$. Let $\sigma_0(\overline{V})$ be the total cross section. Let n_1 be the number molecules/cc. falling on the molecules in d^3r. Total number of molecules in d^3r are nd^3r. Then a moments reflection will suggest that the total number of scattering events of n_1 molecules in d^3r are

$$n_1 \, \overline{V} \, \sigma_0 \, n \, d^3r \tag{19.7}$$

Dividing by number $n_1 d^3r$ of type 1 molecules in d^3r, we get collision probability as

$$w = n \, \overline{V} \, \sigma_0 = \frac{1}{\tau}. \tag{19.8}$$

The mean free path (mfp) is

$$l = \bar{u} \, \tau \;\; = \;\; \frac{\bar{u}}{\overline{V}} \, \frac{1}{n\sigma_0} \tag{19.9}$$

Clearly, $\vec{V} = \vec{u}_1 - \vec{u}_2$ and

$$< V^2 > = < u_1^2 > + < u_2^2 > .$$

Or, approximately writing

$$\overline{V} \approx (\bar{u}_1^2 + \bar{u}_2^2)^{1/2} \;\approx\; \sqrt{2}\,\bar{u}.$$

Thus, using equation (9), the mfp is,

$$l = \frac{1}{\sqrt{2}\,n\,\sigma_0} \tag{19.10}$$

The result for mfp as given by equation (10) turns out to be exact result for hard sphere gas with proper Maxwell velocity distribution.

Orders of magnitudes: Consider one atmospheric pressure ($\sim 10^6$ dynes/$cm^2 \sim$ 101.33 kN/m^2) and $T = 300K$. Take nitrogen molecule diameter to be $d = 2 \times 10^{-10}$ m. Then $\sigma = \pi \, d^2 = 3.142 \times 4 \times 10^{-20}$.

$$\therefore \;\; \sigma \approx 12.568 \times 10^{-20} m^2$$

Density is given by

$$n = \frac{P}{k_B T} \approx 24.5 \times 10^{24}/m^3$$

$$\therefore \quad Mean\,free\,path = l = \frac{1}{\sqrt{2}\,n\,\sigma} = 2.3 \times 10^{-7} m$$

Clearly $d << l$. For Nitrogen Gas,

$$m = \frac{Molecular\;weight}{Avogadro\;number} \simeq \frac{28}{6 \times 10^{23}}\;gm.$$

Then

$$v_{rms} = \sqrt{\frac{3k_B T}{m}} \approx 500\;m/s.$$

Corresponding collision time is $\tau = \dfrac{l}{v_{rms}} \simeq 4.6 \times 10^{-10}/s$

Corresponding collision rate is $\sim \dfrac{1}{\tau} \simeq 2.2 \times 10^9\;/s.$

19.3　Calculational Method

To get the correct dependence of physical quantities, the physical process must be correctly understood. Following are the common features of the calculations.

1. We take a hypothetical plane in the system and look at the transport across it.

2. We assume, on average, $n/3$ molecules move along x, y or z axis of which $n/6$ move along positive direction and $n/6$ along negative direction. Here n is a density of the fluid.

3. Two planes at a distance $z + l$ and $z - l$ are drawn parallel to the hypothetical plane at z. Obviously, the molecules in the layer of width l around the hypothetical plane, on an average, just cross it without undergoing collision. This is so because l is mfp.

4. Physical quantity transferred across the hypothetical plane is different for different physical phenomena.

For viscosity, it is the net horizontal component of momentum that is transferred across the hypothetical plane.

For heat conduction it is the energy which is transferred across the hypothetical plane.

For diffusion process, the molecules are transferred across the plane.

External electric field is applied perpendicular to the hypothetical plane across which the charge is transported leading to the electrical conductivity.

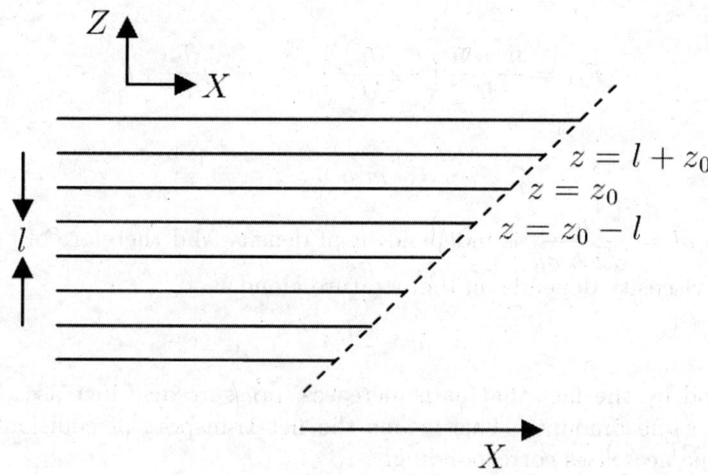

Figure 19.1: u_x as a function of z.

19.4 Viscosity Calculation

It is observed that, as water moves in the canal and when its flow is laminar (i.e. there are no eddies and turbulence), water at the bed is stationary, stuck to the bed. The water at open surface has highest velocity. Thus there is a velocity gradient across z direction. If the flow is along x direction, its velocity u_x is a function of z as shown in figure 1. Thus, $u_x = u_x(z)$. This function can be calculated if one knows the flow properties or it can be found experimentally. Let $z = z_0$ be the hypothetical plane.

All particles in the strip between z_0 and $z_0 + l$ have a larger value of x component of velocity than those on the hypothetical plane. Similarly all those particles which lie in the strip between z_0 and $z_0 - l$ have a lower value of u_x than those on the hypothetical plane. There is no force acting along x direction. As a result, when particles in the plane $z = z_0 + l$ cross the hypothetical plane, it imparts the momentum $m\, u_x(z_0 + l)$. Similarly, when particles from lower strip cross $z = z_0$ plane, the momentum imparted is $m\, u_x(z_0 - l)$. Particle current in the $\pm z$ direction is $\dfrac{1}{6} n\, \bar{u}$ where $\bar{u}$ is a mean speed.

The net transport of momentum across the hypothetical plane induces a stress P_{zx} because of z motion of the molecules. This is empirically related to the velocity gradient by Newton's law as,

$$P_{zx} = -\eta\, \frac{\partial u_x}{\partial z}. \tag{19.11}$$

Here η is called coefficient of viscosity. Its unit is poise=gm/cm/sec in honour of Poiseulle. Thus, because of total transport of flowing molecules across a plane $z = z_0$, the stress is,

$$P_{zx} = \frac{n\bar{u}}{6}\left[-m\, u_x(z + l) + m\, u_x(z - l)\right]\ .$$

Then,

$$P_{zx} = \frac{n\,m\,\overline{u}}{6}\left(-2\frac{\partial u_x}{\partial z}\right) l \equiv -\,\eta\frac{\partial u_x}{\partial z}.$$

Then,

$$\eta = \frac{1}{3} m\,n\,\overline{u}\,l. \tag{19.12}$$

Here the product $nl = \dfrac{n}{\sqrt{2}\,n\,\sigma_0}$ is independent of density and therefore of pressure. Then the coefficient of viscosity depends on temperature alone as,

$$\eta \sim T^{1/2} \tag{19.13}$$

This is understood by the fact that as n increases, pressure also increases but the mfp l decreases by the same amount. This means the net transport of collisionless molecules across $z = z_0$ plane decreases correspondingly.

This situation cannot prevail for an extremely dilute gas such that $l >> L$, where L is the size of container. Then in that case $n \to 0$ and the argument fails. Maxwell verified these assertions.

19.5 Thermal Conductivity Calculation

For a thermal conductivity along the rod, it is observed in a steady state that the heat current is

$$\vec{j}_x = -\kappa\,\frac{\partial T}{\partial x} \tag{19.14}$$

where κ is a coefficient of thermal conductivity. By the arguments given above,

$$j_x = \frac{n\overline{u}}{6}\left[E(z-l) - E(z+l)\right] \tag{19.15}$$

where E is mean energy per molecule transported. Thus, expanding E in Taylor's series,

$$j_x = \frac{1}{6}n\overline{u}\left(-2\frac{\partial E}{\partial z}\right) l \tag{19.16}$$

$$= -\frac{1}{3}n\,\overline{u}\,c\,\frac{\partial T}{\partial z}\,l$$

where c is a specific heat per molecule. Here we used $E = cT$. Then,

$$\kappa = \frac{1}{3}\,n\,\overline{u}\,c\,l. \tag{19.17}$$

19.6 Diffusion

For diffusion of the particles one needs inhomogeneity in the density distribution of them. The process of diffusion leads to a uniform distribution. Diffusion current is given by Fick's law as,

$$j_z = -D\,\frac{\partial n}{\partial z}. \tag{19.18}$$

The partial derivative $\dfrac{\partial n}{\partial z}$ is called as concentration gradient. D is called as a diffusion coefficient. The diffusion current across the plane is

$$j = \frac{1}{6}\,\overline{u}\,[n(z-l) - n(z+l)] = -\frac{\overline{u}\,l}{3}\,\frac{\partial n}{\partial z}.$$

Or,

$$D = \frac{1}{3}\,\overline{u}\,l. \tag{19.19}$$

19.7   Electrical Conductivity

Here we use the fact that during the distance l, the electric field $\mathcal{E}$ simply accelerate the charges. The time between two successive collisions is τ. Beyond the time τ, most likely, the collision randomizes the velocity of the charges. Acceleration of the charge of a value e and mass m is, $a = e\,\mathcal{E}/m$. The distance traveled due to the external electric field $\mathcal{E}$ is,

$$z = \frac{e\,\mathcal{E}}{2m}\,t^2 + v_z(0)\,t.$$

Because the collisions are random, the average distance traveled is

$$\overline{z} = \frac{e\,\mathcal{E}}{2m}\,\overline{t^2}.$$

But,

$$\overline{t^2} = \frac{1}{\tau}\int_0^{\infty} t^2\,e^{-t/\tau} = 2\tau^2$$

Thus, $\overline{z} = \dfrac{e\,\mathcal{E}}{m}\tau^2$.

If there were no collisions the charge will be accelerated without limit. But this does not happen. Finally, in the competition between acceleration in time τ and the collisions, a steady state is reached. The charge then has a constant velocity v_d in the direction of the field. The quantity v_d is called as a drift velocity. Current that crosses the hypothetical plane is proportional to $\mathcal{E}$. It is also found that the drift velocity v_d is also proportional to $\mathcal{E}$. We then write a response in linear regime as

$$j = \sigma_e\,\mathcal{E}. \tag{19.20}$$

and,

$$v_d = \mu\,\mathcal{E}. \tag{19.21}$$

The quantity σ_e is called electrical conductivity and μ is called electrical mobility.

Our claim is that v_d is an average velocity in addition to the random component along the electric field. This can be seen as follows. Because of acceleration $a = e\,\mathcal{E}/m$, the velocity acquired by the charge is

$$v(t) = v(0) + \frac{e\mathcal{E}}{m}t.$$

Because the velocity $v(0)$ is random, its time average is zero. Also, $\bar{t} = \tau$ leading to

$$\bar{v}(t) = v_d = \frac{e\tau}{m}\mathcal{E}.$$

We can get the same drift velocity from previous relations. Or,

$$\mu = \frac{e\,\tau}{m}. \tag{19.22}$$

The current carried across the plane is

$$j_z = n\,e\,v_d = \frac{n\,e^2\,\tau}{m}\,\mathcal{E}$$

Then the conductivity is,

$$\frac{j}{\mathcal{E}} = \sigma_e = \frac{n\,e^2\,\tau}{m}. \tag{19.23}$$

The relationship between σ_e and μ is,

$$\sigma_e = n\,e\,\mu. \tag{19.24}$$

This is a well known relation which connects the electrical conductivity and the mobility.

Illustrative Problems

Problem 1: Let L be the distance between the cathode and anode in a CRO tube. The tube is filled with a gas at a temperature T. Estimate the gas pressure so that 90% of the electrons produced by electron gun in cathode reach anode.

Solution: If $P(x)dx = e^{-x/l}dx/l$ is a probability that the particle traveled distance x, probability of its surviving any distance beyond L centimeter is 0.9 such that,

$$0.9 = l^{-1}\int_{L}^{\infty} e^{-x/l}\,dx = e^{-L/l}.$$

Putting

$$l = \frac{1}{\sqrt{2}n\sigma}$$

we get,

$$n = \frac{\ln{(10/9)}}{L\sqrt{2}\sigma}$$

and the required pressure is $P = nk_BT$ or,

$$P = \frac{k_BT\ln{(10/9)}}{L\sqrt{2}\sigma}.$$

Typically, $L \sim 20cm$, $\sigma \sim 10^{-15}cm^2$. (can you say why this value of σ)? and $T = 300K$. With 1 atmosphere$= 10^6 dyne/cm^2$ the value of pressure is $P = 1.5 \times 10^{-7}$ atmospheres.

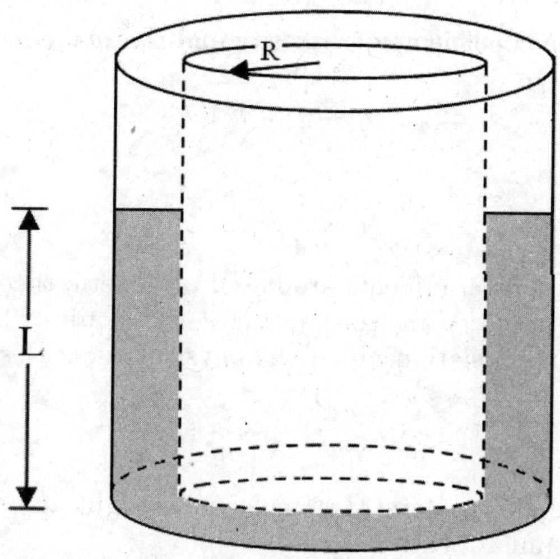

Figure 19.2: Viscometer

Problem 2: Standard method of measuring η is using viscometer. The viscometer has inner stationary cylinder of radius R and outer cylinder of radius $R + \delta$. ($R >> \delta$). The region between two cylinder is filled with a liquid whose value of η is to be found. As shown in figure 2, the liquid is filled up to a height L. The outer cylinder rotates with a small constant angular velocity ω. The inner cylinder experiences a torque G which is measured. Find η.

Solution: $u_{max} = (R + \delta)\omega$. $u_{min} = 0$ which corresponds to the velocity of the liquid stuck to the inner cylinder.

$$\therefore \quad \frac{du}{dr} = \frac{(R + \delta)\omega}{\delta} \approx \frac{R\omega}{\delta}.$$

The stress on the inner cylinder is force/area and it is stress $= \dfrac{-\eta\,R\,\omega}{\delta}$. Then the tangential force on the inner cylinder is, force $= \dfrac{2\pi\,R\,L\,\eta\,R\,\omega}{\delta}$. This gives torque G as

$$G = \frac{2\pi\,R^3\,L\,\eta\omega}{\delta}$$

giving η as G is measured.

To get the feel for the numbers, we take typical experimental parameters for air. $L = 10$ cm, $\delta = 0.1$ cm, $\omega = 2\pi$ radians/sec.($\equiv$ one rotation/sec, R = 2 cm and $\eta = 6 \times 10^{-4}$ gm/cm/s. Then $G \simeq 19.2$ dynes-cm $= 19.2 \times 10^{-7}$ Nm.

19.8   Viscosity and Thermal Conductivity

We have learned that the coefficients of viscosity and thermal conductivity are given by,

$$\eta = \frac{1}{3}\, m\, n\, \overline{u}\, l \tag{19.25}$$

and

$$\kappa = \frac{1}{3}\, c\, n\, \overline{u}\, l. \tag{19.26}$$

In arriving these results, our arguments are based on kinetic theory of gases. Considering the simplistic arguments and non rigorous averages, the factor of 1/3 need not be taken too seriously. However the inter relationship of the physical quantities is correct. The ratio

$$\frac{\kappa}{\eta} = \frac{c}{m} = \frac{N_A\, c}{N_A\, m} = \frac{C_V}{M} \tag{19.27}$$

depends on temperature alone. Here M is molecular weight and C_V is molar specific heat at constant volume. A similar other ratio is

$$\frac{\kappa}{\sigma_e} = \frac{1}{3}\frac{m\, c\, \overline{u}^2}{e^2}. \tag{19.28}$$

For a dilute gas we have $\overline{u} = \sqrt{\dfrac{8}{\pi}\dfrac{k_B T}{m}}$, $c = 3\, k_B/2$. Then the ratio is,

$$\frac{\kappa}{\sigma_e} = \frac{4}{\pi}\frac{k_B^2}{e^2}\, T. \tag{19.29}$$

This ratio is proportional to T. It is known as Wideman-Franz ratio.

19.9   Thermal Conductivity of Metals

For metals the situation is described by Fermi-Dirac distribution. It is known that the electrons are mainly responsible for the transport at low temperature. The quantity in our expression nc is electronic specific heat per unit volume $=C_V$. We have already calculated it when we studied specific heat of electron gas to be

$$C_V = \frac{\pi^2}{2}\frac{N k_B^2}{E_F}\, T. \tag{19.30}$$

Since only the electrons near Fermi surface alone are involved in the transport, we take $\overline{u} = v_F$. It is independent of temperature. At low temperatures, practically very few phonons are excited. The scattering of electrons is then due to impurities only. The mean free path l depends upon $1/n$ and is independent of temperature. This is so because $n = n_i$, the impurity concentration is independent of temperature. From expressions (25) and (30) it is clear that

$$\kappa_i = AT \tag{19.31}$$

where A is a constant. At higher temperatures, the number of phonons produced are proportional to T^3 and thus l is proportional to $(1/T^3)$. n and c that appear in the expression of κ refer to electrons. This factor is proportional to T. This leads to the contribution of phonons to thermal conductivity is,

$$\kappa_p \propto T \times \frac{1}{T^3} \propto \frac{1}{T^2} \ . \tag{19.32}$$

Effective thermal conductivity due to impurity scattering and phonon scattering in metals is,

$$\frac{1}{\kappa} = \frac{1}{\kappa_p} + \frac{1}{\kappa_i}. \tag{19.33}$$

This gives a behavior of κ with temperature for metals as

$$\kappa = A\,T^2 + \frac{B}{T}. \tag{19.34}$$

This kind of behaviour gives rise to a peak in κ which is well observed experimentally.

Short Questions

1. Explain the difference between the steady state and the equilibrium.

2. Why three body collisions are ignored in a dilute gas? At higher densities what is their role? (Think third particle carrying momentum and energy in the collision.)

3. Define collision time. Obtain probability $P(t)$ that the particle has not suffered a collision up to t.

    Ans: $P(t) = e^{-t/\tau}$

4. Estimate the number of atoms of a gas in one cc at STP. (Use 22400 cc at STP contain N_A atoms.)

    Ans: $3 \times 10^{19}/cc.$.

5. It is found that the mfp $l = 10^{-7}\ m$ for one atmospheric pressure. What is the value of l when the pressure is reduced to 10^{-6} atmospheres?

    Ans: 10 cm.

6. Explain in details the mechanism of viscosity of a fluid.

7. Explain in details the mechanism of Thermal conductivity of a fluid.

8. Explain the process of diffusion. Relate diffusion current to the concentration gradient.

9. Define electrical conductivity and mobility. What is the relation between them?

10. Explain the temperature behavior of η. Explain as to why the η is pressure independent?

11. In what way the behavior of κ for the dilute hard sphere gas and metals differ? Why?

Problems

Problem 1: What fraction of cases does an atom travel a distance less than l in a gas?

Ans: $(1 - 1/e)$

(Hint: Required probability $= \displaystyle\int_0^l P(x)dx$)

Problem 2: Consider a non steady case where the temperature of the material is $T(z,t)$. The material has the following properties. Density $= \rho$, specific heat/mass$= c$, thermal conductivity $= \kappa$. Obtain the partial differential equation of $T(z,t)$.

Ans: $\dfrac{\partial T}{\partial t} = \dfrac{\kappa}{\rho c}\nabla^2 T$

(Hint: Consider two planes separated by distance dz. Find a difference between heat fluxes $Q(z)$ and $Q(z+dz)$ and equate it to $\dfrac{\partial}{\partial t}(n\,m\,c\,T(z)))$

Problem 3: A practical method can be devised to measure κ in a following way. There is a long cylindrical wire of radius a and has an electrical resistance $R\ \Omega/cm$. A thin cylindrical shell surrounds the wire and is kept at a fixed temperature T_0. A constant current I is passed through the wire. After a steady state is reached $\Delta T = T_0 - T_{wire}$ is measured. Find κ.

Ans: $\kappa = \dfrac{I^2 R}{2\pi\Delta T}\ln(b/a)$

(Hint: In steady state $\nabla^2 T = 0$. Or, for a given geometry,

$$\frac{d^2 T}{dr^2} + \frac{1}{r}\frac{dT}{dr} = 0$$

Subject to the condition $T(b) = T_0$, Heat flux from surface of inner wire is $\kappa\dfrac{\partial T}{\partial r}$. Solution is

$$T(r) = A\ln r + B$$

$\dfrac{\partial T}{\partial r} = \dfrac{A}{r}$, Put the boundary conditions and get the answer.)

Problem 4: Show that the density distribution $n(\vec{r},t)$ in diffusion process satisfies

$$\frac{\partial n}{\partial t} = D\nabla^2 n$$

(Hint: Use the arguments similar to the problem (2) for mass transport.)

Problem 5: A glass bulb of volume 1 *litre* is filled with the Hydrogen at a temperature $T = 273K$ and at pressure $P = 10^{-4} \, Torr$ (1 *mm* of mercury is 1 Torr). The bulb contains a filament of area $0.2 \, cm^2$ and is suddenly heated to become red hot. Under this condition any hydrogen molecule hitting the filament, dissociates in to atoms. These atoms stick to the wall of the bulb when they hit it. Find mean free path at the starting pressure. Also find the time required for a pressure to drop one order of magnitude.

$$\text{Ans: } l \sim 230cm, \ t = 0.26sec.$$

(Hint: $22400c.c.$ at NTP has $N_A = 6 \times 10^{23}$ molecules.

$$\therefore n = \frac{10^{-4}}{760} \times \frac{6 \times 10^{23}}{22400} \cong 3.5 \times 10^{12}/c.c.$$

For hydrogen molecule $d \sim$ bond length$\sim 2\mathring{A}$

$$\therefore \sigma = \pi d^2 \sim 12.56 \times 10^{-16} cm^2$$

$$l = \frac{1}{\sqrt{2}n\sigma} = \frac{10^4}{1.42 \times 3.5 \times 12.56}$$

Rate at which molecules strike filament $= \dfrac{1}{4}n\bar{u}A.$

$$\therefore \frac{dn}{dt} = -\frac{1}{4}n\bar{u}A, \ \text{or,} \ P = P_0 \exp\left(-\frac{\bar{u}A}{4V}t\right)$$

Put $\dfrac{P}{P_0} = 0.1$ Put the numbers and get the answer.)

Problem 6: By the method of Laplace's transform or otherwise show that the solution of diffusion equation of problem (4) in one dimension is

$$n(z, t) = \frac{1}{\sqrt{4\pi Dt}} \exp\left(-\frac{z^2}{4Dt}\right)$$

Chapter 20

Boltzmann Transport Equation

In the previous chapter we saw that by simple minded arguments and the concept of collision time and mean free path , many non equilibrium phenomenon could be qualitatively understood. However, to go beyond qualitative discussion, and to have a systematic development, we have to find a differential and/or integral equation for a distribution function $f(\vec{r},\ \vec{v};\ t)$. This enables one to carry out systematic approximation schemes. Distribution function is a function of three variables - namely, position $\vec{r}$ of a particle, its velocity $\vec{v}$ and time t. Thus

$$f(\vec{r},\vec{v};t)d^3r d^3v$$

gives number of molecules laying between $\vec{r}$ and $\vec{r}+d\vec{r}$ with their velocity between $\vec{v}$ and $\vec{v}+d\vec{v}$ at time t. Equation of evolution of f in space, velocity and time was discovered by Boltzmann and is known as Boltzmann transport equation. In equilibrium, $f(\vec{r},\vec{v};t)$ goes over to $f(\vec{v})$ and is independent of space and time. This is famous Maxwell-Boltzmann distribution of velocity studied in chapter (9). Study of Boltzmann transport equation (called BTE) also tells how to account for binary collision effects in a detailed way.

20.1 Derivation of Boltzmann Transport Equation without Collision

Consider a situation when collision between the molecules in a gas is ignored. We concentrate on a small configuration space element $d^3r d^3v$, around position $\vec{r}$ and velocity $\vec{v}$. After time dt, the molecules will go into the element $d^3r' d^3v'$ around $(\vec{r'},\vec{v'})$. The new position is

$$\vec{r'} = \vec{r} + \dot{\vec{r}}dt, \ \ \vec{v'} = \vec{v} + \frac{\vec{F}}{m}dt \tag{20.1}$$

where $F(\vec{r},t)$ is an external force acting on the system. Since all the molecules from $d^3r d^3v$ traveled into $d^3r' d^3v'$ in time $t' = t + dt$, we must have

$$f(\vec{r},\vec{v};t)d^3r d^3v = f(\vec{r'},\vec{v'};t')d^3r' d^3v'$$

We claim that $d^3r d^3v = d^3r' d^3v'$. We prove it for 1 dimension, generalization to three dimensions is immediate. Clearly,

$$x' = x + \dot{x}dt; \quad v'_x = v_x + \frac{F_x}{m}dt \tag{20.2}$$

The volume element is related as

$$dx' dv'_x = J dx dv_x$$

where J is a Jacobian of transformation. It is

$$J = \begin{vmatrix} \dfrac{\partial x'}{\partial x} & \dfrac{\partial x'}{\partial \dot{x}} \\[2ex] \dfrac{\partial v'_x}{\partial v_x} & \dfrac{\partial v'_x}{\partial(F_x/m)} \end{vmatrix} = \begin{vmatrix} 1 & dt \\[2ex] \dfrac{1}{m}\dfrac{\partial F_x}{\partial x}dt & 1 \end{vmatrix}$$

$$= 1 + \vartheta(dt^2)$$

Extending the arguments to $d^3r d^3v$, i.e. six components , we arrive

$$d^3r d^3v = d^3r' d^3v' \tag{20.3}$$

This means

$$f(\vec{r}, \vec{v}; t) = f(\vec{r} + \dot{\vec{r}}\, dt, \vec{v} + \frac{\vec{F}}{m}dt; t + dt) \tag{20.4}$$

Using Taylor expansion,

$$\begin{aligned}
f(\vec{r} + \dot{\vec{r}}dt, \vec{v} + \frac{F}{m}dt; t + dt) &= f(\vec{r}, \vec{v}; t) \\[1ex]
&+ \frac{\partial f}{\partial t}dt + \left(\frac{\partial f}{\partial x}v_x + \frac{\partial f}{\partial y}v_y + \frac{\partial f}{\partial z}v_z\right)dt \\[1ex]
&+ \left(\frac{\partial f}{\partial v_x}\cdot\frac{F_x}{m} + \frac{\partial f}{\partial v_y}\cdot\frac{F_y}{m} + \frac{\partial f}{\partial v_z}\cdot\frac{F_z}{m}\right)dt
\end{aligned} \tag{20.5}$$

Using equation (4) we get

$$\frac{\partial f}{\partial t} + \left(v_x\frac{\partial f}{\partial x} + v_y\frac{\partial f}{\partial y} + v_z\frac{\partial f}{\partial z}\right) + \frac{1}{m}\left(F_x\frac{\partial f}{\partial v_x} + F_y\frac{\partial f}{\partial v_y} + F_z\frac{\partial f}{\partial v_z}\right)$$

Or,

$$\frac{\partial f}{\partial t} + \vec{v}\cdot\nabla f + \frac{\vec{F}}{m}\cdot\nabla_{\vec{v}}f = 0 \tag{20.6}$$

Where $\nabla_{\vec{v}}f$ is a gradient of f w.r.t. $\vec{v}$. Equation (6) is known as Boltzmann Transport equation without collision. It is sometimes called as Vlasov equation, derived by him for the collisionless plasma and extensively used by Landau.

Define the operator

$$D = \frac{\partial}{\partial t} + \vec{v}\cdot\nabla + \frac{\vec{F}}{m}\nabla_{\vec{v}} \tag{20.7}$$

for Boltzmann equation. Then collisionless BTE is

$$Df = 0 \tag{20.8}$$

20.2   Boltzmann Transport Equation with Collision

Existence of a collision process implies interaction between the molecules. As the molecules in the element $d^3r\,d^3v$ at a time t travel to $d^3r'\,d^3v'$ at a time $t' = t + dt$, some molecules from outside $d^3r\,d^3v$ are scattered into $d^3r\,d^3v$ and some molecules from $d^3r\,d^3v$ are scattered outside this volume. This difference is denoted by $D_c f$. Then, obviously,

$$f(\vec{r} + \dot{\vec{r}}\,dt, \vec{v} + \frac{F}{m}dt; t + dt) = f(\vec{r}, \vec{v}; t) + D_c f\,dt \tag{20.9}$$

Thus,

$$Df = D_c f = \frac{\partial f}{\partial t} + \vec{v} \cdot \nabla f + \frac{\vec{F}}{m} \cdot \nabla_{\vec{v}} f \tag{20.10}$$

Determination of the operator D_c can be carried out at different levels of complication.

We will be using the ideas of classical scattering theory which are discussed elsewhere.[1] We need to know the ideas of scattering because we have to calculate how many molecules enter in $d^3r\,d^3v$ due to collision and how many molecules leave $d^3r\,d^3v$. Only dominant mechanism is that of binary collisions between the molecules in random motion.

Consider two molecules of masses m_1, velocity v_1 and m_2 with velocity v_2 collide with each other. Then the following are the facts.

(1) total momentum $\vec{P} = m_1\vec{v_1} + m_2\vec{v_2} = constant$

(2) Center of mass has coordinate

$$\vec{R}_{cm} = \frac{m_1\vec{r}_1 + m_2\vec{r}_2}{m_1 + m_2} \tag{20.11}$$

Center of mass has a velocity $\vec{V}_{cm}$ as

$$\vec{V}_{cm} = \frac{m_1\vec{v}_1 + m_2\vec{v}_2}{m_1 + m_2} \tag{20.12}$$

(3) Two body problem can be reduced to a one body problem of effective mass $\mu = \dfrac{m_1 m_2}{m_1 + m_2}$ for the relative motion plus a motion of the center of mass (CM)

(4)
$$\text{Total kinetic energy} = \frac{1}{2}m_1 v_1^2 + \frac{1}{2}m_2 v_2^2 = \frac{1}{2}(m_1 + m_2)V_{cm}^2 + \frac{1}{2}\mu v^2 \tag{20.13}$$

where $\vec{v} = $ relative velocity $= \vec{v}_1 - \vec{v}_2$. Relative distance between the two particles is $\vec{r_1} - \vec{r_2} = \vec{r}$.

(5) In terms of relative velocity,

$$\left.\begin{array}{ccc} \vec{v_1} & = & \vec{V}_{cm} + \dfrac{\mu}{m_1}\vec{v} \\[2mm] \vec{v_2} & = & \vec{V}_{cm} - \dfrac{\mu}{m_2}\vec{v} \end{array}\right\} \tag{20.14}$$

---

[1]See Classical Mechanics by P.V. Panat (Narosa 2005) ch.14

(6) We assume that the scattering is elastic. Let the molecular velocities of particles $1(\vec{v}_1)$ and $2(\vec{v}_2)$ change to $\vec{v}_1'$ and $\vec{v}_2'$ after the collision. Then, because of elastic nature of collision, the magnitudes of two relative velocities $\vec{v} = \vec{v}_1 - \vec{v}_2$ and $\vec{v}' = \vec{v}_1' - \vec{v}_2'$ will not change i.e.

$$|\vec{v}| = |\vec{v}_1 - \vec{v}_2| = |\vec{v}'| = |\vec{v}_1' - \vec{v}_2'| \tag{20.15}$$

(7) Scattering process is described by a quantity b the impact parameter. We can consider the scattering in a center of mass frame. In the process of scattering, the direction of relative velocities change through an angle θ, called scattering angle and is $\cos\theta = \dfrac{\vec{v} \cdot \vec{v}'}{v^2}$

From the number of particles of initial flux, a fraction $d\sigma$ is scattered in a solid angle $d\Omega$. Quantity $d\sigma/d\Omega$ is known as differential scattering cross-section and is given by

$$\frac{d\sigma}{d\Omega} = \frac{b}{\sin\theta}\left|\frac{db}{d\theta}\right| \tag{20.16}$$

Total scattering cross -section is $\displaystyle\int d\sigma = \sigma_{Tot}$.

This differential scattering cross section is clearly a function of initial and final velocities and we will write, from henceforth, as, $\sigma(\vec{v}_1, \vec{v}_2 \to \vec{v}_1', \vec{v}_2')$ as a cross section for scattering of two particles with velocity state $\vec{v}_1$ and $\vec{v}_2$ into a velocity state $\vec{v_1}'$ and $\vec{v_2}'$

(8) Cross section has following properties.

  (a) Because equations of motion at microscopic scale are time reversal invariant (i.e. when $t \to -t$)

$$\sigma(\vec{v}_1, \vec{v}_2 \to \vec{v_1}', \vec{v_2}') = \sigma(-\vec{v_1}', -\vec{v_2}' \to -\vec{v}_1, -\vec{v}_2)$$

  (b) Parity invariance gives

$$\sigma(\vec{v}_1, \vec{v}_2 \to \vec{v_1}', \vec{v_2}') = \sigma(-\vec{v}_1, -\vec{v}_2 \to -\vec{v_1}', -\vec{v_2}')$$

  (c) Successive application of time reversal and space reflection gives,

$$\sigma(\vec{v}_1, \vec{v}_2 \to \vec{v_1}', \vec{v_2}') = \sigma(\vec{v_1}', \vec{v_2}' \to \vec{v}_1, \vec{v}_2)$$

We now calculate $D_c f$, the collision term. In its calculation a dilute gas assumption is made so that only the binary collisions are considered. Another assumption regarding $f(\vec{r}, \vec{v}; t)$ is made. The distribution function $f(\vec{r}, \vec{v}; t)$ is assumed to be unchanged during the time of collision, and $f(\vec{r}, \vec{v}; t)$ is taken to be same over a spatial distances of the order of range R of the molecular interaction. Moreover, it is assumed that the initial velocities $\vec{v}_1$ and $\vec{v}_2$ of the two molecules have no correlation between them. This is possible when $l >> R$. This assumption is called assumption of 'molecular chaos' and was introduced by Boltzmann.

We calculate the loss terms namely, we want to find out, because of collision, how many molecules of velocity $\vec{v}_1$ are thrown out of d^3r around $\vec{r}$ in time dt. We call such

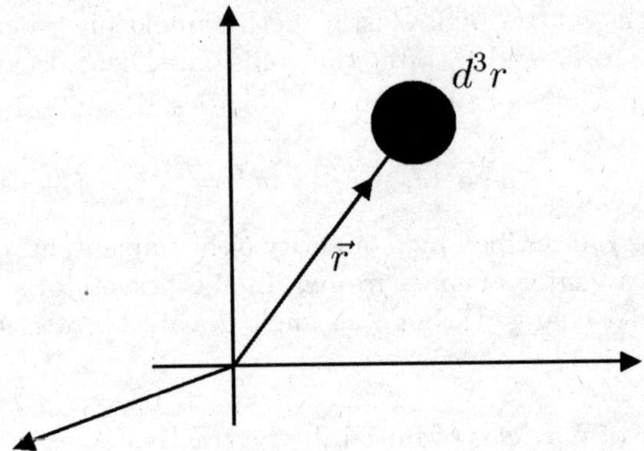

Figure 20.1: Infinitesimal volume element d^3r at the position $\vec{r}$.

molecule as A_1 type. This process will take place because of collision with molecules of velocity $\vec{v_2}$ (call them as A_2 type) in in d^3r. Process of collision leads to $\vec{v_1}$ and $\vec{v_2}$ changing to $\vec{v_1'}$ and $\vec{v_2'}$ with a probability

$$\sigma(\vec{v}_1, \vec{v}_2 \to \vec{v}\,'_1, \vec{v}\,'_2)d^3v_1' d^3v_2'$$

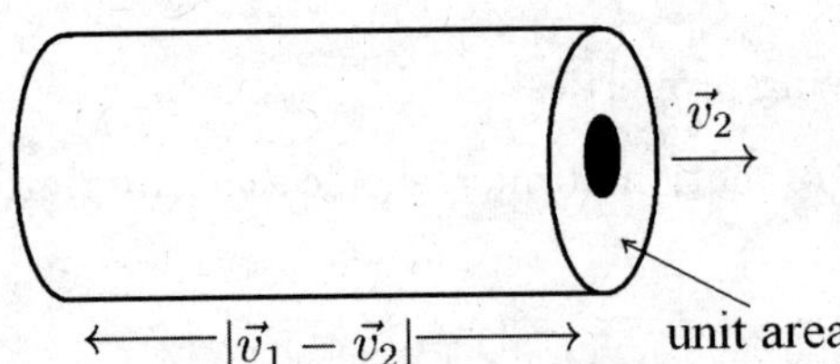

Figure 20.2: Collision cylinder.

Fix now a single molecule of A_2 type. Then, all the A_1 type molecules in the cylinder of volume $|\vec{v_1} - \vec{v_2}|dt$ will be incident on the single molecule of type of A_2 in d^3r. Number of A_2 type molecules in d^3r are $f(\vec{r}, \vec{v_2}; t)d^3r d^3v_2$. Number of A_1 type of molecules in the cylinder are $f(\vec{r}, \vec{v_1}; t)|\vec{v_1} - \vec{v_2}|dt$. Taking into account scattering probability, we see that the number of particles of A_1 type scattered out is

$$f(\vec{r}, \vec{v_2}; t)d^3r d^3v_2 \times f(\vec{r}, \vec{v_1}; t)|\vec{v_1} - \vec{v_2}|dt \times \sigma(\vec{v_1}, \vec{v_2}; \vec{v_1'}, \vec{v_2'})d^3v_1' d^3v_2' \qquad (20.17)$$

Total number of A_1 type particles thrown out by particles of all possible velocities $\vec{v_2}$ in to $\vec{v_1'}$ and $\vec{v_2'}$ is the loss term. Denoting the loss term by $D_c^{(-)}f$, we see

$$D_c^{(-)} f(\vec{r}, \vec{v_1}\,; t)dt \;=\; \int_{v_2} \int_{\vec{v_1'}} \int_{\vec{v_2'}} dt\, |\vec{v_1} - \vec{v_2}| f(\vec{r}, \vec{v_1}\,; t)$$

$$\times \; f(\vec{r}, \vec{v_2}\,; t)d^3v_2 \; \sigma(\vec{v_1}, \vec{v_2} \to \vec{v_1}', \vec{v_2}') \; d^3v_1' \; d^3v_2' \qquad (20.18)$$

We can argue out the calculation of a gain term in exactly similar manner. However, we can think of gain term as identically due to inverse scattering. Here scattering takes place between molecules of velocities $\vec{v_1}'$ and $\vec{v_2}'$, so that

$$
\begin{aligned}
f(\vec{r}, \vec{v}_1; t) &\rightarrow f(\vec{r}, \vec{v_1}'; t) \\
f(\vec{r}, \vec{v}_2; t) &\rightarrow f(\vec{r}, \vec{v_2}'; t) \\
\sigma(\vec{v_1}, \vec{v_2} \rightarrow \vec{v_1}', \vec{v_2}') &\rightarrow \sigma(\vec{v_1}', \vec{v_2}' \rightarrow \vec{v_1}, \vec{v_2})
\end{aligned}
$$

Also by our collision considerations,

$$
|\vec{v}_1 - \vec{v}_2| = |\vec{v_1}' - \vec{v_2}'|
$$

and

$$
d^3 v_1 d^3 v_2 = d^3 v_1' d^3 v_2'
$$

We then immediately write

$$
D_c^{(+)} f(\vec{r}, \vec{v}_1 \; ; t) dt = \int_{\vec{v}_2} \int_{\vec{v_1}'} \int_{\vec{v_2}'} dt |\vec{v}_1 - \vec{v}_2| f(\vec{r}, \vec{v}\,'_1 \; ; t) d^3 v_1' \times
$$
$$
f(\vec{r}, \vec{v_2}'; t) \; d^3 v_2' \; \sigma(\vec{v_1}', \vec{v_2}' \rightarrow \vec{v_1}, \vec{v_2}) d^3 v_2 \tag{20.19}
$$

Write

$$
f(\vec{r}, \vec{v}_1; t) = f_1; \quad f(\vec{r}, \vec{v}_2; t) = f_2
$$
$$
f(\vec{r}, \vec{v_1}'; t) = f_1'; \quad f(\vec{r}, \vec{v_2}'; t) = f_2'
$$

and

$$
\vec{v}_{rel} = |\vec{v}_1 - \vec{v}_2| = |\vec{v_1}' - \vec{v_2}'|
$$

and

$$
D^c f = \left(D_c^{(+)} - D_c^{(-)} \right) f
$$

Boltzmann transport equation in its full glory is then,

$$
\frac{\partial f}{\partial t} + \vec{v}_1 \cdot \nabla f + \frac{\vec{F}}{m} \cdot \nabla_{v_1} f
$$
$$
= \int \int \int |\vec{v_{rel}}| (f_1' f_2' - f_1 f_2) \sigma(\vec{v}_1 \vec{v}_2 \rightarrow \vec{v}_1', \vec{v}_2') \times d^3 v_2 d^3 v_1' d^3 v_2' \tag{20.20}
$$

This is an integro differential equation. Interaction between the molecules is taken care of by calculation of scattering cross-section σ. The collision term is very complex and can be explicitly calculated only in simple cases such as a hard sphere interactions. This equation cannot be applied if collision are not elastic, as can happen in case of internal excitation due to collisions.

In case of metals, considerations are slightly different. Here, electrons are treated with crystal wave vector $\vec{k}$ rather than with velocity $\vec{v}$. The electron distribution taken here is a Fermi distribution. The transitions from $\vec{k}$ to $\vec{k}'$ are affected due to various processes. Let $Q(\vec{k}, \vec{k}')$ be the probability that an electron makes a transition from $\vec{k} \rightarrow \vec{k}'$. Let f_k be the probability that the electron is in the state $\left|\vec{k}\right\rangle$. It will be able to make a transition to $\left|\vec{k}'\right\rangle$ if, by Pauli's principle, it is empty. Thus total transitions from $\left|\vec{k}\right\rangle$ irrespective of a state $\left|\vec{k}'\right\rangle$ is

$$D_c^{(-)} f_k = \int f_{\vec{k}}(1 - f_{\vec{k}'}) Q(\vec{k}, \vec{k}') d^3 k'$$

By similar arguments, the gain term is

$$D_c^{(+)} f_k = \int f_{\vec{k}'}(1 - f_{\vec{k}}) Q(\vec{k}', \vec{k}) d^3 k'$$

Thus, BTE for solid is

$$\frac{\partial f_k}{\partial t} + \frac{\hbar \vec{k}}{m} \cdot \nabla f_k + \frac{\vec{F}}{m} \cdot \nabla_{\vec{k}} f_k = \int \left[f_{\vec{k}}(1 - f_{\vec{k}'}) Q(\vec{k}, \vec{k}') - f_{\vec{k}'}(1 - f_{\vec{k}}) Q(\vec{k}', \vec{k}) \right] d^3 k' \quad (20.21)$$

Here we take f_k as $f(\vec{r}, \vec{k}; t)$ and $\vec{F}$ is written as $\dfrac{\hbar \dot{\vec{k}}}{m}$. $Q(\vec{k}, \vec{k}')$ is a transition probability from electron momentum $\vec{k}$ to electron momentum $\vec{k}'$. This transition could be due to many possible processes in solid such as electron-phonon, electron-electron scattering etc. For elastic scattering $Q(\vec{k}, \vec{k}') = Q(\vec{k}', \vec{k})$, and the collision term simplifies

$$D_c f_k = \int Q(\vec{k}', \vec{k})(f_{k'} - f_k) d^3 k'$$

For metals, f_k is Fermi distribution function.

20.3   Relaxation Time (RT) Approximation

Collision term is a most complicated term. There are many approximation methods. Simplest is the relaxation time approximation.

We assume that the system has a <u>local</u> equilibrium. It is assumed that the collisions restore local equilibrium. The scattering term vanish in equilibrium. Any deviation in distribution function from equilibrium value $f^{(0)}$ should be such that system returns to it in a relaxation time exponentially. To see this, let

$$f(\vec{r}, \vec{v}; t) = f^{(0)}(\vec{r}, \vec{v}) + f^{(1)}(\vec{r}, \vec{v}; t)$$

where $f^{(1)} << f^{(0)}$. Obviously, the relaxation time τ will be different for different velocities of the molecules. We write, in RT approximation,

$$\left. \frac{\partial f}{\partial t} \right)_{coll} = D^{(c)} f = -\frac{f(\vec{r}, \vec{v}; t) - f^{(0)}(\vec{r}, \vec{v}; t)}{\tau_v} = -\frac{f^{(1)}}{\tau_v} \quad (20.22)$$

Consider a case of absence of external fields. Then

$$\frac{\partial f(\vec{r}, \vec{v}; t)}{\partial t} = -\frac{f^{(1)}(\vec{r}, \vec{v}; t)}{\tau_{\vec{v}}} \quad (20.23)$$

But $f^{(0)}(\vec{r}, \vec{v})$ does not have explicit dependence on t since it is an equilibrium distribution. A question may come to your mind that the equilibrium distribution should be velocity dependent only. It should not depend upon space. This is true if there is no external potential. In the presence of external potential, e.g. gravity, one has $f^{(0)}(\vec{r}, \vec{v})$ as an equilibrium distribution. In presence of gravity, the equilibrium distribution is a well known barometric distribution.

Thus, $\partial f/\partial t = \partial f^{(1)}/\partial t$. Distribution function is then given as

$$f(\vec{r}, \vec{v}; t) = f^{(0)}(\vec{r}, \vec{v}) + f^{(1)}(0)e^{-t/\tau_v}$$

Thus the fluctuation dies out exponentially. Then, BTE in RT approximation is,

$$\frac{\partial f}{\partial t} + \vec{v}.\nabla f + \frac{\vec{F}}{m}.\nabla_{\vec{v}} f = -\frac{f - f^{(0)}}{\tau_v} \tag{20.24}$$

The job now is to find $f^{(1)}$ and evaluate the transport coefficient by taking averages of the requisite physical quantities with $f^{(1)}$.

20.4 Calculation of Relaxation Time

We argued that relaxation time is given as $\tau = \dfrac{1}{n\,\sigma\,\overline{v}}$. This expression is obtained more like on a dimensional argument. Exact expression for τ is given by

$$\tau^{-1}(\vec{v}_1) = \int_{\vec{v}} \int_{\Omega'} |\vec{v}_{rel}|\, \sigma(|\vec{v}_{rel} \to \vec{v}\,'_{rel}|) f(\vec{v}) d^3v\, d\Omega' \tag{20.25}$$

where, as discussed,

$$|\vec{v}_{rel}| = |\vec{v} - \vec{v_1}| = |\vec{v}\,'_{rel}| = |\vec{v}\,' - \vec{v_1}'|$$

$$\sigma\left(|\vec{v}_{rel}| \to |\vec{v}\,'_{rel}|\right) = \sigma\left(\vec{v}, \vec{v_1} \to \vec{v}\,', \vec{v_1}'\right)$$

is a differential scattering cross section; $d\Omega'$ is a solid angle $\sin\theta' d\theta' d\phi'$. Here, the direction of $\vec{v}\,'_{rel}$ w.r.t. $\vec{v}_{rel}$ is (θ', ϕ'). $f(\vec{v})d^3v$ is number of molecules/c.c. with velocity between $\vec{v}$ and $\vec{v} + d\vec{v}$. The result of equation (25) is easy to understand. Mean number of molecules/c.c. of velocity between $\vec{v}$ and $\vec{v} + d\vec{v}$ are $f(\vec{v})d^3v$. Relative flux of molecule with velocity $\vec{v}_1$ towards a molecule of velocity $\vec{v}$ is $|\vec{v} - \vec{v}_1|f(\vec{v}_1)d^3v_1$. The number of scattering events when $(\vec{v} - \vec{v}_1)$ goes to $(\vec{v}\,' - \vec{v}\,'_1)$ are

$$|\vec{v} - \vec{v}_1|f(\vec{v}_1)d^3v_1 \cdot \sigma(\vec{v}, \vec{v}_1 \to \vec{v}\,', \vec{v}\,'_1)d\Omega'$$

We want total scattering from all molecules of velocity $\vec{v}$ in d^3r. It is

$$\int_{\vec{v}} \int_{\Omega'} f(\vec{v}_1)|\vec{v} - \vec{v}_1|d^3v_1\sigma(\vec{v}, \vec{v}_1 \to \vec{v}\,', \vec{v}\,'_1)f(\vec{v})d^3v\, d^3r\, d\Omega' \tag{20.26}$$

Scattering rate, which is inverse of the lifetime is obtained by dividing equation (26) by number of molecules of velocity between $\vec{v}_1$ and $\vec{v}_1 + d\vec{v}_1$ in d^3r. This gives result (25)

To evaluate transport coefficients in RTA , one has to find $\tau^{-1}(\vec{v})$ from equation (25). This is not so easy in many cases. Thus the problem of difficulty of solving BTE, is shifted to finding $\tau^{-1}(\vec{v})$.

20.5   Flux Transported Across a Surface

In a transport process, there is a flow of physical quantity giving rise to currents. The distribution function $f(\vec{r}, \vec{v}; t)$ is not an equilibrium distribution i.e. Maxwell distribution. Therefore $\langle \vec{v} \rangle \neq 0$. This non zero net velocity is called hydrodynamic velocity. We devote it by $\vec{u}$. Velocity $\vec{v}$ of a molecule w.r.t. $\vec{u}$ is called 'peculiar velocity' $\vec{U}$ such that

$$\vec{U} = \vec{v} - \vec{u}$$

and

$$\left\langle \vec{U} \right\rangle = \langle \vec{v} \rangle - \vec{u} = 0$$

Mean total density $n(\vec{r}, t)$ at $\vec{r}$ at time t is given by

$$n(\vec{r}, t) = \int f(\vec{r}, \vec{v}; t) d^3 v$$

Let $\chi(\vec{r}, \vec{v}; t)$ be a physical quantity of the molecule which is transported. Then average value of χ between $\vec{r}$ and $\vec{r} + d\vec{r}$ at time t is

$$\langle \chi(\vec{r}, t) \rangle = \frac{\int \chi(\vec{r}, \vec{v}; t) f(\vec{r}, \vec{v}; t) d^3 v}{\int f(\vec{r}, \vec{v}; t) d^3 v}. \tag{20.27}$$

The quantity χ is energy in case of thermal conductivity, is a component of momentum in case of viscosity and so on.

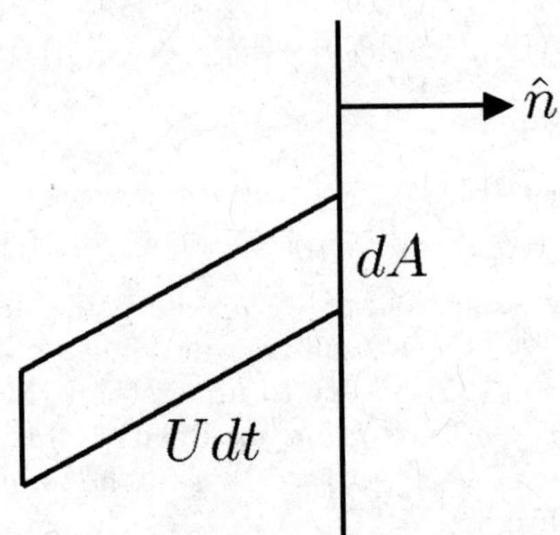

Figure 20.3: Collision cylinder of peculiar velocity $\vec{U}$.

We now calculate net amount of χ per unit time per unit area. Let $\hat{n}$ denote normal to the area dA. Consider a cylinder of volume $dA(\hat{n} \cdot \vec{U})dt$. Number of molecules of velocity $\vec{v}$ in the cylinder is $f(\vec{r}, \vec{v}; t) \, d^3 v \, dA \, (\hat{n}.\vec{U}) \, dt$. Each of the molecule carries a physical quantity χ with it.

Thus, amount of χ transported from left to right per unit area across dA in unit time is,

$$\int_{\hat{n}.\vec{U} > 0} f(\vec{r}, \vec{v}; t) \chi(\vec{r}, \vec{v}; t) \left| \hat{n}.\vec{U} \right| d^3 v.$$

It is clearly a flux of χ passing across the base of the cylinder from left to right. Similar flux of χ passing from right to left is

$$\int_{\hat{n}.\vec{U}<0} f(\vec{r},\vec{v};t)\chi(\vec{r},\vec{v};t)\left|\hat{n}.\vec{U}\right| d^3v.$$

Clearly the net flux $\mathcal{F}_n(\vec{r},t)$ in the direction $\hat{n}$ is

$$\mathcal{F}_n(\vec{r},t) = \int f(\vec{r},\vec{v};t)\chi(\vec{r},\vec{v};t)\left(\hat{n}\cdot\vec{U}\right) d^3v$$

Thus the vector flux is

$$\begin{aligned}\vec{\mathcal{F}} &= \int f(\vec{r},\vec{v};t)\chi(\vec{r},\vec{v};t)\vec{U}\; d^3v \\ &\equiv n\left\langle\chi\vec{U}\right\rangle.\end{aligned} \tag{20.28}$$

This is the principle result of this section. Identification of χ for required physical process will give various coefficients.

Illustrative Problem

Problem 1: Find the momentum flux of a gas moving along x direction that need to be averaged to get viscosity.

Solution: Suppose the flow is in x direction. Then momentum transported across the field surface is mv_x. The value of $\hat{n}.\vec{U}$ is U_z where $\hat{n}$ is a z direction showing the orientation of the surface. Then $\chi = mv_x$. The momentum flux or stress is

$$P_{zx} = mn\left\langle u_x U_z\right\rangle = mn\left\langle v_x v_z\right\rangle \tag{20.29}$$

Problem 2: Express heat flux in terms of velocities and density of a gas to find thermal conductivity.

Solution: In thermal conduction there is a energy transport and not a mass transport. Therefore the peculiar velocity $\vec{u}=0$. From equation (27) the heat flux is

$$Q_x = n\left\langle v_x \frac{1}{2}mv^2\right\rangle = \frac{1}{2}mn\left\langle v_x v^2\right\rangle \tag{20.30}$$

20.6   Calculation of Thermal Conductivity

We show here a typical method of RTA to find a transport coefficient κ.

1 We make the calculations in a steady state.

2 There is no external force leading to $\vec{F}\cdot\nabla_v f/m = 0$.

3 There is a temperature variation with position. Thus $T = T(\vec{r})$.

BTE under above conditions is

$$\vec{v} \cdot \nabla f = -\frac{f - f^{(0)}}{\tau} \tag{20.31}$$

As T is a local temperature, the equilibrium distribution g depends on a position through $T(x)$. Also, locally the distribution of velocities is Maxwellian. We assume that <u>locally</u> f and $f^{(0)} = g(\vec{r}, \vec{v})$, the Maxwell distribution do not differ significantly. Let

$$f(\vec{r}, \vec{v}; t) = f^{(0)}(\vec{r}, \vec{v}) + f^{(1)}(\vec{r}, \vec{v}; t) \equiv g + f^{(1)} \tag{20.32}$$

Then, from equation (31), it is clear that for a flow along x axis,

$$\begin{aligned}
f^{(1)}(\vec{r}, \vec{v}; t) &= -\tau v_x \frac{\partial g}{\partial x} \\
&= -\tau v_x \frac{\partial g}{\partial T} \cdot \frac{\partial T}{\partial x}
\end{aligned}$$

Then,

$$Q_x = -\kappa \frac{\partial T}{\partial x} = \left[-\frac{1}{2} mn\tau \int d^3v \, v_x^2 \, v^2 \frac{\partial g}{\partial T} \right] \cdot \frac{\partial T}{\partial x} \tag{20.33}$$

Thus, from equation (32)

$$\kappa = \frac{1}{2} mn\tau \frac{\partial}{\partial T} \int d^3v \, v_x^2 \, v^2 \, g(v) \tag{20.34}$$

But

$$g(v) = n \left(\frac{m}{2\pi k_B T} \right)^{3/2} e^{-mv^2/2k_B T} \tag{20.35}$$

Angular integral in equation (34)is trivial. We get (see appendix 2)

$$\begin{aligned}
\kappa &= \frac{2\pi}{3} mn\tau \frac{\partial}{\partial T} \int_0^\infty v^6 \, g(v) \, dv \\
&= \frac{2\pi}{3} mn\tau \frac{\partial}{\partial T} \left\{ n \left(\frac{m}{2\pi k_B T} \right)^{3/2} \left(\frac{2k_B T}{m} \right)^{7/2} \cdot \frac{1}{2} \Gamma(7/2) \right\} \\
&= \frac{5 \, nk_B^2 T}{m} \tau \tag{20.36}
\end{aligned}$$

Here, strictly, we should find velocity dependent τ of equation (26). This is a formidable task requiring a knowledge of molecular interaction to calculate σ. Instead we have cheated by taking some average value of τ.

　　Following exactly similar procedure, we can calculate other transport coefficients. We will not be finding $(\partial f/\partial t)_{coll}$ by more exact method of expanding f in terms of Sonine polynomials which are the eigenfunctions of Boltzmann operator for $1/r^4$ potential[2].

---

[2]See 'The Mathematical Theory of Non-uniform Gases' by S. Chapman, T. G . Cowling, 'Cambridge University Press' (1991)

Short Questions

1. Why do we need to study a BTE when we already have a more intuitive method based on mean free path?

2. Show that, in an elastic scattering process, the magnitude of relative velocity remains same.

3. Explain why $f\left(\vec{r} + \vec{v}dt, \vec{v} + \dfrac{\vec{F}}{m}dt; t + dt\right)$ is same as $f(\vec{r}, \vec{v}; t)$ in a collisionless BTE?

4. State the symmetry properties of σ

5. While calculating $D^{(c)}f$; what approximations are made?

6. Explain what is RTA? Show that $\langle v_x \rangle$ is zero with equilibrium distribution but is non zero when $f^{(1)}(\vec{r}, \vec{v}; t)$ is used for its evaluation.

7. What are the fluxes of physical quantities transported for viscosity, thermal conductivity and electrical conductivity? Why?

Problems

Problem 1: Show by using RTA; the electrical conductivity can be written as

$$\sigma_{el} = \frac{8\,n\,e^2}{3m\sqrt{\pi}} \int_0^\infty d\xi \; \xi^4 \; \tau(\xi\bar{v})e^{-\xi^2}$$

where we write $\bar{v} = \sqrt{\dfrac{2k_B T}{m}}$ as the most probable speed in equilibrium. (Hint: Here $\dfrac{\partial f}{\partial t} = \vec{v} \cdot \nabla f = 0$ and

$$\frac{F}{m}\frac{\partial f}{\partial v_z} = \frac{eE}{m} \cdot \frac{\partial f}{\partial \mathcal{E}} \cdot \frac{\partial \mathcal{E}}{\partial v_z} = eEv_z\frac{\partial f}{\partial \mathcal{E}}$$

Here $\mathcal{E} = \dfrac{1}{2}mv^2$ and E is electrical field in z direction. BTE in RTA is,

$$eEv_z\frac{\partial f}{\partial \mathcal{E}} = -\frac{-f^{(1)}}{\tau}$$

Then,

$$j_z = e \int v_z f(\vec{r}, \vec{v}; t)d^3v \cong -e^2 E \int \tau v_z^2 \frac{\partial g}{\partial \mathcal{E}}d^3v.$$

But $\dfrac{\partial g}{\partial \mathcal{E}} = -\beta g$, $\displaystyle\int_0^\pi v^2 \cos^2\theta \, d\theta = 2v^2/3$; $\sigma_{el} = \dfrac{j_z}{E}$, $\dfrac{v}{\bar{v}} = \xi$.

Then

$$\sigma_{el} = \frac{4\pi \, ne^2 \cdot \overline{v}^2}{3k_B T \pi^{3/2}} \int_0^\infty \tau \, (\xi \overline{v}) \xi^4 e^{-\xi^2} \, d\xi.)$$

Problem 2: Show by using RTA and solving BTE, the coefficient of viscosity η is given by

$$\eta = \frac{16 n k_B T}{15\sqrt{\pi}} \int_0^\infty e^{-\xi^2} \xi^6 \tau(\overline{v}\xi) d\xi$$

(Hint: BTE is $v_z \dfrac{\partial f}{\partial z} = -f^{(1)}\tau \approx v_z \dfrac{\partial g}{\partial U_x} \cdot \dfrac{\partial U_x}{\partial z}.$
Then using the illustrative problem (1),

$$P_{zx} = mn \, \langle v_x U_z \rangle .$$

Use $\dfrac{\partial g}{\partial v_x} = -\beta m v_x g;$

$$\int_0^{2\pi} \cos^2 \phi \, d\phi = \pi; \quad \int_0^\pi \sin^3 \theta \cos^2 \theta d\theta = \frac{4}{15}.$$

Evaluate $\displaystyle\int v_x^2 v_z^2 \, v^2 \, g(v) \, dv \sin\theta \, d\theta \, d\phi$ and get the answer.)

Problem 3: Using the BTE and RTA find diffusion coefficient D and compare it with $D = \dfrac{1}{3}\overline{v}l.$

(Hint:

$$\mathcal{F}_n = J_n = \int d^3 v f(\vec{r}, \vec{v}; t)\hat{n} \cdot \vec{U}\chi.$$

Take $\chi = 1$. The transport equation is

$$v_z \frac{\partial g}{\partial z} = -\frac{f^{(1)}}{\tau}.$$

$$J_z = -\int \tau v_z^2 \frac{\partial g}{\partial z} d^3 v = -\left(\int \tau v_z^2 \frac{\partial g}{\partial n} d^3 v \right) \frac{\partial n}{\partial z}$$

This gives

$$D = \int \tau v_z^2 \left(\frac{m}{2\pi k_B T} \right)^{3/2} e^{-mv^2/2k_B T} 2\pi \, v^2 \, dv \sin\theta d\theta$$

Carrying out the integration, and assuming τ to be independent of velocity we get $D = \dfrac{\tau k_B T}{2m}$. With $l = \overline{v}\tau$, $D = \dfrac{8 k_B T}{3\pi m}\tau$. This value is about half of what we get by the hand waving arguments.)

Problem 4: Show that the number of collisions happening per second per unit volume at $\vec{r}$ in a gas as

$$\int d^3v_1 \int d^3v_2 \sigma_{tot} |\vec{v}_1 - \vec{v}_2| f(\vec{r}, \vec{v}_1; t) f(\vec{r}, \vec{v}_2; t)$$

where σ_{tot} is a total cross section.

Problem 5: If $\chi(\vec{r}, \vec{v})$ is a conserved quantity then show that

$$\int d^3v \chi(\vec{r}, \vec{v}) \left[\frac{\partial f(\vec{r}, \vec{v}; t)}{\partial t} \right]_{coll} = \int d^3v \chi(\vec{r}, \vec{v}) D^{(c)} f = 0$$

(Hint: If $\chi_1 + \chi_2 = \chi_1' + \chi_2'$ in collision process,

$$\int d^3v \chi D^{(c)} f = \int d^3v_1 \int d^3v_2 \int \sigma_{tot} |\vec{v}_1 - \vec{v}_2| \chi_1 (f_1' f_2' - f_1 f_2)$$

By interchanging $\vec{v}_1 \leftrightarrows \vec{v}_2$; $\vec{v}_1 \leftrightarrows \vec{v}\,'_1$; $\vec{v}_2 \leftrightarrows \vec{v}\,'_2$, $\vec{v}_1 \leftrightarrows \vec{v}\,'_2$ and $\vec{v}_2 \leftrightarrows \vec{v}\,'_1$
and using symmetry property of σ we get

$$\int \chi D^{(c)} f d^3v$$
$$= \frac{1}{4} \int d^3v_1 \int d^3v_2 \sigma_{tot} |\vec{v}_1 - \vec{v}_2| (f_1' f_2' - f_1 f_2)(\chi_1 + \chi_2 - \chi_1' - \chi_2')$$
$$= (0)$$

Chapter 21

Brownian Motion

A topic of Brownian motion is historically and also physics wise a very important topic. When thermodynamics was in its hey days, the molecular ideas, kinetic theory and a microscopic view of the interaction of heat with matter were not accepted and even ridiculed. Stalwarts like Lord Kelvin, himself an eminent thermodynamist, and Oswal were the major critics. Their main objection was, if we accept $\frac{1}{3}mn\overline{v^2} = P$, how can we measure $\overline{v^2}$? Or, $\frac{1}{2}m\overline{v^2} = \frac{k_B T}{2}$, what is the value of k_B ? Or of $\overline{v^2}$? In effect, they were asking the estimate of Avogadro number.

The total acceptability to molecular theory of gases (in effect Statistical Mechanics) came after Einstein's famous explanation of Brownian motion based on statistical mechanical ideas. This paper of Einstein (Ann.Physik $\underline{17}$, 549 (1905)) is cited more than his relativity papers in those days. Ideas and equation of Einstein were experimentally verified by Perrin (Comp. Rend. Paris $\underline{146}$, 976 (1908)) for which he got Nobel Prize in 1926. Actually, Perrin also determined Avogadro's number with an accuracy of about 15%.

In 1827, Robert Brown a British botanist and a close friend of Darwin discovered that the pollen grains floating in a liquid were undergoing incessant random motion. These observations were made by the high magnification microscope available then. This was puzzling to Brown. He first thought that, perhaps, he had discovered a clue of life. To prove his ideas, he carried out the same experiment with suspension of powders which were non living. He saw the same motion. Many explanations such as internal currents, uneven evaporation, surface tension force etc. were tried but failed. Einstein conclusively suggested that fluctuations in the molecular collisions with the Brownian particles such as the pollen grains are responsible for the motion. He also gave experimentally verifiable relationship between $< x^2 >$, the mean square displacement of the Brownian particle (B.P.) and t the time of motion. This relationship was well verified by Perrin in his famous experiments. This was the earlier macroscopic manifestation of microscopic forces in the thermal phenomena. It is interesting to note that it is easier to measure $< x^2 >$ than the velocity of the B.P. For the velocity measurement, we have to measure $[x(t + \tau) - x(t)]/\tau]$ in a small time interval τ. The value of $x(t)$ changes very fast. It is mathematically described that the path of the

B.P. is such that it is continuous every where with derivative no where.

Probably Brown might have seen the motion as shown in figure 1.

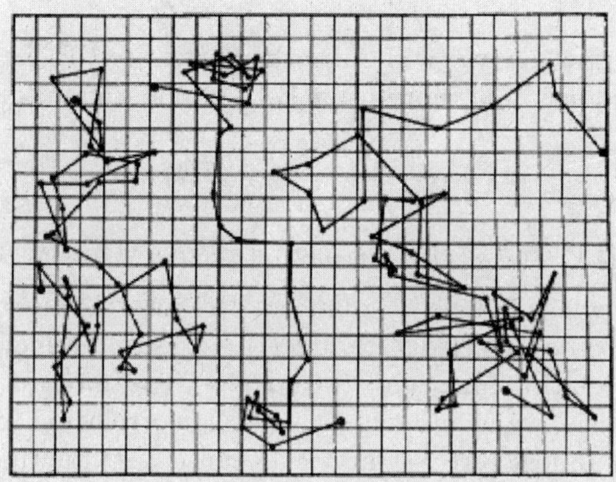

Figure 21.1: Motion of three Brownian particles as observed by Perrin.

21.1 Brownian Motion

A Brownian motion is a motion of a floating particle of mass M floating in a liquid of its molecular mass m such that $m << M$. However the mass M is not so large that its thermal motion is unobservable. It is to be noted that the liquid in which the B.P. is floating has viscosity. The B.P. and the liquid are in equilibrium at a temperature T.

Because of a temperature T of the liquid, its molecules are in incessant random motion. We concentrate a motion of B.P. floating in it. The liquid molecules randomly hit the B.P. from all sides. This implies that $< \vec{r} >$ of the B.P. will be zero. However because of random discrete hits, the B.P. also affects the local equilibrium. This gives rise to some nonzero mean square displacement $< \vec{r}^{\,2} >$ proportional to t. This means the B.P. is subjected to a random force $K(t)$. It would be naive to conclude that $\overline{K(t)} = \langle K(t) \rangle = 0$. Instead we would expect the force to be such that, it would give a small damped motion due to the viscosity of the liquid. Moreover, the force will have another component such that its time average over a long time is zero. Simplest frictional force on the body in liquid is a viscous force, proportional to its velocity. A well known relation for the frictional force $\vec{F}$ is that due to Stokes for a sphere moving in a liquid of coefficient of viscosity η

$$\vec{F} = 6\,\pi\,\eta\,a\,\vec{v}$$

where a is a radius of the body moving with velocity $\vec{v}$ in the liquid of coefficient of viscosity η. We assume that the random force acting on the B.P. due to surrounding liquid is,

$$K(t) = -\frac{v(t)}{\mu} + F(t) \tag{21.1}$$

where μ is called mobility. Then keeping in mind the above discussion, and taking the time average,

$$\overline{K(t)} = -\frac{\overline{v(t)}}{\mu} + \overline{F(t)} = -\frac{\overline{v(t)}}{\mu} \qquad (21.2)$$

with $\overline{F(t)} = 0$. From now on we will take the mass of B.P. as m rather than M for convenience. Then the equation of motion for B.P. in one dimension is,

$$m\ddot{x} = m\dot{v} = -\frac{v(t)}{\mu} + F(t) \qquad (21.3)$$

where m is the mass of B.P. and $v(t)$ is its instantaneous velocity.

Equation(3) is known as Langevin equation. Because $F(t)$ is random, this is a stochastic differential equation. The knowledge that $\overline{F(t)} = 0$ is not sufficient. We must know the variance of $F(t)$ or a correlation function $< F(t)F(t+\tau) >$ in terms of experimental parameters. On the physical basis, we would expect value of correlation function to be significant for small values of τ and almost zero for large τ because $F(t)$ fluctuates very fast. At $\tau = 0$, this correlation function is expected to have a maximum value and should decay at least exponentially for $\tau \gg 0$. The Langevin equation can be solved as,

$$\dot{v} = -\frac{v}{m\mu} + \frac{F(t)}{m}$$

$$\text{Thus,} \qquad x\ddot{x} = -\frac{x\dot{x}}{m\mu} + \frac{xF(t)}{m}$$

$$\text{Or,} \qquad \frac{d}{dt}(x\dot{x}) - \dot{x}^2 = -\frac{x\dot{x}}{m\mu} + \frac{xF(t)}{m}$$

$$(21.4)$$

We now take an ensemble average. We then have,

$$\overline{\frac{d}{dt}(x\dot{x})} - \overline{\dot{x}^2} = -\frac{\overline{x\dot{x}}}{m\mu} + \frac{\overline{xF(t)}}{m}$$

Our claim is that $\overline{xF(t)} = 0$. This can be seen as follows. Fix x and look for $\overline{xF}$. Since x is fixed and $\overline{F} = 0$, it is zero. Repeat this procedure at all values of x. What it means is that the variables $x(t)$ and $F(t)$ are independent. Langevin equation then becomes,

$$\frac{d}{dt}(\overline{x\dot{x}}) = \frac{k_B T}{m} - \frac{\overline{x\dot{x}}}{m\mu} \qquad (21.5)$$

where we used equipartition theorem to write $\overline{\dot{x}^2} = < \dot{x}^2 > = \frac{k_B T}{m}$. This equation is a first order linear differential equation with constant coefficients. Its solution is,

$$\overline{x\dot{x}} = C \, exp \, (-t/m\mu) + \mu k_B T. \qquad (21.6)$$

The constant C can be found by applying initial condition. Let the B.P. be at the origin at $t = 0$. Then,

$$\overline{x\dot{x}} = \mu \, k_B \, T[1 - exp(-t/m\mu)] \qquad (21.7)$$

Now $x\dot{x} = \dfrac{1}{2}\dfrac{d}{dt}x^2$. Thus, further integration gives the average of x^2. Or,

$$\frac{d}{dt}\overline{x^2} = 2\mu\, k_B\, T[1 - exp(-t/m\mu)]$$

Or,

$$\overline{x^2}(t) = 2m\mu^2 k_B T\left(\frac{t}{m\mu} - 1 + \exp\left(-t/m\mu\right)\right) \tag{21.8}$$

This is a most important result because it connects experimentally observed quantities $\overline{x^2}(t)$ with the elapsed time t, mobility μ and an absolute temperature T. This enables to estimate the Boltzmann constant k_B and with it the Avogadro number $N_A k_B = R$.

We will subsequently prove an important relation due to Einstein between the diffusion coefficient D and the mobility μ as

$$D = \mu k_B T \quad\text{or,}\quad \frac{D}{m\mu} = \frac{k_B T}{m} = \overline{v^2} \tag{21.9}$$

It is customary to express equation (8) in terms of D and μ. Then,

$$\overline{x^2}(t) = 2Dm\mu\left(\frac{t}{m\mu} - 1 + \exp\left(-t/m\mu\right)\right) \tag{21.10}$$

It is clear that the slowing time is $\tau = m\mu$. If $t << \tau$, i.e. we have a motion of the B.P. just started from initial condition $t = 0$ and $x = 0$ then,

$$\begin{aligned}
\overline{x^2}(t) \;&\cong\; \frac{D}{m\mu}t^2 \qquad t << m\mu \\
&=\; \frac{k_B T}{m}t^2 = \overline{v^2}t^2 \tag{21.11}
\end{aligned}$$

This motion is called ballistic because $\overline{x^2} \propto t^2$. This is also called the free motion.

On the other hand, when $t >> \tau$, we have

$$\overline{x^2}(t) \simeq 2Dt - 2Dm\mu \qquad \tau << t \tag{21.12}$$

From equation (12), it is clear that $\sqrt{\overline{x^2}} \propto t$ and not t^2. Such a motion is called diffusive motion. Generally, we write

$$\sqrt{\overline{x^2}} \propto t^\alpha$$

Whenever $\alpha = 1$ we have a diffusive motion. Whenever, $0 < \alpha < 1$, we have a sub diffusive motion. When, $1 < \alpha < 2$ the motion is called super diffusive. For $\alpha = 2$ we have a ballistic motion.

From equation (1), it is clear that the frictional force is proportional to v, with proportionality constant as $1/\mu$. Consider a B.P. of a spherical shape of radius R floating in a liquid of viscosity η. Then by Stoke's law, $1/\mu = 6\pi\eta R$. Using equation (9), $D = k_B T/(6\pi\eta R)$. Thus,

$$\langle x^2\rangle = \overline{x^2} = \frac{k_B T}{3\pi\eta R}t \tag{21.13}$$

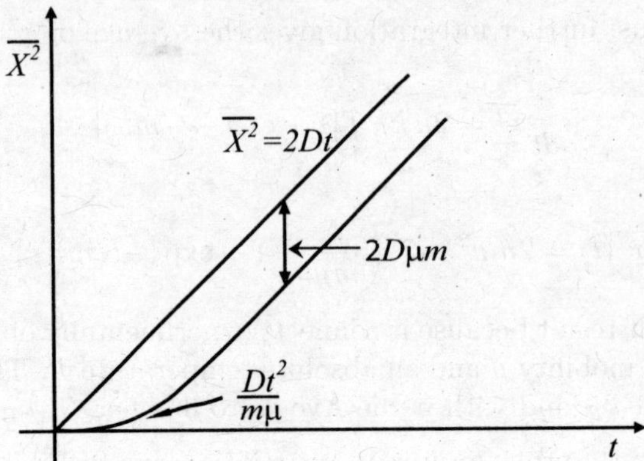

Figure 21.2: Solution of equation (10). Asymptotics is clearly seen for $t << \tau$ and $t >> \tau$.

This relation, discovered by Einstein is experimentally well verified by Perrin. He observed a single B.P. meticulously under a high power microscope noting its positions $x_1, x_2, x_3....x_N$ at successive times $t_1, t_2, t_3, ...t_N$. Then,

$$\overline{x^2} = \frac{1}{N} \sum_{k=1}^{N} x_k^2$$

same kind of observation is done on another B.P. for the same time and another value of $\overline{x^2}$ is obtained. This process is repeated many times. The average of these quantities can be found by adding all x^2 and dividing them by the number of B.P. observed. This is an ensemble average. Perrin observed that

1. $\overline{x^2}(t)$ increases linearly with t.

2. Increase in the temperature of the bath (liquid) increases the value of $\overline{x^2}$

3. Larger the size of B.P. more sluggish it is and therefore, $\overline{x^2} \propto 1/R$.

4. In the medium of more viscosity, $\overline{x^2}(t)$ of B.P. is less as compared with the medium of lower viscosity.

21.2   Mobility and Diffusion

In a fluid, if there is some lump of particles then, obviously it is not uniformly distributed. Fluid molecules then randomly hit the particles in the lump making eventually, their distribution uniform. This process is called diffusion. We call the particle of the lump as a labeled particle. Thus the labeled particles are distributed uniformly in the fluid upon attainment of equilibrium. During the process of attainment of equilibrium, there is a current of the labeled particles. This current of the labeled particles is directly proportional to their density gradient. Thus,

$$\vec{j} = -D\nabla n \tag{21.14}$$

Here $n(\vec{r}, t)$ is a local density of the labeled particles and D is the diffusion coefficient. This equation is experimentally well verified and is known as Fick's law. Equation of continuity is given by

$$\frac{\partial n}{\partial t} = -\nabla \cdot \vec{j} \tag{21.15}$$

Combining Fick's law and the equation of continuity we get the diffusion equation as

$$\frac{\partial n(\vec{r}, t)}{\partial t} = D\nabla^2 n(\vec{r}, t) \tag{21.16}$$

We consider the one dimensional motion of the diffusing particle. This is so because we need not go in to the special function solutions of this partial differential equation. Generalization of our solution to three dimensions is easy.

Using the Laplace's transform we solve equation (16) in one dimension[1]. We take initial conditions as $n(x, t = 0) = N\delta(x)$. Then, the solution of equation (16) is,

$$n(x, t) = \frac{N}{\sqrt{4\pi Dt}} \exp\left(-\frac{x^2}{4Dt}\right) \ . \tag{21.17}$$

The quantity $n(x, t)dx/N$ can be thought of as a probability of finding a labeled particle between x and $x + dx$. Because $n(x, t)$ is an even function of x (Gaussian in nature), $\overline{x} = 0$. However,

$$\begin{aligned}
\overline{(x^2)} &= \frac{1}{\sqrt{4\pi Dt}} \int_{-\infty}^{\infty} x^2 \exp\left(-\frac{x^2}{4Dt}\right) dx \\
&= 2Dt
\end{aligned} \tag{21.18}$$

Thus, comparing the equation (18) with equation (12), we see that the process of diffusion is a Brownian motion of the labeled particles. Another way of looking at it is as follows.

Consider,

$$\begin{aligned}
\frac{d}{dt}\overline{x^2} &= \frac{d}{dt}\frac{\int_{-\infty}^{\infty} x^2 n(x, t)dx}{\int_{-\infty}^{\infty} n(x, t)dx} \\
&= \int_{-\infty}^{\infty} x^2 \frac{\partial n}{\partial t} dx \ .
\end{aligned} \tag{21.19}$$

Here, $\dfrac{d}{dt}\displaystyle\int_{-\infty}^{\infty} ndx = \dfrac{d}{dt}N = 0$ since the number of the labeled particles is fixed. Using Fick's law

$$\begin{aligned}
\frac{d}{dt}\overline{x^2} &= D\int_{-\infty}^{\infty} x^2 \frac{\partial^2 n}{\partial x^2} dx \\
&= 2D
\end{aligned} \tag{21.20}$$

---

[1] See 'Transforms for Engineers' by Andrews and Shivamoggi; Bellingham, WA: SPIE Optical Engineering Press, 1999

Then,

$$\overline{x^2} = 2Dt \tag{21.21}$$

The problem of random walk can be mapped on the problem of diffusion. We want to now establish the Einstein's relation

$$D = \mu k_B T \ .$$

Consider a stationary barometric distribution.

$$n(x) = n(0) \exp\left(-\frac{mgx}{k_B T}\right) \ .$$

Here, the particles are distributed according to their height and the inhomogeneity is sustained even in equilibrium.

Then,

$$\frac{\partial n}{\partial x} = -\frac{mg}{k_B T} n \tag{21.22}$$

This equation is a result of the balance between the diffusive tendency and the gravitational attraction. The diffusion current is

$$j_{diff} = -D\frac{\partial n}{\partial x} = \frac{mgD}{k_B T} n \ . \tag{21.23}$$

Let μ be the mobility when a drift is developed in the particles. This happens because the frictional force balances the downward force due to gravity. By definition of the drift velocity, $v_d = \mu mg$. The drift current is

$$j_{drift} = nv_d = \mu nmg \ . \tag{21.24}$$

Equating the two currents j_{diff} and j_{drift} we get the famous Einstein relation

$$D = \mu k_B T \tag{21.25}$$

Strictly speaking, equating the two currents is a nontrivial statement and is hard to justify. This is so because they are a result of the same effect of the collision of liquid molecules with the labeled particle. None the less, Einstein's relation is true. Barometric distribution can be replaced by another distribution of the type $n = n_0 \exp\left(-\frac{\phi(r)}{k_B T}\right)$. Here $\phi(r)$ is an external potential acting on all the particles and is conservative leading to $\vec{F} = -\nabla\phi$. Then, the entire derivation of equation (24) straight way goes through. Therefore, $D = \mu k_B T$ is a very general relation.

21.3   Correlation Functions and Fluctuation-Dissipation Relations

In previous sections we encountered functions like $\overline{x\dot{x}} = \langle x(t)\dot{x}(t)\rangle$ and $\overline{x^2} = \langle x(t)x(t)\rangle$ or, $\overline{v^2} = \langle v(t)v(t)\rangle$. These are correlation functions. However, general functions like $\overline{x(t)x(t+\tau)}$, $\overline{F(t)F(t+\tau)}$ are also called correlation functions. These are related to the process of dissipation where the system under study looses its energy to its reservoir. In these functions, $x(t)$, $v(t)$ or $F(t)$ are random variables. These random variables are of the special type where their first moment is zero but their second and higher moments may be finite. The correlation function is to be provided in the equation of random variables.

We now discuss how we arrive at these correlation functions. Consider a Brownian particle (B.P.) in a fluid. From equation (1) we have two parts of the external force $K(t)$. The time average of one part is non zero where as, the time average of the other is zero.

Consider an original equation of motion in one dimension for the B.P. in a liquid medium.

$$m\frac{dv}{dt} = K(t) + F_{ext} . \tag{21.26}$$

Here, F_{ext} could be a gradient of a trap potential applied for all B.P. $K(t)$ is a random force. As the random force $K(t)$ will act on B.P. , the Brownian particle will get an extra energy. It will move until it gets another hit by the particles of the reservoir (here the liquid). In the process, on an average, to reach to an equilibrium will require some time say τ^*. The time τ^* is roughly of the order of mean molecular separation ($10^{-7}\ cm$) divided by the mean speed of the molecules at a temperature T ($\sim 5 \times 10^4 cm/sec$). Thus the τ^* is of the order of $10^{-12} sec$. Just as the $K(t)$ has a nonzero average part, so also to the velocity of B.P. We integrate equation (26) over a time τ such that $\tau > \tau^*$ but is a macroscopically a small time within which the B.P. gets many hits from the liquid particles . While integrating it we take an ensemble average. More over we assume that the F_{ext} does not vary violently over a position so that we can write

$$\left\langle m \int_t^{t+\tau} \frac{dv}{dt} dt \right\rangle = m \langle v(t+\tau) - v(t) \rangle \simeq F_{ext}\tau + \int_t^{t+\tau} \langle K(t') \rangle\, dt' . \tag{21.27}$$

If B.P. does not interact with the surrounding liquid then $v(t + \tau)$ is same as $V(t + \tau) = v(t) + F_{ext}\tau$. But this is not so, because the local equilibrium is disturbed by the each hit. Suppose that τ' is a time less than $\tau*$. Then, $\langle K(t') \rangle$ is not zero. This is a subtle point. If the B.P. was fixed in its position, then, the random hits of the molecules of the liquid will not move it. Then the time average of the force of random hits (viz $\overline{K}(t)$ will be zero. But the Brownian particle is not fixed in the liquid. As a result, locally, an equilibrium is disturbed. The B.P. can move a certain distance without collision. It can leave a local non uniform density as well as a back flow for times of the order of τ^*. This means velocity of the B.P. measured is not $V(t + \tau)$but something more due to the integration of the $K(t)$ term. This means that the motion of the B.P. affects its cause which is a random force $K(t)$. But after a time $\tau > \tau^*$, the interaction between the molecules will have reestablished equilibrium conditions. The velocity $v(t + \tau)$ of the B.P. is different from $V(T + \tau)$. This means the approximation $< K(t) >= 0$ is inadequate. What we need is the force experienced by the B.P. should be such that it should yield a slowly varying velocity tending to restore an equilibrium between the B.P. and the liquid.

Here we can treat the B.P. as a small system A and the liquid as its heat bath B. The two together are in equilibrium. Let A be in the state r with probability $P_r{}^0$ evaluated in equilibrium. Corresponding accessible states to B (liquid) are $\Omega(E)$. After collision let the change in velocity be $\Delta v(\tau)$. Here $\tau > \tau^*$ so that we get a new equilibrium position. In the process the heat bath B changes its energy to $E + \Delta E$, with consequent accessible states $\Omega(E + \Delta E)$. Corresponding probability is denoted as $P_r(t + \tau)$. Then,

$$\frac{P_r(t + \tau)}{P_r^{(0)}(t)} = \frac{\Omega(E + \Delta E)}{\Omega(E)} \tag{21.28}$$

With $\beta = \dfrac{\partial ln\Omega}{\partial E}$ as a temperature of the bath, then,

$$P_r(t + \tau) \approx P_r^0(1 + \beta\Delta E) \tag{21.29}$$

This is a probability of the bath in state r at a time $(t + \tau)$ and is needed to evaluate $\overline{K}$. Thus,

$$\overline{K} = \sum_r P_r(t + \tau)K = \sum_r P_r^{(0)}(1 + \beta\Delta E)K = \beta < K\Delta E >_0 \tag{21.30}$$

which in general does not vanish. Here $\sum_r P_r^{(0)}K(t)$ is an average of K over an equilibrium distribution and is in general zero. Energy change ΔE of B in time τ is,

$$\Delta E = - \int_t^{t+\tau} v(t')\ K(t')dt'. \tag{21.31}$$

However $K(t)$ changes much faster than $v(t)$. This implies,

$$\Delta E \approx -v(t) \int_t^{t'} K(t'')dt''. \tag{21.32}$$

Using equations (30), (31) and (27) we get,

$$m \langle v(t + \tau) - v(t) \rangle \approx F_{ext}(t)\tau - \beta v(t) \int_t^{t+\tau} dt' \int_t^{t'} dt'' \langle K(t')K(t'') \rangle_0 \tag{21.33}$$

Write $t'' - t' = s$. Then the limits of integration of t'' change from $t - t'$ to 0. Then,

$$m \langle v(t + \tau) - v(t) \rangle \approx F_{ext}(t)\tau - \beta v(t) \int_t^{t+\tau} dt' \int_{t-t'}^0 ds \langle K(t')K(t' + s) \rangle_0 \tag{21.34}$$

The entity $G(s) = \langle K(t)K(t + s) \rangle_0$ is called as a correlation function of the random force $K(t)$ It is related with μ (compare with equation (1)) i.e. to the dissipation. Thus the fluctuation is connected with energy dissipation. If $s > \tau^*$, there are many collisions leading to

$$\lim_{s \to \infty} G(s) \to \langle K(t) \rangle_0 \langle K(t + s) \rangle_0 = 0$$

$G(s)$ is expected to have a highest value when $s = 0$ In general, $|G(s)| \leq G(0)$. We note here the important point that the average of $G(s)$ is taken as an ensemble average. Here large number of similar systems are looked in the same slice of the time. As a result it is independent of initial time but depends upon the time difference s. This is why we have written it as $G(s)$ rather than $G(t, S)$.

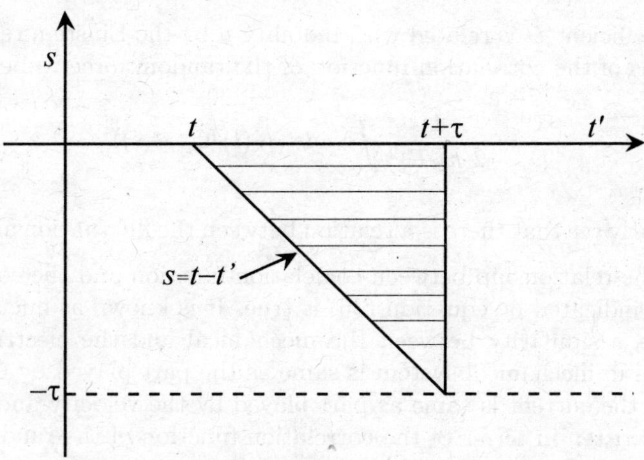

Figure 21.3: Domain of integration of the double integral.

We now consider the evaluation of equation (32). The domain of this integral is as shown in figure (3). Here the integral is of the type,

$$I = \int_t^{t+\tau} dt' \int_{t-t'}^0 ds\, G(s).$$

This can be written as (see the figure)

$$I = \int_{-\tau}^0 ds \int_{t-s}^{t+\tau} dt'\, G(s).$$

Since $G(s)$ is independent of t, the integral is,

$$I = \int_{-\tau}^0 ds\, G(s) \cdot (\tau + s).$$

The integral will have a significant contribution when s is close to 0. We approximate the integral as,

$$I \approx \tau \int_{-\tau}^0 G(s)ds = \frac{\tau}{2} \int_{-\infty}^{\infty} G(s)ds.$$

Here we extended the limits of integration to ∞ because there, $G(s)$ is zero, anyway! We also note that the correlation function $G(s)$ is symmetric around $s = 0$. Then the equation (32) becomes,

$$m\,\langle v(t+\tau) - v(t)\rangle \approx F_{ext}(t)\tau - (\beta v(t)\tau/2) \int_{-\infty}^{\infty} ds\,\langle K(t)K(t+s)\rangle_0 \qquad (21.35)$$

This is exactly the first integral of the Langevin equation. Comparing it with equation (3), we identify the mobility as,

$$\frac{1}{\mu} = \frac{1}{2k_BT} \int_{-\infty}^{\infty} ds\,\langle K(t)K(t+s)\rangle_0 \qquad (21.36)$$

As the diffusion coefficient D is related with mobility μ by the Einstein relation $D = \mu k_B T$, we write it in terms of the correlation function of the random force to be,

$$\frac{1}{D} = \frac{1}{2(k_B T)^2} \int_{-\infty}^{\infty} ds \, \langle K(t) K(t+s) \rangle_0 \tag{21.37}$$

We thus clearly see that there is a relation between the fluctuation and the dissipation.

In general the relationship between correlation function and the corresponding dissipation function as indicated by equation (36) is true. It is known as fluctuation dissipation theorem. There is a similarity between the mechanical and the electrical system. Part played by the mass in mechanical system is same as the part played by the the inductance L. Part played by the current is same as part played by the velocity and so on. Then, the resistance can be written in terms of the correlation function of the random voltages as

$$R = \frac{1}{2k_B T} \int_{-\infty}^{\infty} \langle V(0) V(s) \rangle_0 \, ds$$

The random voltage $V(t)$ is a result of interaction of electrons in the conductor with the degrees of freedom other than with the external applied voltage. It is like a random force $K(t)$.

In all these quantities we did not mention how the averages are taken and what is the probability distribution. This distribution is given by Fokker-Planck equation. We will not deal with the advanced topic. We also do not address the spectral analysis of the correlation function and a general proof of a fluctuation dissipation theorem.

Short Questions

1. What was the controversy between the thermodynamists and the ideas of kinetic theories of gases? How Einstein resolved the controversy by studying the Brownian Motion?

2. Explain the process by which a random force of hit $K(t)$ can be separated in to two parts - one with a non average force proportional to velocity and the other whose time average is zero.

3. Explain why $\overline{xF(t)} = 0$.

4. Explain a difference between the ballistic, diffusive, sub diffusive and super diffusive motion.

5. Explain the significance of

$$\overline{x^2} = \frac{k_B T}{3\pi \eta R} t \ .$$

6. State how Perrin determined Avogadro number.

7. Prove that the Fick's law and the equation of continuity lead to a diffusion equation.

8. By equating the diffusion current and the drift current show that

$$D = \mu k_B T$$

9. State the significance of the relation like that given by equation (37) between D and the correlation function in forces.

Problems

Problem 1 Perrin and Chaudesalgue[2] found the mean square displacement $\overline{x^2}$ for gamboge particles of radius 0.212μ. They made the following table

Period in sec	$\overline{x^2}$ in $10^{-12}\ m^2$		$x^2/t\ cm^2/s$	Average
30	45.0	Temperature	1.50×10^{-8}	
60	86.5	$13^{\circ}C$	1.44×10^{-8}	$1.55 \times 10^{-8} cm^2/s$
90	140.0	and	1.55×10^{-8}	
120	195.0	$\eta = 0.012\ poise$	1.62×10^{-8}	

What is Perrin's estimate of N_A from this data?

Problem 2 Millikan determined the Avogadro number by the same method as determination of the basic electric charge e. His argument was observing a Brownian motion of single oil drop of charge e in the air will give more accurate value of $(\Delta x)^2$. Explain this statement.

He observed the steady oil drop between two parallel capacitor plates and at right angles to the direction of gravity trough a high power microscope. Because the air molecules between the plates constantly hit it the Brownian motion of the drop results. He observed $(\Delta x)^2$ in time τ. Then he let the drop fall a certain distance to get its terminal velocity v_1. If μ is mobility, $mg = \mu v_1$. Then he applied some extra electric field E and the terminal velocity of the same drop was found to be v_2. He obtained a value of $N_A e$ as

$$N_A e = \frac{2RT}{E}\frac{\tau(v_1 - v_2)}{(\Delta x)^2} = \text{Faraday Number}.$$

Get this equation.

N_A thus found by Millikan is $N_A = 6.06 \times 10^{23}$
(Hint: Show that the mobility $\mu = \frac{eE}{v_1 - v_2}$ and put it in Einstein formula.)

Problem 3 Consider a Langevin equation

$$\dot{v} + \frac{v}{m\mu} = \frac{F(t)}{m}$$

---

[2]See J.Perrin "Brownian Motion and Molecular Reality", Taylor and Francis, London, 1910

(a) Show its Solution as

$$v(t) = v_0 \exp\left(-\frac{t}{m\mu}\right) + \frac{1}{m} \exp\left(-\frac{t}{m\mu}\right) \int_0^t \exp\left(\frac{t'}{m\mu}\right) F(t')\, dt'$$

$$\equiv v_0 \exp\left(-\frac{t}{m\mu}\right) + \exp\left(-\frac{t}{m\mu}\right) G(t)$$

(b) Assume that $G(t)$ is distributed as Gaussian, i.e.

$$W\left(G(t)\right) = \frac{\exp\left(-G^2(t)/\overline{G^2}\right)}{\sqrt{2\pi\overline{G^2}}}$$

Show

$$W\left(G(t)\right) = \frac{\exp\left[-\frac{m(v-v_0 e^{-t/m\mu})^2}{2k_B T(1-e^{-2t/m\mu})}\right]}{\left\{2\pi \frac{k_B T}{m}\left(1-e^{-2t/m\mu}\right)\right\}^{1/2}}$$

where v_0 is initial velocity of the labeled particle.

(Hint: (a) is obvious, for (b),

$$\overline{(v(t)-v_0 e^{-t/m\mu})^2} = e^{-2t/m\mu} \int_0^t dt'\, e^{t'/m\mu} \int_0^t e^{t''/m\mu}\, \langle F(t')F(t'')\rangle\, dt''$$

Only $t' = t''$ term will contribute to the average. This has an average $= \dfrac{k_B T}{m}\left(1-e^{-2t/m\mu}\right)$.)

Comment: As $t \gg m\mu$, $W(G) \to$ M.B. distribution indicating a transition to equilibrium. It is actually a solution of Fokker-Planck equation.

Problem 4 Thermal noise in an electrical circuit can be understood by the following arguments. Consider a high capacity capacitance connected to a resistor R in contact with a heat bath at temperature T as shown. Let the charge e be kept at a distance z from one of the plates. Show that charge induced on plate I is $-e(d-z)/d$ where as the charge induced in plate II is $-ez/d$. Suppose the charge moves with a velocity $v = \dot{z}$. Show that current passing through circuit is $I = ev/d$. Find potential difference (V) across R.

$$\text{Ans: } V = \frac{evR}{d}$$

Show that the force is $K = -eV/d = -e^2 Rv/d^2$. Then what is the mobility?

$$\text{Ans: } \mu = \frac{e^2 R}{md^2}$$

This is a frictional force and gives rise to the noise in a circuit. Note- We have calculated the result in low frequency limit. The amount depends upon temperature because $\frac{1}{2}C\overline{V^2} = \frac{k_B T}{2}$. Can you see why many accurate detectors are maintained at low temperatures?

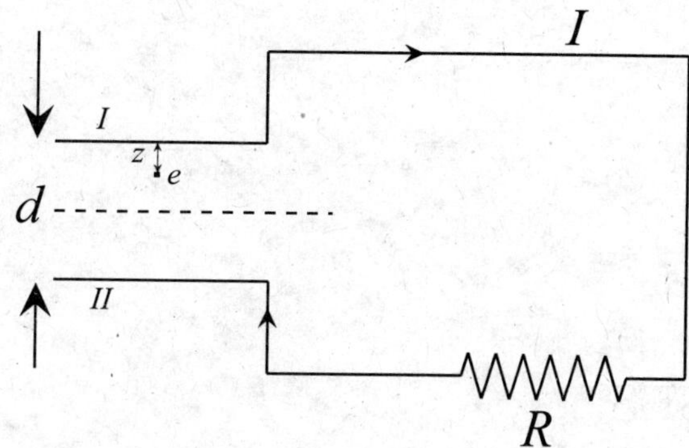

Figure 21.4: Charge e creating a thermal noise- problem 4

Chapter 22

REAL GASES

When the gas is dilute and at high enough temperature, it acts almost like an ideal gas. Molecules of an ideal gas do not interact with each other. They have a kinetic energy only. As seen earlier, equation of state of an ideal gas is

$$P\,V = N\,k_B\,T. \qquad (22.1)$$

Ideal gas does not condense but the real gas does. This is so because, the molecules of a real gas interact with each other through an interatomic potential $V(|\vec{r_i} - \vec{r_j}|)$ between i^{th} and j^{th} molecule. A schematic shape of the potential is as shown in figure(1).

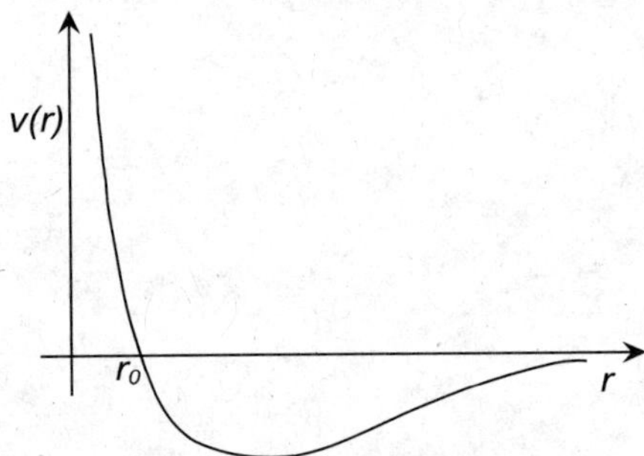

Figure 22.1: Intermolecular two body potential.

The typical characteristic of it is, (a) It has a short range, (b) There is a vehement repulsion when the two molecules come closer than a distance say r_0. (c) There is a weak attraction between the two molecules when a distance between them is larger than r_0. Such potentials are called two body potentials and have many forms. Typical famous once are (1) Morse Potential and (2) Lennard - Jones Potential. Such potentials can be obtained from first principle by using Density Functional Theory. It is assumed that the molecules of real gases interact among each other two at a time. We neglect three body and tensorial forces from our discussion. Historically many equation of states were proposed for the real gases. The van der Waals equation is prominent and simplest of them all. A general form of equation of state was suggested by Kamerling Onnes as

$$\frac{P}{nk_BT} = 1 + B_2(T)\,n + B_3(T)\,n^2 + \dots\dots \tag{22.2}$$

Here, $n = N/V$ is density, and $B_2(T)$, $B_3(T)$ etc. are known as virial coefficients and equation (2) is known as virial expansion of pressure in terms of powers of density. Virial expansion can easily be cast in to van der Waals form. Task of statistical mechanics is to evaluate $B_2(T)$, $B_3(T)$ etc in terms of microscopic parameters of the real gas. There is a systematic method of evaluation of virial coefficients due to Mayer and goes as far back as 1937. It is known as Mayer's Cluster expansion.

We will not discuss real Quantum gases because it requires tools of field theory and therefore is beyond our scope.

22.1   Mayer Cluster Expansion

We will consider a real gas of one component confined in a volume Ω and at a temperature T. To get the equation of state, we have to evaluate the partition function. Let $\vec{r}_1, \vec{p}_1; \vec{r}_2, \vec{p}_2; \vec{r}_3, \vec{p}_3; \dots\vec{r}_N, \vec{p}_N$ be phase space coordinates of N molecules of our real gas. Its N particle Hamiltonian with two body short range potential is,

$$(H\{\vec{r},\vec{p}\}) = \sum_{i=1}^{N} \frac{p_i^2}{2m} + \frac{1}{2}\sum_{i=1}^{N}\sum_{j=1}^{N} V\left(|\vec{r}_i - \vec{r}_j|\right). \tag{22.3}$$

Here $i \neq j$ is to be taken to avoid double counting. The argument of Hamiltonian for $6N$ degrees of freedom is written for brevity as $\{\vec{r},\vec{p}\}$.

Then the partition function is given by,

$$Z = \frac{1}{N!} \int\int\int \dots \int e^{-\beta\,H(\vec{r},\vec{p})}\, d^3p_1\dots d^3p_N \cdot d^3r_1\dots d^3r_N \tag{22.4}$$

The kinetic energy part of the equation (4) can be integrated immediately. We term it as Z_0 where,

$$Z_0 = \frac{1}{N!}\left(\frac{\Omega}{\lambda^3}\right)^N. \tag{22.5}$$

The potential is denoted by V. To avoid the confusion, we write Ω for volume in this chapter. Here λ is defined as a thermal de Broglie wavelength as $\lambda = \dfrac{h}{\sqrt{2\pi m k_B T}}$. From equation (5) we immediately get ideal gas pressure as,

$$P = \frac{1}{\beta}\cdot\frac{\partial \ln Z_0}{\partial \Omega} = \frac{N}{\beta\Omega}.$$

Note, while obtaining equation (5), we have counted Ω^N extra to confirm Z_0 to its ideal gas value. We then write,

$$e^{-\beta F} = Z = \frac{Z_0}{\Omega^N}\int\int\dots\int exp\left(-\frac{\beta}{2}\sum_{i,j} V\left(|\vec{r}_i - \vec{r}_j|\right)\right)\, d^3r_1\dots d^3r_N. \tag{22.6}$$

Equation (6) can be rewritten as

$$Z = Z_0 \; <e^{-\beta \, V}> . \tag{22.7}$$

Here, $(1/\Omega)$ as a uniform probability distribution and $V = \frac{1}{2}\sum'_{i,j} V\left(|\vec{r}_i - \vec{r}_j|\right)$. Prime on the summation means $i = j$ term is avoided. With F_0 and F as the free energies of ideal and real gas respectively, we write,

$$-\beta\Delta F = -\beta(F - F_0) = \ln\left\langle e^{-\beta \, V} \right\rangle . \tag{22.8}$$

Systematic evaluation of ΔF is due to Mayer. Expressing equation (8) in terms of cumulants M_n (as obtained in the appendix 3), we write,

$$-\beta\Delta F = \ln\left\langle e^{-\beta \, V} \right\rangle = \sum_{n=1}^{\infty} \frac{(-\beta)^n \, M_n}{n!} . \tag{22.9}$$

Clearly,

$$M_1 = \langle V \rangle .$$

Thus,

$$M_1 = \frac{1}{2} \cdot \frac{1}{\Omega^N} \sum_i \sum_j \int \cdots \int d^3r_1 \, d^3r_2 ... d^3r_N \; V\left(|\vec{r}_i - \vec{r}_j|\right) . \tag{22.10}$$

There are $N(N-1)$ pairs of dummy indices i and j . Then the first cumulant M_1 becomes,

$$M_1 = \langle V \rangle \; = \; \frac{N(N-1)\,\Omega^{N-2}}{2} \frac{1}{\Omega^N} \int V\left(|\vec{r}_1 - \vec{r}_2|\right) \, d^3r_1 \, d^3r_2$$

$$\cong \; \frac{1}{2} \cdot \frac{N^2}{\Omega} \int V(r) \, d^3r = \frac{1}{2} N\rho \int V(r) \, d^3r \tag{22.11}$$

Here we have neglected N as compared N^2. Thus the cumulant per particle (note its extensivity) is,

$$\frac{M_1}{N} = \frac{1}{2} \, \rho \int V(r) \, d^3r . \tag{22.12}$$

Next cumulant M_2 is given by

$$M_2 = <V^2> - <V>^2$$

$$= \sum_{i<j} \sum_{k<l} \left(\left(< V(|\vec{r}_i - \vec{r}_j|)V(|\vec{r}_k - \vec{r}_l| >)\right) - < V(|\vec{r}_i - \vec{r}_j|) >< V(|\vec{r}_k - \vec{r}_l|) > .$$

From now on, we will abbreviate $V\left(|\vec{r}_i - \vec{r}_j|\right)$ as V_{ij} etc. M_2 has various terms with common and uncommon indices.

a) No common indices or unlinked terms. These terms are of the type

$$< V_{ij}V_{kl} >=< V_{ij} >< V_{kl} > .$$

Here the indices are all distinct, $i \neq j \neq k \neq l$. We diagrammatically represent it

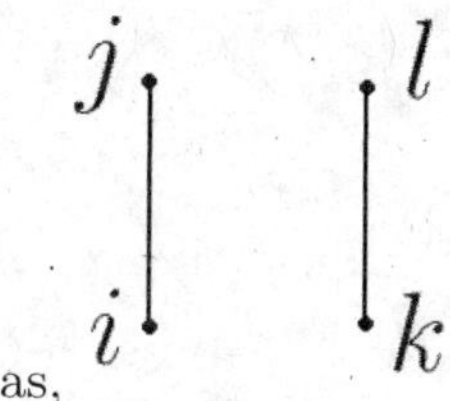

as,

Thus,

$$M_2 = < V_{ij} V_{kl} > - < V_{ij} >< V_{kl} >= 0$$

for all unlinked terms.

b) One index common, or "reducibly linked" terms are of the type

$$< V_{12} V_{23} >= \frac{1}{\Omega^N} \int V(|\vec{r_1} - \vec{r_2}|) V(|\vec{r_2} - \vec{r_3}|) d^3 r_1 \, d^3 r_2 ... d^3 r_N = \frac{1}{\Omega^3} \int V_{12} \, V_{23} \, d^3 r_1 \, d^3 r_2 \, d^3 r_3.$$

With transformation

$$\vec{r} = \vec{r_1} - \vec{r_2}; \quad \vec{R} = \frac{\vec{r_1} + \vec{r_2}}{2}$$

We get $d^3 r_1 d^3 r_2 = d^3 r d^3 R$.

We again write

$$\vec{R} + \frac{\vec{r_1}}{2} - \vec{r_3} = \vec{r'} \quad \text{and} \quad \vec{R} + \frac{\vec{r_1}}{2} + \vec{r_3} = 2\vec{R'} \, ,$$

and get $d^3 r \, d^3 R \, d^3 r_3 = d^3 r \, d^3 r' \, d^3 R'$. Then,

$$\langle V_{12} V_{23} \rangle \;\; = \;\; \frac{1}{\Omega^2} \int V(|\vec{r}|) V(|\vec{r}\,'|) d^3 r d^3 r'$$

$$= \;\; \langle V_{12} \rangle \langle V_{23} \rangle$$

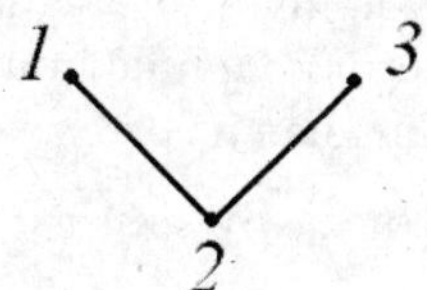

contributing zero to M_2. Diagrammatically we write it as

c) Both indices are common or "irreducible" terms. These terms are of the type

$$\langle V_{12}^2 \rangle - \langle V_{12} \rangle^2 \; .$$

The corresponding diagram is

Thus,

$$\langle V_{12}^2 \rangle = \frac{1}{\Omega^2} \int \int V^2(|\vec{r_{12}}|) d^3r_1 \, d^3r_2$$

$$= \frac{1}{\Omega} \int V^2(r) d^3r$$

$$= \frac{\rho}{N} \int V^2(r) d^3r$$

$$= \vartheta \left(\frac{1}{N} \right).$$

The other term is

$$\langle V_{ij} \rangle^2 = \frac{1}{\Omega^2} \left(\int V(r) \, d^3r \right)^2 = \frac{\rho^2}{N^2} \left(\int V(r) d^3r \right)^2$$

Then,

$$M_2 = \sum_{i<j} \left[\langle V_{ij}^2 \rangle - \langle V_{ij} \rangle^2 \right]$$

$$M_2 \cong \frac{N^2}{2} \left[\frac{\rho}{N} \int V^2(r) d^3r - \frac{\rho^2}{N^2} \left(\int V(r) d^3r \right)^2 \right]$$

Or, in the thermodynamic limit,

$$\frac{M_2}{N} = \frac{1}{2} \rho \int V^2(r) d^3r. \tag{22.13}$$

By now, you might have guessed that only the linked or irreducible graphs contribute to M_n in the thermodynamic limit. This is so because, unlinked as well as reducibly linked terms cancel. This has a correct consequence that ΔF is extensive.

This can be easily seen as follows. Consider the following term of M_3.

$$\langle V_{ij} V_{kl} V_{mn} \rangle = \langle V_{ij} \rangle \langle V_{kl} \rangle \langle V_{mn} \rangle \quad \text{if } ij \neq kl \neq mn.$$

Now each $\langle V_{ij} \rangle \sim \frac{1}{\Omega}$. Then $\langle V_{ij} V_{kl} V_{mn} \rangle$ of above term $\sim \frac{1}{\Omega^3}$. Then,

$$\sum_{ijklmn} \langle V_{ij} V_{kl} V_{mn} \rangle \sim \frac{N^6}{\Omega^3} = N^3 \rho^3.$$

Thus in thermodynamic limit this terms of M_3 gives, $\dfrac{M_3}{N} \to \vartheta\left(N^2\right)$.

To make ΔF extensive, this term must cancel with the other terms of the cumulant. This is possible for all orders of the expansion. This in a nut shell is a statement of the linked cluster theorem which states that only the irreducible diagrams contribute to the cumulant in the thermodynamic limit. All the other diagrams which are unlinked as well as irreducibly linked cancel each other. We have not proved but made this theorem plausible for you.

Irreducible contribution to M_3 comes from

$$M_3(irreducible) = \sum_{ij} \left[\langle V_{ij}^3 \rangle - 3 \langle V_{ij} \rangle \langle V_{ij}^2 \rangle + 2 \langle V_{ij} \rangle^3 \right]$$

But

$$\sum_{ij} \langle V_{ij}^3 \rangle \sim \frac{N^2}{2\Omega} \int V^3(r) d^3 r$$

$$\sum_{ij} \langle V_{ij}^2 \rangle \sim \frac{N^2}{2\Omega} \int V^2(r) d^3 r$$

$$\sum_{ij} \langle V_{ij} \rangle \sim \frac{N^2}{2\Omega} \int V(r) d^3 r$$

$$(22.14)$$

Then,

$$\lim_{\substack{N \to \infty \\ \Omega \to \infty \\ N/\Omega = \rho}} \frac{M_3(irreducible)}{N} \sim \frac{1}{2}\rho \int V^3(r) d^3 r - \frac{3}{4}\frac{\rho}{\Omega}\left(\int V(r) d^3 r\right)\left(\int V^2(r) d^3 r\right)$$

$$+ \ 2\frac{\rho}{\Omega^2}\left(\int V(r) d^3 r\right)^3$$

$$= \ \frac{1}{2}\rho \int V^3(r) d^3 r \qquad (22.15)$$

in the thermodynamic limit. Diagrammatically it is shown as

There is one more diagram of M_3 which is fully connected and irreducible. It is and contributes to M_3 to an order ρ^2 in the thermodynamic limit. Such graphs are known as "ring graphs" about which we will say more later.

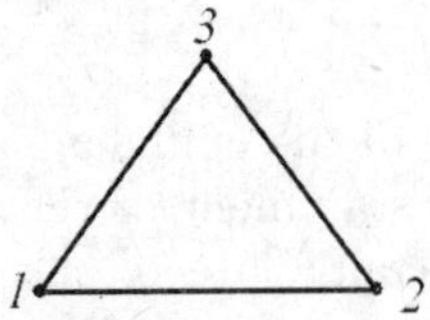

We retain only the irreducible graphs and use the following rules.

I   We define n^{th} order graph as the one where we have n interaction lines. Each interaction line means a factor of $(-\beta V(r))$.

II n^{th} order graph may have ν vertices meaning ν particles participate in the contribution the cumulant.

III   Volume factor for a graph with ν vertices is $\dfrac{1}{\Omega^{\nu-1}}$.

IV   Multiply the diagram contribution by a combinatorial factor

$$\binom{N}{\nu} = \frac{N^{\nu}}{\nu!}.$$

This multiplication ensures sum over indices.

A moment's reflection will reveal that the diagram with ν vertices contribute to the order of $\rho^{\nu-1}$ in the calculation of $\Delta F/N$.

With these rules in mind, we draw possible irreducible graphs as shown in figure (2).

We see then that the diagrams in first column are proportional to the density ρ, diagrams in the second column contribute to the order of ρ^2 and so on.

22.2   Summation of the Diagrams

There are many types of diagrams. They have many groups . Each group has a topologically similar structure. If one is interested in obtaining a virial expansion of pressure in the powers of ρ, we sum the diagrams column wise. The summation of diagrams in column 1 of figure (2) gives $B_2(T)$. Summation of the diagrams in the second column are of the order of ρ^2 and give $B_3(T)$. When the potential is of a short range, i.e. $\int r^2 V(r)\, dr$ converges, we can sum the diagrams of definite powers of density. On the other hand, if the potential is of an infinite range, (such as Coulomb potential) one sums the ring diagrams as shown in the figure (3).

A ring graph of ν vertices behaves as $\rho^{\nu-1}$. Summation of the ring graphs involves summation of a series of increasing powers of ρ.

Diagram		Volume dependence	N dependence	M_3/N	Remark
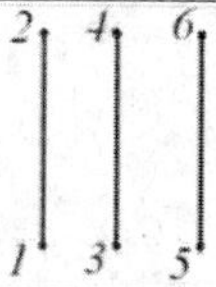		$\dfrac{1}{\Omega^3}$	N^6	$N^2\rho^3$	Cancells
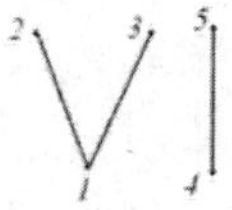	$\langle V_{12}V_{13}V_{45}\rangle$ $=\langle V_{12}V_{13}\rangle\langle V_{45}\rangle$ $=\langle V_{12}\rangle\langle V_{13}\rangle\langle V_{45}\rangle$	$\dfrac{1}{\Omega^3}$	N^5	$N\rho^3$	Cancells
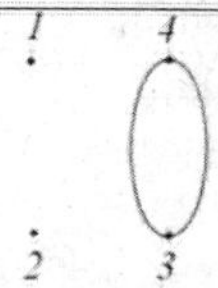	$\langle V_{12}V_{34}^2\rangle=\langle V_{12}\rangle\langle V_{34}^2\rangle$	$\dfrac{1}{\Omega^2}$	N^4	$N\rho^2$	Cancells
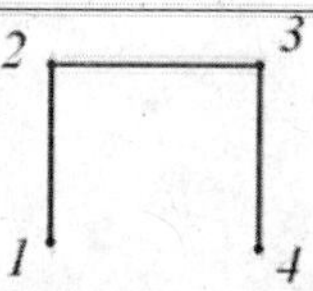	$\langle V_{12}V_{23}V_{34}\rangle=\langle V_{12}\rangle\langle V_{23}\rangle\langle V_{34}\rangle$	$\dfrac{1}{\Omega^3}$	N^4	ρ^3	reducibly linked Cancells
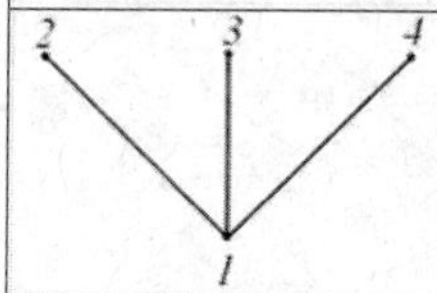	$\langle V_{12}V_{13}V_{14}\rangle=\langle V_{12}\rangle\langle V_{13}\rangle\langle V_{14}\rangle$	$\dfrac{1}{\Omega^3}$	N^4	ρ^3	reducibly linked Cancells
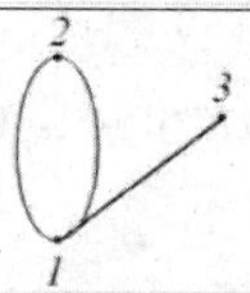	$\langle V_{12}^2V_{13}\rangle=\langle V_{12}^2\rangle\langle V_{13}\rangle$	$\dfrac{1}{\Omega^2}$	N^3	ρ^2	reducibly linked Cancells
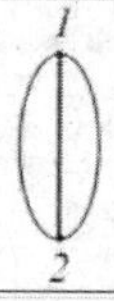	$\langle V_{12}^2\rangle$	$\dfrac{1}{\Omega}$	N^2	ρ	irreducibly linked useful
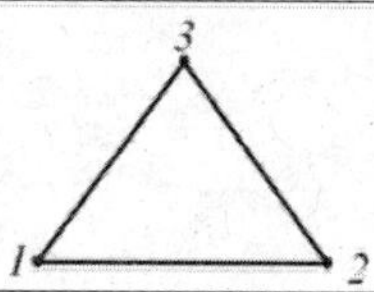	$\langle V_{12}V_{23}V_{31}\rangle$	$\dfrac{1}{\Omega^2}$	N^3	ρ^2	irreducibly linked useful

Table 22.1: Possible graphs with three interaction lines.

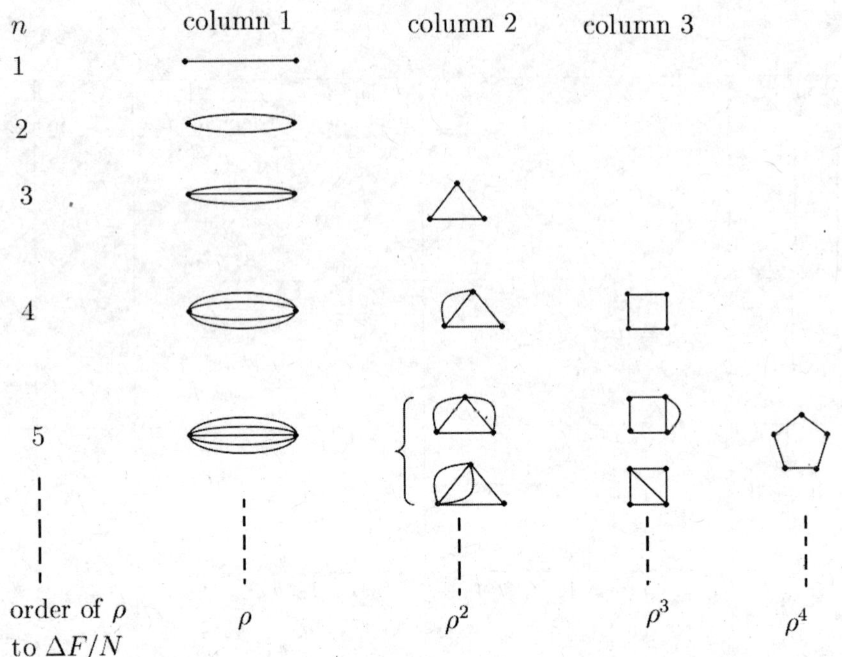

Figure 22.2: Various graphs contributing to M_n

22.3   Calculation of $B_2(T)$ and $B_3(T)$

a Calculation of $B_2(T)$: Second virial coefficient is found by summing the graphs shown in figure (4).

Figure 22.3: Diagrams to be summed for $B_2(T)$

By the rules given previously, we get,

$$
\begin{aligned}
-\beta(F - F_0) &= \frac{N\rho}{2} \sum_{n=1}^{\infty} \frac{(-\beta)^n}{n!} \int V^n(r) \, d^3r \\
&= \frac{N\rho}{2} \int (e^{-\beta V(r)} - 1) \, d^3r
\end{aligned}
$$

$$\tag{22.16}$$

Thus,

$$F(\Omega, T) = F_0(\Omega, T) - \frac{Nk_BT\rho}{2}\int (e^{-\beta V(r)} - 1)\, d^3r \; + \; ... \qquad (22.17)$$

But the pressure is, $P = \left(\dfrac{-\partial F}{\partial \Omega}\right)_T$

Then,

$$\frac{P\Omega}{Nk_BT} = 1 - \frac{1}{2}\int (e^{-\beta V(r)} - 1)\, d^3r + ... \qquad (22.18)$$

Clearly the second virial coefficient is,

$$B_2(T) = -\frac{1}{2}\int (e^{-\beta V(r)} - 1)\, d^3r \qquad (22.19)$$

b Calculation of $B_3(T)$: For this evaluation, we sum the graphs in the third column of figure(3). A typical three vertex graph is

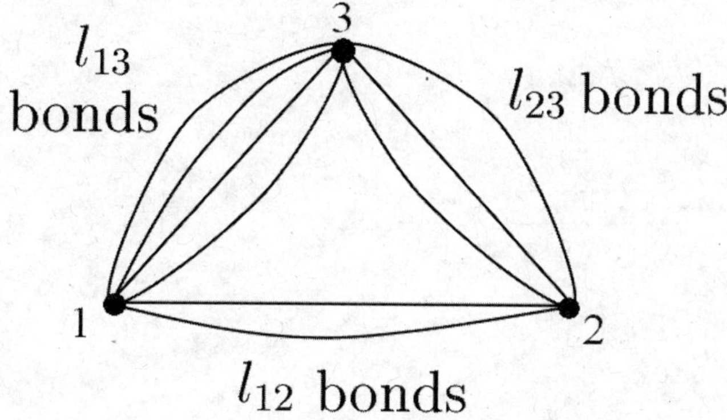

Here l_{13} bonds (i.e. $(-\beta V(r_{13}))^{l_{13}}$ power) connects vertices 1 and 3, l_{12} bonds connect points 1 and 2 and l_{23} bonds connect points 2 and 3. These bonds must observe the constraint

$$l_{12} + l_{13} + l_{23} = n$$

which is the order of the potential. Thus applying the rules given, we see that the contribution to M_3 is,

$$\frac{N^3}{3!\,\Omega^2}\sum_{n=1}^{\infty}\left\langle V_{12}^{\,l_{12}} V_{13}^{\,l_{13}} V_{23}^{\,l_{23}}\right\rangle \times (\text{Number of possible ordering}) \,. \qquad (22.20)$$

Here, number of possible arrangements of n objects in to $(12), (13), (23)$ classes with a constraint $l_{12} + l_{13} + l_{23} \equiv l = n$ is a multinomial coefficient $\dfrac{n!}{l_{12}!\, l_{13}!\, l_{23}!}$. All this type of diagrams have three vertices and hence the above sum will begin with $n = 3$. Thus,

$$
\begin{aligned}
M_3 &= \frac{N^3}{3!\,\Omega^2} \sum_{n=3}^{\infty} \frac{(-\beta)^l}{n!} \cdot \frac{n!}{l_{12}!\,l_{13}!\,l_{23}!} \left\langle V_{12}^{\,l_{12}} V_{13}^{\,l_{13}} V_{23}^{\,l_{23}} \right\rangle \\
&= \frac{N\rho^2}{3!} \left\langle \left(\sum_{l_{12}=1} \frac{(-\beta)^{l_{12}} V_{12}^{\,l_{12}}}{l_{12}!} \right) \left(\sum_{l_{13}=1} \frac{(-\beta)^{l_{13}} V_{13}^{\,l_{13}}}{l_{13}!} \right) \left(\sum_{l_{23}=1} \frac{(-\beta)^{l_{23}} V_{23}^{\,l_{23}}}{l_{23}!} \right) \right\rangle
\end{aligned}
$$

$$(22.21)$$

Here we divided each $n \geq 3$ into possible l_{12}, l_{13} and l_{23} bonds maintaining the constraint. Clearly, for $n = 3$, $l_{12} = l_{13} + l_{23} = 1$. For $n = 4$ there are three possibilities like, $l_{12} = l_{13} = 1$, $l_{23} = 2$ and similar other ones. Thus we get

$$
\frac{M_3}{N} = \frac{\rho^2}{3!} \int \left(e^{-\beta V_{12}} - 1 \right) \left(e^{-\beta V_{13}} - 1 \right) \left(e^{-\beta V_{23}} - 1 \right) d^3 r_{12} d^3 r_{13} d^3 r_{23}
\qquad (22.22)
$$

Then, equation(17) modifies to

$$
\begin{aligned}
F(\Omega, T) &= F_0(\Omega, T) \\
&\quad - \frac{N k_B T \rho}{2} \int \left(e^{-\beta V(r)} - 1 \right) d^3 r \\
&\quad - \frac{N k_B T \rho^2}{3!} \int \left(e^{-\beta V_{12}-1} \right) \left(e^{-\beta V_{13}-1} \right) \left(e^{-\beta V_{23}-1} \right) d^3 r_{12} d^3 r_{13}
\end{aligned}
$$

$$(22.23)$$

This gives

$$
B_3(T) = -\frac{1}{3} \int \int \int \left(e^{-\beta V_{12}-1} \right) \left(e^{-\beta V_{13}-1} \right) \left(e^{-\beta V_{23}-1} \right) d^3 r_{12} d^3 r_{31}
\qquad (22.24)
$$

Clearly the evaluation of B_3, B_4...etc. progressively becomes more and more difficult even for the usual potentials.

Problems

Problem 1: Take the potential energy function between the two molecules as

$$
V(r) = \begin{cases} \infty & r < r_0 \\[2mm] -V_0 \left(\dfrac{r_0}{r} \right)^s & r > r_0 \end{cases}
$$

and calculate $B_2(T)$. Simplify $B_2(T)$ in a dilute gas limit and obtain van der Waals equation.

Solution:

$$B_2(T) = -2\pi \int_0^\infty r^2 \left(e^{-\beta V(r)} - 1\right) dr .$$

$$= -2\pi \int_0^{r_0} r^2 dr - 2\pi \int_{r_0}^\infty r^2 \left(e^{-\beta V(r)} - 1\right) dr$$

If temperatures is high enough so that $V_0/k_B T << 1$, then $e^{\beta V} \cong 1 + \beta V$.

$$\therefore B_2(T) \cong -\frac{2\pi r_0^3}{3} - \frac{2\pi V_0}{k_B T} \int_{r_0}^\infty \frac{r_0^s}{r^{s-2}} dr$$

$$= -\frac{2\pi r_0^3}{3} \left(1 - \frac{3}{3-s}\frac{V_0}{k_B T}\right)$$

$$\equiv b' - \frac{a'}{k_B T}$$

where $b' = -\frac{2\pi}{3} r_0^3$ and $a' = \frac{3 V_0 b'}{s-3}$.

Equation of state is

$$\frac{P}{\rho k_B T} \cong 1 + \rho B_2(T) + \ldots$$

$$= 1 + \left(b' - \frac{a'}{k_B T}\right)\rho$$

Or, $\left(P + a'\rho^2\right) = \rho k_B T \left(1 + b'\rho\right) \cong \frac{\rho k_B T}{1 - b'\rho}$.

For a gas of ν number of moles, $\rho = \frac{N}{\Omega} = \frac{\nu N_A}{\Omega}$. More over $v =$ molar volume $=\Omega/\nu$ and $N_A k_B = R$. Also write $a = N_A{}^2 a'$ and $b = N_A b'$. Then above equation gets a more familiar form of van der Waals equation as

$$\left(P + \frac{a}{v^2}\right)(v - b) = RT$$

22.4   Comment About Classical Coulomb Gas

We assume that we have an electrically neutral system of positive and negative charges confined in a volume Ω. Normally we have a neutral mixture of electrons of mass m and positive and negative ions of mass M with $m/M << 1$. In such a mixture called plasma electrons are much more mobile than the ions due to their small mass. The velocities of electrons are distributed according to Maxwell - Boltzmann distribution if the Coulomb interaction is ignored. This is possible if the plasma is "dilute". There are many interesting phenomena related with plasma.

One of the important phenomenon is that of a screening of the strength of a charge takes place in plasma. If we have an impurity in the form of a charge $Q = Ze$, then the electrons in the plasma will relocate around the impurity. Specifically, suppose that the impurity has a positive charge. Then the electrons would flock around it and as a result the impurity potential $\phi(r)$ would be almost zero far away from the impurity. On the other hand, in the absence of plasma, its potential is of a long range $\sim 1/r$. Debye solved this problem of a potential produced by the impurity in plasma and showed that

$$\phi(r) = -\frac{Q}{r}\exp\left(-r/\lambda_D\right) \tag{22.25}$$

where λ_D is a screening length and also known as the Debye length. If ρ is a uniform density of the plasma(i.e. density far away from the impurity) then with the electron temperature T, λ_D is given by

$$\frac{1}{\lambda_D^2} = \frac{4\pi Z e^2 \rho}{k_B T} \tag{22.26}$$

This theory holds when the average thermal energy of electrons is much larger than average inter particle potential energy. Within this limit, as plasma becomes denser, the screening length becomes smaller.

When we have to deal with plasma as a non-ideal gas the terms like 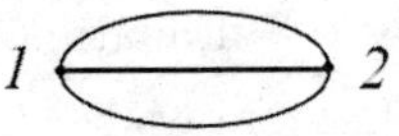 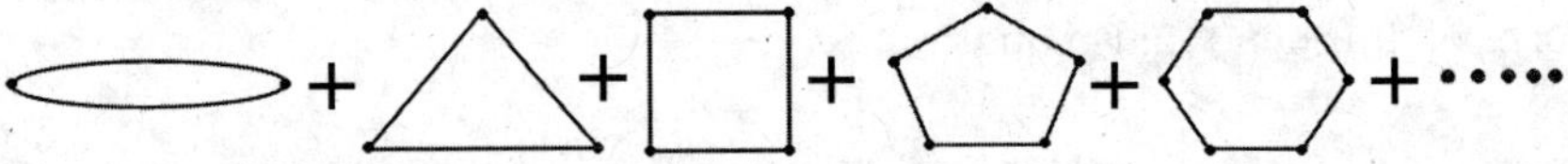 diverge. For

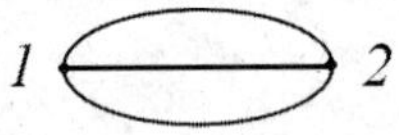

$$1 \bigcirc 2 \quad \sim \int_0^\infty v^3(r)d^3r \sim \int_0^\infty \frac{r^2 dr}{r^3} \sim e^6 \ln r\big|_0^\infty \to \infty$$

This is true of calculation of any diagram with long range potential like Coulomb potential. The way out of this situation is to sum all the ring graphs of figure (3). All these graphs

Figure 22.4: Ring graphs

have $2n$ fold symmetry (n rotations and n reflections.)

Denoting ΔF_{ring} as a change in free energy from its ideal gas value, and $v(q)$ as a Fourier transform, one can show,

$$-\frac{\beta \Delta F_{ring}}{N} = \frac{1}{2\Omega}\sum_q \sum_{n=2}^\infty \frac{(-\beta\rho v(q))^n}{n}$$

$$= \frac{1}{2N}\sum_q \left(\beta\rho v(q) - \rho\ln(1 + \beta\rho v(q))\right) \tag{22.27}$$

One can obtain the interaction energy ΔE_{ring} as

$$\frac{\Delta E_{ring}}{N} = \frac{\partial(\beta \Delta F_{ring})}{\partial \beta}$$

$$= -\frac{1}{2N} \sum_q \left(\rho v(q) - \frac{\rho v(q)}{1 + \beta \rho v(q)} \right). \tag{22.28}$$

$(\Delta E)_{ring}$ is called correlation energy. Here we use $v(q) = \left(4\pi e^2 Z/q^2\right)$. Also, we know that

$$\sum_q f(q) = \frac{\Omega}{(2\pi)^3} \int f(q) d^3q.$$

We can then easily evaluate the sum in equation (28) and obtain

$$\frac{\Delta E_{ring}}{N} = \frac{3}{2} k_B T - \frac{Ze^2}{2\lambda_D}. \tag{22.29}$$

This selective summing of topologically similar but divergent diagrams is common in field theory and is known as renormalization. Using the properties of Fourier transform, one can obtain equation (27). In fact defining the Fourier transform,

$$v(\vec{r}_1 - \vec{r}_2) = \frac{1}{(2\pi)^3} \int d^3q e^{i\vec{q} \cdot (\vec{r}_1 - \vec{r}_2)} \, v(q). \tag{22.30}$$

Then it is easy to show that,

$$\int v_{12} v_{23} \ldots v_{n1'} d^3r_1 \, d^3r_2 \ldots d^3r_n = \int \frac{1}{(2\pi)^3} \, d^3q \, e^{i\vec{q} \cdot (\vec{r}_1 - \vec{r}_2)} \, (v(q))^n.$$

Then the equation (27) is immediate after applying the diagram summation rules.

22.5 Behavior of $B_2(T)$ with Temperature

Nature of he potential is given in figure (1). The second virial coefficient depends upon temperature T only. Thus,

$$B_2(T) = -\frac{1}{2} I = -\frac{1}{2} \int_0^\infty \left(e^{-\beta V(r)} - 1 \right) d^3r.$$

Clearly, for $r < r_0$, $V(r)$ is large and positive leading to a positive contribution to $B_2(T)$. For larger values of r, $V(r)$ is negative. Then, for lower values of temperature, $\left(e^{-\beta V(r)} - 1 \right)$ becomes large and positive. This makes $B_2(T)$ negative. If the potential has a form $V(r) = -V_0 \, f(r)$, then $\dfrac{V_0}{k_B T} \gg 1$ in the low temperature regime. When $\dfrac{V_0}{k_B T} \ll 1$, it is a high temperature regime. In the high temperature regime, $B_2(T)$ slightly decreases. We have seen in the study of phase transition that van der Waals equation can be cast in reduced form. $B_2(T)$ also possesses this property. We write, $B_2' = \dfrac{B_2(T)}{r_0^3}$ and $T' = T/V_0$. Also

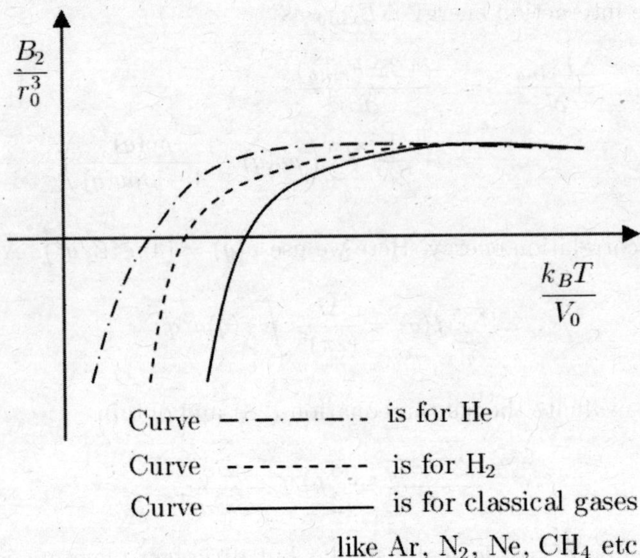

Figure 22.5: B_2/r_0^3 versus reduced temperature $k_B T/V_0$ for the classical gases as well as the deviant gases like Hydrogen and Helium

the potential of the figure (1) has a form $V(r) = V_0 \phi\left(\dfrac{r}{r_0}\right)$, the scaling is possible. Such a scaled coefficient B_2 versus the scaled temperature is shown in the figure (6).

Thus,

$$B_2(T) = -2\pi \, r_0^3 \int r'^{\,2} dr' \left(e^{-\frac{V_0}{k_B T}\phi(r')} - 1 \right) = -2\pi \, r_0^3 \int r'^{\,2} dr' \left(e^{-\frac{\phi(r')}{k_B T'}} - 1 \right).$$

For different gases, the values of V_0 and r_0 can be estimated. A graph of $B_2'(T')$ versus T' is shown in figure(4)and a universal behavior is seen. Universal behavior means we have a same curve for all gases. Two gases, namely the hydrogen and the helium are odd men out. This is so because the quantum effects predominate over the classical effects for H_2 and He.

Short Questions

1. Distinguish between a real gas and the ideal gas as regards (i) inter atomic potential (ii) equation of state (iii) range of the potential and (iv) partition function.

2. Show that the Fourier transform of the Coulomb potential $V(r) = e^2/r$ is $v(q) = 4\pi e^2/q^2$.

3. Find $B_2(T)$ for hard sphere gas of density ρ where $V(r) = \infty$ for $r < r_0$ and $V(r) = 0$ for $r > r_0$.

$$\text{Ans: } 2\pi r_0^3/3$$

4. Verify volume dependence and N dependence of various diagrams that appear in figure(2) for three interaction lines.

5. Find dependence on density and a type of divergence for 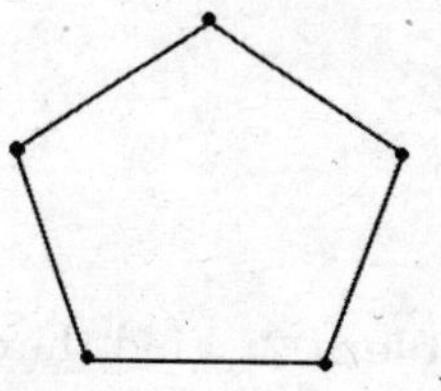 with Coulomb interaction.

$$\text{Ans: } \left(\frac{\rho^4 (4\pi e^2)^5}{q^7} \right) \Big|_{q_0}^{\infty}$$

Problems

Problem 1: A gas of N particles are confined in a volume Ω. $\phi(r)$ is an interatomic potential. Find the expression of $B_2(T)$ using virial theorem. Use also the Boltzmann factor $e^{-\beta\phi(r)}$. (Hint: The Virial theorem states that

$$\langle T \rangle = \frac{1}{2} \left\langle \sum_i \vec{F}_i \cdot \vec{r}_i \right\rangle.$$

and is applicable to periodic and/or bounded systems. Here T is a kinetic energy, $\vec{F}_i$ is a total force acting on the ith particle at the position $\vec{r}_i$. The average indicated is a time average, but we can take it as an ensemble average. Contribution to virial comes from surface and from intermolecular forces. Pressure contributes to the virial as follows. Because of the surface, particle is reflected which means it exerts a pressure. Its contribution to virial is,

$$\frac{1}{2} \int P\, \vec{r} \cdot \vec{n} dS = \frac{1}{2} P \int \nabla \cdot \vec{r} d^3 r = \frac{3}{2} P\, V.$$

We find the contribution of intermolecular force.

$$\vec{F} = -\nabla \phi(r) = -\frac{d\phi}{dr}.$$

Consider one molecule at $\vec{r}$ and the other at the origin. Then,

$$-\frac{1}{2} \left\langle \sum_i \vec{F}_i \cdot \vec{r}_i \right\rangle = -\frac{1}{2} \int \langle \vec{r} \cdot \vec{F}(r) \rangle.$$

As the particles are distributed according to a Boltzmann factor,

$$-\frac{1}{2} \int \langle \vec{r} \cdot \vec{F}(r) \rangle \rightarrow \frac{N^2}{4\Omega} \int r \frac{d\phi}{dr} e^{-\beta\phi(r)} d^3 r$$

$$\therefore \langle T \rangle = \frac{3}{2} PV \;+\; \frac{4\pi}{4\Omega(-\beta)} \int r^3 \frac{d}{dr}\left(e^{-\beta\phi(r)} - 1\right) dr.$$

Note that the addition of a constant -1 does not alter the result since its derivative is zero. We write $\langle T \rangle = \frac{3}{2} N k_B T$. After simplification we get

$$\frac{PV}{N k_B T} = 1 - \frac{2\pi\, N}{\Omega} \int r^2 \left(e^{-\beta\phi(r)} - 1\right) dr$$

Problem 2: Find third virial coefficient for a hard sphere gas with a potential

$$V(r) = \infty \quad r < r_0$$
$$\quad\;\; = 0 \quad r > r_0.$$

(Hint:

$$B_3(T) = -\frac{1}{3} \int \int \int \left(e^{-\beta V_{12}} - 1\right) \left(e^{-\beta V_{13}} - 1\right) \left(e^{-\beta V_{23}} - 1\right) d^3 r_{12} d^3 r_{31}.$$

Note that $\left(e^{-\beta V} - 1\right) = -1$ for $r < r_0$ and is zero for $r > r_0$. Fix $r_{12} < r_0$ Then the particle 3 takes all possible values. The integrand $= -1$ for all values of $r_{13} < r_0$ and $r_{23} < r_0$ For all other values of r_{13} and r_{23} it is zero.

$$\therefore \quad B_3(T) = \frac{1}{3} \int_{r_{12}=0}^{r_0} d^3\, r_{12} \left(\int d^3\, r_{13}\right)_c$$

where $\left(\int d^3\, r_{13}\right)_c$ is a common volume between the spheres of radius r_0 at points 1 and 2. It is a volume of rotation of a shaded region with AB as an axis. A strip MN of width dy

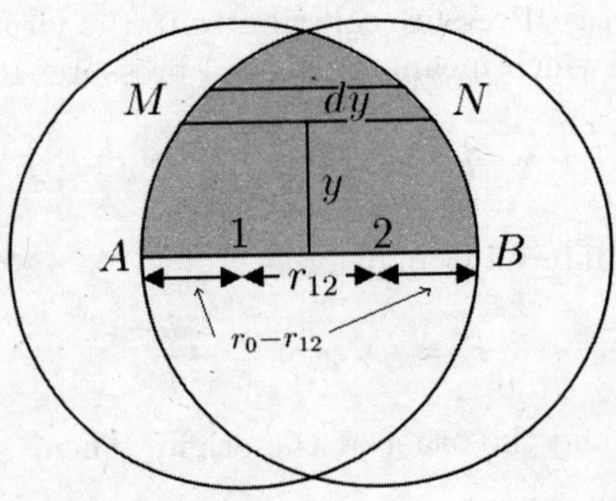

at y has a length $\left(2\sqrt{r_0^2 - y^2} - r_{12}\right)$. Then,

$$\left(\int d^3\, r_{13}\right)_c = \int_0^{\sqrt{r_0^2 - r_{12}^2/4}} \left(2\sqrt{r_0^2 - y^2} - r_{12}\right) 2\pi\, y \, dy$$

The final integration gives,

$$B_3(T) = \frac{5\pi^2 r_0^6}{18}.$$

From the short question 3, $B_2(T) = \dfrac{2\pi r_0^2}{3}$ giving $B_3(T) = \dfrac{5}{8} B_2^2(T)$. q.e.d.)

Problem 3: A Sutherland potential is given as

$$\phi(r) \;=\; \infty \qquad \text{for} \quad r < r_0$$
$$\;=\; -\varepsilon \left(\frac{\sigma}{r}\right)^6 \quad \text{for} \quad r > r_0.$$

Obtain the second virial coefficient of the Sutherland gas. Also find the correction to the ideal gas.

$$\text{Ans: } B_2(T) = -\frac{2\pi r_0^3}{3}\left[1 + \frac{\varepsilon}{k_B T}\left(\frac{\sigma}{r_0}\right)^6\right]$$

Problem 4: Molecules adsorb on a surface. An adsorbed surface layer (film) of area A consists of N molecules. These N molecules freely move on the surface and can be considered as a two dimensional molecular gas with a film pressure P i.e. the average force per unit length. $V(r)$ is a potential energy between the adsorbed molecules. Find equation of state up to $B_2(T)$.

$$\text{Ans: } \frac{P\,A}{Nk_B T} = 1 - \frac{\pi\,N}{A}\int_0^\infty r\left(e^{-\beta V(r)} - 1\right) dr$$

Chapter 23

ISING MODEL, RENORMALIZATION GROUP, ETC.

So far we studied the phase transition phenomenon from mean field angle. We also studied solid-liquid transition, critical exponents and the Kadanoff construction where, given two critical exponents, all others could be estimated. Using the theory of renormalization group due to K. G. Wilson we can estimate even these two exponents. To discuss a complete theory of the Renormalization Group(RG) is beyond the scope of the present book. However, a specific discussion of RG theory as applied to the one and the two dimensional Ising model will be presented. This is so because the essence of RG can be conveyed much more simply in Ising model due to its discreet nature.

Study of model systems is important in statistical mechanics. Many of the models can be solved exactly. These models have similarity with some realistic systems. There are many models such as Heisenberg model, lattice-gas model, Toda model, Ising model, spherical model, X-Y model etc. Of these, Ising model is a most well known of them all. Originally Ising proposed this model to explain magnetic phenomenon. Ising wrote the Hamiltonian of magnetic system as

$$H_N = -J \sum_{n.n} \sigma_i . \sigma_j - \mu B \sum_{i=1}^{N} \sigma_i \tag{23.1}$$

where $\sigma_i = \pm 1$, N is number of spins, μ is Bohr magneton, and B is external magnetic field in Z direction. Note, we have <u>not</u> written as $\vec{\sigma_i}.\vec{\sigma_j}$. It means, only the Z component of $\vec{\sigma}$ i.e. σ_z is used. Thus, actually, $\sigma_i \sigma_j$ is $\sigma_{iz} \sigma_{jz}$. J is called coupling constant. Positive $J(J > 0)$ gives ferromagnetism. Negative J gives anti-ferromagnetism. $\mu B \sigma_i$ is an alignment energy of magnetic moment $\mu \sigma_i$ along B. The term $-J \sum_{n.n} \sigma_i . \sigma_j$ is evaluated when i and j are nearest neighbors. Moreover, we use a periodic boundary condition. This means σ_N has σ_1 and σ_{N-1} as nearest neighbors. Thus, the Ising Hamiltonian is,

$$H_N = -J \sum_{i=1}^{N} \sigma_i \sigma_{i+1} - \frac{1}{2} \mu B \sum_{i=1}^{N} (\sigma_i + \sigma_{i+1}). \tag{23.2}$$

Ising model is solved exactly in one and two dimensions. Ising himself solved 1-dimensional model. Exact solution of Ising model in two dimensions (2-D) is due to Onsagar(1944) and is a mathematical triumph. We will not give Onsagar solution but will present his result. Solution in 3-D is available only numerically. If all the spin components σ_x, σ_y and σ_z are considered then the Hamiltonian has a term like $-J\vec{\sigma_i}\vec{\sigma_j}$. this is completely quantum mechanical because various components of σ do not commute. This quantum mechanical model is known as Heisenberg model and much less is known about it.

23.1 Ising Model in One Dimension

The Hamiltonian that we consider is,

$$H_N = -J\sum_{i=1}^{N}\sigma_i\sigma_{i+1} - \frac{1}{2}\mu B \sum_{i=1}^{N}(\sigma_i + \sigma_{i+1}) \tag{23.3}$$

with $\sigma_{N+1} = \sigma_1$.

We now define a matrix P such that

$$\langle \sigma'|P|\sigma\rangle = \exp\left(\beta J\sigma\sigma' + \frac{\beta\mu B}{2}(\sigma + \sigma')\right). \tag{23.4}$$

The Partition function for the problem is

$$Z_N(J,\beta) = \sum_{\sigma_1}\sum_{\sigma_2}....\sum_{\sigma_N}\exp\left\{\beta\sum_{i=1}^{N}\left(J\sigma_i\sigma_{i+1} + \frac{\mu B}{2}(\sigma_i + \sigma_{i+1})\right)\right\} \tag{23.5}$$

where each σ_i takes two values $+1$ and -1. From the definition of matrix P in equation (4), we get its explicit form as

$$P = \begin{bmatrix} e^{\beta(J+\mu B)} & e^{-\beta J} \\ e^{-\beta J} & e^{\beta(J-\mu B)} \end{bmatrix} \tag{23.6}$$

Partition function then becomes,

$$\begin{aligned}
Z_N(J,B,T) &= \sum_{\sigma_1}\sum_{\sigma_2}....\sum_{\sigma_N}\exp\left[\beta J(\sigma_1\sigma_2 + \sigma_2\sigma_3...\sigma_N\sigma_1) + \frac{\beta\mu B}{2}(\sigma_1 + \sigma_2 + ...\sigma_N)\right] \\
&= \sum_{\sigma_1}\sum_{\sigma_2}....\sum_{\sigma_N}\langle\sigma_1|P|\sigma_2\rangle\langle\sigma_2|P|\sigma_3\rangle...\langle\sigma_N|P|\sigma_1\rangle \\
&= \sum_{\sigma_1}\langle\sigma_1|P^N|\sigma_1\rangle \\
&= Tr\left(P^N\right) \tag{23.7}
\end{aligned}$$

This result is obtained by using the completeness relation of the states $\{\sigma_i\}$ which is $\sum_{\sigma_i}|\sigma_i\rangle\langle\sigma_i| = 1$. But the trace of a matrix is invariant under a similarity transformation. Therefore we choose a representation of P such that it is diagonal with the eigen value λ_1 and λ_2. Then,

$$Z_N(J, B, T) = \lambda_1^N + \lambda_2^N. \tag{23.8}$$

The eigen values are obtained by solving

$$\begin{vmatrix} e^{\beta(J+\mu B)} - \lambda & e^{-\beta J} \\ e^{-\beta J} & e^{\beta(J-\mu B)} - \lambda \end{vmatrix} = 0$$

Or solving with some algebra we get ,

$$\lambda_{\frac{1}{2}} = e^{\beta J}\left[\cosh(\beta\mu B) \pm \sqrt{\cosh^2(\beta\mu B) - 2e^{-2\beta J}\sinh(2\beta J)}\right] \tag{23.9}$$

Clearly, $\lambda_1 > \lambda_2$. Then, the partition function is

$$Z_N(J, B, T) = \lambda_1^N\left[1 + \left(\frac{\lambda_2}{\lambda_1}\right)^N\right] \xrightarrow[N\to\infty]{} \lambda_1^N \tag{23.10}$$

Thus,

$$\begin{aligned} \ln(Z_N) &= N\beta J + N\ln\left[\cosh(\beta\mu B) + \sqrt{\cosh^2(\beta\mu B) - 2e^{-2\beta J}\sinh(2\beta J)}\right] \\ &= -\beta F(B, T) \end{aligned} \tag{23.11}$$

Thus, free energy per particle is

$$\frac{F(B, T)}{N} = -J - \frac{1}{\beta}\ln\left[\cosh(\beta\mu B) + \sqrt{\cosh^2(\beta\mu B) - 2e^{-2\beta J}\sinh(2\beta J)}\right] \tag{23.12}$$

Once we now the free energy we can calculate all thermodynamic quantities. Thus the internal energy is ,

$$E = F + TS = -T^2\frac{\partial}{\partial T}\left(\frac{F}{T}\right) \tag{23.13}$$

Substituting from equation (12) and some simplification yields

$$\begin{aligned} \frac{E(B, T)}{N} &= -J - \frac{\mu B \sinh(\beta\mu B)}{\left[\exp(-4\beta J) + \sinh^2(\beta\mu B)\right]^{1/2}} \\ &+ \frac{2Je^{-4\beta J}}{\left[\cosh(\beta\mu B) + \left\{e^{-4\beta J} + \sinh^2(\beta\mu B)\right\}^{1/2}\right]\left[e^{-4\beta J} + \sinh^2(\beta\mu B)\right]^{1/2}} \end{aligned} \tag{23.14}$$

This expression for internal energy of 1-D Ising model is not very illuminating. However,

$$M(B,T) = -\left(\frac{\partial F}{\partial B}\right)_T = \frac{N\mu B \sinh(\beta\mu B)}{\left[\exp\left(-4\beta J\right) + \sinh^2(\beta\mu B)\right]^{1/2}} \tag{23.15}$$

indicates no ferromagnetism for any $T \neq 0$. Actually, what one needs is a non zero M

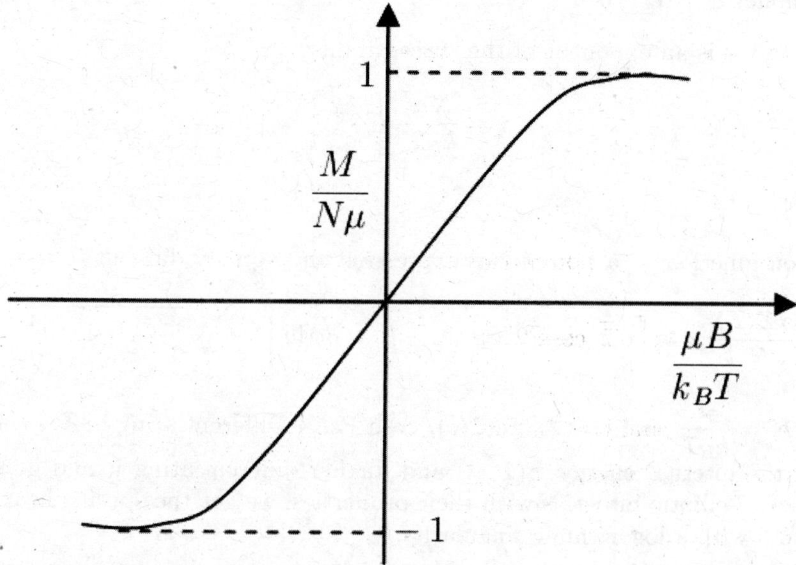

Figure 23.1: Magnetization versus B/T for $1 - D$ Ising model.

at a finite temperature with $B = 0$ to get ferromagnetism. From equation (15) it is clear that $M \to 0$ when $B \to 0$. Thus 1-D Ising model never yields ferromagnetism at a finite temperature. In 1-D Ising system we see that the thermal agitations win over spin-spin interaction at any temperature because there are too few neighbors.

23.2 Ising Model in Two Dimensions

Exact solution of the Ising model in two dimensions on a square lattice was obtained by Onsagar.

1. Onsagar showed that the ferromagnetic order was possible at a nonzero temperature for 2-D Ising model on a square lattice.

2. The critical temperature T_c below which the 2-D Ising model becomes ferromagnetic is,

$$\frac{J}{k_B T_c} = 0.4407 \qquad \text{(Square lattice)}$$

$$\frac{J}{k_B T_c} = 0.2747 \qquad \text{(triangular lattice)}$$

3. Specific heat singularity is,

$$C(T) = -\frac{8}{\pi} N k_B \left(\frac{J}{k_B T_c}\right)^2 \ln|1 - T/T_c|.$$

This implies $\alpha = \alpha' = 0$

4. $\gamma = \gamma'$ and it is an exponent of the susceptibility.

$$\chi = \frac{N\mu^2}{k_B T_c} \begin{cases} c_+ \ t^{-7/4} \\ c_- \ t^{-7/4} \end{cases}$$

5. $\beta = 1/8$.

6. Partition function - a horrendous expression and equally difficult to use. It is,

$$\frac{\ln Z(T)}{N} = \ln\left[\sqrt{2} \ \cosh(2K)\right] + \frac{1}{\pi} \int_0^{\pi/2} d\phi \ln\left[1 + \sqrt{1 - \kappa^2 \ \sin^2 \phi}\right] \qquad (23.16)$$

where $K = \dfrac{J}{k_B T}$ and $\kappa = 2\sinh(2K)/\cosh^2(2K)$ Differentiating $\ln Z(T)$ w.r.t. $(-\beta)$, we get the internal energy $E(T, A)$ and further differentiating it and going through the maze of elliptic integrals with their properties, we get the specific heat expression given in 3 with a logarithmic singularity.

Illustrative Problem

Problem 1: Consider a spin 1 Ising model in zero magnetic field whose Hamiltonian is,

$$H_N(\sigma_i) = -J\sum_i \sigma_i \ \sigma_{i+1} \qquad \sigma_i = \pm 1 \ \text{and} \ 0.$$

Show that the free energy is given by

$$\frac{F(T)}{N k_B T} = -\ln\left\{\frac{1}{2}\left[(1 + 2\cosh(\beta \ J)) + \sqrt{8 + (2\cosh\beta J - 1)^2}\right]\right\}$$

and find limiting forms of F when $T \to 0$ and $T \to \infty$.

Solution: Here,

$$Z = \sum_{\sigma_1}\sum_{\sigma_2}\cdots\sum_{\sigma_N} e^{\beta J(\sigma_1\sigma_2 + \sigma_2\sigma_3 + \cdots\cdots\sigma_N\sigma_1)}.$$

The matrix P has elements $\langle\sigma|P|\sigma'\rangle = e^{\beta J\sigma\sigma'}$. The matrix P is,

$$P = \begin{bmatrix} e^{\beta J} & 1 & e^{-\beta J} \\ 1 & 1 & 1 \\ e^{-\beta J} & 1 & e^{\beta J} \end{bmatrix}.$$

Eigen values of this matrix are $\lambda_1 = 2\sinh(\beta J)$ and

$$\lambda_{\frac{2}{3}} = \frac{1}{2}\left\{[1 + 2\cosh(\beta J)] \pm \sqrt{8 + (2\cosh\ \beta J - 1)^2}\right\}.$$

Clearly λ_2 is largest eigen value. Then,

$$Z_N(J,T) = \lambda_1^N + \lambda_2^N + \lambda_3^N.$$

In the limit $N \to \infty$, $\qquad Z_N(J,T) = \lambda_2^N$. Then,

$$\ln Z_N = -\frac{F}{k_B T} = N\ \ln\lambda_2.$$

Thus, the free energy is,

$$F = -Nk_B T \ln\left(\frac{1}{2}\left\{[1 + 2\cosh(\beta J)] + \sqrt{8 + (2\cosh\ \beta J - 1)^2}\right\}\right).$$

When $T \to \infty$, $\beta \to 0$ and in this limit, $F = -Nk_B T \ln 3$. In the low temperature limit, $T \to 0$ and $\beta \to \infty$ giving $F = -NJ$=constant.

23.3 Renormalization Group Analysis of 1-D Ising Model

For this and the next section, we will heavily lean upon the treatment due to Maris and Kadanoff[1]. Understanding of the R-G analysis is based upon the idea, that the correlation length diverges as $T \to T_c$. Under this situation, if there is a change of length scale, then the partition function must be insensitive to it. To illustrate this idea, consider a square lattice and set spin $\hbar/2$ at each lattice point. Suppose that the system undergoes a phase transition at T_c. Now delete alternate points of the lattice. Call this transformation as τ. Since the physics of the original lattice is same as that of the lattice under τ, we should have a similar partition function.

The quantity $K = \dfrac{J}{k_B T}$ in the Ising model acts as a coupling constant. It couples neighboring spins. The concept of a coupling constant is a very general one. For every problem where the interactions are involved, there is a coupling constant. Fine structure constant $\alpha = \dfrac{e^2}{\hbar c}$ is a well known coupling constant in the quantum e.m. interactions.

Under the transformation τ, the original partition function $Z(K, N) \to Z(K', N/2)$. It means the coupling constant K changes to K'. In a technical jargon, we say that K is "renormalized" to K'. A repeated application of τ will produce, $K \to K' \to K''$ etc. Then in the phase transition case, a stage comes when further application of τ does not renormalize the coupling constant. This is called as a fixed point and is denoted as K_c. The idea of renormalization came from the field theory (originally from QED). Its group structure and equation of renormalization of the coupling constant was developed by Bogoliubov.

---

[1] H J Maris and L P Kadanoff "Teaching of Renormalization Group" Am. J. Phys. *46*, 652 (1978)

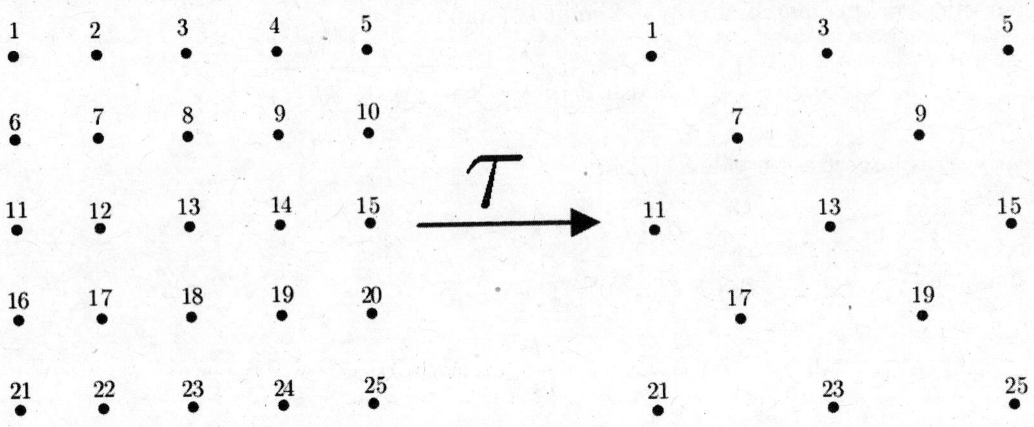

Figure 23.2: Labeled square lattice and the same lattice where the alternate lattice points are missing. Clearly, under τ the lattice looks similar to the original one albeit rotated by 45° .

To illustrate these ideas we apply them to one dimensional Ising model where in we know that there is no phase transition at a finite T. To simplify further, we take a field free Ising model. Its partition function is,

$$Z_N(K,N) = \sum_{\sigma_1} \sum_{\sigma_2} \cdots \sum_{\sigma_N} e^{K(\sigma_1\sigma_2 + \sigma_2\sigma_3 \ldots \sigma_N\sigma_1)}. \tag{23.17}$$

where $K = J/k_B T$ and each σ takes $\pm$ values. Make a sum over all the even numbered spins i.e. on $\sigma_2, \sigma_4, \sigma_6\ldots$ etc. Then the remaining spins are $N/2$ in number. Thus, the "new" partition function is,

$$Z(K,N) = \sum_{\sigma_1} \sum_{\sigma_3} \sum_{\sigma_5} \cdots \left\{ e^{K(\sigma_1+\sigma_3)} + e^{-K(\sigma_1+\sigma_3)} \right\} \times \left\{ e^{K(\sigma_3+\sigma_5)} + e^{-K(\sigma_3+\sigma_5)} \right\} \cdots .$$
$$\tag{23.18}$$

By performing this sum, we are eliminating half of the degrees of freedom. We now insist that the terms like $e^{K(\sigma_1+\sigma_3)} + e^{-K(\sigma_1+\sigma_3)}$ in equation (18) must look like $e^{K'\sigma\sigma'}$. Then,

$$e^{K(\sigma_1+\sigma_3)} + e^{-K(\sigma_1+\sigma_3)} = f(K)e^{K'\sigma_1\sigma_3}. \tag{23.19}$$

Equation (18) becomes

$$Z(K,N) = \sum_{\sigma_1} \sum_{\sigma_3} \cdots f(K)e^{K'\sigma_1\sigma_3} f(K)e^{K'\sigma_3\sigma_5} \cdots$$
$$= [f(K)]^{N/2} \, Z(K',N/2). \tag{23.20}$$

We know that $\ln Z = -F/k_B T$ and the free energy is an extensive quantity. We then write,

$$\ln Z(K,N) = Ng(K)$$
$$= \frac{N}{2}\ln f(K) + \frac{N}{2}g(K') \tag{23.21}$$

Equation (21) is one renormalization equation which connects new partition function with the old one. We need a second relation connecting K to K'. This is found by evaluating eq.(19) for the values of σ_1 and σ_3. We take $\sigma_1 = \sigma_3 = 1$ and get,

$$e^{2K} + e^{-2K} = f(K)e^{K'}.$$

Another choice is $\sigma_1 = 1$ and $\sigma_3 = -1$ giving,

$$2 = f(K)e^{-K'}.$$

Solving we get,

$$K' = \frac{1}{2}\ln\left[\cosh(2K)\right] \tag{23.22}$$

and,

$$f(K) = 2\sqrt{\cosh(2K)} \tag{23.23}$$

The renormalization equations to be solved self consistently are,

$$g(K') = 2g(K) - \ln\left[2\sqrt{\cosh(2K)}\right] \tag{23.24}$$

and

$$K' = \frac{1}{2}\ln[\cosh(2K)]. \tag{23.25}$$

Maris and Kadanoff solved eqs. (24) and (25) numerically. They took initial value of $K' = 0.01$. This means we have almost non interacting spins leading to $Z(0.01, N) \approx 2^N$ or, $g(0.01, N) = \ln 2$. Then

$$g(K) = \frac{1}{2}g(K') + \frac{1}{2}\ln 2 + K'/2 \tag{23.26}$$

and,

$$K = \frac{1}{2}\cosh^{-1}(e^{2K'}). \tag{23.27}$$

Following table for K, $g(K)$ by RG and its exact value is provided by Maris and Kadanoff.

Table 1: Non-Convergent Behaviour of the Coupling Constant K for the 1-D Ising Model.

K	$g(K)$ R.G.Calculation	Exact
0.01	$\ln 2$	0.693197
0.100334	0.698147	0.698172
0.327447	0.745814	0.745827
1.662637	1.697968	1.697968
2.702146	2.706633	2.706634

From Table 1 it is clear that from small K, each renormalization increases it leading to no

$$\times \longrightarrow \longrightarrow \longrightarrow \longrightarrow \longrightarrow \times$$
$$K = 0 \qquad\qquad\qquad\qquad K = \infty$$

Figure 23.3: Flow diagram of Ising model

fixed point. The way the coupling constant goes on changing is known as a flow diagram. Thus the flow diagram for 1-D Ising model is as shown in figure (3).

For 1-D Ising model, as K goes on increasing, after each scaling (renormalization), there is no phase transition at any $T \neq 0$. Obviously at $T = 0$ all the spins are aligned and ferromagnetism results. Technically then there are two fixed points for 1-D Ising model, namely, $K = 0$ and $K = \infty$ and they are trivial. At $K = 0$ i.e. at $T = \infty$, the lattice is totally disordered. At $K = \infty$ i.e. at $T = 0$, the lattice is completely ordered. For 2-D Ising model we have a nontrivial fixed point, as seen in the next section.

23.4   Renormalization Group Analysis of 2-D Ising Model

We take a field free model. Like a 1-D model, we sum alternate degrees of freedom. From figure (2) it is clear that Z has the form,

$$Z = \sum_{\sigma_1} \sum_{\sigma_2} .... e^{K\sigma_8(\sigma_7+\sigma_3+\sigma_9+\sigma_{13})} \cdot e^{K\sigma_{12}(\sigma_7+\sigma_{11}+\sigma_{17}+\sigma_{13})} ..... \qquad (23.28)$$

Here we eliminate σ_8 and σ_{12} by summing over them. Thus

$$Z = \sum_{\sigma_1} \sum_{\sigma_2} .... \left\{ e^{K(\sigma_7+\sigma_3+\sigma_9+\sigma_{13})} + e^{-K(\sigma_7+\sigma_3+\sigma_9+\sigma_{13})} \right\}$$
$$\times \left\{ e^{K(\sigma_7+\sigma_{11}+\sigma_{17}+\sigma_{13})} + e^{-K(\sigma_7+\sigma_{11}+\sigma_{17}+\sigma_{13})} \right\} ... \qquad (23.29)$$

Here comes the difficulty. Can we write the following expression?

$$e^{K(\sigma_1+\sigma_2+\sigma_3+\sigma_4)} + e^{-K(\sigma_1+\sigma_2+\sigma_3+\sigma_4)} = f(K)e^{K'(\sigma_1\sigma_2+\sigma_2\sigma_3+\sigma_3\sigma_4+\sigma_4\sigma_1)}. \qquad (23.30)$$

This should hold for all nonequivalent choices of σ. These nonequivalent choices are,

1. $\sigma_1 = \sigma_2 = \sigma_3 = \sigma_4 = \pm 1$

2. $\sigma_1 = \sigma_2 = \sigma_3 = \pm 1 , \sigma_4 = \mp 1$

3. $\sigma_1 = \sigma_2 = \pm 1, \sigma_3 = \sigma_4 = \mp 1$ and,

4. $\sigma_1 = \pm 1, \sigma_2 = \sigma_3 = \sigma_4 = \mp 1$

These will give us four equations, where as, from eqn.(30) we have two unknown, namely, $f(K)$ and K. That means we must introduce two more degrees of freedom and modify the eqn.(30) in a physically meaningful way. Simplest possibility is,

$$e^{K(\sigma_1+\sigma_2+\sigma_3+\sigma_4)} + e^{-K(\sigma_1+\sigma_2+\sigma_3+\sigma_4)}$$
$$= f(K)\left\{ e^{\frac{1}{2}K_1(\sigma_1\sigma_2+\sigma_2\sigma_3+\sigma_3\sigma_4+\sigma_4\sigma_1)} \times e^{K_2(\sigma_1\sigma_3+\sigma_2\sigma_4)} \times e^{K_3 \ \sigma_1\sigma_2\sigma_3\sigma_4} \right\}. \qquad (23.31)$$

We have four unknown, namely, $f(K)$, K_1, K_2 and K_3. We have four nonequivalent choices of σ. Thus,

1. For $\sigma_1 = \sigma_2 = \sigma_3 = \sigma_4 = 1$ we get,
$$e^{4K} + e^{-4K} = f(K)\, e^{(2K_1 + 2K_2 + K_3)}$$

2. For $\sigma_1 = \sigma_2 = \sigma_3 = 1, \sigma_4 = -1$, we get,
$$e^{2K} + e^{-2K} = f(K)\, e^{-K_3}.$$

3. For $\sigma_1 = \sigma_2 = +1$ and $\sigma_3 = \sigma_4 = -1$ we get,
$$2 = f(K)e^{K_3 - 2K_2}$$

4. For $\sigma_1 = \sigma_3 = +1$ and $\sigma_2 = \sigma_4 = -1$, we get,
$$2 = f(K)\, e^{K_3 + 2K_2 - 2K_1}$$

These equations can easily be solved to give

$$K_1 = \frac{1}{4} \ln\left[\cosh(4K)\right] \tag{23.32}$$

$$K_2 = K_1/2 \tag{23.33}$$

$$K_3 = K_2 - \frac{1}{2}\cosh(2K) \tag{23.34}$$

$$f(K) = 2\left[\cosh(2K)\right]^{1/2}\left[\cosh(4K)\right]^{1/8} \tag{23.35}$$

Structure of these equations is, that the two terms do not have a standard nearest neighbor form. $K_2(\sigma_1\sigma_3 + \sigma_2\sigma_4)$ gives next nearest neighbor interaction and $K(\sigma_1\sigma_2\sigma_3\sigma_4)$ gives closed looping around the square. It seems that by removing some degrees of freedom, and due to high connectivity of 2-D system, effective interaction is more complicated than that of the original problem. To get the same sort of terms in the transformed Z as those in the original Z, we must somehow get rid of K_2 and K_3 terms. This means, we must approximate equation (31) in such a way that the old Z and the new Z look similar. There are two possibilities.

1) Ignore K_2 and K_3. Then, $K_1 = K'$ and $K' = \frac{1}{4}\ln(\cosh(4K))$

and writing a similar equations to (21) and (24).

$$g(K') = 2g(K) - \ln\left\{2\sqrt{\cosh(2K)} \cdot (\cosh(4K))^{1/8}\right\}$$

These equations give the same flow pattern as that of $1 - D$ Ising model leading to no fixed point.

2) Other possibility is to look at the terms involving K_1 and K_2. Both terms are positive and hence have a tendency to enhance the alignment of the spins. We try the following approximation. Ignore K_3. Then drop K_2 but increase K_1 to a new value K' so that the 'alignment tendency' remains same. This is similar to the mean field theory which,

here, takes in to account the next nearest neighbor interaction.

Thus,

$$\frac{1}{2}K_1\left(\sigma_1\sigma_2 + \sigma_2\sigma_3 + \sigma_3\sigma_4 + \sigma_4\sigma_1\right) + K_2\left(\sigma_1\sigma_3 + \sigma_2\sigma_4\right).$$

$$= K'(K_1, K_2)(\sigma_1\sigma_2 + \sigma_2\sigma_3 + \sigma_3\sigma_4 + \sigma_4\sigma_1). \tag{23.36}$$

The maximum value of K_1 term in the equation(36) is $2K_1$ and that of K_2 is $2K_2$. We then approximate

$$
\begin{aligned}
K' &= K_1 + K_2 \\
&= \frac{3}{8}\ \ln[\ \cosh\ (4K)].
\end{aligned}
\tag{23.37}
$$

The partition function can be written as

$$Z(K, N) = [f(K)]^{N/2} Z(K'(K_1, K_2); N/2). \tag{23.38}$$

Then writing $\ln Z = Ng(K)$ etc. we get,

$$g(K') = 2g(K) - \ln\left\{2[\cosh\ (2K)]^{1/2}[\cosh\ (4K)]^{1/8}\right\}. \tag{23.39}$$

and,

$$K' = \frac{3}{8}\ \ln[\ \cosh\ (4K)]. \tag{23.40}$$

Recurrence relations of eq.(39) and (40) has a new feature, namely, a nontrivial fixed point. Using computer, we solve these equations and get a fixed point at $K = K_c \cong 0.50698$. The flow pattern starts from K_c in such a manner that for $K < K_c$, K' goes on decreasing at each renormalization. For $K > K_c$, the K' goes on increasing after each RG transformation. The fixed point K_c is unstable and gives a non analytic behaviour to $Z(K_c)$. This fixed point

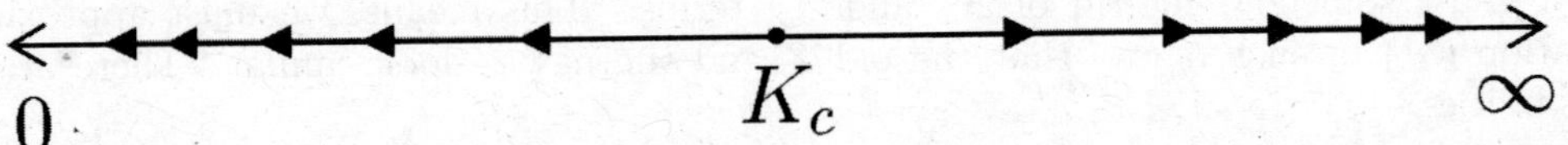

Figure 23.4: Fixed point in $2 - D$ R.G. Ising model

is associated with the phase transition. Exact value is given by Onsagar as $[J/(k_BT_c) = 0.44069]$. With our crude model calculation, our value turns out to be $K_c = 0.507$. Not bad!

Illustrative Problem

Problem 2: Show that the power law singularity of the specific heat

$$C \propto |T - T_c|^{-\alpha}$$

is such that

$$\alpha = 2 - \ln 2 / \ln\left[\left(\frac{dK'}{dK}\right)_{K_c}\right].$$

Solution: Recall that the most divergent part of the specific heat will be proportional to

$$C \propto \left|1 - \frac{T}{T_c}\right|^{-\alpha} \propto \frac{d^2 g(K)}{dK^2}$$

(can you see why?)

Thus the most divergent part of the free energy is proportional to

$$a|K - K_c|^{2-\alpha}.$$

From equation of RG, we have,

$$g(K') = 2g(K) - \ln\left\{2[\cosh\ (2K)]^{1/2}[\cosh\ (4K)]^{1/8}\right\}.$$

At $K = K_c$,

$$g(K'_c) = 2g(K_c) - \ln\left\{2[\cosh\ (2K_c)]^{1/2}[\cosh\ (4K_c)]^{1/8}\right\}.$$

We expand around K_c as,

$$K' = K_c + (K' - K_c)\left(\frac{dK'}{dK}\right)_{K_c}.$$

Near the critical point,

$$a|K' - K_c|^{2-\alpha} = 2a|K' - K_c|^{2-\alpha}\left(\frac{dK'}{dK}\right)_{K_c}^{2-\alpha}$$

Taking the logarithm of both sides, we get,

$$\alpha = 2 - \ln 2 / \ln\left[\left(\frac{dK'}{dK}\right)_{K_c}\right].$$

With $\dfrac{dK'}{dK} = \dfrac{3}{2}\tanh(4K_c)$, and after some simplification and substitution of $K_c = 0.507$ we get, $\alpha = 0.131$. Actual value of α is zero giving a logarithmic singularity.

23.5 Epilogue

Aim of this chapter was to give a glimpse of one of the most modern topics, namely the Renormalization group approach to the critical phenomena. Many ideas were evolved from earlier work of Widom as well as of Kadanoff (Scaling hypothesis) and also from the work of

Fisher. These earlier approaches did not provide a systematic method to evaluate the critical exponents. These deficiencies were remedied by Wilson by introducing the renormalization group (RG) concept which existed in the field theory earlier. Wilson got a Nobel Prize in physics for RG in 1982.

We have not discussed RG in its full details and we have not done a continuous scaling. We also did not discuss these ideas in a general $d-$ dimensions. We refer to a book by Ma[2] for these interesting details.

Ideas of RG are not restricted to the magnetic system alone but are applicable to any second order phase transition such as superconductivity or even superfluidity.

Short Questions

1. Give in short the difference between the Heisenberg model and the Ising model. Explain how the nature of magnetism depends upon the sign of J.

2. Consider a field free Ising model in 1-D for spin 1/2 system. Find corresponding P matrix and its eigenvalues.

   Ans:

   $$P = \begin{bmatrix} e^{\beta J} & e^{-\beta J} \\ e^{-\beta J} & e^{\beta J} \end{bmatrix}$$

   and

   $$\lambda_1 = 2\cosh(\beta J) \text{ and } \lambda_2 = 2\sinh(\beta J)$$

3. Explain as to why the 1-D Ising model does not have a ferromagnetic transition at a finite value of T?

4. Explain qualitatively, what steps of RG are taken in the analysis of 1-D Ising model.

Problems

Problem 1: Consider a one dimensional Ising chain where the nearest neighbors and the next nearest neighbors interactions are taken in to account. Hamiltonian for this problem is,

$$H = -J_1 \sum \sigma_i \sigma_{i+1} - J_2 \sum \sigma_i \sigma_{i+2}.$$

Find partition function and other properties.

(Hint: use $\sigma_i \sigma_i = 1$ and define the operator $\tau_i = \sigma_i \sigma_{i+1} = \pm 1$. Then

$$H = -J_1 \sum_i \tau_i - J_2 \sum_i \tau_i \tau_{i+1}$$

which is isomorphic to the Ising model. Clearly, $J_1 = \mu B$ and $J = J_2$.)

---

[2]S.K. Ma, *Modern Theory of Critical Phenomena*, Benjamin, (1976)

Ans: $\ln Z_N = N\beta J_2 + N\ln\left[\cosh(\beta J_1) + \sqrt{\cosh^2(\beta J_1) - 2e^{-2\beta J_1}\sinh(2\beta J_1)}\right]$

Problem 2: Obtain the RG equations for the $1-D$ Ising model with an external[3] magnetic field B.

Problem 3: Consider a $1-D$ ladder with a Hamiltonian,

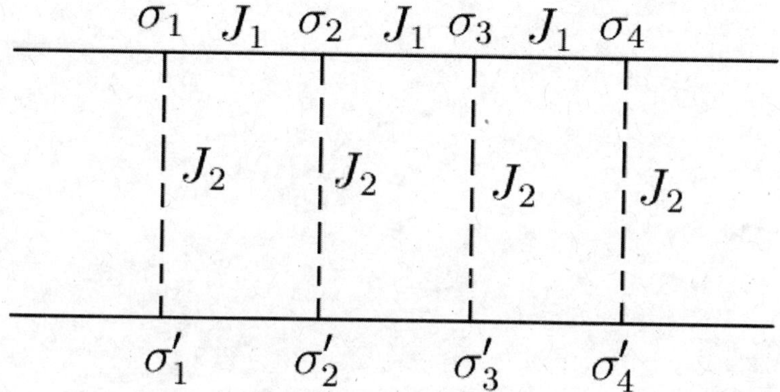

Figure 23.5: $1-D$ ladder with the interaction J_1 between two σs and J_2 between σ and σ'

$$H = -J_1\sum(\sigma_i\sigma_{i+1} + \sigma'_i\sigma'_{i+1}) - \frac{J_2}{2}\sum(\sigma_i\sigma'_i + \sigma_{i+1}\sigma'_{i+1})$$

Show that the partition function is

$$\ln Z = N\ln\left[2\cosh K_2\left\{\cosh 2K_1 + \sqrt{1 + \sinh^2 2K_1\tanh^2 K_2}\right\}\right]$$

Here $\beta J_1 = K_1$ and $\beta J_2 = K_2$

(Hint: This is a difficult problem. The transfer matrix elements are

$$\langle\sigma\mu|P|\sigma'\mu'\rangle = e^{\beta J_1(\sigma\sigma' + \mu\mu') + \frac{\beta J_2}{2}(\sigma\mu + \sigma'\mu')}$$

There will thus be 16 matrix elements and the matrix is 4×4 . It is

$$
(\sigma\mu)\equiv
\begin{array}{c}
(1\ 1)\\
(1\ -1)\\
(-1\ 1)\\
(-1\ -1)
\end{array}
\begin{bmatrix}
e^{2K_1+K_2} & 1 & 1 & e^{-2K_1+K_2}\\
1 & e^{2K_1-K_2} & e^{-2K_1-K_2} & 1\\
1 & e^{-2K_1-K_2} & e^{2K_1-K_2} & 1\\
e^{-2K_1+K_2} & 1 & 1 & e^{2K_1+K_2}
\end{bmatrix}
\begin{array}{c}
(11)\ (1-1)\ (-11)\ (-1-1) \equiv (\sigma'\mu')
\end{array}
$$

(column headers: (11) $(1-1)$ (-11) $(-1-1)$ $\equiv (\sigma'\mu')$)

---

[3]This problem is completely solved in "Statistical Mechanics", Second Edition by R K Patharia(Butterworth and Heinemann), 1996

Find largest eigen value. The eigen value equation will be of fourth order, get $\ln Z = N \ln \lambda_{max}$.)

Chapter 24

SIMULATION METHODS

The focus of this chapter is computer simulation methods in statistical physics. We first present a somewhat wider perspective on modeling, simulation, computation, and the two principle simulation methodologies of statistical physics. In the rest of the chapter, we discuss these two methodologies in their most basic form, and illustrate them through exploratory case studies.

The bibliography section at the end of this book lists select resources that may be useful as more comprehensive introductions to the topics discussed in this chapter. This list is in no way representative of the enormous literature on simulation in physics. We would also like to point the interested reader to the enormous number of resources on the topic, such as tutorials, lecture notes, codes, etc., that is available in the public domain over the internet.

24.1 Simulation in Statistical Physics

24.1.1 Reality, Models, Mathematics

Simply stated, a *model* is a description of a relevant aspect of the system or phenomenon of interest. The key step of the modeling process is that of making bold, meaningful, and useful abstractions of reality — This is the step that separates what is of interest from what is not, and consequently simplifies the description and the analysis that follows. Familiar abstractions in Physics include, e.g., the notion of a point particle. The scientific methodology is driven by an urge to describe reality rationally through the use of such models, the need to make predictions about it in a *quantitative* fashion, and the process of hard experimental validation of all models against reality. Indeed, a model or a theory is considered scientific only if it is manifestly *falsifiable* — Falsifiability of a model means that it is possible to make experimentally testable predictions from the model and, if these predictions turn out to be inconsistent with reality, then that could, in principle, invalidate the model.

If we are willing to ignore deep philosophical questions about the nature of reality or the nature of understanding, we might even go as far as saying that models represent the state of our understanding of reality. For example, celebrated theories of Physics that represent our current state of understanding about the physical reality could be seen as mathematical models of physical reality. These theories have historically evolved through the same process of falsification, and through a tight handshake between the *theoretical* and *experimental* approaches.

Mathematics turns out to be a powerful, expressive, and economical language for describing reality in most branches of the scientific enterprise. In any case, the need for a *quantitative* understanding makes the use of Mathematics inevitable. As a result, we end up with mathematical models of reality.

24.1.2   Computation and Simulation

Once a mathematical model of a system under study is formulated, the next and inevitable task is to extract meaningful information from the model and to make predictions about the behaviour of the underlying system. The purely analytical approach to elucidating mathematical properties of a model is conventionally called *theory* (as opposed to *experiment*). In this approach, one invariably needs to resort to *computation* of some sort or the other, either to gain insights for analytical work, or because purely analytical treatment of a model becomes increasing unwieldy as a function of the complexity and sophistication of the model.

A *computational simulation* is a form of computation which one could meaningfully think of as *imitating* the behaviour of the underlying system, and not just as the solution of, say, a bunch of ordinary differential equations.

The use of computation-based methodologies has been on the rise over the past few decades because of the ever-increasing availability of inexpensive computing power. Because of its somewhat specialized nature, computation and simulation are sometimes considered a third scientific methodology, besides theory and experiment. Indeed, computation can be seen to extend the power of theory beyond what is possible by purely analytical means, and to help bridge the gap between theory and experiment. Moreover, in problems where an accurate, well-validated model of reality is available, simulation can, at times, almost entirely replace an actual experimental investigation. Theory, computation, and experiment thus form a tightly-linked trio of scientific methodologies.

24.1.3   Principal Simulation Methods of Statistical Physics

To a good approximation, it could be said that simulation methodologies in Statistical Physics come in two flavours, *deterministic* and *stochastic*. Molecular dynamics (MD) simulations belong, predominantly, to the deterministic category. At the other end of the spectrum of simulation methods are what are called the Monte Carlo (MC) methods, which are *stochastic* by nature. These two methodologies in their most basic form will be discussed in the rest of this chapter.

24.2   Molecular Dynamics (MD)

The basic idea of MD is to explore the behaviour of a physical system by computing, with sufficient accuracy, the trajectories of individual particles constituting the system. Along the way, relevant information about the system is accumulated, which is used for computing physical quantities as statistical averages along the trajectory. The formal basis of why MD works for statistical mechanics problems is the (often assumed) *ergodic hypothesis*, i.e., equivalence of ensemble and time averages, which allows computing ensemble average of physical quantities of interest as time averages along a sufficiently long phase space trajectory of the system. As a side remark, note that the meaning of *particle*, as always, depends on the context and on how the system is being modeled: It could mean an individual atom, a molecule, a group of atoms that form a chemical group that is part of an even larger molecule (a protein, e.g.), or an entire star.

In this section, we will explore the MD method in its most basic form. With a view of understanding the behaviour of the method itself as thoroughly as possible, we have restricted the discussion in this section to elementary constant energy MD simulations. We note here in passing that constant energy MD (together with a constant number of particles) corresponds to a microcanonical ensemble, whereas constant temperature MD (again with a constant number of particles) corresponds to a canonical ensemble. Many important directions (e.g., constant temperature MD simulations) are not even touched upon in this introductory chapter – we refer the reader to more comprehensive texts such as those listed in the bibliography section at the end of this book.

24.2.1 Equations of Motion

The molecular dynamics method attempts to numerically integrate the equations of motion of a system of N interacting particles. In usual notation, let us write the equations of motion as

$$m_i \frac{d^2 \vec{r}_i}{dt^2} = \vec{F}_i, \tag{24.1}$$

where $i = 1, \ldots, N$ is the particle index. This Newtonian form of the equations of motion is not binding; any other convenient form (say, Lagrangian or Hamiltonian) of the equations of motion is equally alright to work with. We restrict ourselves to conservative systems and velocity-independent forces for the sake of this chapter, which makes the force $\vec{F}_i$ a function of only the particle positions $(\vec{r}_1, \vec{r}_2, \ldots, \vec{r}_N)$. To make things simpler, we will also assume that no external forces act on this system.

The Lennard-Jones Potential

For concreteness, let us consider a system of particles interacting with one another via the well-known Lennard-Jones potential. A conventional forms for this potential is

$$V(\vec{r}_1, \vec{r}_2, \ldots, \vec{r}_N) = \sum_{i=1}^{N} \sum_{j=1}^{i-1} v(r_{ij}),$$

$$\text{where } v(r_{ij}) = 4\epsilon \left\{ \left(\frac{\sigma}{r_{ij}} \right)^{12} - \left(\frac{\sigma}{r_{ij}} \right)^{6} \right\}, \tag{24.2}$$

and $r_{ij} = |\vec{r}_i - \vec{r}_j|$ is the distance between particles i and j.

	ϵ	σ_*	m	ρ	$\sqrt{m\sigma^2/4\epsilon}$
	K	Å	amu	g/cm^3	s
He	10.80	2.57	4.002602	0.1785	8.58×10^{-13}
Ne	36.68	2.79	20.1797	0.901	1.13×10^{-12}
Ar	120.0	3.38	39.948	1.784	1.07×10^{-12}
Kr	171.0	3.6	83.8	3.74	1.38×10^{-12}
Xe	221.0	4.1	131.29	5.8971	1.73×10^{-12}

Table 24.1: Lennard-Jones parameters for noble gas systems. The last two columns list, respectively, the density ρ at 293 K, and the characteristic time scale associated with each system. The units used in this table do not conform to a single system of units (such as SI), but are commonly used because of their convenience in the atomic domain.

A number of characteristics of this potential are worthy of note. First of all, this potential has a pairwise additive form. Further, it depends on particle positions $(\vec{r}_1, \vec{r}_2, \ldots, \vec{r}_N)$ only through the $N(N-1)/2$ pairwise distances $|\vec{r}_i - \vec{r}_j|$. As a consequence, it respects the homogeneity (translational invariance) and isotropy of the space in which the particles lives. For any pair of particles, the potential has a shallow minimum at distance

$$\sigma_* = 2^{\frac{1}{6}}\sigma. \tag{24.3}$$

For distances less than this, the interactions become sharply repulsive, whereas for distances larger than this, the interactions are weakly attractive. The characteristic energy and length scales in this potential are, respectively, 4ϵ and σ; values of these parameters for a number of noble gases are given in Table 24.1.

The Lennard-Jones force $-\nabla_i V(\vec{r}_1, \vec{r}_2, \ldots, \vec{r}_N)$ on the i particle can be written as

$$\vec{F}_i(\vec{r}_1, \vec{r}_2, \ldots, \vec{r}_N) = \sum_{j(\neq i)=1}^{N} f(r_{ij})(\vec{r}_i - \vec{r}_j),$$

$$\text{where} \quad f(r_{ij}) = 6\frac{4\epsilon}{\sigma^2}\left\{2\left(\frac{\sigma}{r_{ij}}\right)^{14} - \left(\frac{\sigma}{r_{ij}}\right)^{8}\right\}. \tag{24.4}$$

System of Units for Simulation

When setting up a MD simulation, it is not very convenient to work in standard measurement units (such as Kelvin, Joule, meter, or kilogram) for at least two reasons:

1. From a computational perspective: Real numbers (*real* as in mathematics) with a fractional part are represented in typical modern digital computers as *floating-point* numbers that allow a finite and fixed amount of storage space per number, both in the *significand* (also called *mantissa*) and the *exponent*. The set floating-point numbers (and the resulting finite-precision arithmetic) has many peculiar characteristics (more about this later in the chapter).

If we choose to stick to, say, SI units, then numerical magnitudes of most quantities are likely to be either too small or two large for the finite-precision arithmetic to handle in a meaningful fashion. This usually results in a catastrophic loss of precision, making the numerics unreliable.

2. From a purely physical perspective, it is always a good idea to work with a set of units that is based on the natural length/time/energy/mass scales in the problem. This way, the results of an analysis or computation become independent of the specific details of a system. The mathematical reason for this is that such choice of units transforms the equations of motion into a generic, dimensionless form.

For the Lennard-Jones potential (as applied to noble gases, for example), this means that if the set of simulation units is based on the natural scales of 4ϵ (energy), σ (length), and m (mass), then one single analysis or computation would become applicable equally to all noble gases (it would just need to be rescaled for the specific values of these quantities for a specific system).

Example. For a Lennard-Jones system, the above choice of amounts to setting $4\epsilon = \sigma = m = 1$ in Eq. 24.1–24.4. This choice also implies $\sqrt{m\sigma^2/4\epsilon}$ as the unit of time (See Table 24.1 for typical values of this time scale). Another conventional choice of units is $m = 1, \epsilon = 1$, and $\sigma_* = 1$.

Example. Consider the equation of motion of a one dimensional simple harmonic oscillator

$$\frac{d^2 x}{dt^2}(t) = -\frac{k}{m}x(t). \tag{24.5}$$

There is no natural length scale in this problem. However, the natural time scale in this system is ω^{-1}, where $\omega = \sqrt{k/m}$ is the frequency of oscillation of the simple harmonic oscillator.

24.2.2 Integrating Equations of Motion: the Verlet Algorithm

Having obtained the equations of motion of a system of particles, the next task is to solve them numerically to get particle trajectories. One of the simplest, robust, and hence popular, method for numerically integrating equations of motion (Eq. 24.1) is the Verlet algorithm. In the present context, the term "numerical integration of a differential equation" simply means obtaining the values of positions $\vec{r}_i$ at time $t + \delta$ from their values at time t. The Verlet scheme is obtained by Taylor-expanding a particle's trajectory $\vec{r}(t+\delta)$ both forward and backward in time upto order 3 in δ; i.e.,

$$\vec{r}(t+\delta) = \vec{r}(t) + \frac{\delta}{1!}\vec{v}(t) + \frac{\delta^2}{2!}\vec{a}(t) + \frac{\delta^3}{3!}\vec{b}(t) + O(\delta^4)$$

$$\vec{r}(t-\delta) = \vec{r}(t) - \frac{\delta}{1!}\vec{v}(t) + \frac{\delta^2}{2!}\vec{a}(t) - \frac{\delta^3}{3!}\vec{b}(t) + O(\delta^4),$$

where $\vec{v}$ and $\vec{a}$ stand, respectively, for the velocity and the acceleration the particle, and $\vec{b}$ for the third derivative of $\vec{r}$ with time. With some rearrangement after adding the above two expansions, we obtain the expression for the Verlet iterative solver:

$$\vec{r}(t+\delta) = 2\vec{r}(t) - \vec{r}(t-\delta) + \delta^2 \vec{a}(t). \tag{24.6}$$

It is easy to see that the error (per time step of δ) in the resulting trajectory is $O(\delta^4)$, that is, of the order of the first *neglected* term in the expansions above.

To use this recurrence operationally, one needs to specify two initial position values $\vec{r}(0)$ and $\vec{r}(\delta)$ along a trajectory, from which the trajectory is computed forward in time using the above recurrence relation given a fixed and sufficiently small value of the *time step* parameter δ. The meaning of "sufficiently small" needs to be decided on a case by case basis by assessing the numerical stability of the Verlet method (see Figure 24.1 and the accompanying discussion), and from physical considerations (such as the allowable extent of numerical non-conservation of total energy).

Example. Consider the one dimensional simple harmonic oscillator (Eq. 24.5). Given two initial values at $t = 0$ and δ, the corresponding Verlet recurrence takes the form

$$x(t + \delta) = \left(2 - \delta^2 \frac{k}{m}\right) x(t) - x(t - \delta),$$

and generates the trajectory values $x(0), x(\delta), x(2\delta), \ldots$. Velocities would need to be computed, for $t = \delta, 2\delta, \ldots$ as

$$v(t) = \frac{x(t + \delta) - x(t - \delta)}{2\delta}.$$

The Velocity Verlet Algorithm

Often times one is interested in computing the values of particle velocity along the trajectory. The Verlet algorithm in the form developed above does not directly incorporate this computation in its structure. Instead, one has to resort to some sort of finite-difference approximation for the velocity, such as

$$\vec{v}(t) = \frac{\vec{v}(t + \delta) - \vec{v}(t - \delta)}{2\delta}, \tag{24.7}$$

in order to compute velocities.

A variant of the Verlet algorithm, called the velocity Verlet method, resolves this situation by setting up a double recurrence in position and velocity. Given initial conditions $\vec{r}(0)$ and $\vec{v}(0)$, this modified recurrence takes the form

$$\begin{aligned}
\vec{r}(t + \delta) &= \vec{r}(t) + \delta \vec{v}(t) + \frac{\delta^2}{2} \vec{a}(t) \\
\vec{v}(t + \delta) &= \vec{v}(t) + \frac{\delta}{2} \left(\vec{a}(t) + \vec{a}(t + \delta)\right).
\end{aligned} \tag{24.8}$$

Here, $\vec{a}(t + \delta)$ is computed from the updated positions $\vec{r}(t + \delta)$.

Example. Consider the one dimensional simple harmonic oscillator (Eq. 24.5) again. Given initial values of x and v at $t = 0$, the corresponding Velocity Verlet recurrences take the form

$$\begin{aligned}
x(t + \delta) &= \left(1 - \frac{\delta^2}{2} \frac{k}{m}\right) x(t) + \delta v(t) \\
v(t + \delta) &= v(t) - \frac{\delta}{2} \frac{k}{m} \left(x(t) + x(t + \delta)\right).
\end{aligned} \tag{24.9}$$

and generates the phase-space trajectory $(x(0), v(0)), (x(\delta), v(\delta)), (x(2\delta), v(2\delta)), \ldots$.

The Verlet Method: Energy Conservation and Numerical Stability

The energy conservation behaviour of the Verlet method (in its numerical stable regime) is a consequence of the time-reversal invariance of the Verlet recurrence; i.e., the form of the Verlet recurrence (Eq. 24.6) remains unchanged under the transformation $\delta \rightarrow -\delta$.

Example. Consider the velocity Verlet method (Eq. 24.9) as applied to the one dimensional simple harmonic oscillator (Eq. 24.5) again. The behaviour of the Velocity Verlet solver as a function of the time step parameter δ is illustrated in Figure 24.1 for the simple harmonic oscillator with $k = m = 1$. For $\delta \leq 0.1$, we see that the computed solution is quite close to the true solution $\cos(t)$, the phase-plane curve traced by the oscillator is indeed a circle with unit radius, and the total energy $(x^2 + v^2)/2$ is nicely conserved (within numerical fluctuations that are not visible on the scale of the y-axis), We see that the numerical solution, although it remains oscillatory, starts differing from the exact solution increasingly as δ increases. This is also reflected as a progressively shrinking and distorting phase-plane curve, and progressively larger fluctuations in the total energy (which should ideally remain constant at the value of $1/2$). Beyond $\delta = 2$, the Verlet recurrence (Eq. 24.9) produced unphysical, non-oscillatory solutions (not shown in the figure). Whether this behaviour with respect to the time step is a numerical artifact can only be answered through a formal stability analysis; see below.

It is important to assess the numerical stability of a differential equation solver with reference to the differential equation of interest. Such analysis provides, in the least, an upper bound on the allowable time step δ. A full treatment of this topic is not possible within the scope of this chapter; however, we discuss below how this could be done for the Verlet solver (Eq. 24.6) as applied to the simple harmonic oscillator.

Example. Let us write the simple harmonic oscillator equation in the form

$$\frac{d^2 x}{dt^2}(t) = -\omega^2 x(t).$$

Since we are interested in oscillatory solutions to this equation ($\omega^2 > 0$), let us write an *ansatz* of the form

$$y(t) = y_0 e^{i\alpha t}$$

for the discrete solution $y(t)$ of the Verlet recurrence (Eq. 24.6)

$$y(t + \delta) = (2 - \delta^2 \omega^2)y(t) - y(t - \delta).$$

This will give us a handle over how far the solution y of the Verlet recurrence differs from the true oscillatory solution (of the equation of motion) of the form $x(t) = e^{i\omega t}$, as a function of the time step δ. Here, ω is the frequency of oscillation of the true solution $x(t)$ to the original equation of motion, and α is to be interpreted as the frequency of the solution $y(t)$ to the Verlet recurrence. Substituting this *ansatz* in the Verlet recurrence, we get

$$y_0 e^{i\alpha t} \left(e^{i\alpha \delta} - (2 - \omega^2 \delta^2) + e^{-i\alpha \delta} \right) = 0,$$

which implies

$$e^{i\alpha \delta} - (2 - \omega^2 \delta^2) + e^{-i\alpha \delta} = 0.$$

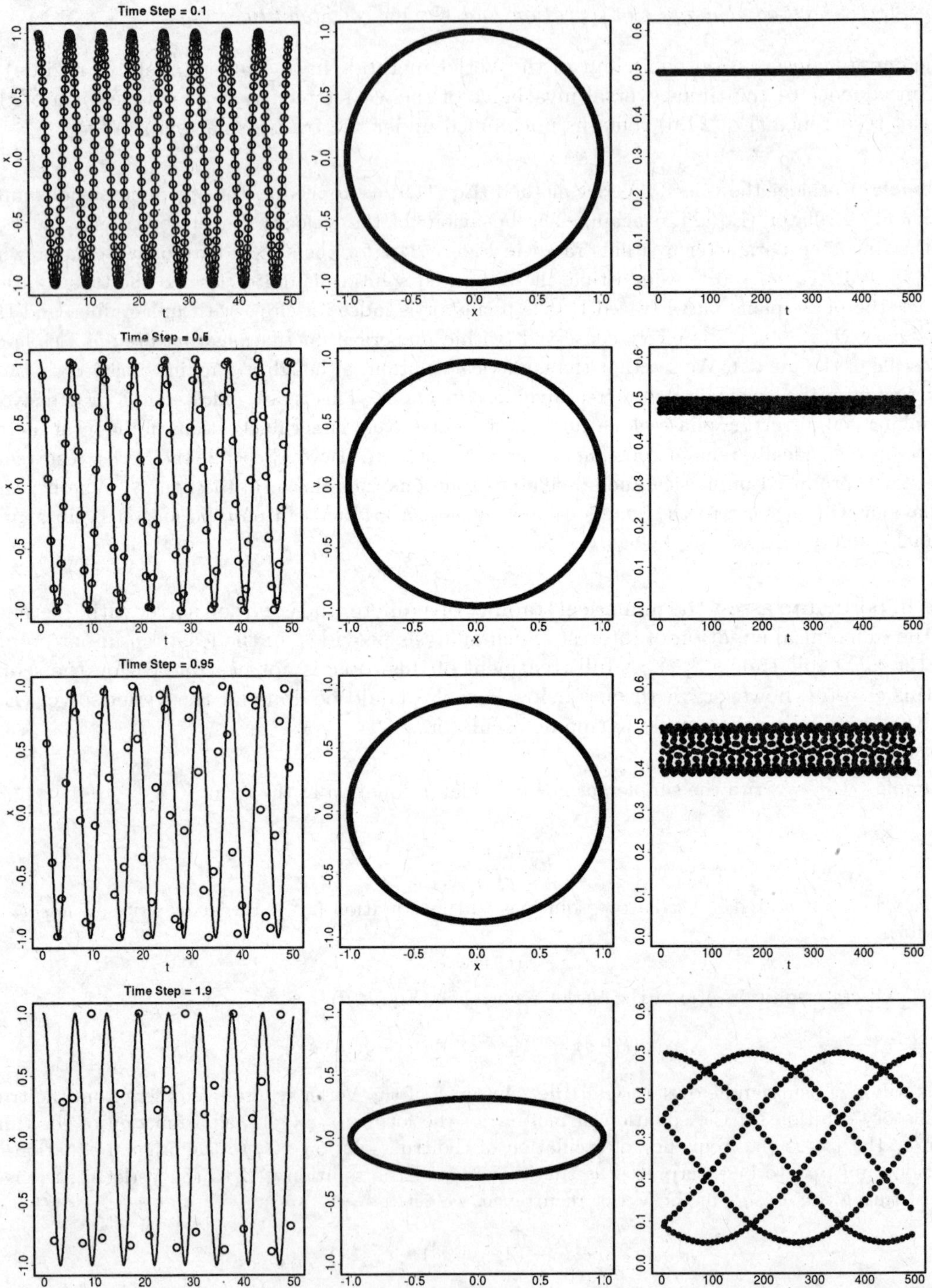

Figure 24.1: Behaviour of the velocity Verlet algorithm for the simple harmonic oscillator. The four rows correspond to $\delta = 0.1, 0.5, 0.95, 1.9$ respectively. For each row, the first column compares the computed solution (circles) for $x(t)$ with the exact solution $\cos(t)$ (continuous curve), the second column plots the energy ellipse in the x–v phase plane, and the third column plots total energy $(x^2 + v^2)/2$ as a function of t.

With some algebraic manipulations, this reduces to the identity

$$\omega^2 \delta^2 = 4 \sin^2 \frac{\alpha \delta}{2}.$$

Clearly, there is no real value of α that would satisfy the above identity when $\delta^2 > 4/\omega^2$. This analysis shows that

1. as observed in the numerical results, $\omega \delta = 2$ represents the boundary between physical (oscillatory) and unphysical (non-oscillatory) solutions of the Verlet recurrence, and

2. this observed behaviour is not an artifact of the finite-precision numerics that was used to compute the solutions.

From purely physical considerations it could be argued that the choice of the time step parameter δ should be such that the fastest motions in the system are adequately represented in the numerical solution. Thus, e.g., when choosing a value for δ for the velocity Verlet algorithm, the typical velocity magnitudes expected to occur in the system being simulated should also be considered. These magnitudes would depend on, amongst other things, the initial velocities, and the typical magnitudes of forces/accelerations expected to occur in the system.

24.2.3 Schematics of a Bare-Basics MD Simulation

Having discussed the key elements of a MD simulation, let us return once again to a system of interacting particles (e.g., via the Lennard-Jones potential). Assuming velocity Verlet, the schematic structure of a bare-basics MD simulation is as follows:

1. Choose a convenient system of units; see discussion in Sec. 24.2.1. Choose parameters that define the system, such as the dimensionality D of the space, number N of particles in the system. Also choose values of simulation parameters such as the time step size δ, the total number N_δ of such steps, etc.

2. Specify the initial conditions for this system of N particles: For velocity Verlet, this means the initial positions $\vec{r}_i(0)$ and the initial velocities $\vec{v}_i(0)$, $i = 1, \ldots, N$.

3. For $t = \delta, 2\delta, \ldots, N_\delta \delta$:

   - compute new positions $\vec{r}_i(t)$ and velocities $\vec{v}_i(t)$ of all the particles $i = 1, \ldots, N$ from the previous positions $\vec{r}_i(t - \delta)$ and velocities $\vec{v}_i(t - \delta)$ using the velocity Verlet recurrence (Eq. 24.8).

   - store this updated information for later analysis, or compute any relevant quantities such as the total kinetic and potential energies, etc.

By construction, this simulation conserves the number of particles N and the total energy E. The volume of the system is not conserved in the present bare-basic scheme. Constant volume simulations need the system to be enclosed in a fixed-volume enclosure; this feature will be added in Sec. 24.2.5 using a device called periodic boundary conditions.

This all needs to be implemented using an appropriate programming language (such as `Fortran, Scheme, Haskell, C, C++, Java, ...`) or a scripting/computing environment (such as `python`, `matlab`, `R`, ...). The key to good computational work is the spirit of self-learning and exploration. We thus leave the programming-related details for the reader to fill-in. A variety of books on this subject already include enough details on programming and, oftentimes, actual codes[1]. A selection of such resources is included in the bibliography section at the end of this book.

24.2.4 Breathing-Mode Oscillations of Highly Symmetric Clusters

Having gained some insight into the behaviour of the Verlet solvers, let us explore the dynamics of a bit more complex system, i.e., highly symmetric Lennard-Jones clusters. Such systems whose physics is quite well-understood and where one knows what to expect at least qualitatively, are highly valuable as test cases for establishing the sanity of the simulation system and for validating the computer programs for the simulation. They often provide unexpected insights into the behaviour of the simulation system (the computer, programming language used, simulation codes, etc., taken as a whole).

Specifically, let us consider a cluster of 4 particles that are confined to the x–y plane, and interact with one another via the Lennard-Jones potential (Eq. 24.2). Our choice of simulation units is defined by $4\epsilon = \sigma = m = 1$ (see Sec. 24.2.1 and Table 24.1). For initial positions, let us place them at the corners of a square with side length $a = 0.89 \times \sigma_*$, centred at the origin, and sides parallel to the coordinate axes. Let us set the initial velocities to zero. For this problem, let us choose a time step $\delta = 0.05$ (in units of $\sqrt{m\sigma^2/4\epsilon}$).

As noted before, the Lennard-Jones potential is invariant under translations and rotations of the system, i.e., it respects these two symmetries of the space in which the system lives. It is clear from the symmetry of our initial state that all forces in this system act along the diagonals of the square. As such, our four particles will always be placed at the corners of a square at all times, and the dynamics of this system is completely determined by the initial size a of this square. For example, if $a \ll \sigma$, then all pairwise interactions will be highly repulsive, leading to a sharp explosion, with the four particles flying away to infinity along the four diagonal directions. If $a \gg \sigma$, then the weakly attractive interaction would lead to an initial collapse of the system followed possibly by an explosion. For $a \approx \sigma_*$, we expect to see stable breathing-mode oscillations of this square configuration. Indeed, these expectations are fulfilled as seen in a series of simulation snapshots in Figure 24.2 for the oscillatory regime.

Our Visualization Scheme. Our convention for visualizing instantaneous configurations of Lennard-Jones particles confined to a two-dimensional plane is as follows: we represent each Lennard-Jones particle by a circle of radius σ_*. With this convention, overlapping and non-overlapping pairs of circles respectively represent pairwise repulsive and attractive interactions. The centre of each circle is emphasized with a black dot, and the gray trail represents a short-time trace of this black dot over the immediate past. Depending on the particle trajectories, this trace gets smeared or erased.

---

[1]A suite of `Fortran` codes that accompanied the famous book *Computer Simulations of Liquids* by Allen and Tildesley (Oxford University Press, 1987) is available at the website `http://www.ccp5.ac.uk/librar.shtml#ALLENTID`

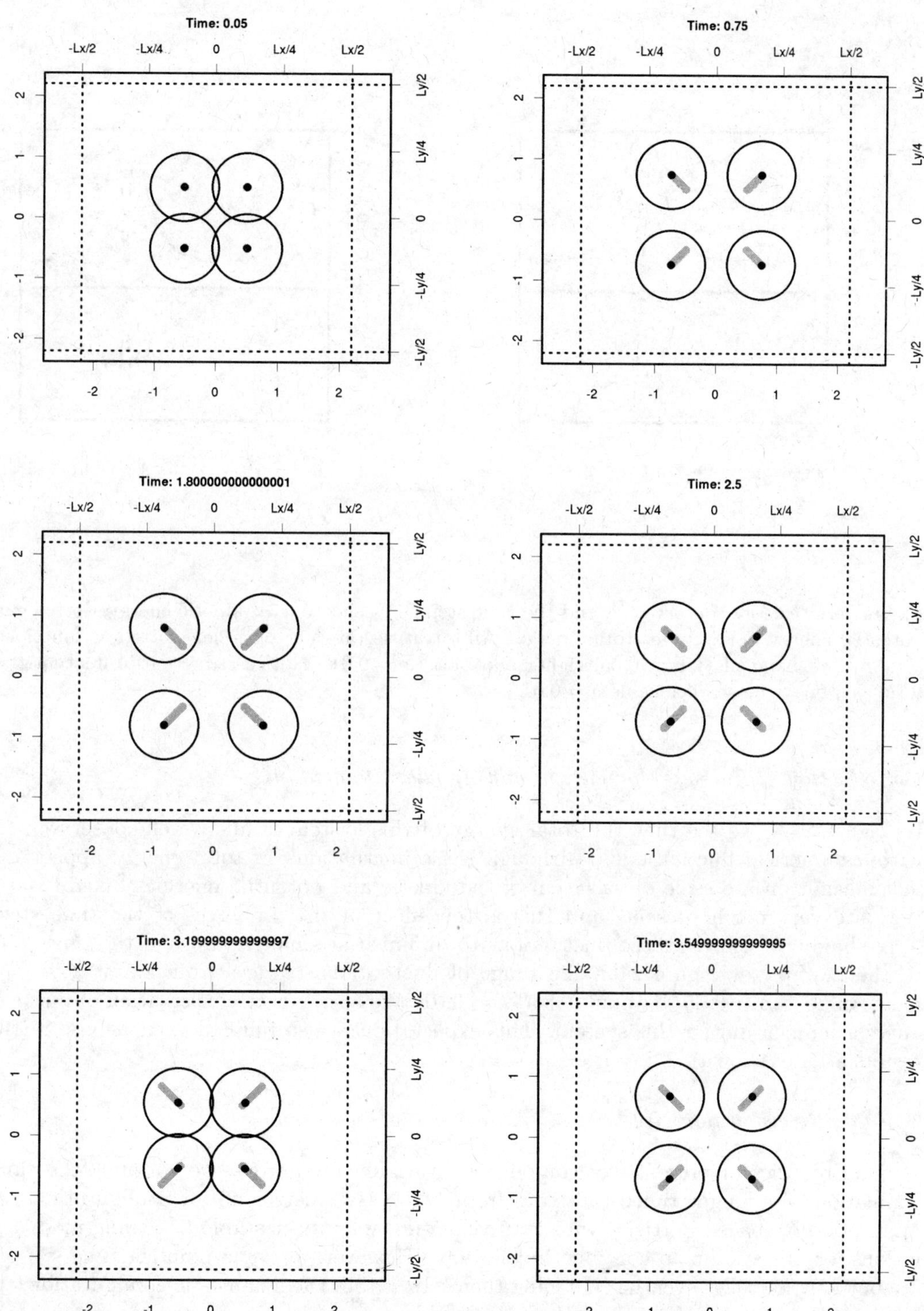

Figure 24.2: Snapshots of an oscillating Lennard-Jones square at $t = 0.05, 0.75, 1.8, 2.5, 3.2, 3.55$ (rowwise from top left to bottom right). See discussion in the text and the note on the visualization scheme used.

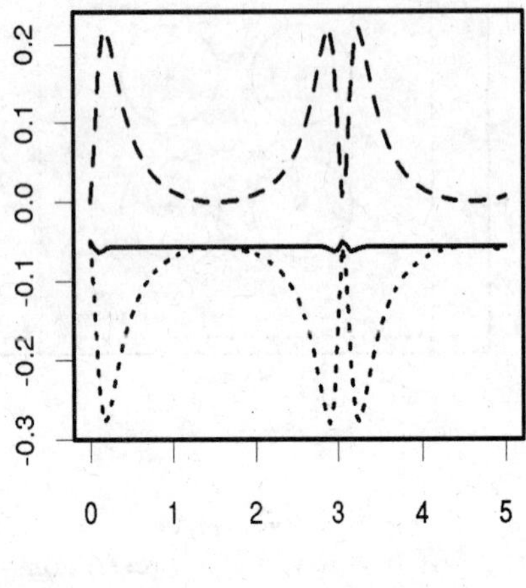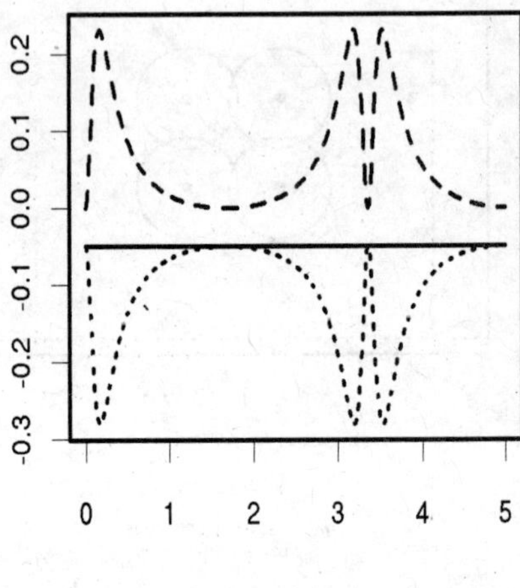

Figure 24.3: Energy conservation and the time step δ. Dashed curve: kinetic energy, dotted curve: potential energy, solid curve: total energy. All quantities are per particle, plotted as functions of time t. Left-hand plot: $\delta = 0.05$, right-hand plot: $\delta = 0.01$. Notice reduction of fluctuations in total energy at the smaller value $\delta = 0.01$.

Conservation of Energy, Momentum, and Angular Momentum

In Figure 24.3, we see that the total energy of this system is fairly well-conserved – i.e., within numerical fluctuations. Although these fluctuations in total energy appear quite insignificant on the scale of variation of the kinetic and potential energies taken together, they are very much present, and this is the effect of the finiteness of the time step δ. Accordingly, we expect these fluctuations to diminish at smaller values of the time step δ. In the example of Figure 24.3, the range of fluctuations reduces from about 25% of the mean value at $\delta = 0.05$ to about 1% at $\delta = 0.01$. We also expect conservation of linear and angular momentum for this system. This expectation is also fulfilled extremely well (these results are not shown).

A Finite-Precision Surprise

Ground realities of present-day computing invariably hit hard if we continue the similar simulations for longer time. Starting from highly symmetric initial configurations with appropriately spaced particles with zero velocities, we expect stable breathing-mode oscillations for our system to continue indefinitely. However, at some point in time, we often see our four particles eventually freeing themselves from the symmetric configuration, their motions becoming randomized and unpredictable.

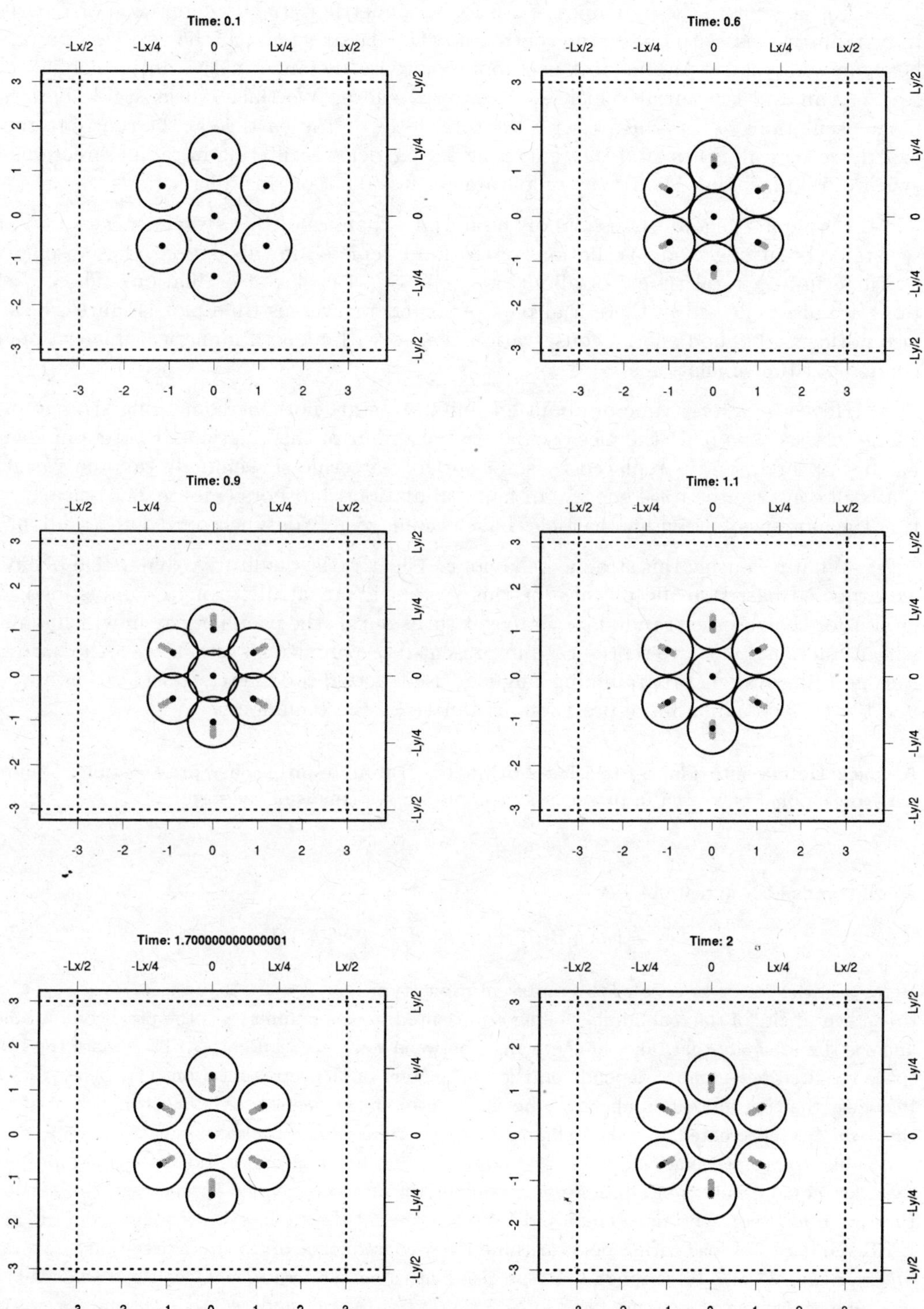

Figure 24.4: Snapshots of a hexagonal oscillating Lennard-Jones cluster at $t = 0.1, 0.6, 0.9, 1.1, 1.7, 2.0$ (rowwise from top left to bottom right).

Let us explore the dynamics of a 6-fold symmetric initial configuration of 7 particles in two dimensions, with one at the centre and the rest six sitting at the vertices of a regular hexagon. Let us set the nearest-neighbour pairwise distance in the initial configuration to $1.2\sigma_*$ and all the initial velocities to zero as before. We thus expect stable breathing-mode oscillations of this cluster, with the total force on the particle at the centre to remain exactly zero at all times, and the remaining six particles oscillating in radial directions in a synchronized fashion. The time step parameter $\delta = 0.05$ for this illustration.

Figure 24.4 shows snapshots of the initial dynamics of this system, where we indeed see stable breathing-mode oscillations of the kind depicted in this figure. These oscillations continue upto a total time t of about 20 units, or 14 cycles of oscillation. These oscillations are also reflected in the initial part of the energy versus time plot (Figure 24.6). We also notice fairly good energy conservation (i.e, to within small numerical fluctuations) all throughout the simulation.

However, after a time of about 24 units, we start noticing something strange in the energy versus time plot: the nice oscillatory behaviour of the kinetic and potential energies has disappeared, and is replaced by some sort of patternless, seemingly random variations that still conserve the total energy. In fact, simulation snapshots (Figure 24.5) clearly show that the motions of individual particles have become visibly desynchronized and randomized.

What is causing this strange behaviour? Clearly, the qualitative dynamical behaviour we expected based on the physics for this system is not at all wrong or unjustified. The culprit for the observed randomization, as it turns out, is the peculiar way in which numbers with a fractional part are represented in present-day computing systems (i.e., the hardware, the operating system, programming language used, actual codes, etc., considered as a whole), and the corresponding finite-precision arithmetic of these numbers.

A Quick Detour into Finite-Precision Arithmetic The finite-precision representation of numbers with a fractional part used in present-day computing systems is of the form

$$d_0.d_1 \ldots d_{p-1} \times \beta^e,$$

which stands for the real number

$$\left(d_0 + d_1\beta^{-1} + \ldots + d_{p-1}\beta^{-(p-1)}\right)\beta^e.$$

Here, β is the (integer) base of the representation (typically $\beta = 2$), $0 \leq d_0, d_1, \ldots, d_{p-1} < \beta$ are the integer digits of the real number being represented in this manner, p is the precision parameter, and e is the *exponent* that takes integer values between some e_{min} and e_{max}. The size of the storage space required per number depends on the parameters of the representation, i.e., $\beta, p, e_{min}, e_{max}$. It is clear that not all real numbers can be represented in this fashion. The resulting set of numbers that *can* be represented in this fashion, called *floating-point numbers*, is necessarily a *discrete* set, unlike the (mathematical) set of real numbers, and has non-uniform distribution (specifically, there are equal numbers of floating-point numbers in intervals defined by successive *powers* of β). Floating-point representations employed by most present-day computing systems conform to the IEEE standard 754 for floating-point arithmetic. A consequence of the finite precision is the *round-off error* that propagates through finite-precision arithmetic operations. Because of this, the result of a computation – e.g., potential and force computation, Verlet update, etc. – depends on the order

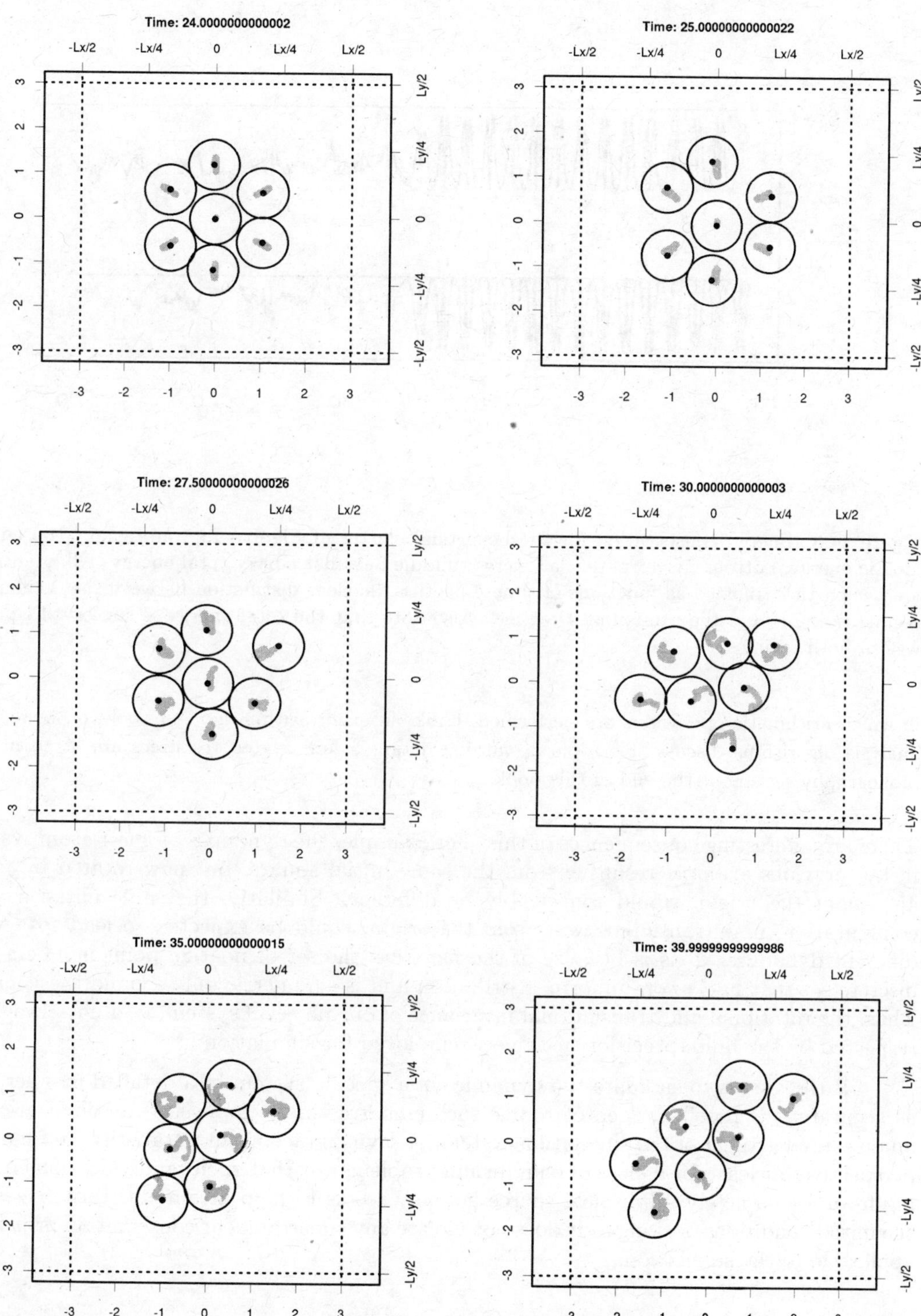

Figure 24.5: Snapshots of the (randomized) Lennard-Jones hexagon at $t = 24, 25, 27.5, 30, 35, 40$ (rowwise from top left to bottom right).

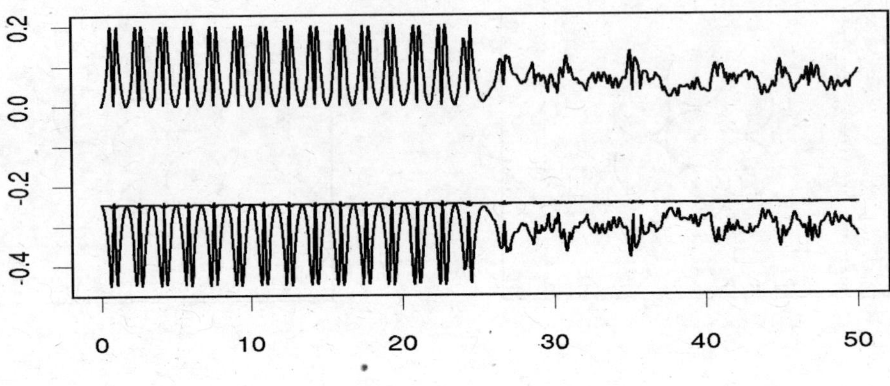

Figure 24.6: Energy conservation for the hexagonal cluster of Figures 24.4 and 24.5. Top curve: kinetic energy, bottom curve: potential energy, middle near-flat curve: total energy. All quantities are per particle, plotted as functions of time t. Notice the clear distinction between the oscillatory regime ($t < 24$) and the randomized regime. Also note that the total energy is conserved equally well in both regimes.

in which arithmetic operations are performed, unlike in exact arithmetic. All books on numerical analysis/algorithms discuss this topic in varying detail; a few related resources are listed in the bibliography section at the end of this book.

There are interesting consequences of this. For example, the dynamics of our 4-atom system in the previous example resulting from the same initial square, but now rotated by (say) 45° about the origin, would conceivably be different. Similarly, the same initial square configuration, now translated away from the origin, could be expected to lead to a very different dynamics: This is because of the fact that the set of floating-point numbers is a discrete set that has a non-uniform distribution and is symmetric only around the value 0. Thus, the rotational and translational invariance of our physical system need not always be respected by the finite-precision arithmetic employed for simulation.

If the idea is to simulate the dynamics of a specific system in a detailed manner and as accurately as possible, then of course such embarrassingly unphysical randomization is an unwanted artifact of the simulation system. A saving grace, from a statistical mechanical perspective (at least for a class of equilibrium problems), is that such randomization emerging from a completely unphysical source actually seems to help attain equilibrium *despite* the initial conditions, so long as it does not violate any constraints or conservation principles applicable to the simulation.

It must be understood that such artifacts of the finite-precision arithmetic are highly dependent on how the simulation is implemented in a programming language or a computing environment, initial conditions, the value of simulation parameters such as the time step δ, etc. The possibility of encountering such seemingly strange artifacts must always be kept in mind.

24.2.5 Simulating Extended Systems

From a statistical mechanical perspective, the interest is often focussed on understanding (or predicting) properties of *extended* systems consisting of large (technically, infinite) number of particles, unlike the small finite systems we explored in the previous section. On the other hand, simulations can only be done for a *finite* number of particles. The grand challenge in the simulation domain, then, is to infer the behaviour of a system in the thermodynamic limit from finite-size simulations. This is usually a hard and a tedious task.

Eliminating Surfaces Using Periodic Boundary Conditions

The least that could be done to finite system to make it resemble an infinite system is to eliminate all surfaces – Extended systems have no surfaces. The simplest way of incorporating this characteristic of extended systems into a simulation is a device called *periodic boundary conditions* (PBC).

For example, a line segment of (finite) length L has two boundaries – its left and right end points. PBC turn this line segment into a circle of circumference L by joining together the two end points. Mathematically, this is equivalent to a x *modulo* L operation, i.e., remainder of the division of x by L, which effectively maps all real numbers x onto points on this circle. The same transformation, when applied to a square, turns it into a two-dimensional torus. It could be extended to rectangular regions in any number of dimensions by applying the *modulo* L transformation to each orthogonal coordinate direction.

As the space in which a physical system lives gets wrapped around in this fashion, the interactions between particles also need to be wrapped around. This is especially true when the range of interactions (e.g., a few σ for Lennard-Jones particles) is comparable to the length L of the "simulation box" – This possibility is typically realized when the density ρ of the physical system being simulated is high enough, and the simulation box size L is chosen to match this density value for a fixed number N of particles in the simulation.

A conventional way of visualizing this wrapping-around is to consider the simulation box and its "images" which are translated from itself by integer multiples of L along all coordinate directions (see Figure 24.7). In two dimensions, e.g., the first shell of images consists of 8 images which are translations of the simulation box by translation vectors $(-L, -L)$, $(-L, 0)$, $(-L, +L)$, $(0, +L)$, $(+L, +L)$, $(+L, 0)$, $(+L, -L)$, and $(0, -L)$. The wrapping-around of interparticle interactions is then taken into account by considering the interaction of each particle in the simulation box not only with all other particles in the simulation box but also with the images of all particles in a shell of images of the simulation box.

PBC are typically implemented in MD simulations by identifying, at each time step,

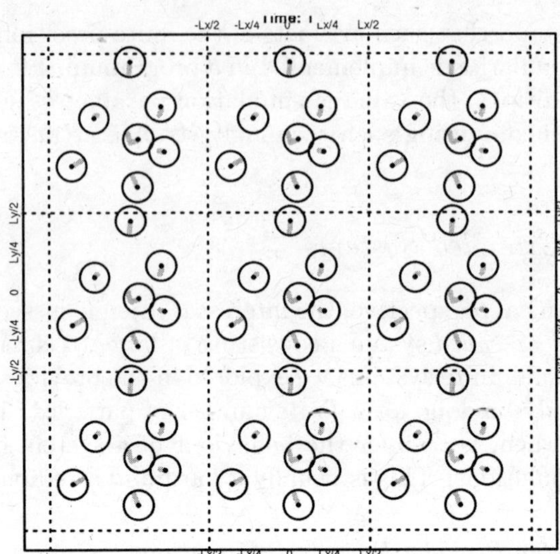

Figure 24.7: Periodic boundary conditions (PBC) visualized as the simulation box (central square) and the first shell of its images. Under PBC, particles leaving the simulation box from the right (top) wall re-enter from the left (bottom) wall and vice versa. A particle is considered to interact not only with all particles in the simulation box, but with all image particles as well. By construction, all image particles mimic their counterpart in the simulation box at the centre.

the particles that crossed the boundaries of the simulation box during that time step, and bringing them back from the opposite face. Specifically, if coordinate values within the simulation box lie between $-L/2$ and $L/2$ along each coordinate direction, then this prescription amounts to replacing, e.g., x by either $x - L$ or $x + L$ depending on whether $x > L/2$ or $x < -L/2$. This way of implementing PBC (instead of the formal *modulo L* operation) assumes that the time step δ is reasonably small enough; i.e., the fastest-moving particle in the system travels at most a small fraction of the size L of the simulation box.

To accommodate PBC in the schematics of Sec. 24.2.3, one needs to specify additionally the simulation box size L and the number of image shells around the simulation box. Typically, the density ρ of the physical system being simulated (see, e.g., Table 24.1) is used to determine L by fixing the number of particles N in the simulation.

With the inclusion of PBC, our MD simulations will now conserve the volume V of the system, in addition to the particle number N and the total energy E. Such microcanonical simulations are sometimes referred to as constant NVT simulations.

Periodic Boundary Conditions Violate Isotropy

A side-effect of incorporating PBC in a simulation must be noted: While the translational invariance in a system remains unbroken in presence of PBC, the PBC do break the isotropy (or rotational invariance) of a system. As a consequence, unlike in the simulations of the previous section which did not incorporate PBC, the total angular momentum is not a

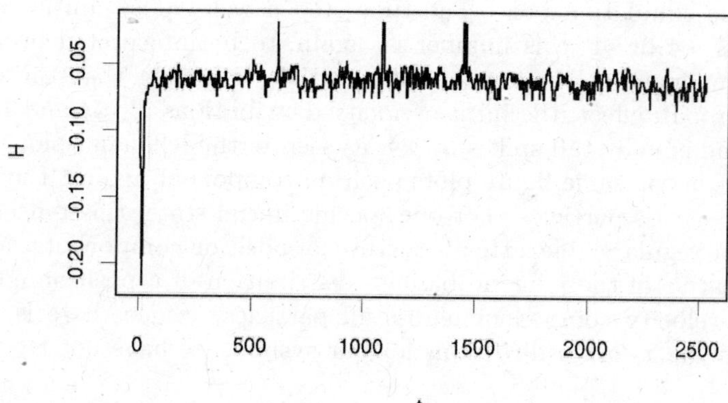

Figure 24.8: Approach to equilibrium as seen through the H function (Eq. 24.10).

conserved quantity any more (even if the original physical system is isotropic). We thus expect to see large fluctuations in it. However, over length scales far smaller than the simulation box length L, we could expect that the system would still appear *locally* isotropic.

Approach to Equilibrium

We end this section with the example of a simulation of an extended system as implemented through PBC. Broadly speaking, thermodynamic equilibrium is characterized by *stationarity* (i.e, time-independence) of the probability distributions describing the system. This implies that static properties of the system also become time-independent. In simulations, whether a system has attained equilibrium or not can thus be judged by monitoring the time evolution of a static property of the system. For example, for a two-dimensional system, we may define a quantity of the form

$$h = \frac{1}{2}(h_x + h_y), \text{ where}$$

$$h_x = \int P(v_x)\log(P(v_x))dv_x, \text{ and}$$

$$h_y = \int P(v_y)\log(P(v_y))dv_y, \tag{24.10}$$

and $P(v_x)$ is, e.g., the probability distribution function (PDF) of the x-components of particlewise velocities. Figure 24.8 shows a possible behaviour of this quantity as the system attains equilibrium. From the figure, it is clear that this quantity becomes stationary fairly quickly for the initial conditions used (see below), after just about 100 time steps. It must be understood that numerical simulations with finite systems will always involve fluctuations.

This particular simulation consisted of $N = 100$ Lennard-Jones particles confined to a two-dimensional box with PBC. The simulation box length L was determined from this N and an arbitrary density value of $\rho = 0.75$ in simulation units (i.e, on an average 0.75 particles per unit σ^2). As initial positions, these 100 particles were placed on a regular square grid over the simulation box. The time step δ was 0.005. Initial velocities were randomly chosen as +4 or -4 (this number is, again, to be interpreted in our simulation units) for each component of velocity of each particle, ensuring that the centre-of-mass velocity remains zero. In effect, the initial velocity distributions $P(v_x)$ and $P(v_y)$ have the shape of discrete and equally tall spikes at ± 4, as seen in the left-hand plot in Figure 24.9. The right-hand plot in the same figure plots position component against the corresponding velocity component across particles. For our specific initial state, this reflects the velocity spikes at ± 4, and the regular cubic lattice structure for position components. More generally, such a plot is indicative of the joint probability distribution of a position component with the corresponding velocity component across all particles. Since there is no conceptual distinction between the x and y directions for our system, we have not treated or plotted them separately.

Boltzmann's H Theorem Boltzmann's H theorem states that the approach of the system to equilibrium is accompanied by minimization of what is called the H function, which is defined as

$$H(t) = \int f(\vec{r}, \vec{v}, t) \log(f(\vec{r}, \vec{v}, t)) d\vec{r} d\vec{v}.$$

Once equilibrium is attained, this quantity remains stationary at its minimum value. Here, f is the single-particle number distribution function over the phase space and a solution of the Boltzmann transport equation. A 1971 paper by E.T. Jaynes[2] argues that the H theorem can be violated by systems with appreciable potential energy starting from specific classes of initial conditions and, in fact, the approach to equilibrium here is accompanied by a *maximization* of the H function. These classes of initial conditions are common enough to be encountered in experiments and simulations. Lennard-Jones systems, especially at the high density value (0.75) and initial conditions used for the simulation behind Figure 24.8 are also known to violate the H theorem[3]. The quantity h that we defined (Eq. 24.10) is a kind of H function that is restricted to velocities.

It is interesting to look at how the velocity distributions themselves evolve to stationarity as the system approaches equilibrium. Starting from a velocity distribution with sharp peaks at ± 4 (Figure 24.9), Figure 24.10 shows the distribution of velocity components (bottom left) and componentwise position-velocity correlations (bottom right) as before. This figure is a snapshot of these quantities over the first 50 times steps, that is for $0 \leq t < 50$. In principle, the same figure could have been done for each time step separately; however, such pooling together of data across time steps help reduce the fluctuations to some extent and make the generic features of the distributions stand out clearly. In this figure, we also see the distribution of velocity *magnitudes* (top left), and and position-velocity *magnitude* correlations (top right).

Notice how the original sharp peaks at ± 4 at $t = 0$ got smeared out by the process of equilibration. This smearing effect is also clearly seen in both the right-hand plots when

---

[2]E.T. Jaynes, *Violation of Boltzmann's H Theorem in Real Gases*, Phys. Rev. A$\underline{4}$, 747–750 (1971).

[3]V. Romero-Rochin and E. Gonzalez-Tovar, *Comments on Some Aspects of Boltzmann H Theorem Using Reversible Molecular Dynamics*, J. Stat. Phys. $\underline{89}$, 735–749 (1997).

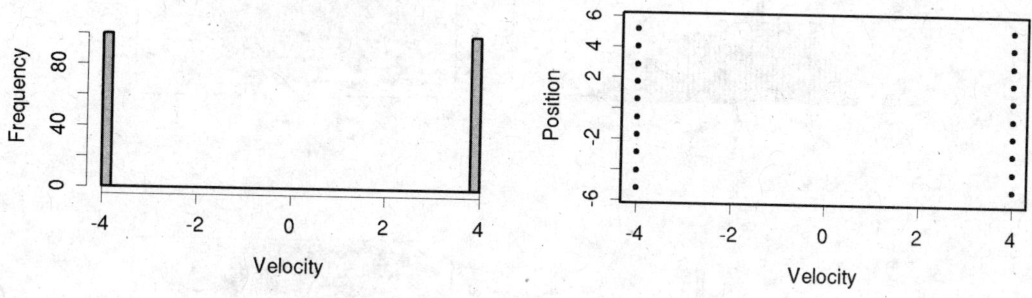

Figure 24.9: Left: Initial distribution of velocity components. Right: Componentwise position-velocity correlations.

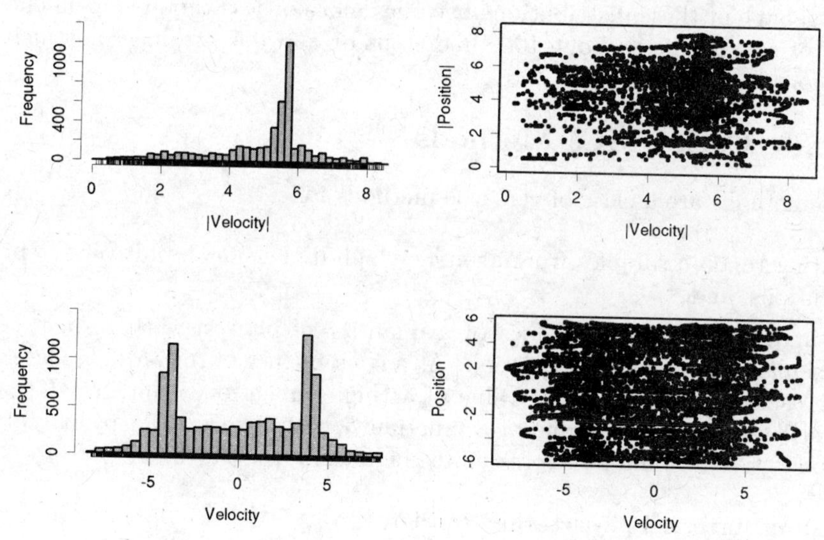

Figure 24.10: $t = 0.25$ (time step 50). Top left: Velocity *magnitude* distribution. Top right: position-velocity *magnitude* correlations. Bottom left: Velocity component distribution. Bottom right: Componentwise position-velocity correlations.

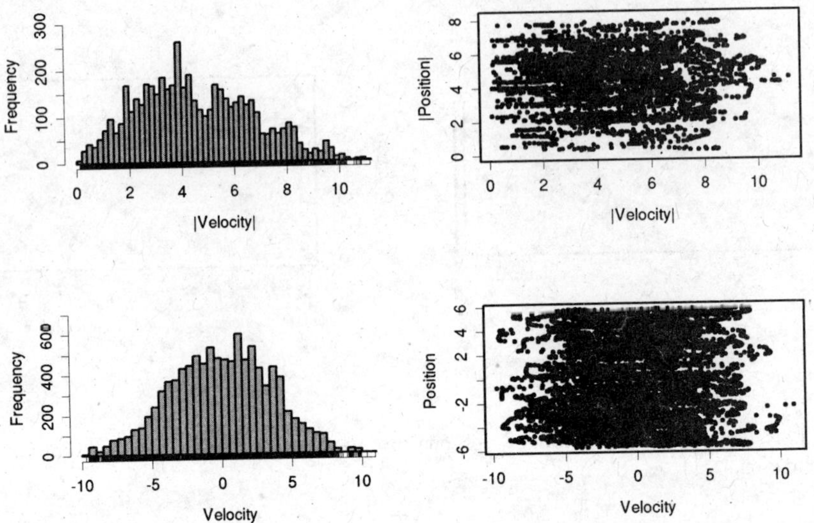

Figure 24.11: $t = 0.5$ (time step 100). Conventions same as in Figure 24.10.

compared to the original discrete, ordered picture at $t = 0$ in Figure 24.9. We just state here that the shape of these distributions becomes more or less time-independent (barring fluctuations, of course) after about 100 time steps or $t = 0.5$ (we have not included any figure to demonstrate this).

24.3   Monte Carlo (MC) Methods

Monte Carlo methods are a class of versatile methods for

- generating random samples from arbitrary high-dimensional multivariate probability distributions, and
- for estimating expectation values of a quantity of interest with respect to a high-dimensional multivariate distribution. As a consequence of this, Monte Carlo methods can be used for estimating the value of a (high-dimensional) integral I that can be expressed as expectation value of a function h with respect to a probability density function f (i.e., I could be expressed in the form $\int h(x)f(x)dx$).

In the context of statistical physics, this translates to

- randomly sampling configurations (or phase space points) of a system in such a way that their distribution follows the probability density function of an appropriate ensemble (e.g., the canonical density function $f(\cdot) = \exp\left(-\beta E(\cdot)\right)/Z$, and
- estimating expectation values of physical properties of interest with respect to this ensemble-specific probability density function f.

It is perhaps no wonder that Monte Carlo methods originated (in the form of the celebrated 1953 paper by Metropolis et al.[4]) in the realm of statistical physics which is replete with high-dimensional distributions and integrals. What makes these methods immensely useful tools for exploring probability distributions is the fact that these methods do not require that the normalization constant (e.g., $1/Z$) be known. The formal basis of why (and under what conditions) Monte Carlo methods work is founded in the theory of Markov chains.

24.3.1 The Metropolis Algorithm

Consider a system, at a constant temperature T, consisting of N particles interacting via a potential $V(\vec{r}_1, \vec{r}_2, \ldots, \vec{r}_N)$. The probability distribution function (PDF) that describes the *spatial* structure of the system has the form

$$P(\vec{r}_1, \vec{r}_2, \ldots, \vec{r}_N) = \frac{1}{Z} \exp\left\{ -\frac{V(\vec{r}_1, \vec{r}_2, \ldots, \vec{r}_N)}{kT} \right\} \equiv P(\{\vec{r}\}). \tag{24.11}$$

A *configuration*[5] $\vec{R}$ of this system is specified by the positions $(\vec{r}_1, \vec{r}_2, \ldots, \vec{r}_N)$ of all particles. We wish to generate a sample of such configurations $\vec{R}_1, \vec{R}_2, \ldots, \vec{R}_M$, of the system in such a fashion that their probabilities of occurrence are in accordance with the PDF (Eq. 24.11). In addition, we may wish to compute expectation values, with respect to this PDF, of quantities of interest (such as, average kinetic and potential energies, a quantity called the *radial distribution function* that describes spatial correlations between particles, etc.)

The Metropolis algorithm for this task, in its conventional form, is as follows. Specify an arbitrary configuration $\vec{R}_0$. If possible, choose this from a high probability density region of the PDF (Eq. 24.11). Suppose we have generated a sequence of such configurations $\vec{R}_0, \vec{R}_1, \ldots, \vec{R}_{n-1}$ using the Metropolis algorithm (to be described shortly). The next configuration $\vec{R}_n$ is generated by repeating the steps 1–4 below N times (giving each particle a chance to get displaced on an average once):

1. Choose particle i at random (with uniform probability) from the N particles of the system.

2. Propose a change in its position by giving it a random displacement, i.e.,

$$\vec{\rho}_i = \vec{r}_i + \delta \vec{u}.$$

Here, δ is a parameter that determines the maximum displacement along each coordinate. $\vec{u}$ is a random vector (with the same dimensionality D as that of $\vec{r}_i$), with each of its coordinates being picked with uniform density over the interval $(-1, 1)$. This is typically accomplished by using uniform (pseudo-)random number generators available in most programming languages, computing environments, and numerical/statistical libraries. In effect, we have picked $\vec{\rho}_i$ a random point in a D-dimensional hypercube of length δ centered on $\vec{r}_i$.

---

[4]Metropolis, Rosenbluth, Rosenbluth, Teller, and Teller, *Equation of State Calculations by Fast Computing Machines*, J. Chem. Phys. <u>21</u>, 1087–1092 (1953).

[5]The notation $\vec{R}$ in this chapter has the same meaning as $\{\vec{r}\}$ in the rest of this book. In this chapter, we have use both the notations as per convenience.

3. Compute the *acceptance probability* $a(\vec{\rho}_i, \vec{r}_i)$ for the displaced position $\vec{\rho}_i$ given the current position $\vec{r}_i$, where

$$a(\vec{\rho}_i, \vec{r}_i) = \min\left\{ 1, \frac{P(\ldots, \vec{\rho}_i, \ldots)}{P(\ldots, \vec{r}_i, \ldots)} \right\}. \tag{24.12}$$

4. Accept the proposed position $\vec{\rho}_i$ with probability $a(\vec{\rho}_i, \vec{r}_i)$. This is accomplished as follows: generate a uniform random number u from the interval $(0, 1)$. If $u < a(\vec{\rho}_i, \vec{r}_i)$, then accept the proposed change and displace the ith particle to its new position $\vec{\rho}_i$. If $u > a(\vec{\rho}_i, \vec{r}_i)$, reject the proposed change, and the ith particle remains in its present position $\vec{r}_i$.

Steps 1–4 above are often collectively referred to as a *Monte Carlo step*. Because of the special forms of the Boltzmann PDF $P(\vec{R})$ (Eq. 24.11) and the acceptance probability (Eq. 24.12), the Metropolis algorithm accepts all "downhill" moves that result in the lowering of the potential energy $V(\vec{R})$ of the system. These are precisely the moves that take the system "uphill" in the PDF $P(\vec{R})$. Proposed moves that tend to increase the potential energy of the system are not rejected outright – they are accepted in a probabilistic fashion in step 4 above. This feature allows, in principle, the Metropolis method to move from local peaks in the PDF (equivalently, local minima in the potential energy) to other or higher peaks in the PDF (i.e., other or lower potential energy configurations) to eventually sample the configuration space as described probabilistically by the PDF.

No Need to Know the Normalization Factor

A highly desirable feature of the Metropolis algorithm is that the PDF $P(\vec{R})$ need not be known precisely all the way upto the normalization factor $1/Z$. This is because the acceptance probability involves only a *ratio* of PDF values (proposed change against the current position), which makes the normalization factor $1/Z$ redundant for the purpose of this algorithm. This is a very useful feature because such normalization factors are usually extremely hard or impossible to compute.

Step Size δ and the Acceptance Ratio

This algorithm, in the form above, needs the step size parameter δ to be specified. A standard prescription for choosing the right value of δ is to monitor the average *acceptance ratio* α, defined as the ratio of the number of accepted changes to the number of proposed changes, and adjust the step size δ such that α remains within a pre-specified range around 0.5, say between 0.4 to 0.6.

Why Metropolis Works

This algorithm, the way it is described here, may sound rather *ad hoc*. However, it has a well-established theoretical justification in the theory of Markov chains. Specifically, provided that the step size δ is not unreasonably large or small, the chain of configurations thus generated turns out to be an *ergodic* Markov chain. A Markov chain is called ergodic if, in our context, it is possible to go from every configuration to every other configuration (not necessarily in a single step). Such Markov chains have a unique stationary distribution (i.e., one that does not change once it is attained). Furthermore, it is guaranteed to be attained eventually. The form of the acceptance probability above guarantees, through what is called the *condition of detailed balance*, that this stationary distribution will indeed be our desired PDF $P(\vec{R})$.

Monitoring Equilibriation and Computing Averages

All this boils down to mean that this algorithm will *eventually* start producing configurations of the system that are indeed distributed according to $P(\vec{R})$. Practically, the problem of how long one needs to wait for this to happen needs to be addressed in an empirical fashion, i.e., by trial and error. The evolution of the Markov chain to the desired PDF is often called *thermalization* or *equilibriation* by analogy to the process of equilibriation of a physical system.

Expectation values of physical quantities computed are thus to be computed after rejecting the initial transient portion of the chain of configurations, i.e., after the Markov chain has attained the desired PDF. Expectation values of a quantity $A(\vec{R})$ of interest, which is typically a function of the configuration $\vec{R}$, is computed as a simple average over a sufficiently long equilibrium portion of the Markov chain, i.e.,

$$\langle A \rangle = \frac{1}{M} \sum_{i=1}^{M} A(\vec{R}_i).$$

This is indeed an estimate of the true expectation value of A because the PDF $P(\vec{R})$ is now represented in the sample of configurations itself.

24.3.2 The Lennard-Jones Fluid Again

Let us illustrate these ideas again with the example of a two-dimensional Lennard-Jones system. As an example of a physically-relevant quantity to be estimated through the simulation, let us consider the radial distribution function $g(r)$.

The Radial Distribution Function

The spatial structure of an isotropic fluid (such as Lennard-Jones) is often characterized by what is variously known as the *radial distribution function* or the *pair correlation function*. This is formally defined as

$$g(\vec{r}_1, \vec{r}_2) = \frac{N(N-1)}{\rho^2} \frac{\int d\vec{r}_3 \ldots \vec{r}_N \exp\left(-\beta V(\vec{r}_1, \vec{r}_2, \ldots, \vec{r}_N)\right)}{\int d\vec{r}_1 \ldots \vec{r}_N \exp\left(-\beta V(\vec{r}_1, \vec{r}_2, \ldots, \vec{r}_N)\right)}.$$

For an isotropic and translationally invariant system, $g(\vec{r}_1, \vec{r}_2)$ depends only on $|\vec{r}_1 - \vec{r}_2|$ and not on $\vec{r}_1$ and $\vec{r}_2$ separately. We thus denote it by $g(r)$. This function serves as a link between simulations and experiments, because its Fourier transform

$$S(k) = 1 + 4\pi\rho \int \frac{\sin(kr)}{kr} g(r) r^2 dr$$

is an experimentally measurable quantity (e.g., through X-ray scattering). Furthermore, for a system of particles interacting with a pairwise additive potential (such as Lennard-Jones) $g(r)$ has the form $\exp(-\beta v(r))$ in the dilute gas limit ($\rho \to 0$). Here, $v(r)$ is the potential energy of a single pair of particles (see Eq. 24.2).

For a solid in a perfectly crystalline state, the $g(r)$ comprises of sharp peaks because crystalline order allows only a discrete set of pair distances to occur. In the liquid state, depending on the temperature, these sharp peaks get smeared out because of the mobility of particles in the liquid state. However, smeared-out relics of the first few peaks still survive, because particle-particle correlations at short distances still exist in the liquid state. In the gas phase, only the first peak survives in a much broadened form. For a Lennard-Jones fluid at constant temperature, so long as the potential energy dominates (i.e., at low enough temperatures) it would be difficult to find pairs of particles that are closer than σ. We therefore expect that $g(r) = 0$ for $r \lesssim \sigma$ for a Lennard-Jones fluid.

Operationally, $g(r)$ is estimated by counting $n(r)$, the average number of particle pairs with distances in the range $r - \frac{1}{2}\Delta r$ and $r + \frac{1}{2}\Delta r$. The estimate of $g(r)$ is simply this number suitably normalized and further averaged over all angular variables to account for isotropy:

$$
\begin{aligned}
g(r) \;&=\; \frac{V}{\frac{1}{2}N(N-1)}\frac{n(r)}{2\pi r \Delta r}\ \text{(two dimensions)}\\[2mm]
&=\; \frac{V}{\frac{1}{2}N(N-1)}\frac{n(r)}{4\pi r^2 \Delta r}\ \text{(three dimensions).}
\end{aligned}
\tag{24.13}
$$

Here, V stands for the volume of the system in three dimensions and area of the system in two dimensions.

A consequence of PBC is that the maximum possible separation between any pair of particles along any one coordinate direction is $L/2$. Geometrically, this could be understood as follows: PBCs turn a line segment into a circle. On a circle, the distance between a pair of points is not unique (i.e., the clockwise and anticlockwise distances match only for pairs of diametrically opposite points). This ambiguity is resolved by taking the shorter of these two distances. In simulations with PBC, this is done in a coordinatewise fashion. Thus the maximum possible separation between any pair of particles in D dimensions is $L\sqrt{D}/2$. This must be taken into consideration when computing $g(r)$ in a configuration R of the system.

The quantity $n(r)$ in the definition of $g(r)$ (Eq. 24.13) is estimated by dividing the interval $0 \leq r \leq L\sqrt{D}/2$ into N_g intervals of equal length $\Delta r = L\sqrt{D}/(2N_g)$, and counting the number of pairs with distances in each of these intervals.

Illustrative Results for $g(r)$

The MC simulation of a Lennard-Jones system presented here was performed at a fairly low temperature of $kT = 0.1$ ($\beta = 10$). The value of the step size parameter for the Metropolis algorithm was $\delta = 0.15$, chosen by trial-and-error so as to keep the average acceptance ratio α in the range $0.4 \lesssim \alpha \lesssim 0.6$.

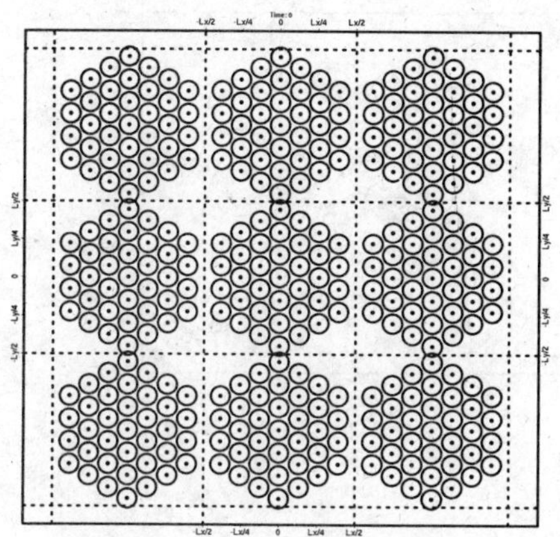
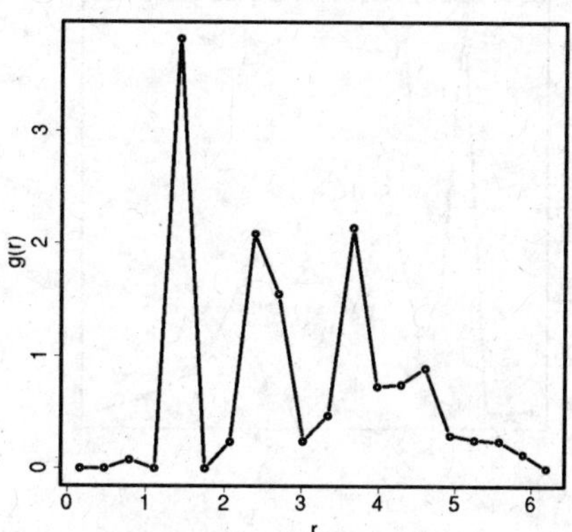

Figure 24.12: Initial configuration for MC simulation at $\beta = 10$ and the corresponding $g(r)$. Notice the sharp peaks characteristic of solid-like order.

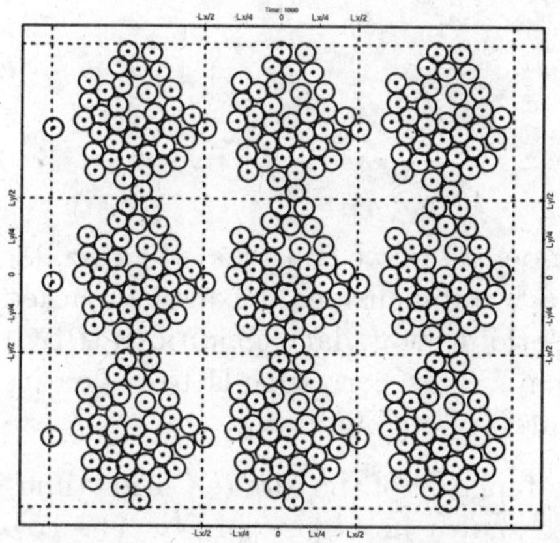
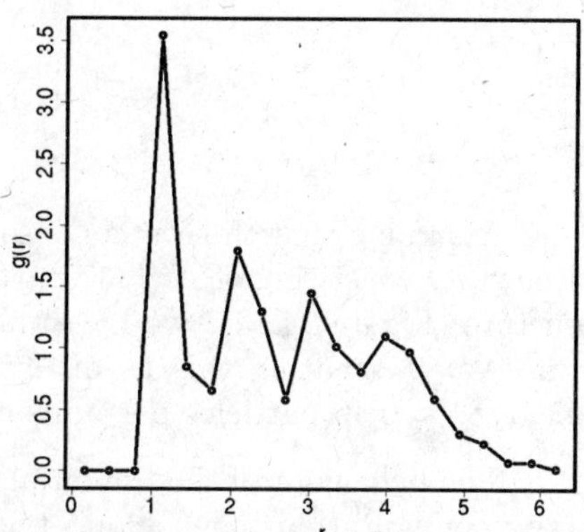

Figure 24.13: A typical Configuration generated by the Metropolis algorithm ($t = 1000, \beta = 10$), and the corresponding $g(r)$.

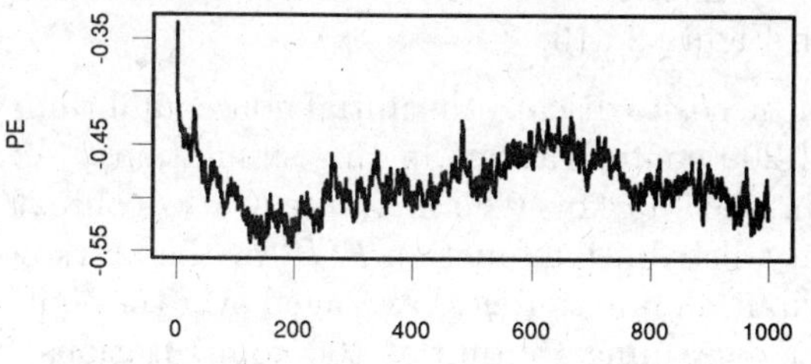
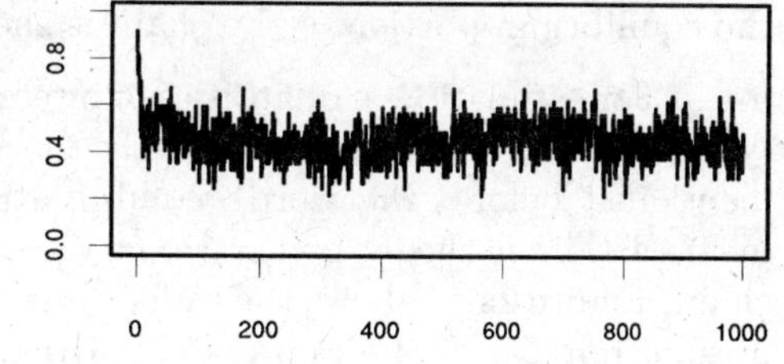

Figure 24.14: Potential energy V and the average acceptance ratio α as functions of the Monte Carlo step index t ($\beta = 10$).

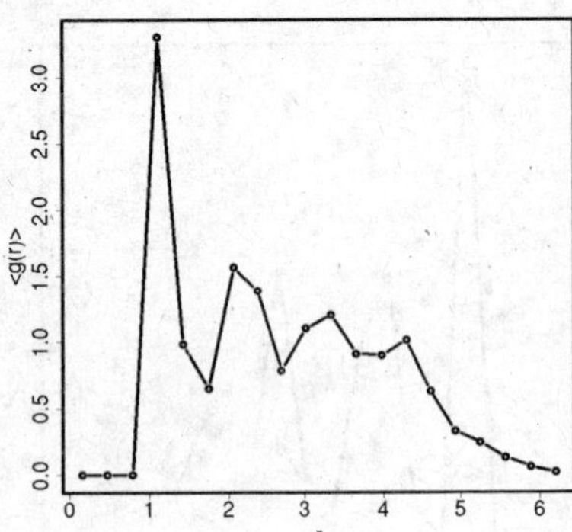 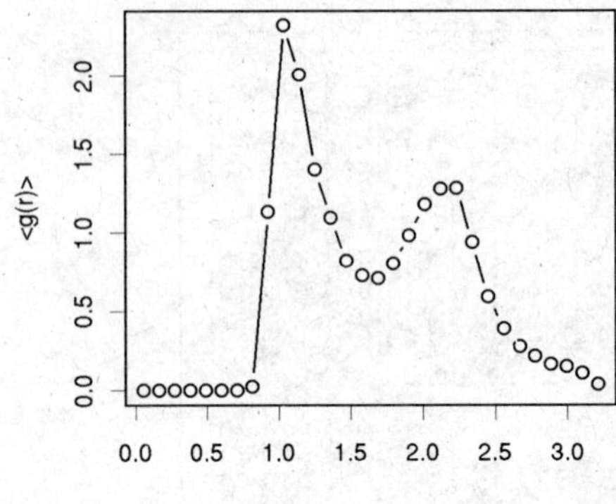

Figure 24.15: Left-hand plot: Monte Carlo averaged $g(r)$ at $\beta = 10$. Right-hand plot: the same quantity for a 16-particle system at density $\rho = 0.75$ and $\beta = 1$, for comparison.

Assuming that the system is solid-like at this temperature, we constructed a peculiar initial state consisting of $N = 37$ particles arranged in a two-dimensional hexagonal packed structure (Figure 24.12). we have thus employed the periodic boundary conditions (PBC) with $L \approx 9$, treating this as an extended fluid system. This corresponds to a density $\rho \approx 37/81 \approx 0.46$ particles per unit area (density measured in σ^{-2} units).

The behaviour of the potential energy $V(\vec{R}_t)$ as a function of the Monte Carlo "time" t (i.e., configuration index along the Markov chain) is shown in Figure 24.14. The corresponding behaviour of the Boltzmann PDF $P(\vec{R})$ (Eq. 24.11) is not shown separately; it follows from the behaviour of the potential energy. Notice the distinction between the non-equilibrium and the equilibrium portions of the Markov chain of configurations. Figure 24.14 also shows the variation of the average acceptance ratio α. A typical configuration in the equilibrium portion of the chain is shown in Figure 24.13.

For estimation of quantities of interest, one needs to discard the initial non-equilibrium portion of the chain (approximately the initial 200 configurations in this simulation). As mentioned before, the word "equilibriation" in the Markov Chain Monte Carlo context means attaining the desired stationary probability distribution function $P(\vec{R})$ by the Markov chain. Figure 24.15 shows the radial radial distribution function $g(r)$ averaged over the equilibrium portion of the Markov chain (i.e., after discarding the initial 200 configurations). The presence of multiple sharpish peaks are indicative of a solid-like phase. For comparison, the same figure also includes a similarly averaged-out $g(r)$ at $\beta = 1$ for a Lennard-Jones system consisting of 16 particles at density $\rho = 0.75$. Here, the third and higher peaks are absent, which indicates a liquid-like phase.

Appendix A

COUNTING OF STATES

We repeatedly used following relations as

$$\sum_k f(k) = \frac{V}{(2\pi)^3} \int f(k)\, d^3k \quad (3-D) \tag{A-I-1}$$

$$\sum_k f(k) = \frac{A}{(2\pi)^2} \int f(k)\, d^2k \quad (2-D) \tag{A-I-2}$$

$$\sum_k f(k) = \frac{L}{(2\pi)} \int f(k)\, dk \quad (1-D) \tag{A-I-3}$$

$$\sum_k f(k) = \frac{L^d}{(2\pi)^d} \int f(k)\, d^dk \quad (d-D) \tag{A-I-4}$$

Since this does not form a main part of the text, we thought that the discussion could go in the Appendix. The factor in front of the integrals is called density of states for free particle states.

Here $\vec{k}$ is a wave vector. Corresponding momentum is $\vec{p} = \hbar\vec{k}$. Consider a particle confined in a rectangular parallelepiped of sides L_x, L_y and L_z. A particle momentum is such that its corresponding de Broglie wavelength $\lambda = h/p$ is much smaller than any of the lengths L_x, L_y and L_z. This means the details of the walls are unimportant as far as 5the counting of states of $\vec{k}$ is concerned. Let us limit our discussion to one dimension only. The co-ordinate of the particle is restricted in the range $0 \leq x \leq L_x$. We put an infinite potential at $x = 0$ and at $x = L_x$. This is a motion of a particle in 1-D box of a Quantum Mechanics course. The Schrödinger equation to be solved is

$$\frac{-\hbar^2}{2m} \frac{d^2\psi}{dx^2} = E\psi \qquad 0 \leq x \leq L_x \tag{A-I-5}$$

with a boundary condition $\psi(0) = \psi(L_x) = 0$.

Wave functions which satisfy the boundary condition are eigen functions and are,

$$\psi_{n_x}(x) = A_x \sin(k_x\, x) \tag{A-I-6}$$

with $k_x = \dfrac{\pi n_x}{L_x}$, $\quad n_x = 1, 2, 3......$ Here $\{n_x\}$ are taken as positive, because negative values of n_x lead to a state, $-\psi$ which is a same state in Q.M. Then corresponding energy eigen value is,

$$E_{n_x} = \frac{\hbar^2 \pi^2}{2L_x^2}\frac{n_x{}^2}{m}. \tag{A-I-7}$$

Similarly, a particle in box moving along y direction gives,

$$\psi_{n_y}(y) = A_y \sin(k_y\, y). \tag{A-l-8}$$

Here $k_y = \dfrac{\pi n_y}{L_y}$, $\quad n_y = 1, 2, 3.....$, and similarly for the z motion. Thus, the particle confined in a box of volume $V = L_x L_y L_z$ has energy,

$$E_{n_x, n_y, n_z} = \frac{\hbar^2 \pi^2}{2m}\left[\frac{n_x^2}{L_x^2} + \frac{n_y^2}{L_y^2} + \frac{n_z^2}{L_z^2}\right] \tag{A-I-9}$$

and corresponding eigenfunction is,

$$\psi_{n_x, n_y, n_z}(x, y, z) = A \sin\left(\frac{\pi x n_x}{L_x}\right) \sin\left(\frac{\pi y n_y}{L_y}\right) \sin\left(\frac{\pi z n_z}{L_z}\right) \tag{A-I-10}$$

with $\{n_x, n_y, n_z\}$ are all positive integers. This means, (a) k_x, k_y and k_z are restricted in the first quadrant of the $\vec{k}$ space. (b) Energy becomes almost continuous as the box size becomes larger and larger. In this limiting case, $E_{n_x} - E_{n_x-1} = \Delta E_{n_x}$ become very small. This means we can consider E as a continuous variable of the quantum numbers.

For a fixed values of k_y and k_z, number Δn_x of possible integers when k_x lies between k_x and $k_x + dk_x$ equals,

$$\Delta n_x = \frac{L_x}{\pi} dk_x \, . \tag{A-I-11}$$

Each integer in Δn_x represent a state of k_x. In general, number states between k_x and $k_x + dk_x$, number states between k_y and $k_y + dk_y$ and number states between k_z and $k_z + dk_z$ are,

$$\begin{aligned}\Delta n_x \Delta n_y \Delta n_z &= \frac{L_x L_y L_z}{\pi^3} dk_x\, dk_y\, dk_z \\[2mm] &= \frac{V}{\pi^3} d^3 k \, . \end{aligned} \tag{A-I-12}$$

Let $\rho(k)$ be the density of states. Then from the equation (12), number of states between $\vec{k}$ and $\vec{k} + d\vec{k}$ is $\rho(\vec{k})\, d^3 k$ where

$$\rho(\vec{k})\, d^3 k = \frac{V}{\pi^3}\, d^3 k \, . \tag{A-I-13}$$

But this expression is <u>wrong</u> by a factor of 8. This is so because it includes both the values of k_x and also of $-k_x$ and so on. However, from our boundary conditions we include only positive components of the wave vector $\vec{k}$. As a result, taking the volume of the first octant in to account, we get,

$$\rho(\vec{k})\, d^3k \;=\; \frac{V}{8\pi^3}\, d^3k \;=\; \frac{V}{(2\pi\hbar)^3}\, d^3p \;.$$
(A-I-14)

Thus, $\rho(\vec{k}) = \dfrac{V}{(2\pi)^3}$. Or,

$$\sum_k f(\vec{k}) \;=\; \int \rho(\vec{k}) f(\vec{k}) d^3k$$

$$=\; \frac{V}{(2\pi)^3} \int f(\vec{k}) d^3k \;.$$
(A-I-15)

Generalization of this formula to 1-D;2-D and d-D as given at the beginning of this appendix is immediate.

Conversion of $\rho(k)$ to $\rho(\omega)$ is done using the dispersion relation between ω and k. Consider a case of photon for illustration. Photon obeys $\omega = c\, k$. Then,

$$\rho(k)d^3k = 4\pi\rho(k)\, k^2\, dk = \rho(\omega)\, d\omega \;.$$
(A-I-16)

Thus,

$$\rho(\omega) = \frac{4\pi V}{(2\pi)^3}\, \frac{\omega^2}{c^3}$$
(A-I-17)

This is the expression which we used in the theory of a Black Body Radiation.

Calculation of $\rho(k)$ given here is applicable to free particle system only. This need not be true for an electron in a periodic potential of a crystal. Symmetry of the crystal group is reflected in to the wave function of electron (Bloch state) As a result the energy function $\epsilon(k)$ is a complicated function of k and is different in different symmetry directions. This is reflected in the band structure diagrams. Since many properties of the solid depend on the density of states, calculations of the band structure and $\rho(\vec{k})$ was an important industry in 60's and 70's.

Appendix B

Gamma Function and Useful Integrals

Sometimes Gamma function is also called as a factorial function. It is defined as

$$\Gamma(\mu) = (\mu - 1)! = \int_0^\infty x^{\mu-1} \, e^{-x} \, dx \quad : \quad \mu > 0. \tag{A-II-1}$$

Clearly,

$$\Gamma(\mu + 1) = \mu\Gamma(\mu) \tag{A-II-2}$$

The argument of Gamma function , namely μ can be a rational number (fraction) and has a form,

$$\Gamma(\mu + 1) = \mu(\mu - 1)(\mu - 2).....(1 + p)p\Gamma(p)$$

where $0 < p \leq 1$. Clearly, for integer $\mu = n$, $\Gamma(n + 1) = n!$. Useful value of $\Gamma(1/2) = \sqrt{\pi}$. Thus,

$$\Gamma(1/2) = \int_0^\infty x^{-1/2}e^{-x} \, dx = \sqrt{\pi}.$$

This is obtained as follows. Put $\sqrt{x} = \xi$; $d\xi = \dfrac{1}{2\sqrt{x}} \, dx$. Then,

$$I = 2\int_0^\infty e^{-\xi^2} d\xi = \int_{-\infty}^\infty e^{-\xi^2} d\xi.$$

Then clearly

$$I^2 = \int_{-\infty}^\infty \int_{-\infty}^\infty e^{-(\xi^2+\eta^2)} \, d\xi \, d\eta.$$

With $\xi = r\cos\theta$, $\eta = r\sin\theta$, $d\xi \, d\eta = r \, dr \, d\theta$. Then,

$$I^2 = \int_0^\infty r \, e^{-r^2} \, dr \int_0^{2\pi} d\theta = \frac{2\pi}{2} = \pi$$

leading to $I = \sqrt{\pi}$.

In general it is easy to see by transformation of variables and integration by part that, for $\nu > -1$,

$$\int_0^\infty x^\nu \, e^{-\alpha x^2} \, dx = \frac{1}{2} \frac{1}{\alpha^{(\nu+1)/2}} \, \Gamma\left(\frac{\nu+1}{2}\right) \tag{A-II-3}$$

In general the integrals satisfy,

$$\frac{\partial}{\partial\alpha} I_\nu = -I_{\nu+2} \, . \tag{A-II-4}$$

The useful integrals in the kinetic theory are,

1.
$$I_0 = \int e^{-\alpha x^2} \, dx = \frac{1}{2} \left(\frac{\pi}{\alpha}\right)^{1/2}$$

2.
$$I_1 = \int x e^{-\alpha x^2} \, dx = \frac{1}{2\alpha}$$

3.
$$I_2 = \int x^2 e^{-\alpha x^2} \, dx = \frac{1}{4} \left(\frac{\pi}{\alpha^3}\right)^{1/2}$$

4.
$$I_3 = \int x^3 e^{-\alpha x^2} \, dx = \frac{1}{2\alpha^2}$$

5.
$$I_4 = \int x^4 e^{-\alpha x^2} \, dx = \frac{3}{8} \left(\frac{\pi}{\alpha^5}\right)^{1/2}$$

Appendix C

MOMENT GENERATING FUNCTION AND CUMULANT

For any random variable x with probability distribution $P(x)$, various moments of x are defined as,

$$\bar{x} = \langle x \rangle \;=\; \int x \, P(x) \, dx$$

$$\bar{x}^2 = \langle x^2 \rangle \;=\; \int x^2 \, P(x) \, dx$$

Or, in general,
$$\bar{x}^n = \langle x^n \rangle \;=\; \int x^n \, P(x) \, dx \qquad \text{(A-III-1)}$$

These integrals are defined on the domain of x. Then the moment generating function of a random variable x is defined as

$$\begin{aligned}
\phi(t) \;&=\; \langle e^{tx} \rangle \\[6pt]
&=\; 1 + t<x> + \frac{t^2 <x^2>}{2!} + \dots\dots + \frac{t^n <x^n>}{n!} + \dots \\[6pt]
&=\; \sum_{k=0}^{\infty} \frac{t^k <x^k>}{k!} \qquad \text{(A-III-2)}
\end{aligned}$$

Thus $\phi(t)$ generates various moments.

$$\ln \, \phi(t) = \ln \, \langle e^{tx} \rangle \qquad \text{is of interest to us.}$$

Thus,

$$\ln \phi(t) = \ln \left(1 \;+\; t<x> \; \frac{t^2 <x^2>}{2!} \;+\; \dots\dots \right) \qquad \text{(A-III-3)}$$

By taking arbitrary t to be sufficiently small, so that $tx \ll 1$, we write,

$$\ln \phi(t) = \left(1 + t<x> \frac{t^2 <x^2>}{2\,!} + \ldots\ldots\right) - \frac{1}{2}\left(t<x> \frac{t^2 <x^2>}{2\,!} + \ldots\ldots\right)^2$$

$$+ \frac{1}{3}\left(t<x> \frac{t^2 <x^2>}{2\,!} + \ldots\ldots\right)^3 + \ldots\ldots \qquad\qquad \text{(A-III-4)}$$

Here we used a standard expansion of a logarithmic function. Thus,

$$\ln(1+x) = x - \frac{x^2}{2} + \frac{x^3}{3} - \frac{x^4}{4} + \ldots.$$

Clearly, $\quad \ln \phi(t) = \sum_{n=1}^{\infty} \frac{t^n M_n}{n\,!}.$

Here the cumulants M_n are,

$$M_1 = \langle x \rangle$$

$$M_2 = \langle x^2 \rangle - \langle x \rangle^2$$

$$M_3 = \langle x^3 \rangle - 3\langle x \rangle\langle x^2 \rangle + 2\langle x \rangle^3$$

Advantage of working with the cumulants is apparent when two independent random variables x and y are in the system. It is left as an exercise to show that

$$M_n(x+y) = M_n(x) + M_n(y).$$

These cumulants are extensively used in the Mayer's cluster expansion theory of the real gases.

Appendix D

EVALUATION OF SOME FERMI-DIRAC AND BOSE-EINSTEIN INTEGRALS

Fermi Dirac integrals

As we have seen previously (chapter ??) the Fermi Dirac distribution has a sharp fall near Fermi energy $E_F = k_B T_F$, leading to the evaluation of thermodynamic functions tricky. Most of the integrals, after some simple transformations, can be cast in a general form

$$F(\alpha) = \int_0^\infty \frac{G(x)}{e^{x-\alpha}+1} \tag{A-IV-1}$$

To give an example, consider a total energy evaluation of an ideal degenerate Fermi gas of free electrons. Here $\epsilon_k = \dfrac{h^2 k^2}{2m}$ and the integral for total energy E is,

$$E = \frac{2 \times 4\pi V}{(2\pi)^3} \int \frac{\dfrac{\hbar^2 k^2}{2m} k^2 dk}{\exp\left[\beta\left(\dfrac{\hbar^2 k^2}{2m} - \mu\right) + 1\right]} \tag{A-IV-2}$$

Writing $\dfrac{\beta h^2 k^2}{2m} = x$, we transform energy integral of equation (2) as,

$$E = \frac{V}{\pi^2 \hbar^3} \sqrt{2m} \, (k_B T)^{5/2} \int \frac{x^{3/2} dx}{e^{x-\alpha}+1} \tag{A-IV-3}$$

Where $\beta\mu = \alpha$.

In general thus, the integral as given in equation (1) need to be evaluated[1]. Usually, $G(x)$ is some polynomial in x. The general method goes as follow. Write $x - \alpha = \xi$. It must

---

[1] Method presented below was communicated to the author by Dr. Hugh E DeWitt

be noted that, when the Fermi gas is degenerate, $\mu \approx E_F$ and $\alpha = \mu/K_B T >> 1$. Equation (1) can be written as

$$
\begin{aligned}
F(\alpha) &= \int_{-\alpha}^{\infty} \frac{G(\xi + \alpha)}{e^{\xi} + 1} d\xi \\
&= \int_{-\alpha}^{0} \frac{G(\xi + \alpha)}{e^{\xi} + 1} d\xi + \int_{0}^{\infty} \frac{G(\xi + \alpha)}{e^{\xi} + 1} d\xi
\end{aligned}
$$

But

$$
\frac{1}{e^{\xi} + 1} = 1 - \frac{1}{e^{-\xi} + 1} \tag{A-IV-4}
$$

Thus

$$
\begin{aligned}
F(\alpha) &= \int_{-\alpha}^{0} G(\xi + \alpha) d\xi - \int_{-\alpha}^{0} \frac{G(\xi + \alpha)}{e^{-\xi} + 1} d\xi + \int_{0}^{\infty} \frac{G(\xi + \alpha)}{e^{\xi} + 1} d\xi \\
&= \int_{0}^{\alpha} G(\xi) d\xi - \int_{0}^{\alpha} \frac{G(\alpha - \xi)}{e^{\xi} + 1} d\xi + \int_{0}^{\infty} \frac{G(\xi + \alpha)}{e^{\xi} + 1} d\xi \\
&= \int_{0}^{\alpha} G(\xi) d\xi + \int_{0}^{\infty} \frac{G(\alpha + \xi) - G(\alpha - \xi)}{e^{\xi} + 1} d\xi + \int_{\alpha}^{\infty} \frac{G(\alpha - \xi)}{e^{\xi} + 1} d\xi
\end{aligned}
$$

$$\tag{A-IV-5}$$

But $\alpha >> 1$, $\int_{\alpha}^{\infty} \frac{G(\alpha - \xi)}{e^{\xi} + 1} d\xi \sim \vartheta\left(e^{-\alpha}\right)$ and can be neglected.

Now $G(\alpha \pm \xi)$ can be expanded around α and we get

$$
G(\alpha + \xi) - G(\alpha - \xi) = 2\xi G'(\alpha) + \frac{\xi^3}{3} G'''(\alpha) + \dots
$$

The integral of the equation (1) is,

$$
F(\alpha) = \int_{0}^{\alpha} G(\xi) d\xi + 2G'(\alpha) \int_{0}^{\infty} \frac{\xi d\xi}{e^{\xi} + 1} + \frac{G'''(\alpha)}{3} \int_{0}^{\infty} \frac{\xi^3 d\xi}{e^{\xi} + 1} + \dots
$$

Integrals like $\int_{0}^{\infty} \frac{\xi^n d\xi}{e^{\xi} + 1}$ for $n > 1$ are evaluated as follows.

$$
\begin{aligned}
\int_{0}^{\infty} \frac{z^{x-1} dz}{e^{z} + 1} &= \int_{0}^{\infty} z^{x-1} e^{-z} \sum_{n=0}^{\infty} (-1)^n e^{-nz} dz \\
&= \Gamma(x) \sum_{n=1}^{\infty} (-1)^{n+1} \frac{1}{n^x} \\
&= \left(1 - 2^{1-x}\right) \Gamma(x) \sum_{n=1}^{\infty} \frac{1}{n^x} \\
&= \left(1 - 2^{1-x}\right) \Gamma(x) \zeta(x) \tag{A-IV-6}
\end{aligned}
$$

where $\zeta(x) = \sum_{n=1}^{\infty} \frac{1}{n^x}$ is known as Riemann Zeta function.

The values of second and third definite integral is $\pi^2/72$ and $7\pi^4/120$ respectively. Then

$$F(\alpha) = \int_0^\alpha G(\xi)d\xi + \frac{\pi^2}{6}G'(\alpha) + \frac{7\pi^4}{120}G'''(\alpha) + \dots \tag{A-IV-7}$$

This is a general expression. We can use this expression to find temperature dependence of highly degenerate Fermi-Dirac systems. Recall , total number of noninteracting Fermions of spin $1/2$ have a density n given by

$$n\pi^2 = \frac{\pi}{2\lambda^3}\int_0^\infty \frac{z^{1/2}}{e^{z-\alpha}+1}dz \tag{A-IV-8}$$

where λ is de-Broglie thermal wavelength given by

$$\lambda = \frac{h}{\sqrt{8\pi^2 m k_B T}}$$

We can now apply the formula (7) to evaluate n of equation (8) as, $(G(z) = z^{1/2})$

$$2n\pi^2\lambda^3 = \frac{2}{3}\alpha^{3/2}\left(1 + \frac{\pi^2}{8\alpha^2}\right) \tag{A-IV-9}$$

Taking approximate value of $\alpha \approx E_F/K_B T$, equation (9) can be further simplified to

$$\mu = E_F\left(1 - \frac{\pi^2}{12}\left(\frac{K_B T}{E_F}\right)^2 + \dots\right) \tag{A-IV-10}$$

giving a temperature dependence of μ for highly degenerate electron gas. For calculation of energy , we can use the expression (see equation 3).

$$\frac{E}{V} = \frac{\hbar^2}{4m\pi^2\lambda^5}\int_0^\infty \frac{z^{3/2}}{e^{z-\alpha}+1}dz \tag{A-IV-11}$$

with $G(z) = z^{3/2}$, and using equation(7), we get

$$\frac{E}{V} = \frac{\hbar^2}{4m\pi^2\lambda^5}\left\{\frac{2}{5}\alpha^{5/2} + \frac{\pi^2}{4}\alpha^{1/2}\right\}. \tag{A-IV-12}$$

Using Equation (10), we can write ,

$$\alpha^{1/2} \simeq \left(\frac{E_F}{k_B T}\right)^{1/2}\left\{1 - \frac{\pi^2}{24}\left(\frac{k_B T}{E_F}\right)^2\right\} \tag{A-IV-13}$$

Substituting equation (14) in (13), we get,

$$\frac{E}{V} = \frac{\hbar^2}{2m\pi^2}\frac{1}{5\lambda^5}\left(\frac{E_F}{k_B T}\right)^{5/2}\left\{1 + \frac{5\pi^2}{12}\left(\frac{k_B T}{E_F}\right)^2 + \dots\right\} \tag{A-IV-14}$$

With $\left(3\pi^2 n\right)^{2/3}\lambda^2 \simeq \alpha = E_F/k_B T$, equation (14) becomes

$$\frac{E}{N} = \frac{3}{5}E_F \left\{ 1 + \frac{5\pi^2}{12}\left(\frac{k_B T}{E_F}\right)^2 + ... \right\} \tag{A-IV-15}$$

and $\dfrac{C_V}{N k_B}$ = electronic specific heat$= \dfrac{\pi^2}{2}\left(\dfrac{k_B T}{E_F}\right)$

In general the Fermi-Dirac integrals are traditionally defined as

$$f_\nu(\alpha) = \frac{1}{\Gamma(\nu)}\int_0^\infty \frac{x^{\nu-1}}{e^{x-\alpha}+1}dx$$

For small e^α

$$f_\nu(z) = f_\nu(e^\alpha) = \sum_{k=1}^\infty (-1)^{k-1}\frac{e^{\alpha k}}{k^\nu}$$

The earlier discussion gives expansion of $f_\nu(e^\alpha)$ when $e^\alpha >> 1$ ie. at low temperatures.

Bose- Einstein Integrals

While dealing with Bose-Einstein condensation or involving the Bose function in calculation of various physical properties, we come across the integrals of the type,

$$I_\nu(\alpha) = \int_0^\infty \frac{x^{\nu-1}}{e^{x+\alpha}-1} \tag{A-IV-16}$$

in the range $0 \le e^{-\alpha} < 1$, $\nu > 0$ and $\alpha = 0$, $\nu > 1$.

As far as Bose gas is concerned, μ is negative and as $T >> 0$, $\mu \to -\infty$. Also at the point of B-E condensation, i.e. at finite $T = T_c$, $\mu = 0 \therefore -\infty \le \beta\mu \le 0$. Thus writing $\beta\mu = -\alpha$ we obtain the kind of integral given in (16)

Asymptotic properties

Far away from T_c, and when , $e^{-\beta\mu} >> 1$ the integral (16) reduces to

$$\begin{aligned}
I_\nu(\alpha) \quad &\to \quad \int_0^\infty e^{-\alpha}x^{\nu-1}e^{-x}dx \\
&= \quad e^{-\alpha}\Gamma(\nu) \tag{A-IV-17}
\end{aligned}$$

We then redefine the integral (16) in light of equation (17) as,

$$\mathcal{J}_\nu(\alpha) = \frac{1}{\Gamma(\nu)}\int_0^\infty \frac{x^{\nu-1}dx}{e^{x+\alpha}-1} \tag{A-IV-18}$$

For $e^\alpha >> 1$, we can expand the integrand of equation (18) as

$$\frac{x^{\nu-1}dx}{e^{x+\alpha}-1} = x^{\nu-1}e^{-x-\alpha}(1-e^{-x-\alpha})^{-1} = x^{\nu-1}e^{-x-\alpha}\sum_{l=0}^\infty e^{-lx-l\alpha}$$

The integral (18) then becomes,

$$\mathcal{J}_\nu(\alpha) = \frac{1}{\Gamma(\nu)} \sum_{l=1}^{\infty} \int_0^{\infty} x^{\nu-1} e^{-lx-l\alpha} dx$$

$$= \sum_{l=1}^{\infty} \frac{e^{-l\alpha}}{l^\nu} \tag{A-IV-19}$$

Thus, for $e^{-\alpha} << 1$, $\mathcal{J}_\nu(\alpha)$ behaves as $e^{-\alpha}$ for all ν. Clearly, when $\alpha \to 0$, $\mathcal{J}_\nu(\alpha)$ has a largest value for $\nu > 1$ as

$$\mathcal{J}_\nu(\alpha) = \frac{1}{1^\nu} + \frac{1}{2^\nu} + \frac{1}{3^\nu} + \dots \tag{A-IV-20}$$

This is nothing but a Riemann Zeta function mentioned earlier and which is defined as

$$\xi(\nu) = \sum_{n=1}^{\infty} \frac{1}{n^\nu} \quad (\nu > 1)$$

Typical values of Riemann zeta function are,

$$\left.\begin{aligned} \xi(2) &= \frac{\pi^2}{6} \approx 1.645 \quad ; \quad \xi(4) = \frac{\pi^4}{90} \approx 1.082 \\ \xi(3/2) &\approx 2.612 \quad ; \quad \xi(5/2) = 1.341 \end{aligned}\right\} \tag{A-IV-21}$$

For $\alpha \neq 0$, and $\nu = 1$, $\mathcal{J}_1(\alpha)$ has a closed analytical form as

$$\mathcal{J}_1(\alpha) = \frac{1}{\Gamma(1)} \int_0^{\infty} \frac{dx}{e^{\alpha+x} - 1} = -\ln\left(1 - e^{-\alpha}\right) \tag{A-IV-22}$$

$\mathcal{J}_1(\alpha)$ diverges logarithmically when $\alpha \to 0$ as $\ln(1/\alpha)$.

The region of Bose-Einstein condensation is provided by $\mu \to 0$,   i.e. $\alpha \to 0$. Consider a case when $\alpha \to 0$ and $0 < \nu < 1$. Then

$$\mathcal{J}_\nu(\alpha) = \frac{1}{\Gamma(\nu)} \int_0^{\infty} \frac{x^{\nu-1} dx}{e^{x+\alpha} - 1}$$

$$\approx \frac{1}{\Gamma(\nu)} \int_0^{\infty} \frac{x^{\nu-1} dx}{x + \alpha} \tag{A-IV-23}$$

This integral can be cast in the beta function to yield

$$\mathcal{J}_\nu(\alpha) \simeq \frac{\Gamma(1-\nu)}{\alpha^{1-\nu}} \quad (0 < \nu < 1). \tag{A-IV-24}$$

In general, Robinson[2] has shown that for $\alpha \to 0$, and non integral $\nu > 1$,

$$\mathcal{J}_\nu(\alpha) = \frac{\Gamma(1-\nu)}{\alpha^{1-\nu}} + \sum_{l=0}^{\infty} \frac{(-1)^l}{l!} \xi(\nu - 1) \alpha^l \tag{A-IV-25}$$

---

[2] J. E. Robinson Phys. Rev. **83**, 678(1951)

Equation (25) is the required result, useful in the calculation of thermodynamic function involving Bose- Einstein condensation.Its particular forms are,

$$\left.\begin{array}{rcl} \mathcal{J}_{1/2}(\alpha) & = & \dfrac{1.77}{\sqrt{\alpha}} - 1.46 \ + \ 0.208 \ \alpha + ... \\[2mm] \mathcal{J}_{3/2}(\alpha) & = & -3.54 \ \sqrt{\alpha} \ + \ 2.61 + 1.46 \ \alpha + ... \\[1mm] \mathcal{J}_{5/2}(\alpha) & = & 2.36 \ \alpha \ \sqrt{\alpha} + 1.34 - 2.61 \ \alpha + ... \end{array}\right\} \qquad \text{(A-IV-26)}$$

In general, for $\nu > 1$,

$$\frac{d}{d\alpha} \mathcal{J}_\nu(\alpha) = \mathcal{J}_{\nu-1}(\alpha). \qquad \text{(A-IV-27)}$$

Robinson showed that equation (27) can be taken as a definition of a function $\mathcal{J}_\nu(\alpha)$ valid for all ν

Appendix E

Mean Field Solution of Ising Model

The Ising Hamiltonian is,

$$H = -J \sum_{n.n} \sigma_i \sigma_j - \mu_B B \sum_i \sigma_i \qquad \text{(A-V-1)}$$

We will use a mean field method to get the ferromagnetism from this hamiltonian. It turns out that the mean field procedure is such that the dimensionality is not involved in it.

We say that σ_i sees not only the magnetic field due to its nearest neighbors but also due to all other spins in an average way. This average field has fluctuations. In the mean field theory these fluctuations are ignored.

Consider a quantity (called fluctuation)

$$(\sigma_i - \langle \sigma \rangle)(\sigma_j - \langle \sigma \rangle) = \sigma_i \sigma_j - (\sigma_i + \sigma_j)\langle \sigma \rangle + \langle \sigma \rangle^2 .$$

Thus Ising Hamiltonian

$$H = -J \langle \sigma \rangle \left(\sum \sigma_i + \sum \sigma_J \right) + J \langle \sigma \rangle^2 \sum_{i,j} 1$$

$$-J \sum_{ij} (\sigma_i - \langle \sigma \rangle)(\sigma_j - \langle \sigma \rangle) - \mu_B B \sum \sigma_i \qquad \text{(A-V-2)}$$

One must be careful in counting. It is obvious that $\sum_i \sigma_i$ is same as $\sum_j \sigma_j$. We assume that for each σ_i there are even nearest neighbors. Let them be q. Then total number of nearest neighbor pairs of N spins is $Nq/2$. Taking these considerations H is approximated as H_{mf} to be

$$H_{mf} = -Jq \langle \sigma \rangle \sum_{i=1}^{N} \sigma_i + \frac{JqN}{2} \langle \sigma \rangle^2 - \mu_B B \sum \sigma_i \qquad \text{(A-V-3)}$$

Comparing the first and the third term of this mean field Hamiltonian we can say that the spins are collectively producing an equivalent mean magnetic field as

$$B_{mf} = Jq \langle \sigma \rangle / \mu_B \qquad \text{(A-V-4)}$$

The partition function is

$$Z(N,T,B\langle\sigma\rangle) = \sum_{\sigma_1=\pm}\sum_{\sigma_2=\pm}---\sum_{\sigma_N=\pm}\left[e^{-\beta JqN/2\langle\sigma\rangle^2+\beta\mu_B(B_{mf}+B)\sum\sigma_i}\right]$$

$$\text{(A-V-5)}$$

$$= \exp\left[-\beta JqN\langle\sigma\rangle^2/2\right]\cdot 2^N\cdot\cosh^N\left(\mu_B\beta(B_{mf}+B)\right)\qquad\text{(A-V-6)}$$

The free energy is

$$F = -k_B T\ln Z \qquad\text{(A-V-7)}$$

The average magnetic moment is

$$\begin{aligned}
M &= N\mu_B\langle\sigma\rangle\\
&= -\frac{\partial}{\partial B}F(N,B,T\langle\sigma\rangle)\\
&= N\mu_B\tanh\left[\mu_B(B+B_{mf})/k_B T\right] \qquad\text{(A-V-8)}
\end{aligned}$$

This is the principal result of mean field theory.

This can be compared with the Brillouin function that was obtained for a general spin S in the chapter on Para and ferromagnetism. Following exactly the same analysis as we did in case of a Weiss theory we get $T_c = \dfrac{Jq}{k_B}$. Let $x = \dfrac{Jq\langle\sigma\rangle}{k_B T}$. For $T < T_c$ we have to solve

$$\frac{T}{T_c}x = \tanh x \approx x = \frac{x^3}{3}.$$

Then the behavior of the order parameter M is

$$M = N\mu_B\langle\sigma\rangle \approx \left[3\left(1-\frac{T}{T_c}\right)\right]^{1/2}$$

giving the critical index to be 1/2. This is consistent with the Landau theory. The low temperature behavior can be found by expanding $\tanh x$ for large x as

$$\tanh x = 1 - 2e^{-2T_c/T}$$

This low temperature behavior does not agree with low temperature behavior of $\langle\sigma\rangle$ as,

$$\langle\sigma\rangle = 1 - \text{ constant }\times T^{3/2}$$

This is because the low energy excitations are the spin waves which cannot be deduced from Ising model. A simple calculation will show that the magnetic susceptibility is

$$\chi = \frac{(N\mu_B^2/k_B)}{T_c - T} \qquad\text{(A-V-9)}$$

giving the critical exponent $\gamma = 1$.

Comment

The Heisenberg Hamiltonian is loaded with complications. It is totally quantum mechanical and the components of spin do not commute with each other. However if we ignore fluctuations and take averages of x and y components of σ then we get the Ising model. The mean field result of both the models is same.

Problem

Show that

1. $T_c = \dfrac{Jq}{k_B}$ for the mean field theory of $S = 1/2$ system.

2. The internal energy of the Ising lattice for $B = 0$ is $E_0 = \dfrac{-JqN}{2}\langle\sigma\rangle^2$.

3. $\langle\sigma\rangle = \tanh\left\{\dfrac{T_c}{T}\langle\sigma\rangle\right\}$

4. $\chi = \dfrac{(N\mu_B^2/k_B)}{T_c - T}$

Bibliography

Books on Thermodynamics

1. Heat and Thermodynamics by M.W. Zemansky and Richard Ditman 7th ed. McGraw Hill, NewYork 1997.

   Excellent introductory book with careful expose of the fundamental ideas. Many numerical problems in the book unveil amazing reality in Physics. Connection to experiments is interesting.

2. The Elements of Classical Thermodynamics by A B Pippard, Cambridge University Press, Cambridge 1957. This book is an outstanding experimental physicists' way of understanding of Thermodynamics.

3. Thermodynamics by H B Callen, John Wiley and sons Inc. New York 1960. This is a modern presentation that includes application of Thermodynamics to the difficult topics like elasticity.

4. A Treatise on Heat by M N Saha and B N Srivastava 3rd ed. Indian Press limited Allahabad and Calcutta 1950 A Flawless old style discussion of many topics not found in modern books. Experimental techniques of last century can be found here.

5. Thermodynamics by E Fermi, Dover Publication, New York 1957

Books on Statistical Mechanics

1. Fundamentals of Statistical and Thermal Physics by F Reif, McGraw Hill International Editions 1985. Excellent treatment of the subject by a well known low temperature physicist. Level of this book is similar to ours.

2. Theory of Heat by Richard Becker, Springer Verlag Berlin 1955. This book gives unusual insight in to the subject. Becker plays many tricks using Carnot cycle.

3. Thermodynamics and Statistical Mechanics by A H Wilson, Cambridge University Press, Cambridge 1966.

   Wilson develops subject from historical point of view. Many applications to Solid State Physics and Physical Chemistry problems are discussed.

4. Statistical Mechanics by R K Patharia, Butterworth and Heinemann, Oxford 2nd ed. 1996. This is an excellent book where many modern topics are treated. Method of calculations can be learnt from here.

5. Statistical Mechanics and Thermodynamics by Claud Garrod, Oxford University Press Oxford, 1995. Unusual text where Statistical Mechanics is treated as an application of the Probability theory. Problems and supplementary sections are refreshing.

6. Fundamentals of Statistical Mechanics by Felix Bloch, Stanford University Press, Stanford 1989. Book represents Bloch's way of looking at the subject.

7. Thermodynamics and Statistical Mechanics by Greiner, Neise and Stöcker, Springer, 1994. A good treatment of Statistical Mechanics with many new applications. This is one of the volumes of Greiner on theoretical physics in the tradition of Sommerfeld.

8. Statistical Mechanics by K Huang , John Wiley and sons, Inc., New York 1963. Professor Huang treats many advanced and research topics of Statistical Mechanics.

9. Statistical Physics by L D Landau and E M Lifshitz, Addison-Wesley Pub. Co. Reading, Mass. 1959. Typical Landauian way of looking at the subject. Landau Lifshitz is simply great! Every serious student must contemplate on the material presented in this book.

10. Statistical Mechanics by Shang-Keng Ma, World Scientific, Singapore and Allied Publishers New Delhi 2000. This book serves both, as a text and as a reference. Professor Ma treats the subject in a provocative manner. Many research topics are discussed.

11. Introduction to phase transition and critical phenomena by H E Stanley, Oxford University Press, Oxford,1971. Stanley treats most of the topics of critical phenomena of pre renormalization group.

12. Fundamentals of Statistical Mechanics by B B Laud, New Age International (P) Ltd.New Delhi 1998. This slender volume is simply great for its clarity.

Simulation Methods

Computational Physics

1. Harvey Gould, Jan Tobochnik, and Wolfgang Christian, *Introduction to Computer Simulation Methods* (Addison-Wesley, 2006).

2. Franz Vesely, *Computational Physics: an Introduction* (Kluwer Academic/Plenum Publishers, Second Edition, 2001).

3. J.M. Thijssen, *Computational Physics* (Cambridge University Press, 1999).

4. M.P. Allen and D.J. Tildesley, *Computer Simulations of Liquids* (Oxford University Press, 1989).

Molecular Dynamics

1. D.C. Rapaport, *The Art of Molecular Dynamics Simulation* (Cambridge University Press, 1995).

2. J.M. Haile, *Molecular Dynamics Simulation: Elementary Methods* (John Wiley & Sons, 1997).

3. Daan Frenkel and B. Smit, *Understanding Molecular Simulation* (Academic Press, Second edition, 2001).

Monte Carlo Methods

1. M.H. Kalos and P.A. Whitlock, *Monte Carlo Methods. Vol. 1: Basics* (Wiley Interscience, 1986).

2. M.E.J. Newman and G.T. Barkema, *Monte Carlo Methods in Statistical Physics* (Oxford University Press, 1999).

3. D.P. Landau, *A Guide to Monte Carlo Simulations in Statistical Physics* (Cambridge University Press, Second edition, 2005).

4. K.P.N. Murthy, *Monte Carlo Methods in Statistical Physics* (University Press, 2004).

5. Gilks, Richardson, and Spiegelhalter, *Markov Chain Monte Carlo Methods in Practice* (Chapman and Hall, 1996).

Numerics

1. H.M. Antia, *Numerical Methods for Scientists and Engineers* (Hindusthan Book Agency, Second edition, 2002).

2. W.H. Press, S.A. Teukolsky, W.T. Vetterling, and B.P. Flannery, *Numerical Recipes in C: The Art of Scientific Computing* (Cambridge University Press, Second edition, 1992).

Index

$P - V - T$ surface for CO_2, 73
$P - V - T$ surface of water, 73

Absolute temperature scale, 4
Adiabatic demagnetization, 63
Adiabatic process, 28
adiabatic process, 33
Amagat, 71
Andrews, 71
Average speed, 158
Avogadro Number, 12
Axiomatic treatment, 35

Barometric distribution, 225
Black-body radiation, 258
Bloch's equation, 143
Boltzmann constant, 114
Boltzmann distribution, 136
Boltzmann Transport equation
    with collision, 304
    without collision, 303
Bose distribution, 149
Bose system, 134
Bose- Einstein Integrals, 401
Bose-Einstein condensation, 273
    recent experiments, 279
Bose-Einstein gas
    thermodynamic functions, 274
Boyle's law, 3
Brillouin function, 169
Brown
    Robert, 316
Brownian motion, 156, 317
Brownian particle, 316

canonical distribution, 141
Carnot cycle

petrol engine, 65
photon gas, 269
    to solve thermodynamic problems, 57
Carnot engine
    efficiency, 41
    power output, 45
Carnot's theorem, 30, 31
Charles law, 3
Chemical equilibrium constant, 236
Chemical potential, 12
    ideal classical gas, 214
Classical Coulomb gas, 341
Clausius-Clapeyron Equation, 79
Closed system, 3
co-existence curve, 72
Collision time, 290
Collisions
    wall of container, 161
compressibility, 13
Concave function, 77
Convex function, 77
Correlation function, 88
Cosmic radiation background, 264
Critical exponents, 86
Critical isotherm, 75, 82
Critical point, 73
Cumulant, 396
Curie's law, 170
Curie-Weiss law, 174

de Hass-van Alphen effect, 197
Debye, 254
Debye temperature, 258
Debye's theory, 255
Degeneracy of levels, 191
Density function, 108
Density matrix, 128

Density of states, 391
Diamagnetism, 167
Differential
    exact, 15
    inexact, 15
    perfect, 17
Diffusion, 294, 320
Dipole layer, 184
Distribution
    component of velocities, 159
Doppler cooling of atoms, 283
    limit, 285

Effusion, 162
Ehrenfest classification, 78
Einstein relation
    between mobility and diffusion coef-
        ficient, 322
Einstein's model, 255
Electrical Conductivity, 295
Electrolytic battery, 1
Electron gas
    strongly degenerate, 179
Energy
    free, 38
    Gibbs free, 39
    internal, 14
Ensemble, 107
    microcanonical, 3
    Canonical, 140
    canonical, 3
    grand canonical, 3, 145
    isothermal-isobaric, 228
    microcanonical, 110
        quantum gases, 132
    quantum statistical, 127
Enthalpy, 38
Entropy, 5, 112
    universe, 47
    various statistics, 136
Equipartition theorem, 163
Ergodic hypothesis, 111
Euler summation formula, 201

Fermi Dirac integrals, 398
Fermi distribution, 149

Fermi energy, 177
Fermi gas
    Gibbs energy and enthalpy, 182
    at finite temperature, 180
    Entropy, 182
    free energy, 182
    pressure, 179
    specific heat, 181
Fermi temperature, 179
Fermi wave vector, 177
Ferromagnetism, 167
Fick's law, 321
Filling factor, 193
flow of matter
    equilibrium, 46
Fluctuation dissipation theorem, 326
Fluctuation- Dissipation relation, 322
Fluctuations
    enthalpy, 229
Flux transport
    across surface, 310
Function
    generalized homogeneous, 101
    homogeneous, 101
Fusion curve, 73
Fusion process, 242

Gamma function, 394
gas constant, 12
Gibbs Paradox, 210
Gibbs- Durham relation, 44
Ginzburg-Landau theory, 93
Grand potential, 146
    Bose gas, 150
    Fermi gas, 150

Haber's reaction, 238
Hall coefficient, 205
Hall resistance, 205
Hamilton's equation, 5
Heat, 2, 5
heat
    specific, 13
Heat of reaction, 237
Heisenberg Hamiltonian, 168
Helmholtz coil, 280

anti, 281
Hertzprung Russel diagram, 244
Hooke constant, 9
Hydrostatic system, 8

Ideal Fermi gas at $T = 0$, 176
Ideal gas, 2, 26
Ideal gas relation, 27
Identical particles, 132
integrals in the kinetic theory, 395
integrating factor, 16
Interaction
    dipole-dipole, 167
    exchange, 167
Internal energy, 5
Internal energy relations, 53
Irreversible process, 6
Ising model
    one dimension, 349
    two Dimensions, 351
Isolated system, 2
Isotherm, 9
Isothermal process, 10

Joule-Thomson Process, 59

K.Onnes, 75
kelvin, 3
Kinetic theory of gases, 156
Kirchhoff's law, 259
Klitzing
    standard of resistance, 208

Landau diamagnetism, 195
Landau levels, 191
Landau theory-validity, 98
Langevin equation, 2
    for Brownian motion, 318
Langevin function, 196
Latent heat, 78
Law of corresponding states, 84
Law of mass action, 235
Laws of thermodynamics
    First law, 25
    Second law, 30, 37
    Third law, 40

Zeroth law, 23
Lennard-Jones Potential, 365
Liouville's theorem, 110

Macro state, 5
Magnetic moment, 1
Magnetic response functions, 54
Magnetization, 1
Main sequence, 245
Maxwell Boltzmann distribution, 156
Maxwell's energy distribution function, 165
Maxwell's relations, 39, 40
Maxwell-Boltzmann statistics, 136
Mayer Cluster Expansion, 331
Mean field approximation, 2
Mean field solution
    Ising model, 404
Mean field theory
    Landau theory, 96
Mean free path, 162, 290
mechanical equivalent of heat, 14
Metropolis Algorithm, 385
    acceptance ratio, 386
    and normalization factor, 386
    and step size parameter, 386
    Condition of detailed balance, 386
Micro state, 5
Micro-canonical average, 111
Microstate, 105
Mixed state, 130
Mobility, 320
Molecular Dynamics, 364
    Approach to equilibrium, 381
        and Boltzmann H theorem, 382
    Breathing-mode oscillatory cluster, 372
    Energy conservation, 374
    Finite-precision artifacts, 374
    Schematic Simulation Structure, 371
    Stability Analysis for Harmonic Oscillator, 369
    Verlet Algorithm, 367
        Stability, 369
        Velocity Verlet, 368
    Visualization scheme, 372
Moment generating function, 396
Monte Carlo Simulation, 384

    of Lennard-Jones Fluids, 387
Most probable speed, 157
Multiplicity factor, 193

Nernst theorem, 40

Open system, 3
Order Parameter, 72
Order parameter
    fluctuation, 96

Paramagnetism, 167, 168
Partition function, 143
    diatomic molecule, 216
    elementary excitations, 253
    harmonic oscillator, 212
    homo nuclear molecules, 218
    ideal gas, 209
    thermodynamic quantities, 143
Pascal, 19
Penzias, 264
Perfect Black-body, 259
Perfect gas scale, 4
Perfect white body, 259
Phase space, 106
Phase transition
    first order, 76
phase transition, 71
Photons, 252
Planck's Law, 261
Planck's law
    Einstein derivation, 266
Polarization density, 11
Pressure on the wall, 162
Principle of detailed balance, 152
Proto star, 243
Proton−proton chain reaction, 244
Pure state, 130

Quantum Hall effect, 205
Quantum ideas, 122
Quasi static process, 7, 31

Radial Distribution Function, 387
Rayleigh-Jean's law, 266
Reaction
    chemical, 234

    heterogeneous, 234
    homogeneous, 234
Relaxation time, 308, 309
Renormalization group analysis
    1-D Ising model, 353
    2-D Ising model, 356
Response function, 13
Reversible process, 7
rms speed, 158
Root mean square speed, 158
Rushbrooke inequality, 56

Saha, 238
    ionization formula, 238
Schottky effect, 187
Simulation
    Computation and, 364
    Equations of Motion, 365
    Methods, 363
    of extended systems, 379
    Periodic Boundary Conditions, 379
    Statistical Physics, 363
    System of Units for, 366
Spin polarization, 203
Static scaling hypothesis, 102
Stefen's law, 264
Sterling approximation, 211
Sublimation curve, 73
Susceptibility, 14
    electrical, 14
    isothermal, 14
    Landau diamagnetic, 190
    Pauli paramagnetic, 190, 202

Thermal conductivity calculation, 294
Thermionic current density, 186
Thermionic emission, 184
Thermodynamic equilibrium, 6
Thermodynamic temperature scale, 42
Throttling process, 59
Transport equation, 156
Triple point, 73

van der Waals, 71
van der Waals Equation, 83
van der Waals equation

derivation, 340
van der Waals gas, 11
Vaporization curve, 73
Variable
    extensive, 13
    intensive, 13
Virial coefficient, 68
Viscosity, 6
Viscosity calculation, 293
Vlasov equation, 303
Volume expansion coefficient, 13

Wein's displacement law, 262
Wein's law, 265
Weiss, 71
Weiss theory, 170
White dwarf star, 1, 245
    model, 247
Wilson, 264
Work, 6
    external, 6
    internal, 6
    magnetic, 11
Work function, 185

Young's modulus, 9